Power Electronics and Electric Drives
for Traction Applications

Power Electronics and Electric Drives for Traction Applications

Edited by

GONZALO ABAD
Mondragon University, Spain

Contents

3.6   Operation under voltage and current constraints   115
3.7   Speed control   124
3.8   Sensorless control   125
3.9   Numerical calculation of the steady-state of synchronous machines   140
References   146

4   Control of grid-connected converters   **148**
      Aritz Milicua and Gonzalo Abad
4.1   Introduction   148
4.2   Three-phase grid-connected converter model   149
4.3   Three-phase grid-connected converter control   175
4.4   Three-phase grid-connected converter control under unbalanced voltage conditions   185
4.5   Single-phase grid-connected converter model and modulation   207
4.6   Single-phase grid-connected converter control   212
References   220

5   Railway traction   **221**
      Xabier Agirre and Gonzalo Abad
5.1   Introduction   221
5.2   General description   221
5.3   Physical approach   248
5.4   Electric drive in railway traction   255
5.5   Railway power supply system   276
5.6   ESSs for railway applications   278
5.7   Ground level power supply systems   332
5.8   Auxiliary power systems for railway applications   338
5.9   Real examples   340
5.10 Historical evolution   351
5.11 New trends and future challenges   351
References   357

6   Ships   **362**
      Iñigo Atutxa and Gonzalo Abad
6.1   Introduction   362
6.2   General description   362
6.3   Physical approach of the ship propulsion system   376
6.4   Variable speed drive in electric propulsion   392
6.5   Power generation and distribution system   409
6.6   Computer-based simulation example   439
6.7   Design and dimensioning of the electric system   448
6.8   Real examples   450
6.9   Dynamic positioning (DP)   455
6.10 Historical evolution   458
6.11 New trends and future challenges   463
References   466

List of contributors

Gonzalo Abad received a degree in Electrical Engineering from Mondragon University, Spain, in 2000, an MSc degree in Advanced Control from the University of Manchester, UK, in 2001, and a PhD degree in Electrical Engineering from Mondragon University in 2008. He joined the Electronics and Computing Department of Mondragon University in 2001 and is currently an Associate Professor there. His main research interests include renewable energies, power conversion, and motor drives. He has co-authored several papers, patents, and books in the areas of wind power generation, multilevel power converters, and the control of electric drives. He has participated in different industrial projects related to these fields.

Fernando Briz received an MS and a PhD from the University of Oviedo, Spain. He is currently a Full Professor in the Department of Electrical, Computer, and Systems Engineering, University of Oviedo. His research interests include control systems, power converters, electric drives, and smart grids. He has written more than 100 journal and conference papers in the field of electric drives and power conversion and been the technical project manager of more than 40 projects, both publicly (European, national and regional) and industry funded. Being the recipient of several IEEE conference and transactions prize paper awards, he served on the organizing and technical program committees of several IEEE conferences. Also he occupied several positions on the Industrial Drives Committee of the IEEE-IAS, of which he is currently Vice Chair, Publications.

Aritz Milicua received a degree in Electrical Engineering and a PhD in Electrical Engineering from Mondragon University, Spain, in 2006 and 2015, respectively. After one year working in the industry, he joined the Electronics and Computing Department of Mondragon University in 2007 and is currently an Associate Professor there. His research interests include power electronics, grid quality, distributed generation, and renewable energy. He has co-authored several papers and patents in the areas of control of grid-connected converters, active filtering, FACTS applications, and modulation strategies for multilevel converters. He has participated in different industrial projects related to these fields.

Xabier Agirre received a degree in Electrical Engineering from Mondragon University, Spain, in 2001. During 2002–2004, he worked as an Associate Professor in the Power Electronics Department of Mondragon University, also he worked in the Power Electronics Department at the IKERLAN research center from 2004 to 2007. Since 2008, he has worked in the Power Electronics R & D Department of CAF P&A. His main areas of interest are the development of magnetic components, high-power DC/DC and DC/AC converters, and energy storage systems based on supercapacitors and batteries.

David Garrido received a degree in Electronic Engineering from the Polytechnic University of Catalonia, Spain. He is an Assistant Professor at Mondragon University. Since October 2009, he has coordinated the research and development of MONDRAGON group electric vehicles in his main interest areas, such as instrumentation and control, electric powertrain, and batteries.

Iñigo Atutxa received a degree in Electrical Engineering in 2001 and in Automation and Control Electronics Engineering in 2004 from Mondragon University, Spain. During 2004–2012, he worked at Ingeteam Power Technology as an R & D Engineer, working in projects related to the design and development of medium-voltage and low-voltage frequency converters for several different applications in the industrial, marine, traction, energy, and grid markets. Since 2012, he has worked at Ingeteam Power Technology as Technical Director for the Industrial & Marine Drives Business Unit. He has contributed to more than 10 research and technical papers, books, and patents in the fields of frequency converters and power electronics.

Ana Escalada received a degree in Electronic Engineering from the University of Basque Country, Spain, in 2001, an MSc in Physics from the University of Cantabria, Spain, in 2003, and a PhD in Electrical Engineering from Mondragon University, Spain, in 2007. During 2001–2007, she worked in the Power Electronics Department of the IKERLAN research center. Since 2008, she has worked at the ORONA Elevator Innovation Center as Senior Engineer Team Leader. Her main research interests include electrical machines, power conversion, and energy storage systems for several different applications in the fields of industrial, elevator, and home energy.

Preface

The work presented in this book offers a practical approach to electric drives. Electric drives are in charge of controlling the movement of devices or appliances that we can find in our daily lives, such as air conditioning systems, washing machines, trains, trams, ships, electric vehicles, hybrid vehicles, elevators, ventilation systems, and wind generators. Thus, the electric drive is part of the electromechanical equipment that enables, for instance, the driver of an electric vehicle to accelerate, decelerate, and maintain a constant speed—ultimately, to drive the car. In a similar way, the electric drive enables an elevator to move from one floor to another as required by its users, while maintaining certain standards of comfort, safety, efficiency, and so on.

This book describes in detail electric drives used in the following extensively used elements and devices: trains, ships, electric and hybrid vehicles, and elevators. In all these elements, and in many others, the electric drive is designed to be able to produce a controlled movement in accordance with the needs and preferences of the user. In essence, the basic electric and electronic working principles and fundamentals of the electric drive for each device are the same. However, for an optimized, safe, efficient, reliable, and comfortable. Performance, the basic fundamental electric drive concept must be adapted to each application or device.

Thus, in this book, the various characteristics of electric drives employed in the above-mentioned applications are described, providing details of how the device itself, with its needs, defines the characteristics of the electric drive. This means, for instance, that the electric drive of a train must be prepared to receive energy from the catenary, transform this electric energy into a controlled movement of the wheels that move the wagon, being able to travel to the different speeds and accelerations required by the driver, and avoiding undesirable and dangerous slipping of the traction wheels so typical in trains.

Structurally, the electric drive is composed of three basic technologies. First, electric machine technology is an important part of the electric drive. The electric machine converts electrical energy into mechanical energy employed to move something. For instance, in an elevator, the electric motor moves the drive sheave. At the same time, this drive sheave moves ropes attached to the car to ferry passengers from floor to floor. In a similar way, the movement of a ship is carried out by the propeller, and the electric machine is in charge of generating the rotatory movement of the blades of the propeller at different speeds. Equally, in road vehicles and trains, the electric motor is in charge of controlling the rotatory movement of the driving wheels.

The power electronic converter technology is in the electric drive. The power electronic converter supplies the electric machine with the necessary electric energy, taken from the energy source. For instance, in an electric vehicle, a power electronic converter supplies the electric motor with energy, typically in AC form, converted from the batteries (the energy source) in DC form.

Third, a control strategy or control algorithm is also necessary in the electric drive. There exist different control philosophies or technologies in electric drives. The strategy controls the movement of the electric motor by sending the necessary orders to the power electronic converter, responding to the demands of the user. For instance, in a ship, the control algorithm, following a demand from the user to travel at a certain speed, controls the speed of the electric machine at a constant speed, which also moves the blades of the propeller. To this end, it sends the appropriate orders to the power electronic converter to provide the required energy to the electric machine. Note that the control must be able to employ the required energy from the energy source, no matter how much the wind is in opposition to the ship or the load it carries, or how rough or calm the sea is.

There is another element that has already been mentioned, which is the energy source. Sometimes, the electric source can be considered a part of the drive itself. This element obviously influences the design and construction of the electric drive and consequently the performance of the ship, train, elevator, and so on. For instance, in electric vehicles, the most commonly employed energy sources are batteries. Depending on the nature and characteristics of these batteries, the electric drive must be accordingly adapted, which is an important part of the global design of the drive.

To use an analogy, the propeller of a ship is like a person's legs. The electric machine in charge of rotating the propeller to move the ship could be the heart and the nervous system. These organs provide blood and nervous stimulus to move the muscles of the legs, thus the energy source of the ship, which is often a combination of diesel engines and batteries, in the person would be the food, water, air, and so on. needed to be able to walk. The power electronic converter that converts the energy in a ship from batteries and diesel engines into electric energy for supplying the electric machine in a human could be the digestive and respiratory systems. Finally, the control system in a ship sends orders to the power electronic converter, to produce movement at the machine and therefore at the propeller. In a human the control could be the brain, which is in charge, among other things, of sending orders to the nervous system to move the legs by means of its muscles. Also, of course, there are many other technologies in electric drives which have not been highlighted, for instance, measurers or sensors of speeds, currents, voltages, etc. necessary for control. In humans, we have, for instance, a vision system, auditory system, olfactory system, etc. which are needed to send information from images, sounds, and smells to the brain to be processed.

Obviously, this comparison, like all analogies, is not perfect, but it gives an idea of the romantic parallelism between humans or animals on the one hand and devices such as vehicles, ships, elevators, and so on, which are created by humans, on the other. It is clear that animals are much more complex than the technology created by humans. Animals and humans are the result of many millions of years of evolution. However, humans started creating technology, according to some anthropologists, only around two or three millions years ago, when one of our "grandfathers", an early hominid, discovered that braking a boulder with another boulder creates broken boulders with an edge, which is a kind of device that allowed early humans to cut meat. From that moment on, technology created by humans has evolved to very sophisticated elements of equipment, such as elevators, vehicles, ships, airplanes, robots, smartphones, rockets; unimaginable to those ancient humans.

Over millions of years, life, whether it be plant or animal, has evolved to adapt to an ever-changing environment. In parallel, technology created by humans is also evolving, trying to adapt to the ever-changing needs of humans. For instance, many concepts employed in shipbuilding that once were useful, even innovatory, have passed by the way, to be replaced by the modern, electrically propelled ship. In a very similar way, many species of animals have disappeared or become extinct, but they were the base or root of the species of animals today. In a similar way, in Nature we can find diversification of life, for instance falcons that can fly very fast and have developed incredibly strong eyesight share the skies with ducks that can walk and swim over the water as well as fly. And so, for instance, with trains: there are trains that specialize in travelling at high speed over long distances, others in carrying heavy and bulky loads, and yet others in travelling at low speeds through the cities, in some cases even disconnected from the catenary or energy source and travelling with the help of batteries. Moreover, there exist some types of trains which do not employ catenary or external energy sources but take their required energy from an engine that is located within the train. These types of trains are, essentially, moved by a tractive electric concept that is very similar to those employed in hybrid-electric vehicles that travel on roads with tires. Thus, this could be understood as an adaptive approach of trains to road vehicles. In Nature, we can also find many equivalent approaches. For example, the dolphin is a mammal with the bones, digestive system, respiratory system, limbic system, social habits, and so on that are very similar to other mammals—such as humans, cows, pigs, and horses—but whose adaptation to its marine environment, and specifically its hydrodynamic requirements, has made it externally in appearance like other fishes, for instance sharks. In a similar way, bats are probably the only mammals that fly, having adapted their forelimbs to wings. Also, having the dense bones of mammals compared to those of birds, they require a huge amount of energy to sustain their flight, and so they need to eat a great many insects in relation to their weight.

In this way, it can be said that human technological developments, which evolve to survive the changing needs of societies, are living entities which adapt to the environment, adopting many different strategies. Just as flowers need to attract bees to aid their pollination and, therefore, their reproduction, elevators created by humans must be attractive to other humans in order to maintain the demand for them so that they will be produced again and, consequently, evolve. Note that it is common that humans adapt themselves to new technological advances rather than advances to humans. Look, for instance, at the Internet, social networks, smartphones, and so on. These have changed the habits and behaviors of humans.

Hence, these "living" technologies, or advances, created in multinational companies, industries, research centers or university departments, and so on, compete with each other and so, in a way, are trying to survive. However, not only the technology created by humans is evolving on the basis of competition, because in parallel and just as interconnected, all the abstract concepts created by humans are also competing to survive: nations, countries, tribes, races, individual careers, clans, societies, religions, ideologies, lobbies, empires, kingdoms, economic concepts, financial interests, forms of economies, forms of consumerism, entertainment industries, and so on.

In this way, it is obvious that over thousands of years, by repeating the same behavioral patterns, humans accumulate abstract images, concepts, or ideas, which are also living entities, evolving and competing to survive. Unfortunately, the history of human civilization shows that competition brings conflict. Inevitably, these concepts create an enormous disorder on earth, with a tremendous accumulated inertia, which unfortunately also creates too much physical violence, non-physical violence, conflicts, suffering, antagonisms, hate, jealousy, overpopulation, sorrow, irrational actions, and so on. on all living entities of earth, including its climate, atmosphere, seas and oceans, mountains, animals, plants, societies, technology, religions, and so on. We do not know whether animals, for instance, can perceive this disorder; humans can, however. These are facts. With these concepts, humans seek security, but we have created division and therefore suffering because the division tries to survive. The burden of these concepts dominates humans' thinking, their relationships, and their daily life. Often, we are so consumed by and accustomed and conditioned to the conflict caused by these concepts that we do not even perceive it and, therefore, do not realize how dangerous it is for the planet, which is a unique living entity of which humans are only a part.

Obviously, it is true that we can also find positive and harmonious evolution, but it is probably too slow and comes at a great cost. Human beings call themselves *Homo sapiens*, but it is not clear whether we behave according to the meaning of that second word, *sapiens* (wise). Unfortunately, neither is it clear how long this behavior can be sustained and whether the planet will be able to resist this tremendous division and disorder that we have created for millennia. As another analogy, again, we can say that humans are now behaving like an uncontrolled plague. Even laboratory experiments with mice, which are mammals, just like humans, have shown that when they obtain food by moving a lever but at the same time see that another mouse receives an electric discharge, they choose not to eat. They prefer not to see another member of their species suffer rather than satisfy their hunger. Humans must recognize this situation and get their own house in order before it is too late. This is probably our greatest global challenge.

Everything written in this book—from this preface to the equations, algorithms, analyses, diagrams—has been inspired by works created with or by others. We just gave a certain structure to contributions of many individuals, many groups, and many multidisciplinary teams. We would like to acknowledge these many and uncountable people who contributed to the concepts contained in this book. In conclusion, we hope that this book will be of interest to and useful for the reader.

Gonzalo Abad
Mondragon University, Spain
June 2016

1

Introduction

Gonzalo Abad

1.1 Introduction to the book

This book is mainly focused on the field of what is commonly known as the electric drive. Electric drives are very prevalent in our lives. They are used in many applications or devices throughout the world. Wherever we use a device or element in which a kind of movement is involved, that movement will probably be governed by an electric drive. Examples of such kinds of devices include trains, trams, ships, electric vehicles, elevators, washing machines, air conditioning systems, wind generators, pumps, and rolling mills and so on. Moreover, in order to be effective and efficient, the specific characteristics of the drive designed, for instance, for controlling the drum of a washing machine, will be quite different compared to the electric drive, for example, employed for controlling the speed of rotation of the blades of a specific wind generator. However, in essence, in terms of what we would call basic technology, all these electric drives employed in various applications share a common technological structure, which, in order to be optimized, is adapted to the specific needs of each application.

This book is mainly focused on describing the electric drives employed in four common applications or devices that one can find in real life: railway traction (trains, trams, locomotives, etc.), ships, electric and hybrid vehicles, and last but not least, elevators. As already noted, in all these traction applications, the main movement must be effectively generated and controlled in order to satisfy standards of performance, efficiency, comfort, safety, reliability, etc. For that purpose, electric drives of different characteristics are developed in each application. It is possible to find AC electric drives or DC electric drives, depending on whether the machine they control is AC or DC. In this book, only AC electric drives are treated, since nowadays AC machines have displaced DC machines, owing to their performance capacity, robustness, cost, etc.

Consequently, this book concentrates on AC drives, dividing its contents into eight chapters. The first four chapters deal with the basic technology comprising the electric drive. And the final four chapters look at how this technology is applied to specific applications.

Power Electronics and Electric Drives for Traction Applications, First Edition. Edited by Gonzalo Abad.
© 2017 John Wiley & Sons, Ltd. Published 2017 by John Wiley & Sons, Ltd.

To be more specific, the introductory chapter anticipates what the rest of the chapters deal with in detail. It sets out to contextualize, and give a general view of, what are the different parts involved in the design of an electric drive, as well as discussing the most common types of electric drives we find in the subsequently described applications or devices.

Then, in Chapter 2 and 3, the control of electric drives oriented to two electric AC machines is described: induction machines and synchronous machines. These two machines, among the existing ones, are the most employed machines in electric drives for the applications described. After that, in Chapter 4, the control of grid-connected converters is addressed, which is an important part of certain sophisticated electric drives required to regenerate energy to the electric grid. In order to describe this control, some other necessary and connected aspects of the electric drive are also studied in these three chapters, such as models of converters, machines, steady-state performance, and so on.

Thus, it is possible to remark that, in general terms, these four first chapters try, on the one hand, to define and describe the most commonly employed drive topologies and their controls in the applications described in subsequent chapters. These topologies have, over many years of industry use, become successfully established as industry standards, and yet they continue to evolve. While, on the other hand, the first part of the book also tries to provide the necessary mathematics, block diagrams, explanation styles, etc. to facilitate an understanding of the concepts described that will be suitable for engineers or people from the industry and postgraduate students.

The second part of the book describes each mentioned device or application in greater detail. There is one chapter for each application: railway traction, ships, electric and hybrid vehicles, and elevators. To avoid duplication, much of what the last four chapters refer to that is described in the first four chapters is not reproduced. These four chapters, from the point of view of exposition, share a common structure. However, the specific particularities of each application necessitate individual chapters addressing certain aspects that are not treated in all chapters. In general terms, we can highlight the most relevant themes:

- A holistic and global introduction to each application, providing a general view and showing the different practical aspects that determine the further performance requirements of the electric drive;
- The physics and mechanics describing the functioning of the applications. This important aspect helps to explain why the way an electric drive is employed varies from application to application. And having knowledge of the different ways a device functions helps us to understand the stages involved in the design of electric drives, such as the characteristics of control, the volume of the drive, and so on;
- The particularities of each application are translated into different functioning or operation conditions of the electric drive, for example dynamic performances, comfort, repetitive operation cycles, power levels, speed ranges, producing torque characteristics, currents, voltages, volume, space, and so on;
- The development and analysis of global simulation models, based on previously developed physic models and electric drive models, showing the behaviors and performances of each specific application;
- Dimensioning examples of each device, providing ideas and procedures of how the different elements of the drive can be dimensioned in order to fulfill different specifications and requirements;
- The representative manufacturers involved in each product, describing some real examples that can be found in the market;
- The technological evolution experienced by each device, showing the past, present, and future of the whole technology involved; and
- An emphasis on the future trends and challenges for each application. As is mentioned in this chapter, it is possible to say that all the applications under discussion present common general future trends, since they share the same basic electric drive technology. But also, the specificity of each application's needs give rise to other, different trends and challenges for each of the devices.

Thus, finally it must be highlighted that the contents of this book are discussed by various academic and industry experts collaboratively. These contributors have come together to give their perspectives on and solutions to the challenges generated by this continuously evolving technology.

1.2 Traction applications

The necessity for an electric drive arises in such applications, products, or equipment where motion is required. Nowadays, it is possible to find a huge amount of applications surrounding us, where motion is required.

Thus, for instance, something so popular and common nowadays, trains, locomotives, trams, or metros employ a typical traction operation. As illustrated in Fig. 1-1, the traction wagon presents at least one traction bogie, where a special arrangement of mechanical transmissions and electric motors produces the traction effort at the traction wheels. The traction effort at the wheels produces the linear movement of the entire train along the railway. A specifically designed electric drive enables features related to the comfort of the users—such as speed, jerk, and slip—to be controlled.

(a)

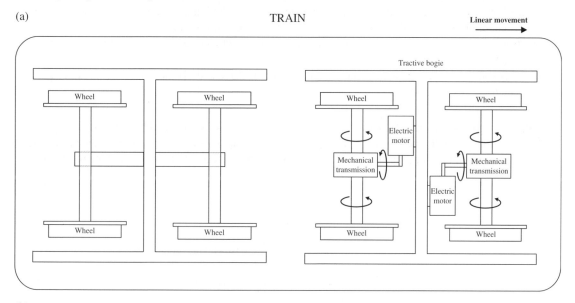

(b)

Fig. 1-1 *(a) Schematic representation of the basic movement operation principle of an electric train, (b) Example of a real tram (Source: CAF. Reproduced with permission of CAF)*

Therefore, it can be noticed that, as in most of the applications where movement must be created and controlled, the movement itself is produced by an electric motor. The rotational movement generated at the motor's shaft is then converted to the movement required by the application, which in the train example is the longitudinal movement of the train itself. Additionally, it must be remarked that the movement of the train must be controlled, guaranteeing some basic performances, such as: smooth and comfortable arrivals and accelerations, minimized energy consumption, and reduced noise levels. In order to achieve this, movement is created by what we call an electric drive. The electric drive is discussed in greater detail later in the chapter, but it can be said here, in a simplified way, that it is composed of:

- an electric motor, which generates the rotational movement;
- a power electronic converter, which supplies the electric motor taking the energy from a specific source of energy, enabling the controlled rotational movement of an electric motor;
- a control algorithm, which is in charge of controlling the power electronic converter to obtain the desired performance of the electric motor; and
- an energy source, which in some cases is part of the electric drive and in other cases is considered an external element.

So, too ship applications. In a modern ship, the advance movement is governed by a thruster or a propeller. The thruster creates a rotational movement of the blades, displacing the water surrounding it and producing the advance movement of the ship. Fig. 1-2 gives a schematic representation of the basic movement principle of a

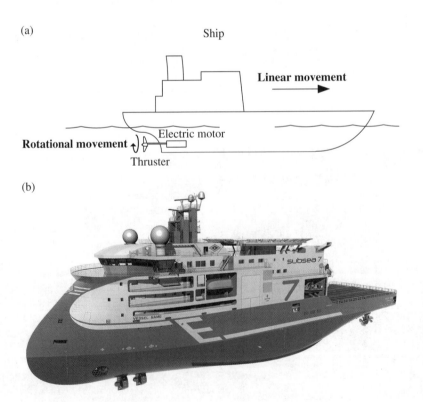

Fig. 1-2   *(a) Schematic representation of the basic movement operation principle of a ship, (b) Example of a real ship (Source: Ulstein. Reproduced with permission of Ulstein)*

ship. In this case again, the element that enables the rotational movement of the blades is an electric motor. Again, in order to obtain reliable and controllable movement, the thruster is controlled by an electric drive specifically designed and optimized for that individual ship, enabling the ship to move at different speeds, under different sea conditions, or to perform dynamic positioning (DP) when performing a specific task. It must be mentioned that, in ship applications, not all the ships utilize an electric motor to move the thruster. Alternatively, for instance, diesel engines can also be employed. However, this book mainly focuses on electric ship propulsion, which is the most commonly used propulsion technology.

On the other hand, we can mention the electric vehicle application. In this case, as schematically illustrated in Fig. 1-3, the linear advance of the vehicle is created by the rotational movement of the traction

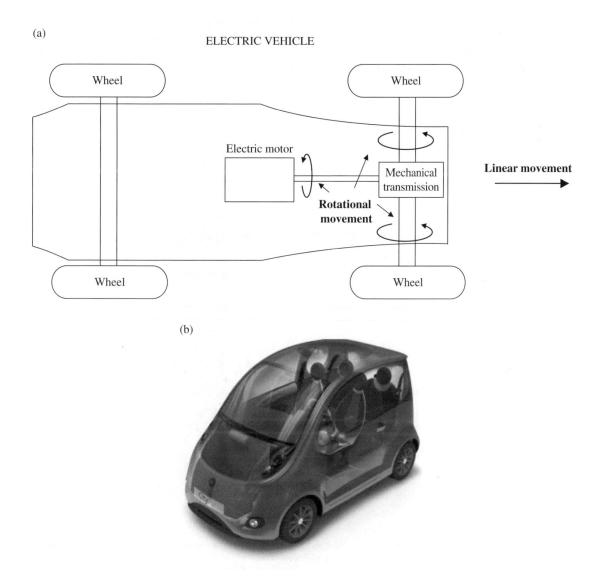

Fig. 1-3 *(a) Schematic representation of the basic movement operation principle of an electric car, (b) Example of a prototype of electric car*

wheels, which are driven by an electric motor. Again, as occurs in the previous two applications described, in order to obtain good longitudinal advance performance, an electric drive is employed, enabling features of the electric vehicle such as: variable speed, electric brake, anti-slip performance, and efficient energy consumption.

In a similar way, it is possible to study vertical transport applications.

Thus, for instance, in many buildings in our cities the elevators have become almost a necessity. In Fig. 1-4, the schematic representation of the basic operation principle of an elevator is depicted. In this case, the car movement between floors must be controlled for the safe and comfortable transportation of the passengers. To this end, the linear movement of the car is generated by the combination of an arrangement of sheave, ropes, and electric motor, which transforms the rotational movement created by the electric motor into the actual linear displacement of the passengers within the car.

The present book analyzes the traction electric drives of the previously introduced four applications; however, it is possible to find a huge number of applications and devices that require an electric drive for the proper control of their movement. Table 1-1 shows some examples of applications governed by an electric drive (the description is not intended to cover all of the possible existing applications).

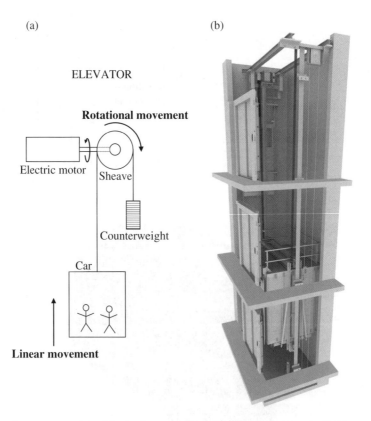

Fig. 1-4 *(a) Schematic representation of the basic operation principle of an elevator, (b) Example of a real elevator (Source: ORONA. Reproduced with permission of ORONA)*

Table 1-1 *Examples of applications and devices governed by an electric drive*

Industry applications	Mining:
	• Conveyors
	• Grinding mills
	• Crushers
	• Shovels
	• Water pumps
	• Ventilation fans
	• Hoists
	Petrochemical:
	• Oil pumps
	• Gas compressors
	• Water-injection pumps
	• Mixers
	Metal:
	• Rolling mills
	• Cooling fans
	• Coilers
	• Extruders
	• Blast furnace blowers
	Paper/Pulp:
	• Grinders
	• Winders
	• Fans
	• Pumps
	Cement:
	• Crushers
	• Draft fans
	• Mills
Home appliances	Washing machines
	Air conditioning systems
	Cooling systems
Energy generation	Wind turbines
	Wave energy
	Tidal energy
	Hydroelectric power conversion plants
	Geothermal energy
Machine tools	Lathes
	Milling machines
	Boring machines
	Drilling machines
	Threading machines
	Grinding machines

(continued overleaf)

Table 1-1   (continued)

Transportation	Machining centers
	Broaching machines
	Sawing machines
	Presses
	Robots and manipulators
	Electric vehicles
	Hybrid vehicles
	Trains
	Trams
	Elevators
	Escalators
	Ship propulsion
	Actuators in aircrafts
	Cranes

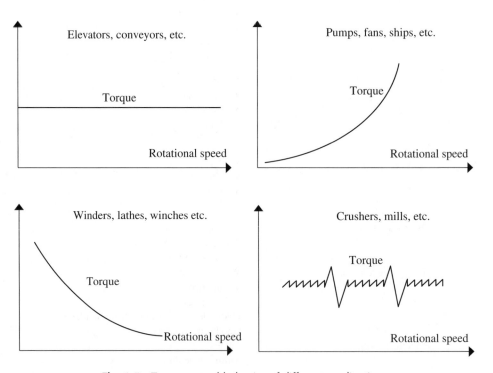

Fig. 1-5   *Torque-speed behavior of different applications*

It can be seen that there are many industry sectors and applications where electric drives of different characteristics and features govern movement.

The reader can intuit that in all of these applications, the movement that is generated needs to overcome a force or torque. Hence, this torque is seen by the electric drive, more specifically by the electric motor that must provide an opposition torque, for the proper operation of the system. Thus, for instance in elevators, the electric drive must be able to provide the required and variable torque at the motor's shaft in order to create an equivalent force to move the car, no matter the number or weight (within reason) of the passengers being transported. Or, for instance, if a train is required to move at a constant speed, the torque that the electric drive must provide to traction the wheels would depend on the force of the air that is in opposition, the slope, the weight of the passengers, etc.

Hence, depending on the nature of the application and the operating conditions of the system itself, the opposition torque that the electric drive must provide often follows a predefined pattern. Fig 1-5 illustrates some typical patterns of torque vs rotational speed of different applications. It can be noticed that the elevators, for instance, operate at constant torque for a fixed number of passengers in the car (constant mass). Or, for instance in ships, that the torque the electric drive must provide to move the thruster at different rotational speeds, and therefore to move the ship at different longitudinal speeds, has an exponential relation to the rotational speed. The physical equations describing these torque vs speed relations of the mentioned four representative applications will be derived in subsequent chapters.

1.3   Electric drives for traction applications

1.3.1   General description

As stated in the previous section, in traction applications or in applications where motion is required, the electric drive is in charge of controlling movement. Depending on the nature of the application and its characteristics in general, the electric drive is specifically designed ad hoc in order to obtain good performances and meet the application's requirements. There are many types of electric drive configurations and it is beyond the scope of this chapter to present all of them. Instead, some of the most representative electric drive configurations will be shown in this section. These representative configurations are the basic drives, which in subsequent chapters are explained and analyzed in detail regarding the following applications or devices: trains, ships, electric vehicles, and elevators. And, as stated previously, we will focus on AC electric drives, which are the most commonly employed drives in these applications.

Hence, Fig. 1-6 shows a general schematic block diagram of an electric drive. As was advanced before and can be seen in the figure, the electric drive is in charge of controlling the movement of the mechanical load. On the other hand, a source of energy that allows the exchange of power required to create and control that movement is necessary. Thus we can distinguish the following main elements [1], [2], [3], [4]:

- Mechanical load: Depending on the application, this is the mechanical compendium of elements which are involved in the movement of the system or device. Thus, for instance, in an elevator, the mechanical load is the compendium of the passengers, the car, the ropes, and the sheave. Or, for instance, in an electric vehicle, the mechanical load is the compendium of the road characteristics, the wheels, the vehicle dimensions and weight, and the number of passengers, etc. As seen in Fig. 1-5, most of the applications require an adjustable speed and torque control.
- Energy source: From this element, the electric drive takes the energy necessary to create the movement. Depending on the nature of the application, the energy flow can be exclusively unidirectional, from the energy source to the load (fans, blowers, etc.), or from the mechanical load to the energy source (wind energy generators, etc.). Alternatively, the power flow can be bi-directional in some applications (electric vehicle, trains, etc.), which means that, depending on the operating conditions, the energy flow can go, at certain times, from the source to the load and, at others, from the load to the source.

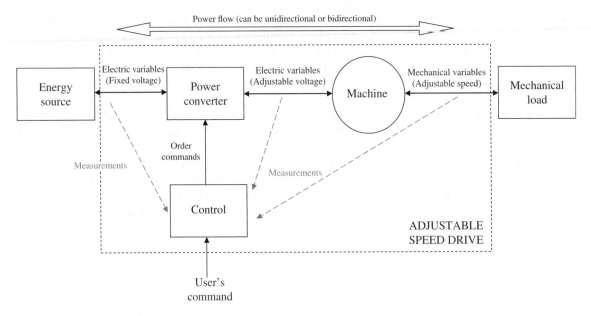

Fig. 1-6　*Schematic block diagram representation of an electric drive*

- Electric drive: As mentioned before, the electric drive in general is composed of many elements; however, probably the most important ones are the following:
  - ○ The electric machine: The electric machine is the element which provides the rotational speed and torque to the mechanical load to assure the correct movement of the system. As will be seen later, there are different types of electric machine configurations (or topologies). In order to be able to obtain adjustable speed performances, the machine must be supplied at adjustable voltage operation conditions.
  - ○ The power electronic converter: The electric machine must be appropriately supplied to be able to provide proper torque and speed at the shaft. The element which is in charge of supplying the electric machine is the power electronic converter and is connected to the energy supply. It is able to supply the required adjustable voltage to the electric machine, converting the voltage from a normally fixed voltage supplied by the energy source. As will be seen later, there are different types of power electronic converter configurations (or topologies).
  - ○ The control: In general, the power electronic converter is governed by a control algorithm (often simply called control) that allows key variables of the electric drive—such as rotational speed, torque, and currents—to be controlled. The control algorithm needs to continuously measure certain variables of the electric drive (electrical and/or mechanicals), in order to be able to control the user's defined command, owing to efficiency or security reasons. Thus for instance in an electric vehicle, the user defines with the pedal the acceleration torque of the vehicle. The control algorithm will be in charge of defining the necessary adjustable voltage that the power converter must provide, so the electric motor responds to the demanded acceleration of the vehicle. As in the previous two elements of the drive, depending on the application (and therefore the machine and converter employed), there are different types of algorithms, which are described later.

Therefore, as has been mentioned repeatedly, the electric machine produces the required rotational speed and torque by the load at the shaft. For that purpose, if the torque and/or speed must vary during the operation, it is necessary to supply the machine with adjustable AC voltage, as depicted in Fig. 1-7. Nowadays, the most efficient, reliable, and cost-effective electric machines are AC machines; therefore, the required adjustable supplying voltage is AC. This means that the supplying voltage can present different amplitude and frequency, as is schematically represented in Fig. 1-7. Note that the electric machine can be conceived as an electromechanical converter, which on one, electrical, side operates with electric variables such as voltages and currents, while on the other, mechanical, side operates with mechanical variables such as torque and rotational speed.

On the other hand, the element that can produce the required adjustable AC voltages for the electric machine is the power electronic converter. As is schematically represented in Fig. 1-8, the power converter in general in an electric drive converts fixed voltage from the energy source into an adjustable AC voltage required by the machine. Depending on the characteristics of the application, it is possible to find energy sources of DC fixed voltage or AC fixed voltage. Thus, depending on the nature of the energy source, the nature or configuration of the power converter would be different as well. To carry the proper conversion from DC to AC, the power converter receives, continuously, order commands from the control algorithm.

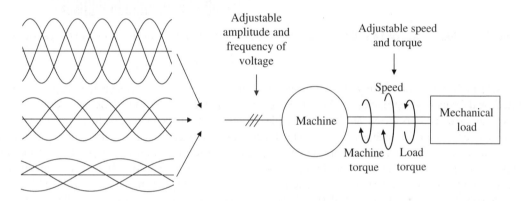

Fig. 1-7   *Electric machine and mechanical load interaction*

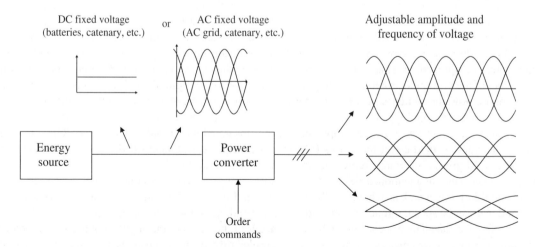

Fig. 1-8   *Interaction between the energy source and the power converter, to supply the electric machine*

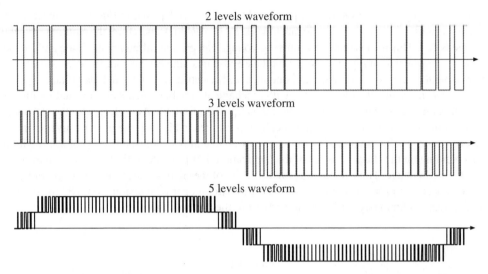

Fig. 1-9 *Output AC voltages obtained by different converter topologies with different voltage levels*

It must be highlighted that, ideally, the AC machines would require sinusoidally shaped voltages of different amplitude and frequency to properly operate. However, at least nowadays, the most efficient and cost-effective power electronic converters are not able to produce ideal sinusoidally shaped voltages. Instead, the power electronic converter at their AC side is able to create staggered or chopped voltages, with a different similitude degree from the sinusoidal ones. As represented, in a simplified form, in Fig. 1-9, depending on the power electronic converter configuration, or topology, it is possible to create voltages of different constant levels that present different similitude appearance from the ideal sinusoidal voltage. Thus, it can be noticed how if the number of voltage levels employed is high (five in this example), the similitude to the sinusoidal shape is closer than if the employed number of levels is low (two in this example). Hence, it can be said that with this type of staged voltage waveforms the electric machines are good enough and therefore useful, providing an acceptable performance of torque and speed behaviors for most of the applications described in this chapter.

The power electronic converter is normally composed of different types and numbers of controlled and uncontrolled switches, also called semiconductors. In addition, passive elements such as, most commonly, capacitors and, less so, inductors are also present at the converter. Depending on the disposition and number of the switches employed, the shape of the output AC voltage that can be obtained is different. By controlling the switches of the power electronic converter, the output voltage waveform of the converter is also controlled. Depending on how the switches of the converter are disposed or arranged, it is said that a different power electronic converter topology is obtained. Typically, the uncontrolled switches employed are the diodes, while the controlled switches employed can be: insulated-gate bipolar transistors (IGBTs), metal-oxide-semiconductor field-effect transistors (MOSFETs), insulated-gate controlled thyristors (IGCTs), and so on. Fig. 1-10 gives a schematic representation of a diode and an IGBT.

On the other hand, as mentioned before, the power electronic converter is continuously governed by the control algorithm. In general, as schematically represented in Fig. 1-11, the control algorithm receives a user's command that can be speed, torque, power, and so on. depending on the nature of the application. This command or reference can change according to the user's needs or the needs of the application at a given moment. Then, considering different measurements taken from the electric drive itself (currents, position, or speed of the shaft, etc.), the control algorithm creates order commands for the controlled switches of the power electronic

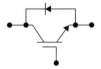

Fig. 1-10 *Schematic representation of a diode and an IGBT*

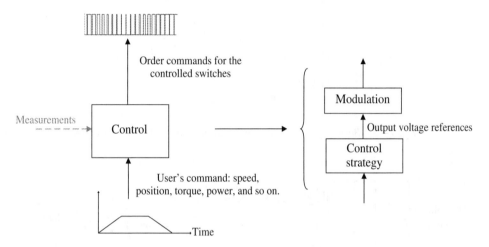

Fig. 1-11 *Schematic representation of the control algorithm of an electric drive*

converter, which produces the necessary output AC voltage to respond to the user's command. There are many different types of control algorithm, adapted to meet the needs of elements such as: the application, converter topology, machine topology, nature of the energy source, etc. However, it can be affirmed that the most popular or commonly employed control algorithms are divided into two different parts, or tasks. The first one can be called the control strategy. In general, by following basic principles of control theories, it creates the voltage references at the AC output, for the second part (i.e. the modulation). The modulation creates the order commands for the controlled switches, based on the voltage references provided by the control strategy.

1.3.2 Different electric drive configurations

The previous section provides an intuitive approach to the electric drive. In this section, a more detailed description of the electric is given, showing the most important configurations that can be found in the applications discussed in this chapter. The electric drive configurations are presented, attending to the type of power electronic converter that is used. They are presented as general electric drives, which are suitable for many applications.

1.3.2.1 Electric drive with a DC/AC power converter

The first electric drive configuration that is presented is depicted in Fig. 1-12. The electric machine is supplied by a DC/AC converter that takes the energy from a fixed DC voltage source. The most common DC sources can be batteries or DC catenaries in railway traction applications.

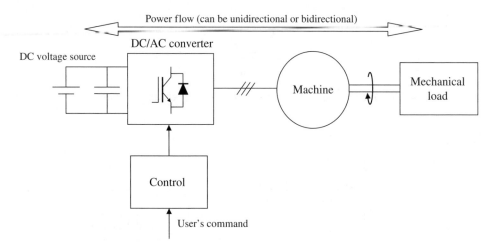

Fig. 1-12   *Basic electric drive configuration with a DC/AC converter*

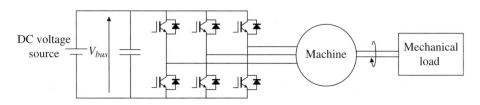

Fig. 1-13   *Basic electric drive configuration with a classic two-level DC/AC converter topology*

Although ideally the DC source is of fixed voltage, it must be said that the real batteries or catenaries vary their voltage, depending on several factors. Because of this, this type of electric drive configuration must be specially prepared to handle DC voltage bus variations that can take extreme values in some cases. On the other hand, with respect to the DC/AC power electronic converter, probably the most common topology employed nowadays is the classic two-level converter, which is presented in Fig. 1-13. In this case, IGBTs have been used as controlled switches, but some other types of controlled switches could optimize the converter performance depending on the needs of the applications where this electric drive is used. It can be seen that, in this configuration, a capacitor is typically located in parallel to the battery (or battery pack), in order to obtain several performance improvements of the drive, which is more deeply described in subsequent chapters.

In principle, depending on the nature of the application (the mechanical load), this configuration can trans-port energy in both directions, from source to load and from load to source—at least if the energy source is prepared to receive energy, since the power converter is bi-directional. With regards to the converter topology, it is also possible to use a more sophisticated power converter topology as, for instance, the three-level neutral point clamped (3L-NPC) topology, the four-level flying capacitor (4L-FC) topology, or the five-level active neutral point clamped (5L-ANPC) topology. One leg of these multilevel topologies is represented in Fig. 1-14. It must be mentioned that this is just an example of the types of modern multilevel converter topologies that can be suitable for this electric drive, and used in some of the applications described in this chapter. By employing multilevel converter topologies, although the complexity of the converter is increased, there are some useful

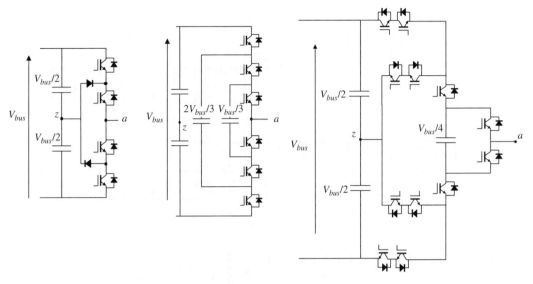

Fig. 1-14 *One arm of the following multilevel converter topologies (from left to right): 3 L-NPC topology, 4 L-FC topology, and 5 L-ANPC topology*

benefits that can also be obtained: they operate at higher voltage to reduce the amount of current employed, improve the output waveform quality, and so on.

Finally, Fig. 1-15 illustrates the shape of the output AC voltages obtained with each of the multilevel converter topologies mentioned.

To conclude, with regards to the machine topology, it can be said that this electric drive allows the use of any kind of AC machine, which obviously requires AC voltages at the input. Subsequent sections of this chapter show that there are different possible machine topologies, but probably the most employed ones are induction machines and synchronous machines as commented before.

1.3.2.2 Electric drive with a DC/DC/AC power converter

The electric drive configuration presented in this section, compared to the one presented in previous section, incorporates a DC/DC converter in order to mainly stabilize the DC voltage at the input of the DC/AC converter. The basic scheme configuration is depicted in Fig. 1-16. Thus, the inclusion of this converter can be useful if the DC source is going to vary its voltage during normal operation. In this way, the DC/DC converter stabilizes the voltage at the output, allowing the operation of the DC/AC converter at constant DC voltage and, therefore, optimizing its performance. In a similar way, the incorporated DC/DC converter can also be useful when the DC source voltage is too low to synthesize the required AC voltage. In this situation, the DC/DC converter increases the DC voltage, thus allowing the proper operation of the system. Here again, at least if the energy source is prepared, the power flow can be bi-directional, depending on the nature of the application. With regards to the DC/AC converter topology, in this case again, it is possible to use a two-level converter topology and a multilevel converter topology as well.

On the other hand, this electric drive configuration needs an extra control for the additional DC/DC converter included, which typically is in charge of ultimately controlling the DC voltage at the input of the DC/AC

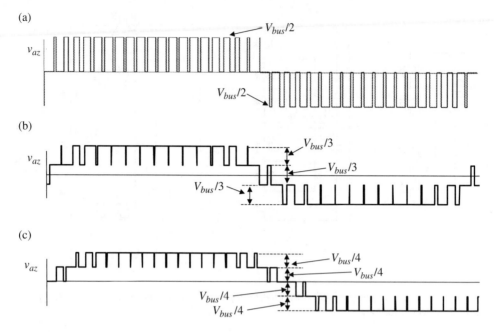

Fig. 1-15   *Output voltages of (a) 3L-NPC, (b) 4L-FC, and (c) 5L-ANPC topologies*

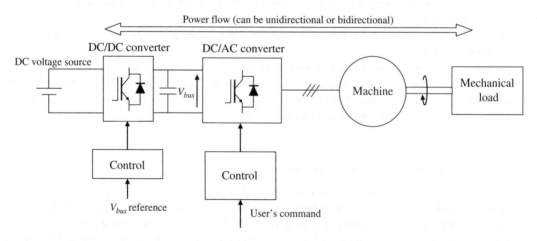

Fig. 1-16   *Electric drive configuration with a DC/DC/AC converter*

converter. One example of a suitable DC/DC converter configuration is depicted in Fig. 1-17 together with a two-level converter as DC/AC converter topology. This configuration allows for the transmission of energy in both directions. Chapter 5, related to railway traction, and Chapter 7, related to electric vehicles, provide more details about DC/DC converters. Finally, with regards to the electric machine topology, this drive is also suitable for electric machines such as induction machines or synchronous machines.

On the other hand, Fig. 1-18 illustrates the auxiliary pre-charging circuit that typically accompanies the DC voltage sources, for instance battery packs. As the reader may notice, the DC source cannot be suddenly connected

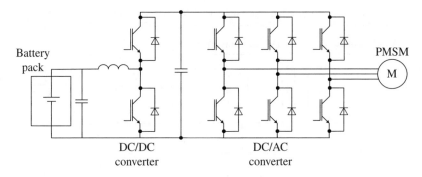

Fig. 1-17   *Example of a possible DC/DC converter*

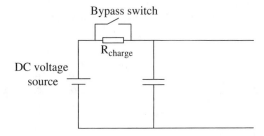

Fig. 1-18   *Charging resistance and bypass switch to avoid damaging the capacitor and batteries*

to the capacitor, because a high current transient would occur. In order to avoid this, a resistance is normally placed when connected, while after a time when the capacitor is charged the resistance is disconnected by means of the bypass switch. In battery packs, as is shown in Chapter 7, related to the electric vehicle, the charging circuit is often incorporated by the manufacturer of the battery pack itself. On the other hand, the necessity of incorporating the capacitor arises to smooth the current shape demanded from the battery pack, therefore increasing its life.

1.3.2.3   *Electric drive with a unidirectional AC/DC/AC power converter*

The next electric drive configuration takes its energy from an AC voltage source. For this purpose, it requires an AC/DC converter at the input. Then, a DC/AC converter is used, as in the previous two electric drive configurations. The schematic block diagram of this drive is depicted in Fig. 1-19. It shows that the two converters employed form an AC/DC/AC converter configuration.

In this case, a passive front end is employed at the input. It is a diode bridge rectifier, which converts the voltage from the AC fixed voltage amplitude of the source to a fixed DC voltage. This converter is composed solely of diodes and so needs no control algorithm. It is not reversible, which means it is able to transmit the energy only from the AC side to the DC side. It can be single-phase or three-phase, depending on the nature of the grid where it is connected. Consequently, this drive configuration is not suitable for generation applications, since it would not be able to take energy, for instance, from a wind turbine's blades and send it to the electric grid. Fig. 1-20 illustrates two possible configurations for the input AC/DC diode bridges. One is used when the applications oblige it to use a single-phase AC source and the other for AC sources which are

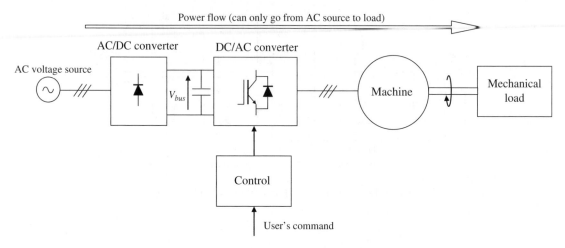

Fig. 1-19   *Electric drive configuration with a uni-directional AC/DC/AC converter*

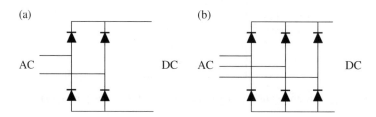

Fig. 1-20   *Two diode bridge converter configurations: (a) single phase diode bridge and (b) three-phase diode bridge*

three-phase. In Chapter 6, which is dedicated to ship propulsion, more diode-based AC/DC converters, which enhance performance, are discussed. On the other hand, it must be remarked that often, a step-down transformer (either single-phase or three-phase) can be employed at the input of the electric drive in order to adapt the voltage level of the AC source and the required electric drive AC input voltage. In this exposition, it has not been represented in Fig. 1-19, for simplicity's sake, and because it is not always needed. In addition, often also in this electric drive configuration, an AC filter is also included at the input of the diode rectifier. The filter topology often consists of a simple inductance per phase. The inclusion of this filter allows a better-quality current exchange with the grid to be obtained. The filter is often not physically placed when a transformer is needed at the input. Instead, the transformer is designed specifically with the necessary leakage inductance, which also provides the filter functionality.

The fixed relation between the AC voltage input (V_{LLrms}) and the obtained DC voltage output of a three-phase diode bridge is:

$$V_{DC} = 1.41 \cdot V_{LLrms} \tag{1.1}$$

This is valid when a sufficiently high capacitance is placed at the DC side, at the input of the DC/AC converter. When the AC source is single phase, the relation follows exactly the same equation, but in this case the AC voltage input is not line-to-line (LL) but is single phase.

With regards to the DC/AC power electronic converter, it is possible to employ all the topologies mentioned in the previous two electric drive configurations (i.e. from the classic two-level converter to the more modern and sophisticated multilevel converters). The electric machine can also be a synchronous machine or an induction machine, depending which fits better in the corresponding application.

It must be highlighted that when the application needs to operate in generator mode, working as an electric brake, it is necessary to incorporate a DC chopper (also called a crowbar) at the DC bus, which allows the regenerative energy in a resistance at the DC side of the AC/DC/AC converter to dissipate. One example of an electric drive configuration with a diode bridge rectifier and a two-level inverter incorporating a DC chopper is depicted in Fig. 1-21. The DC chopper is typically mainly composed of a controlled switch and a resistance. It works in such a way that the controlled switch is activated enabling the regenerative energy that comes from the machine at the resistance to dissipate. This occurs when an increase of the DC voltage of the AC/DC/AC converter is detected. Conversely, it is deactivated when the DC bus voltage goes down again taking normal values, because the machine stops operating as a generator.

In addition, Fig. 1-22 illustrates the DC chopper arrangement needed for the DC bus capacitors of a three-level converter. For multilevel converters of a higher order, with bigger number of capacitors at the DC bus, it would be necessary to place one DC chopper for each capacitor, to ensure the integrity of all the capacitors.

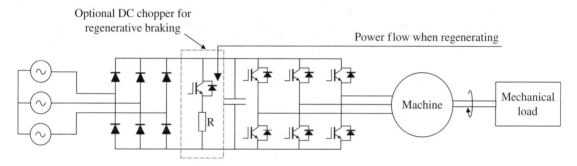

Fig. 1-21 *Electric drive configuration with a uni-directional AC/DC/AC converter (diode bridge rectifier and two-level inverter) incorporating a DC chopper (or crowbar)*

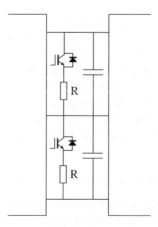

Fig. 1-22 *DC choppers for the DC bus capacitors of a three-level converter*

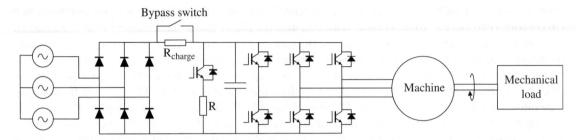

Fig. 1-23   *Electric drive configuration with a unidirectional AC/DC/AC converter (diode bridge rectifier and two-level inverter) incorporating a DC chopper and a charge resistance for the initial charge of the DC bus capacitor (charge resistances can also be placed at the AC side, instead of the DC side)*

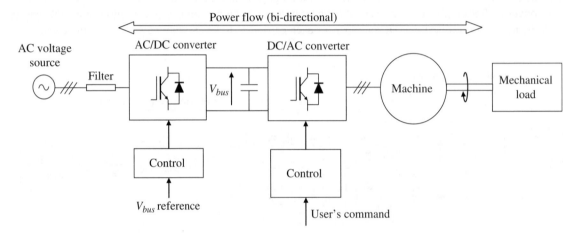

Fig. 1-24   *Electric drive configuration with a bi-directional AC/DC/AC converter (input converter can also be single phase)*

Finally, it must be mentioned that this configuration of electric drive needs an auxiliary circuit for an initial charge of the DC bus capacitor through the diodes of the input rectifier, as occurs in DC/DC/AC converter configurations. Fig. 1-23 illustrates the charging circuit that is simply configured by a charging resistor (R_{charge}) and a bypass switch. Thus, in an initial maneuver, the capacitor is charged from zero to a certain DC voltage through the diodes and the resistance, avoiding a high overcurrent. Then, once the DC bus voltage reaches a certain value, the R_{charge} is bypassed by the bypass switch, allowing the entire AC/DC/AC converter to start its operation in a normal regime. Alternatively, it is possible to find also auxiliary charging circuits that incorporate charging resistances at the AC side of the diode rectifier.

1.3.2.4   *Electric drive with a bi-directional (regenerative) AC/DC/AC power converter*

A more complex and sophisticated electric drive configuration than the one presented in the previous section is the drive that incorporates a bi-directional (also called regenerative) AC/DC/AC converter. A schematic representation of the drive is depicted in Fig. 1-24. As can be seen, an active front end is used as the AC/DC converter, enabling the regeneration mode to the grid. Thus, instead of using a diode rectifier, as in the previous

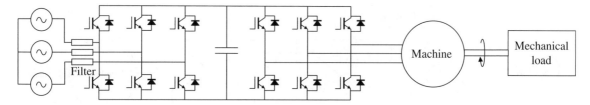

Fig. 1-25 *Electric drive configuration with a bi-directional AC/DC/AC converter using two-level converters as rectifier and inverter with a three-phase grid*

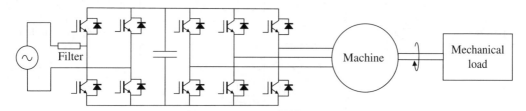

Fig. 1-26 *Electric drive configuration with a bi-directional AC/DC/AC converter using two-level converters as rectifier and inverter with a single-phase grid (the charging auxiliary circuit is often included at the AC side)*

drive, a controlled AC/DC converter is used with its corresponding control algorithm. In general, the topologies of the AC/DC converter and the DC/AC converter are equal. Therefore, if, for instance, a 3L-NPC converter were wanted, it would be employed at both sides of the same topology. In addition, the input AC/DC converter requires a filter at the AC side for the proper operation of the system. The typical filter configurations are: pure inductive filters (L), or combinations of inductances and capacitances (LC or LCL). Hence, the AC/DC converter at the input is able to work in a regenerative mode, regenerating energy from the mechanical system to the grid. Typically, the AC/DC converter at the input is controlled in such a way that it is in charge of controlling the DC bus voltage. A detailed study of this converter and its control is provided in Chapter 4.

Note that, in this configuration, it is not necessary to use the DC chopper, since the regenerative energy is delivered to the grid. Obviously, this configuration of electric drive is suitable when the grid is connected and prepared to receive energy as well as to provide energy. The reader can deduce that this configuration is typical in energy generation applications such as wind turbines, for instance. In this configuration again, the converter at the input can be single-phase or three-phase, depending on the characteristics of the application. Again, a step-down transformer can also be included at the drive, if the voltages of the AC grid source and the voltage of the drive do not match. As in previous configurations, the converter topologies can be a classic two-level or multilevel converter topology. Also with regards to the type of machine, an induction machine or a synchronous machine can be used, depending on the characteristics of the application.

Finally, Fig. 1-25 illustrates an example of electric drive with bi-directional AC/DC/AC converter, using two-level converters as rectifier and inverter and connected to a three-phase grid. In this case, as with the previous drive configuration, the pre-charge of the capacitor is done across the diodes of the input converter and the charging resistances are often located on the AC side.

In some applications, for instance in railway traction with AC catenaries, the grid is not available with three phases. Therefore, in these situations a single-phase input converter is necessary, as depicted in Fig. 1-26.

Alternatively, when the required output power is in the range of multi-megawatts, the parallelization of converters is often used to construct the drive. For instance, in industrial applications and sometimes in ship propulsion, it is quite common. Fig. 1-27 illustrates some examples of converter parallelization.

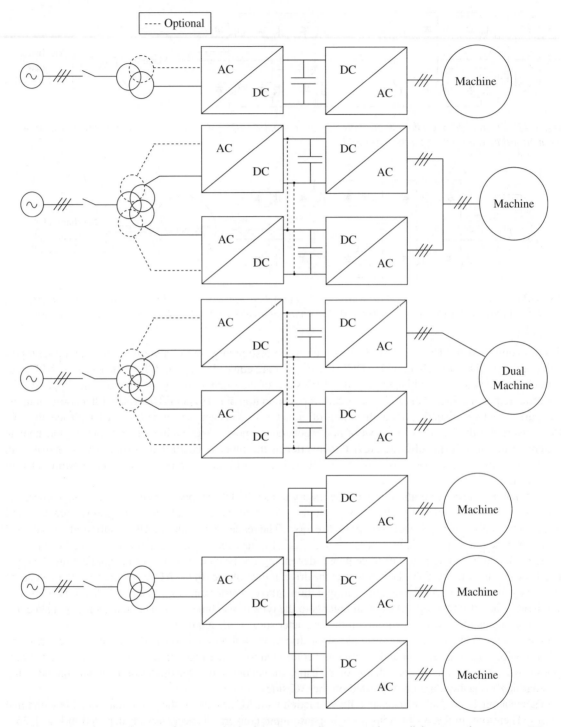

Fig. 1-27   *Electric drive configuration for high power applications with parallelization of converters [5]*

1.3.2.5   *Electric drive with other power converter topologies*

The most employed electric drives are presented in previous sections. This section examines some drives which are not so established in real applications, but could be of interest in the future or simply are used by a low proportion of drive manufacturers in the world. Amongst them, probably the one most studied at research level is the electric drive based on the AC/AC converter without DC-source storage in the middle of the converter (i.e. without a capacitor). These types of power converters are typically called matrix converters. They use in general a higher amount of controlled semiconductors and diodes when arranging different matrix converter topologies. One schematic example of electric drive with an AC/AC matrix converter is depicted in Fig. 1-28.

In general, these types of converters are bi-directional in terms of their power flow transmission capabilities. Since they do not transform the voltage into DC at the middle of the converter stage, they synthesize the output voltage at the machine terminals from the input AC voltage of the energy source. The converter side connected to the AC source, as in previous converter configurations, can be single-phase or three-phase.

There exist many matrix converter topologies that are beyond the scope of this book. The reader can find, in specialized literature, different constructions of matrix converters, as well as the advantages and disadvantages of the more classic converter arrangements shown in previous subsections. Fig. 1-29 shows a schematic representation of a direct matrix converter.

On the other hand, some other electric drive configurations employ a different conversion philosophy and they are quite well established in industry applications, especially in high-power applications. These types of drives are the current source converter (CSC) based electric drives. The electric drives described in the previous sections are commonly called voltage source converters (VSC), since they work and supply the loads as voltage sources. One example of a CSC-based bi-directional electric drive is depicted in Fig. 1-30. It is composed of two two-level CSCs as rectifier and inverter. It will be noted that this converter configuration only uses controlled switches (and not diodes). In this case, they use symmetrical gate-commutated thyristors (GCTs) connected in series, in order to achieve a higher operating voltage and therefore also power. They typically need a capacitor-based filter both at the input and at the output, as well as an inductive filter at the DC-link of the current converters. The reader can find more detailed descriptions of these types of converter configurations in specialized literature.

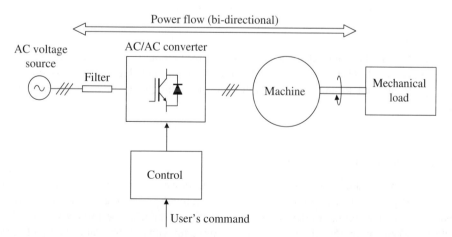

Fig. 1-28   *Electric drive configuration with a bi-directional AC/AC matrix converter*

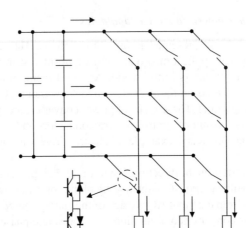

Fig. 1-29 *Schematic representation of a direct matrix converter*

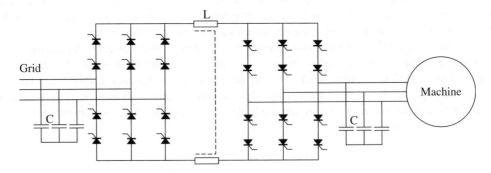

Fig. 1-30 *One example of a CSC-based bi-directional electric drive*

Finally, it must be said that the reader can find in specialized literature many different electric drives based on many different power electronic converter configurations not presented in this chapter—or in this book. It is beyond the scope of this book to summarize all of them. Instead, this chapter gives a short overview of the ones most employed.

1.3.2.6 *Common mode currents*

Common mode voltage results from the operation of the power converter with a switching pulse-width modulation (PWM) pattern. Thus, common mode voltage refers to the voltage between the neutral point of the three-phase star connected load (electric motor) and the potential of the protective earth [1]. Fig. 1-31 illustrates the equivalent circuit for a three-phase power converter with a motor, cable, and grid supply showing the parasitic capacitances. In this way, the appearance of the common mode output voltage of the converter and

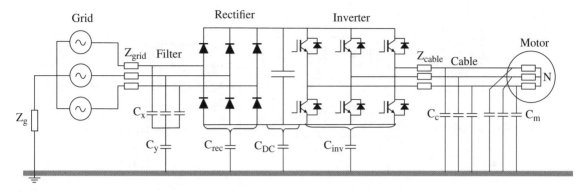

Fig. 1-31   *Equivalent circuit for a three-phase power converter with a motor, cable and grid supply showing the parasitic capacitances [1]*

the presence of parasitic capacitances in the motor and other elements of the electric drive cause the flow of zero-sequence currents.

Therefore, common mode voltages generated by the converter, with high dv/dt, produce common mode currents that often inevitably generate current bearings and shaft voltage in the motor. In general, these shaft and bearing currents through the motor cause accelerated degradation of the motor, among other undesired phenomena. It must be mentioned that there exist several types of bearing currents—such as capacitive bearing current, electric discharge bearing current, circulating bearing current or shaft current—related to the shaft voltage effect and rotor ground current. These currents are related to different physical phenomena.

In order to avoid or at least prevent these damaging currents, it is very important to reduce the influence of common mode voltages by a proper cabling and earthing system. When this is fulfilled, the further reduction of common mode currents can be obtained by increasing the impedance to these currents. Some possible solutions, often incorporated at the electric drive, are listed below [1]:

- ceramic bearings
- common mode passive filters
- systems for active compensation of common mode voltage
- decreasing the converter switching frequency if possible
- motor shaft grounding by using brushes
- conductive grease in the bearings
- common mode choke
- use of one or two insulated bearings
- use of one or two ceramic bearings
- dv/dt filter
- shielded cable for motor supply.

Therefore, in Fig. 1-32, two of those solutions are graphically represented. The common mode choke is typically constructed with three symmetrical coils on a toroidal core, as shown in the figure. The choke presents negligible impedance for differential mode currents, because the total flux in the core is eliminated for three-phase symmetrical currents. However, the impedance seen by the common mode current is important.

(a)   (b)

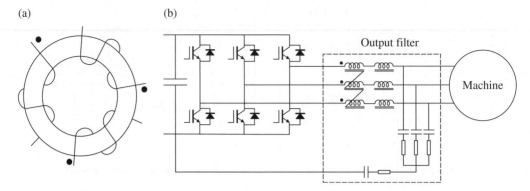

Fig. 1-32   *Two examples of passive solutions for reducing common mode currents: (a) common mode choke and (b) common mode output filter*

On the other hand, the example of common mode passive filter [1] shown in Fig. 1-32(b) is normally based on the combination of elements that give more impedance for the common mode circuit and also create an alternative path for the common mode current bypassing the motor.

Finally, it must be mentioned that there is another way to reduce common mode currents in the electric drive. This alternative way is based on reducing the common mode voltages of the PWM employed. Thus, by reducing the common mode voltage created by the converter, the common mode currents are inherently reduced. Thus, there are several alternative modulation methods oriented to this purpose, and they can be found in specialized literature. In general, they are based on space vector and pulse-width modulation methods [1].

1.4   Classification of different parts of electric drives: converter, machines, control strategies, and energy sources

This section provides a brief perspective of the main elements that can be found in electric drives. Converters, machines, control strategies, and energy sources are briefly introduced. They can be considered the basic electric and electronic technologies employed in electric drives for traction applications or devices.

1.4.1   Converters

By means of controlled and/or uncontrolled switches, such as IGBTs and diodes, we can build up power electronic converters. Such devices can convert voltage from DC to AC or from AC to DC. This subsection enumerates some representative types of converters commonly used nowadays. But, first, some of the commonly employed semiconductors are introduced as well. Table 1-2 shows some of these semiconductor devices. The evolution of power converters and their characteristics is closely related to the evolution of semiconductor devices. In specialized literature, for instance [1]–[4], it is possible to obtain information about the properties of each device and its major characteristics and performance. It can be said it is possible to find in the market semiconductor devices of different rated voltage, current, and many other characteristics, suitable therefore for converters of different characteristics. Thus, probably the most commonly employed semiconductor device is the diode, which is widely used for uncontrolled rectifiers and for controlled power converters. Then there is the thyristor—also known as a silicon-controlled rectifier (SCR). It is line commutated and the

Table 1-2 *Semiconductor devices [1]–[5]*

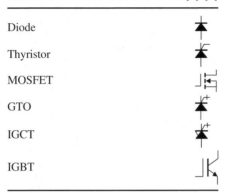

Diode	
Thyristor	
MOSFET	
GTO	
IGCT	
IGBT	

turn on can be controlled by applying a positive gate current. On the other hand, the turn off depends on the load current so it cannot be directly controlled as the turn on.

The MOSFET (metal-oxide semiconductor field-effect transistor) is a voltage-controlled device. It presents controlled turn on and turn off capability. This device is very popular in low-voltage and low-power applications, because it is possible to operate it at hundreds of kHz switching, causing relatively low switching losses.

On the other hand, the gate turn-off (GTO) thyristor is a self-commutated variant of the thyristor. It can be turned off by a negative gate current, but it is a bulky and expensive turn-off snubber circuit with high snubber losses and a complex gate driver. It was very popular in high-power converters, but nowadays GTOs are not used as much as the four applications described in this book.

It is possible to say that the gate-commutated thyristor (GCT) or the integrated gate-commutated thyristor (IGCT) is the successor of the GTO. Thanks to its constructive nature, it presents lower on-state losses. This semiconductor device presents some variants, such as the symmetric gate-commutated thyristor (SGCT) typically used in CSCs.

The insulated-gate bipolar transistor (IGBT) is probably the most employed device in low-voltage and low-power converters but is also much used in high-power converters. It is a voltage-controlled self-commutated device, with a simple gate driver, snubberless operation, and high switching speed.

Now that we have introduced the most commonly used semiconductors, in the following we briefly introduce the most representative power electronic converters employed in electric drives. As depicted in Table 1-3, although there exist many other types of converters, we can say that the most used ones can be divided into three main groups: rectifiers with diodes (or thyristors, but these are not so much used nowadays), AC/DC or DC/AC controlled VSCs, and DC/DC converters.

Hence, most of these converters have already been introduced in previous sections. The machine of the electric drive is supplied by a DC/AC converter that, depending on the characteristics of the application, can be a two-level converter or a multilevel converter. For high-power applications mainly, there are many possible multilevel converter configurations, such as 3L-NPC, 4L-FC, 5L-ANPC. that are shown in previous subsections. In addition, Fig. 1-33 also illustrates some other examples of stacked multi-cell converter topologies, nowadays not commonly employed in commercial drives but they may be employed in the future.

On the other hand, Fig. 1-34 illustrates some examples of commercially available power converters. Finally, diode-based rectifiers and DC/DC converters are also typically found in electric drives. Here also, there is a wide variety of converters, with different characteristics and performances. Examples of them are provided in subsequent chapters, closely related to the applications in which they are typically used.

Table 1-3   *Power electronic converters*

Type	Example
Diode or thyristor bridges or rectifiers	6 pulses
	12 pulses
	18 pulses
	24 pulses
AC/DC or DC/AC Controlled VSCs	Two level converter
	Multilevel converter
	• 3 L-NPC
	• 4 L-FC
	• 5 L-ANPC
DC/DC converters (including or not galvanic isolation)	Flyback
	Forward
	Push/pull
	Dual-active bridge

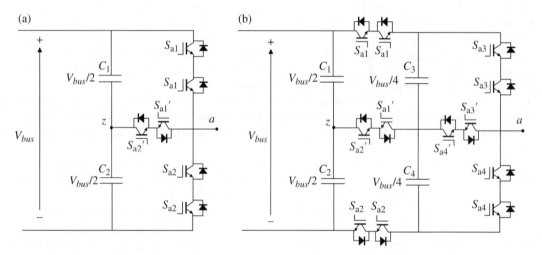

Fig. 1-33   *Stacked multi-cell (SMC) topologies: (a) 3L-SMC, also known as 3L-NPP (neutral point piloted), and (b) 5L-SMC*

1.4.2   Machines

With regards to the AC machines, Table 1-4 shows a simplified classification of machines that can be found in AC electric drives. There are many other type of AC machines; however, this table shows the most representative and used ones in common electric drives. As can be seen, electric machines can be classified into three main groups: induction machines, synchronous machines, and variable reluctance machines. In all these cases, probably the most employed ones are rotating machines with three phases, but it is also possible to find linear

(a)　　　　　　　　　　　(b)

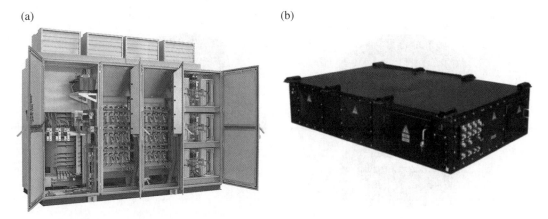

Fig. 1-34 *Examples of commercially available power electronic converters: (a) industry application, MV700: five-level H-bridge neutral point clamped (5 L HNPC) (Source: Ingeteam. Reproduced with permission of Ingeteam) and (b) train application 700 kW/16.5 kV AC Power Converter Box (Source: CAF P&A. Reproduced with permission of CAF P&A)*

Table 1-4　*AC machines*

Type	Example
Induction machines (rotating or linear and three-phase or multiphase)	Cage rotor (asynchronous machines)
	Wound rotor (WRIM) or doubly fed
Synchronous machines (rotating or linear and three-phase or multiphase)	Wound field (WFSM)
	Synchronous reluctance (SyRM)
	Permanent magnet (PM)
	• Axial
	• Transversal
	• Radial
	○ Interior
	○ Surface
	▪ Trapezoidal (brushless DC, BLDC)
	▪ Sinusoidal
Variable reluctance machines (rotating or linear and three-phase or multiphase)	Switched reluctance
	Stepper

machines and also multiphase machines (i.e. those with more than three phases) in both versions (i.e. rotating and linear).

In general, variable reluctance machines are associated with low-power, low-performance applications. This family of machines is cheap and easier to manufacture than the other two groups—induction machines and synchronous machines. In addition, variable reluctance machines require a special converter topology. They are not very common in the applications that are studied in this book, and so are not discussed in any great depth here. On the contrary, the typical AC machines employed in ships, elevators, railway traction, and hybrid and electric vehicles are the induction and synchronous machines. They are constructed in both low-power and

(a)   (b)

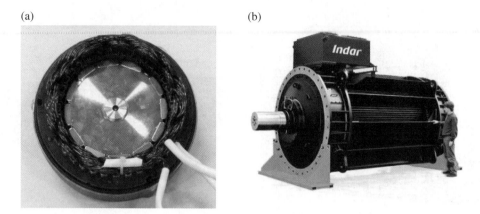

Fig. 1-35 *Machine examples: (a) surface permanent magnet radial synchronous motor for small wind turbines and (b) submersible induction motor for dredge applications (Source: Ingeteam. Reproduced with permission of Ingeteam)*

high-power versions, or in low-voltage or medium-voltage, depending on the needs of the application. In most of their versions, they operate with sinusoidal voltages and currents. Only the brushless DC machine requires non-sinusoidal voltages and currents for the proper operation. The most typical induction machine is the cage rotor induction machine, or asynchronous machine. It is employed in many applications, and the obtained designs are relatively simple and robust. On the other hand, the wound rotor induction machine (WRIM) or doubly fed induction machine is also quite popular. It must be supplied through the rotor and the stator and it is typically employed in wind turbines among other applications.

Finally, with synchronous machines it is possible to obtain a wide spectrum of machine types. Table 1-4 only shows what can be considered the most representative ones. Nowadays, the designs with permanent magnets in the rotor, creating a fixed rotor flux, are very popular. Depending on the direction in which the flux at the rotor is created by the magnets, they can be axial, transversal, or radial synchronous machines. At the time of writing, the most employed design is the radial one. Depending on how the magnets are located at the rotor, they can be interior or surface permanent magnet synchronous machines (PMSM). Finally, depending also on the nature of the stator windings, the machine must be supplied with trapezoidal or sinusoidal voltages. Probably the most employed synchronous machines are radial sinusoidal machines with interior or surface permanent magnets. Fig. 1-35 illustrates some of the above-mentioned machines.

In Chapters 2 and 3, the model and the control of these two machines are analyzed in detail.

1.4.3   Control strategies

This subsection introduces briefly some of the most commonly employed control strategies or philosophies for AC machines. As has been described already, the AC machine of an electric drive must be supplied by the power electronic converter and is typically controlled by a control strategy. Since the beginning of the development of control techniques for AC machines in the early 1970s, many controls have been created. Some of them are based on very simple control approaches; other (extreme) cases are based on really sophisticated control algorithms. This subsection does not intend to classify all existing control strategies. Instead, it offers a simplified view of the most typically employed control strategies for most applications associated with commercially available products or devices. Thus, although it is possible to find a great many control techniques in specialized literature, only a few are mentioned here. Hence, Table 1-5 illustrates a simplified classification of

Table 1-5   *Variable frequency control for AC machines*

Type	Designation
Scalar-based control	V/F control
Vector-based control	Field-oriented
	• Rotor flux oriented vector control
	○ Direct
	○ Indirect
	• Stator flux oriented vector control
	○ Direct
	○ Indirect
	• Airgap flux oriented vector control
	○ Direct
	○ Indirect
	Direct torque control (or direct power control)
	• Direct torque space vector modulation
	• Circle flux trajectory
	○ Constant switching variable hysteresis
	○ Constant switching predictive
	• Hexagon flux trajectory
	Predictive control

variable frequency controls for AC machines. This classification is divided into two groups: scalar-based control and vector-based control. The scalar-based control, in essence, is a very simple and easy-to-implement control philosophy. It does not provide very sophisticated performances in terms of control accuracy, dynamic response, and so on. It was used much more when the control hardware platforms were not as developed as they today and implementation simplicity was a big issue. However, nowadays, thanks to the advance of the control hardware devices such as digital signal processors (DSPs) or microprocessors, vector-based controls are most often used, since it is possible to obtain better control performances with them. Nevertheless, in those applications where simplicity and cost is of great importance, scalar-based control can still be a useful option.

Vector-based control strategies are nowadays the most extensively used ones in electric drives that can be found in commercially available devices or applications. Some of these vector-based control techniques are studied in some depth in Chapters 2 and 3 as they are applied to the control of induction and synchronous machines respectively. Vector-based control techniques, as their name suggests, are based on space vector representation of the most important variables of the machine that is being controlled, such as: voltages, currents, and fluxes. More details about the basic mathematical principles of these controls is provided later. As can be seen in Table 1-5, the first vector-based control categorized group is field-oriented control. Depending on the flux that is utilized as reference for the space vector representation, we can develop rotor, stator, or airgap flux oriented vector control. Additionally, these vector controls can be of direct philosophy and indirect philosophy. In essence, all of these combinations of control techniques provide good performance, but later we will see that probably the most equilibrated and reliable one is the indirect rotor flux oriented vector control, which is probably the one most extensively used by drive manufacturers. Then there is also the direct torque control (or direct power control) philosophy. Within this philosophy, there are also many variants, but Table 1-5 only covers the most popular ones. Direct torque space vector modulation can be said to be very similar to classic vector controls (mentioned earlier), but the main difference is that current loops are omitted. On the other hand, direct torque control with circle flux trajectory is characterized by an absence of modulation, since the pulses for the controlled switches are directly generated by the control itself. This type of control provides very quick dynamic responses, but one drawback, is that they do not make the converter's switches, nor the machine, operate at a constant switching frequency. This has meant that several different control variants that enable one to operate at a constant or near constant switching frequency, such as the variable hysteresis or predictive direct torque controls, have been developed. In addition, hexagon flux trajectory direct torque—better known as direct self-control (DSC)—is another variant of this control family that is traditionally employed in railway traction applications, where the number of commutations of the switches of the converter is limited.

With regards to the modulation techniques which must be employed in communion with the control strategy itself, in order to create the pulses for the controlled switches, as illustrated in Fig. 1-11, it can be affirmed that there exist two main alternative modulation techniques: pulse-width modulation (PWM) and space vector modulation (SVM). Each family of modulation presents different variants and alternatives as well. In specialized literature [1]–[4] and also briefly in subsequent chapters, it is possible to find reach descriptions of such modulations. Nevertheless, it can be said that SVM may be more employed in electric drives, since in general it can obtain better performance with them and be more versatile or adaptable to the needs of the application compared with PWM.

Finally, it is possible to highlight a newer tendency of control philosophies, known as predictive control. By taking advantage of the computational power capacity of newer control hardware devices, this is a more sophisticated control philosophy firmly based on the model of the machine, which achieves the control of the most important magnitudes based on predictions rather than on classic automatic control. Since often this predictive control philosophy is based on space vector representation of the machine, it has been classified as a vector-based control technique.

1.4.4 AC and DC voltage sources

To conclude, this subsection shows some examples of AC and DC voltage sources employed for supplying electric drives. Table 1-6 shows a classification of the most representative voltage sources. For most of the applications, the typical AC voltage source is the electric AC grid, also known as AC utility or AC network. Nowadays, it is created from a combination of AC generator (fixed-speed wind turbines, electric power plants, etc.) and converter-based generation (photovoltaic, thermo-solar, full scale converter based wind turbines, etc.). Depending on the country can reach at the user at different voltage levels and different frequencies.

On the other hand, for railway traction a different voltage source is normally employed which is typically created from the AC grid. To these dedicated voltage sources the name "catenary" is typically used. It is possible to find DC catenaries and AC catenaries of different voltage levels and different frequencies, as is shown in Chapter 5, dedicated to railway traction. In addition, for instance, in hybrid electric vehicles or ships, an alternative AC voltage source is created by one or several AC generators driven by diesel engines. Finally, typically found DC voltage sources are, as mentioned earlier, DC catenaries in railway traction applications and batteries in applications such as electric vehicles.

Table 1-6 *AC and DC voltage sources*

Source	Type
AC	Electric AC grid
	Catenaries for railway traction
	AC generators
DC	Batteries
	Catenaries for railway traction
	Super-capacitors

1.5 Future challenges for electric drives

This section gives a general, and very simplified, view of the challenges facing electric drives and their continued evolution.

As can be deduced from the previous sections, the electric drive technology employed in many traction applications or devices, from an electronic and electric point of view, depends mainly on the evolution of four basic technologies:

- Electric sources: DC (batteries, super-capacitors, etc.) and AC (generators, utility grids, etc.)
- Power electronic converters: rectifiers, inverters, multilevel conversion, etc.
- Electric machines: synchronous, asynchronous, etc.
- Control algorithms or control strategies: vector control, energy management, etc.

Table 1-7 depicts a particular view of the general future challenges of electric drives used in different applications or devices. This table seeks to clarify how the above-mentioned basic technologies should evolve in order to adapt the electric drives' development to the needs and demands of users or society in general. As

Table 1-7 *General future challenges of electric drives used in different applications or devices*

Challenge	Example
Improving performance	Faster dynamics or more adapted to the user's needs
	Improve reliability
	Reduce energy consumption
	Improve efficiency
	Reduce volume, weight, dimensions, etc.
	Improve comfort: Reduce noise and vibrations, improve acceleration and braking, jerk, etc.
	Increase the life of the equipment.
	Reduce maintenance
Adapting to different operating conditions or demands of the specify of the application (versatility)	Be able to adapt to different needs of power, speeds, voltages, energies, etc.
	Be able to move heavier loads and faster in trains for instance
	Improve the autonomy of electric vehicles for instance
	Elevators at higher buildings moving heavier loads
	Trams that can operate without catenary in longer periods
	Approach to the port of ships only moved by batteries reducing the noise.
Reducing costs	Cheaper elements or parts
	Cheaper manufacturing processes
	Faster production times
Being more environmentally friendly	Reduce emissions
	Save energy
	Reduce noise
	Avoid using polluting materials or polluting manufacturing processes
	Avoid damage to animals and plants and Nature in general
Improve working conditions of people involved in doing the electric drives and devices' different parts	Avoid hard/dangerous/stressing working conditions of workers
	Increase: Motivation, identification with challenges, etc.

Table 1-8   *A simplified classification of materials employed in electric drives*

Types	Examples
Semiconductors and microprocessors	Silicon
	Cupper: base plates
	Silver: union between chips, cold plates
	Aluminum oxide (Al_2O_3): isolation between silicon and base plate
Electric machines and inductances for filters	Magnetic cores: silicon steel (for magnetic sheets), soft magnetic materials in small machines
	Electric circuits: copper and aluminum for coils and squirrel cages
	Housing and end-caps: aluminum, iron (cast)
	Steel for shafts
	Materials for electrical insulation: epoxy, varnish, Nomex 410, plastic, polyimide, polyethylene
	Magnets: rare earths and ferrites
Batteries	Lead acid: Pb, H_2SO_4, H_2O_2
	Nickel cadmium: nickel oxide-hydroxide, cadmium, alkaline electrolyte
	Nickel-metal hydride: nickel hydroxide $Ni(OH)_2$, cadmium, KOH electrolyte
	LiOn: lithium cobalt dioxide ($LiCoO_2$), Lithium manganese oxide spinel ($LiMn_2O_4$) Lithium-nickel cobalt manganese
	Polypropylene for cases
	Cathode chemistries: $LiNiCoAlO_2$, $LiNiMnCoO_2$, $LiMn_2O_4$, $LiFePO_4$
	Anode materials: graphite (natural or synthetic), soft or hard carbon
Ultracapacitors	Carbon, aluminum, organic electrolyte
Other electronic elements	Plastic capacitors: polypropylene, polyethylene, aluminum
	Ceramic capacitors
	Transformers: same materials as electric machines and epoxy (dry transformers), mineral oil, ceramics

summarized, the electric drives for tractions applications should try to improve aspects such as: performance, adapting to different operating conditions, being more environmentally friendly, reducing costs, and improving working conditions. To these ends, the specific development of the basic technologies (converters, machines, controls, and supply sources) should involve trying to meet these general challenges. Later chapters look at the specific needs of four specific applications in turn.

Section 1.5 is intended to be a general, and simplified, approach to future challenges.

Finally, the above-mentioned basic technologies for electric drives are made of different materials. Thus, Table 1-8 shows in a simplified manner an enumeration of the materials employed in electric drives and so gives a view of the basic materials needed to create the different elements or parts of an electric drive.

1.6   Historical evolution

Briefly, this section shows some of the most remarkable achievements or developments in electric drive multidisciplinary technology. There have been many advances in the history of electric drives that cannot be included in this section. Instead, a short representative classification of technological inventions and developments is provided in Table 1-9.

Table 1-9 *A summary of the most notable events in the evolution of electric drive technology [6]–[7]*

1837	Davenport patents the DC motor. Some few years earlier, Jacobi, Strating, and Becker also developed several equivalent DC motors and many other inventors also developed more primitive electric motors or electromagnetic arrangements
1859	Gaston Planté invents the first lead acid battery
1882	Development of the New York City DC distribution system by Edison
1882	Jasmin discovers the phenomenon of semi-conductance and proposes its use in AC rectifying
1885	Ferraris develops the rotating magnetic field by polyphase stator windings
1887–1888	Haselwander builds the first three-phase synchronous generator with salient poles
1888	Tesla develops the commercial wound rotor induction motor
1889–1891	Dobrovolsky develops the cage induction motor
1891	Tesla develops the polyphase alternator and Dobrovolsky develops the first three-phase generator and transformer creating the first complete AC three-phase system
1892	Arons develops the first mercury-arc vacuum valve
1897	Graetz develops the three-phase diode bridge rectifier
1899	Jungner invents the first rechargeable nickel–cadmium battery
1902	Hewitt patents the mercury-arc rectifier
1906	Pickard proposes the silicon valve
1929	Park defines the synchronous machine d–q analytical model in synchronous reference frame
1938	Stanley defines the induction motor model in stationary reference frame
1948	Bipolar transistor is invented by Bardeen, Brattain, and Shockley generating an electronic revolution
1952	General Electric manufactures the first germanium diodes
1954	Texas Instruments produces a silicon transistor with high commercial acceptance
1956	Moll, Tanenbaum, Goldey, and Holonyak invents the thyristor, starting the first era of power devices development
1963	Turnbull develops the selective harmonic elimination PWM
1964	Schonung and Stemmler develop the sinusoidal PWM
1968–1972	Hasse and Blaschke pioneer the indirect vector control and direct vector control
1971–1972	Intel introduces the firsts 4-bit and 8-bit microprocessors
1979	Plunkett publishes the hysteresis band PWM for the AC motor drive
1981	Nabae, Takahashi, and Akagi first publish the three-level neutral diode clamped converter
1982	Pfaff, Weschta, and Wick propose the space vector PWM, which is later developed by Van der Broeck, Skudelny, and Stanke
1983	General Electric invents the insulated-gate bipolar transistor (IGBT)
1984–1986	Depenbrock on the one side and Takahashi and Noguchi on the other develop the direct self-control technique and direct torque control technique
1990	Following the development of Ni-Mh (nickel metal hydride) batteries in the second half of the 20th century, the commercialization starts
1991	Sony commercializes the first lithium-based rechargeable battery, improving the technical specifications of the rest of technologies significantly
1997	ABB introduces the insulated-gate controlled thyristors (IGCT)
1990–2000	Introduction of voltage source converter (VSC) drives in many applications and devices
2000–2016	Introduction of permanent magnet-based motors. Introduction of battery-based energy storage systems in many applications
	New semiconductor technologies

References

[1]   Abu-Rub H, Malinowski M, Al-Haddad K. *Power Electronics for Renewable Energy Systems, Transportation and Industrial Application*, Chichester: John Wiley & Sons, 2014.

[2]   Bose BK. *Power Electronics and Drives*, London: Elsevier, 2006.

[3]   Wu B. *High Power Converters and AC Drives*, Piscataway, NJ: IEEE Press, 2006.

[4]   Mohan N, Undeland T, Robbins R. *Power Electronics*, New York: Wiley InterScience, 1989.

[5]   Kouro S, Rodríguez J, Wu BS, et al. Powering the future of industry: High power adjustable speed drive topologies. *IEEE Industry Applications Magazine* August 2012.

[6]   Bose B. Doing research in power electronics. *Industrial Electronics Magazine* March 2015, 1.

[7]   Various Wikipedia articles.

2

Control of induction machines

Fernando Briz and Gonzalo Abad

2.1 Introduction

Induction machines can be considered as the workhorse of industry, as they are rugged and cheap. Before the advent of electronic power converters the use of induction machines was limited to applications operating at nearly constant speed. The advances in the field of power semiconductors, combined with the development of powerful and cheap digital signal processors and adequate control strategies, have resulted in the widespread industrial use of AC variable speed drives (VSDs). VSDs using three-phase induction motors can be found nowadays in practically all industrial and transportation sectors, for low-voltage and medium-voltage applications, and in a wide power range from fractional horsepower to multi-megawatt rating.

This chapter mainly focuses on describing the most important and established control strategy for induction machines (i.e. the vector control). In this chapter, the motor side converter is supposed to be properly supplied, as represented in Fig. 2-1, by a DC supply source or grid converter.

Thus, in this chapter, first the modeling on the induction machine is studied, which later permits us to study and develop the core of the chapter: vector control strategies for induction machines. This control technique is described in detail, analyzing different variants, types of flux estimations, flux-weakening options, and so on. Finally, the most popular and reliable sensorless control strategies are also studied, and the chapter ends with a short section dedicated to the analysis of the steady state of the induction machine.

2.2 Modeling of induction motors

Analysis of induction machines is often realized using per-phase equivalent circuits. While such equivalent circuits are simple and useful for machines operating in steady state, they are inadequate for the analysis of machines operated from electronic power converters, which are frequently subject to changing operating

Power Electronics and Electric Drives for Traction Applications, First Edition. Edited by Gonzalo Abad.
© 2017 John Wiley & Sons, Ltd. Published 2017 by John Wiley & Sons, Ltd.

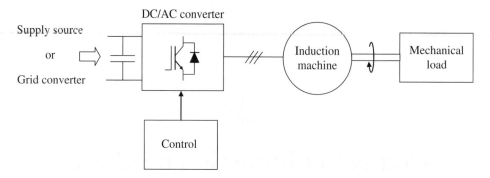

Fig. 2-1   *Basic representation of motor side control of induction machine*

conditions. Complex vector representation provides an alternative mean for the modeling and analysis of three-phase induction machines, and three-phase systems in general. Compared to other notations, complex vectors provide a much simpler and insightful representation of the dynamic effects that physically occur in the machine, i.e. the relationships among voltages, currents and fluxes, as well as electromechanical power conversion.

Dynamic models of three-phase induction machines are presented in this section, and they are used in further sections for the development of high-performance control strategies.

2.2.1   Dynamic model of the induction motor using three-phase variables

Fig. 2-2 shows a simplified winding model of a three-phase, two-pole induction machine, the stator and rotor phase winding being denoted by *as-as'*, *bs-bs'*, *cs-cs'*, and *ar-ar'*, *br-br'*, *cr-cr'* respectively. In this figure, θ_m stands for the angle between the MMF (magneto-motive force) produced by phase *a* of the stator and the MMF produced by phase *a* of the rotor, ω_m being the angular speed in electrical units in rad/s.

If the number of pole pairs of the machine, *p*, is different from 1, the relationship between the mechanical units, θ_{rm}, and electrical units, θ_m, is given by:

$$\theta_{rm} = \frac{\theta_m}{p} \tag{2.1}$$

The stator voltage equation in matrix form is (2.2), R_s being the stator winding resistance and v_{as}, v_{bs}, v_{cs}, i_{as}, i_{bs}, i_{cs}, and ψ_{as}, ψ_{bs}, ψ_{cs} the instantaneous stator voltages, currents and fluxes respectively.

$$\begin{bmatrix} v_{as} \\ v_{bs} \\ v_{cs} \end{bmatrix} = R_s \begin{bmatrix} i_{as} \\ i_{bs} \\ i_{cs} \end{bmatrix} + \frac{d}{dt} \begin{bmatrix} \psi_{as} \\ \psi_{bs} \\ \psi_{cs} \end{bmatrix} \tag{2.2}$$

The stator fluxes can be written in matrix form as function of the stator and rotor currents as (2.3), where L_h is the self-inductance of the stator, which is twice the mutual inductance between stator windings, and $L_{\sigma s}$ is the leakage inductance of the stator windings.

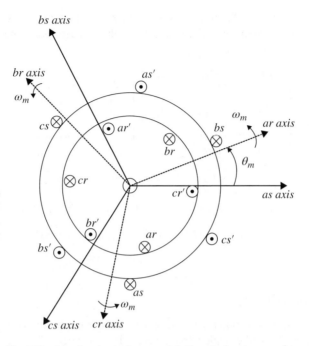

Fig. 2-2 *Idealized winding model of an induction machine*

$$
\begin{bmatrix} \psi_{as} \\ \psi_{bs} \\ \psi_{cs} \end{bmatrix} =
\begin{bmatrix}
L_{\sigma s} + L_h & -\dfrac{1}{2}L_h & -\dfrac{1}{2}L_h \\[2mm]
-\dfrac{1}{2}L_h & L_{\sigma s} + L_h & -\dfrac{1}{2}L_h \\[2mm]
-\dfrac{1}{2}L_h & -\dfrac{1}{2}L_h & L_{\sigma s} + L_h
\end{bmatrix}
\begin{bmatrix} i_{as} \\ i_{bs} \\ i_{cs} \end{bmatrix}
$$

$$
+ L_h
\begin{bmatrix}
\cos\theta_m & \cos(\theta_m + 2\pi/3) & \cos(\theta_m - 2\pi/3) \\[2mm]
\cos(\theta_m - 2\pi/3) & \cos\theta_m & \cos(\theta_m + 2\pi/3) \\[2mm]
\cos(\theta_m + 2\pi/3) & \cos(\theta_m - 2\pi/3) & \cos\theta_m
\end{bmatrix}
\begin{bmatrix} i_{ar} \\ i_{br} \\ i_{cr} \end{bmatrix}
\tag{2.3}
$$

The same procedure can be repeated to derive the rotor electromagnetic equation. The rotor voltage equation is given by (2.4), being R_r the rotor winding resistance and v_{ar}, v_{br}, v_{cr}, i_{ar}, i_{br}, i_{cr} and ψ_{ar}, ψ_{br}, ψ_{cr} the instantaneous rotor voltages, currents and fluxes respectively:

$$
\begin{bmatrix} v_{ar} \\ v_{br} \\ v_{cr} \end{bmatrix} = R_r \begin{bmatrix} i_{ar} \\ i_{br} \\ i_{cr} \end{bmatrix} + \frac{d}{dt} \begin{bmatrix} \psi_{ar} \\ \psi_{br} \\ \psi_{cr} \end{bmatrix}
\tag{2.4}
$$

Similarly, as for the stator circuits, the rotor fluxes can be represented in matrix form as a function of the stator and rotor instantaneous currents as (2.5), where $L_{\sigma r}$ is the leakage inductance of the rotor windings.

$$
\begin{bmatrix} \psi_{ar} \\ \psi_{br} \\ \psi_{cr} \end{bmatrix} = \begin{bmatrix} L_{\sigma r} + L_h & -\dfrac{1}{2}L_h & -\dfrac{1}{2}L_h \\[2mm] -\dfrac{1}{2}L_h & L_{\sigma r} + L_h & -\dfrac{1}{2}L_h \\[2mm] -\dfrac{1}{2}L_h & -\dfrac{1}{2}L_h & L_{\sigma r} + L_h \end{bmatrix} \begin{bmatrix} i_{ar} \\ i_{br} \\ i_{cr} \end{bmatrix}
$$

$$
+ L_h \begin{bmatrix} \cos\theta_m & \cos(\theta_m - 2\pi/3) & \cos(\theta_m + 2\pi/3) \\[1mm] \cos(\theta_m + 2\pi/3) & \cos\theta_m & \cos(\theta_m - 2\pi/3) \\[1mm] \cos(\theta_m - 2\pi/3) & \cos(\theta_m + 2\pi/3) & \cos\theta_m \end{bmatrix} \begin{bmatrix} i_{as} \\ i_{bs} \\ i_{cs} \end{bmatrix} \tag{2.5}
$$

2.2.2   Basics of space vector theory

Given a generic set of three-phase variables, x_a, x_b, and x_c, the corresponding complex space vector, $x_{\alpha\beta}$, is defined as (2.6), a being (2.7). The complex space vector is seen to consist of a real component, x_α, and an imaginary component, x_β.

$$
\vec{x}^s = x_\alpha + j \cdot x_\beta = \frac{2}{3}\left(x_a + a \cdot x_b + a^2 \cdot x_c \right) \tag{2.6}
$$

$$
a = e^{j\frac{2\pi}{3}} \tag{2.7}
$$

Equation (2.6) can be written using a polar form (2.8), X and θ_s being the amplitude and the angle respectively. The constant 2/3 in (2.6) is chosen for convenience to preserve the amplitude of the variables.

$$
\vec{x}^s = X e^{j\theta_s} \tag{2.8}
$$

From (2.6), the relationship between the $\alpha\beta$ components of the complex vector and the *abc* quantities is:

$$
x_\alpha = \mathrm{Re}\{x_{\alpha\beta}\} = \frac{2}{3}\left(x_a - \frac{1}{2}x_b - \frac{1}{2}x_c \right) \tag{2.9}
$$

$$
x_\beta = \mathrm{Im}\{x_{\alpha\beta}\} = \frac{1}{\sqrt{3}}(x_b - x_c) \tag{2.10}
$$

Fig. 2-3 graphically shows the transformation between the *abc* and $\alpha\beta$ systems.

It should be noted that (2.6)–(2.10) do not impose any restriction on the shape of the three-phase quantities, i.e. they do not have to be necessarily sinusoidal nor balanced.

For the particular case of a set of three-phase, sinusoidal, balanced variables of peak value X and frequency ω_s, (2.11), the resulting complex vector $\vec{x}^s$ (2.8) has a magnitude of X and rotates at a constant speed of ω_s, the angle of the resulting complex vector being (2.12).

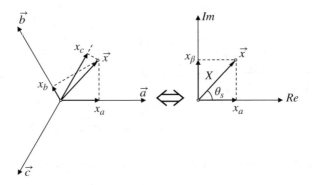

Fig. 2-3   *Three-phase to complex space vector in the αβ axes transformation*

$$x_a = X \cdot \cos(\theta_s + \theta_{s0})$$
$$x_b = X \cdot \cos(\theta_s + \theta_{s0} - 2\pi/3) \tag{2.11}$$
$$x_c = X \cdot \cos(\theta_s + \theta_{s0} + 2\pi/3)$$

$$\theta_s = \int \omega_s dt + \theta_{s0} \tag{2.12}$$

Equations (2.9)–(2.10) can be represented in a matrix form (2.13), with matrix T often being called the direct Clarke transformation. The term x_0 in (2.13) is the zero-sequence component and is needed for the matrix T to be invertible.

$$\begin{bmatrix} x_\alpha \\ x_\beta \\ x_0 \end{bmatrix} = T \begin{bmatrix} x_a \\ x_b \\ x_c \end{bmatrix} \tag{2.13}$$

$$T = \frac{2}{3} \begin{bmatrix} 1 & -\dfrac{1}{2} & -\dfrac{1}{2} \\ 0 & \dfrac{\sqrt{3}}{2} & -\dfrac{\sqrt{3}}{2} \\ \dfrac{1}{2} & \dfrac{1}{2} & \dfrac{1}{2} \end{bmatrix} \tag{2.14}$$

This transformation is schematically represented in Fig. 2-4.
The inverse Clarke transformation is given by:

$$\begin{bmatrix} x_a \\ x_b \\ x_c \end{bmatrix} = T^{-1} \cdot \begin{bmatrix} x_\alpha \\ x_\beta \\ x_0 \end{bmatrix} \tag{2.15}$$

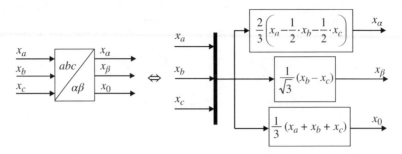

Fig. 2-4   *Transformation from* abc *components to* αβ *components*

$$T^{-1} = \begin{bmatrix} 1 & 0 & 1 \\ -\dfrac{1}{2} & \dfrac{\sqrt{3}}{2} & 1 \\ -\dfrac{1}{2} & -\dfrac{\sqrt{3}}{2} & 1 \end{bmatrix}$$ (2.16)

The space complex vector defined by (2.8) is referred to a stationary reference frame. Complex space vectors that result from applying this transformation to the voltages, currents, and fluxes in induction machines and AC systems in general will rotate in steady state at the fundamental excitation frequency ω_s. It is useful for modeling, analysis, and control purposes to transform the complex vectors from the stationary reference frame to a frame rotating at the same angular frequency as the three-phase variables, which is denoted as the synchronous reference frame. This is done using (2.17), the resulting complex vector in the synchronous reference frame being indicated by an a superscript.

$$\vec{x}^a = \vec{x}^s e^{-j\theta_s}$$ (2.17)

The transformation between the stationary to the rotating reference frame when applied to the Cartesian coordinates is given by (2.18), which is known as the rotational transformation. The components of the resulting complex vector in the synchronous reference frame are indicated by subscripts d and q respectively.

$$\begin{bmatrix} x_d \\ x_q \end{bmatrix} = \begin{bmatrix} \cos\theta_s & -\sin\theta_s \\ \sin\theta_s & \cos\theta_s \end{bmatrix} \cdot \begin{bmatrix} x_\alpha \\ x_\beta \end{bmatrix}$$ (2.18)

The transformation between the stationary and the synchronous reference frames is schematically shown in Fig. 2-5, the corresponding block diagram being shown in Fig. 2-6.

The transformation of a complex vector from the synchronous to the stationary reference frame (2.19) is readily obtained from (2.17). The derivative of the transformation is (2.20). It is used later in this chapter.

$$\vec{x}^s = \vec{x}^a e^{j\theta_s}$$ (2.19)

$$\frac{d\vec{x}^s}{dt} = \frac{d}{dt}\left(\vec{x}^a e^{j\theta_s}\right) = \frac{d\vec{x}^a}{dt} e^{j\theta_s} + j\omega_s \vec{x}^a e^{j\theta_s}$$ (2.20)

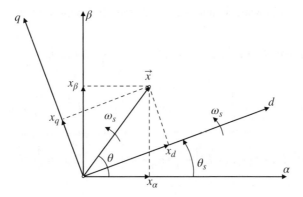

Fig. 2-5   *Transformation between the stationary and the synchronous reference frames*

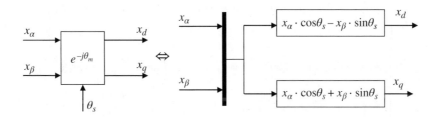

Fig. 2-6   *Transformation from αβ components to dq components*

2.2.3   Dynamic model of the induction machine using complex space vectors

Using the definition of the space complex vectors in (2.6), the corresponding stator voltage, current, and flux linkage space complex space vectors are obtained as (2.21), (2.22), and (2.23) respectively.

$$\vec{v}_s^{\,s} = \frac{2}{3}\left(v_{as} + a \cdot v_{bs} + a^2 \cdot v_{cs}\right) \tag{2.21}$$

$$\vec{i}_s^{\,s} = \frac{2}{3}\left(i_{as} + a \cdot i_{bs} + a^2 \cdot i_{cs}\right) \tag{2.22}$$

$$\vec{\psi}_s^{\,s} = \frac{2}{3}\left(\psi_{as} + a \cdot \psi_{bs} + a^2 \cdot \psi_{cs}\right) \tag{2.23}$$

Combining (2.3) and (2.23), the following expression of the stator flux is obtained [1]:

$$
\begin{aligned}
\vec{\psi}_s^{\,s} =& \left[(L_{\sigma s} + L_h)i_{as} - \frac{1}{2}L_h i_{bs} - \frac{1}{2}L_h i_{cs} + L_h\cos\theta_m i_{ar} + L_h\cos(\theta_m + 2\pi/3)i_{br} + L_h\cos(\theta_m - 2\pi/3)i_{cr}\right] \\
&+ a\left[-\frac{1}{2}L_h i_{as} + (L_{\sigma s} + L_h)i_{bs} - \frac{1}{2}L_h i_{cs} + L_h\cos(\theta_m - 2\pi/3)i_{ar} + L_h\cos\theta_m i_{br} + L_h\cos(\theta_m + 2\pi/3)i_{cr}\right] \\
&+ a^2\left[-\frac{1}{2}L_h i_{as} - \frac{1}{2}L_h i_{bs} + (L_{\sigma s} + L_h)i_{cs} + L_h\cos(\theta_m + 2\pi/3)i_{ar} + L_h\cos(\theta_m - 2\pi/3)i_{br} + L_h\cos\theta_m i_{cr}\right]
\end{aligned} \tag{2.24}
$$

Thus, by considering the following equalities derived from the previous expression:

$$\left[(L_{\sigma s}+L_h)-a\frac{1}{2}L_h-a^2\frac{1}{2}L_h\right]i_{as}=L_h\left[1-a\frac{1}{2}-a^2\frac{1}{2}\right]i_{as}+L_{\sigma s}i_{as}=\frac{3}{2}L_h i_{as}+L_{\sigma s}i_{as} \tag{2.25}$$

$$\left[-\frac{1}{2}L_h+a(L_{\sigma s}+L_h)-a^2\frac{1}{2}L_h\right]i_{bs}=L_h\left[-\frac{1}{2}+a-a^2\frac{1}{2}\right]i_{bs}+aL_{\sigma s}i_{bs}=\frac{3}{2}aL_h i_{bs}+aL_{\sigma s}i_{bs} \tag{2.26}$$

$$\left[-\frac{1}{2}L_h-a\frac{1}{2}L_h+a^2(L_{\sigma s}+L_h)\right]i_{cs}=L_h\left[-\frac{1}{2}-a\frac{1}{2}+a^2\right]i_{cs}+a^2L_{\sigma s}i_{cs}=\frac{3}{2}a^2L_h i_{cs}+a^2L_{\sigma s}i_{cs} \tag{2.27}$$

and

$$L_h[\cos\theta_m+a\cdot\cos(\theta_m-2\pi/3)+a^2\cdot\cos(\theta_m+2\pi/3)]i_{ar}$$

$$=L_h\left[\frac{e^{j\theta_m}+e^{-j\theta_m}}{2}+a\frac{e^{j\left(\theta_m-\frac{2\pi}{3}\right)}+e^{-j\left(\theta_m-\frac{2\pi}{3}\right)}}{2}+a^2\frac{e^{j\left(\theta_m+\frac{2\pi}{3}\right)}+e^{-j\left(\theta_m+\frac{2\pi}{3}\right)}}{2}\right]i_{ar}$$

$$=\frac{L_h}{2}\left[e^{j\theta_m}+e^{-j\theta_m}+a\left(e^{j\theta_m}e^{-j\frac{2\pi}{3}}+e^{-j\theta_m}e^{j\frac{2\pi}{3}}\right)+a^2\left(e^{j\theta_m}e^{j\frac{2\pi}{3}}+e^{-j\theta_m}e^{-j\frac{2\pi}{3}}\right)\right]i_{ar}$$

$$=\frac{L_h}{2}\left[e^{j\theta_m}\left(1+ae^{-j\frac{2\pi}{3}}+a^2e^{j\frac{2\pi}{3}}\right)+e^{-j\theta_m}\left(1+ae^{j\frac{2\pi}{3}}+a^2e^{-j\frac{2\pi}{3}}\right)\right]i_{ar}=\frac{3}{2}L_h e^{j\theta_m}i_{ar} \tag{2.28}$$

$$L_h[\cos(\theta_m+2\pi/3)+a\cdot\cos\theta_m+a^2\cdot\cos(\theta_m-2\pi/3)]i_{br}$$

$$=L_h\left[\frac{e^{j\left(\theta_m+\frac{2\pi}{3}\right)}+e^{-j\left(\theta_m+\frac{2\pi}{3}\right)}}{2}+a\frac{e^{j\theta_m}+e^{-j\theta_m}}{2}+a^2\frac{e^{j\left(\theta_m-\frac{2\pi}{3}\right)}+e^{-j\left(\theta_m-\frac{2\pi}{3}\right)}}{2}\right]i_{br}$$

$$=\frac{L_h}{2}\left[e^{j\theta_m}e^{j\frac{2\pi}{3}}+e^{-j\theta_m}e^{-j\frac{2\pi}{3}}+a\left(e^{j\theta_m}+e^{-j\theta_m}\right)+a^2\left(e^{j\theta_m}e^{-j\frac{2\pi}{3}}+e^{-j\theta_m}e^{j\frac{2\pi}{3}}\right)\right]i_{br}$$

$$=\frac{L_h}{2}\left[e^{j\theta_m}\left(e^{j\frac{2\pi}{3}}+a+a^2e^{-j\frac{2\pi}{3}}\right)+e^{-j\theta_m}\left(e^{-j\frac{2\pi}{3}}+a+a^2e^{j\frac{2\pi}{3}}\right)\right]i_{br}=\frac{3}{2}a\cdot L_h e^{j\theta_m}i_{br} \tag{2.29}$$

$$L_h[\cos(\theta_m-2\pi/3)+a\cdot\cos(\theta_m+2\pi/3)+a^2\cdot\cos\theta_m]i_{cr}$$

$$=L_h\left[\frac{e^{j\left(\theta_m-\frac{2\pi}{3}\right)}+e^{-j\left(\theta_m-\frac{2\pi}{3}\right)}}{2}+a\frac{e^{j\left(\theta_m+\frac{2\pi}{3}\right)}+e^{-j\left(\theta_m+\frac{2\pi}{3}\right)}}{2}+a^2\frac{e^{j\theta_m}+e^{-j\theta_m}}{2}\right]i_{cr}$$

$$=\frac{L_h}{2}\left[e^{j\theta_m}e^{-j\frac{2\pi}{3}}+e^{-j\theta_m}e^{j\frac{2\pi}{3}}+a\left(e^{j\theta_m}e^{j\frac{2\pi}{3}}+e^{-j\theta_m}e^{-j\frac{2\pi}{3}}\right)+a^2\left(e^{j\theta_m}+e^{-j\theta_m}\right)\right]i_{cr}$$

$$=\frac{L_h}{2}\left[e^{j\theta_m}\left(e^{-j\frac{2\pi}{3}}+ae^{j\frac{2\pi}{3}}+a^2\right)+e^{-j\theta_m}\left(e^{j\frac{2\pi}{3}}+ae^{-j\frac{2\pi}{3}}+a^2\right)\right]i_{cr}=\frac{3}{2}a^2L_h e^{j\theta_m}i_{cr} \tag{2.30}$$

the following expression of the stator flux space vector is obtained.

$$\vec{\psi}_s^s = \left(\frac{3}{2}L_h + L_{\sigma s}\right)\vec{i}_s^s + \frac{3}{2}L_h \vec{i}_r^r e^{j\theta_m} \tag{2.31}$$

If a new mutual inductance L_m is defined (2.32), (2.33) is then obtained.

$$L_m = \frac{3}{2}L_h \tag{2.32}$$

$$\vec{\psi}_s^s = (L_m + L_{\sigma s})\vec{i}_s^s + L_m \vec{i}_r^r e^{j\theta_m} \tag{2.33}$$

Defining the stator inductance as:

$$L_s = L_m + L_{\sigma s} \tag{2.34}$$

the following expression of the stator flux is obtained:

$$\vec{\psi}_s^s = L_s \vec{i}_s^s + L_m \vec{i}_r^r e^{j\theta_m} \tag{2.35}$$

The stator voltage equation using complex vector notation (2.36) can be obtained applying (2.6) to (2.2).

$$\vec{v}_s^s = R_s \vec{i}_s^s + \frac{d\vec{\psi}_s^s}{dt} \tag{2.36}$$

And combining (2.35) and (2.36), the following expression for the stator voltage equation is reached.

$$\vec{v}_s^s = R_s \vec{i}_s^s + L_s \frac{d\vec{i}_s^s}{dt} + L_m \frac{d\left(\vec{i}_r^r e^{j\theta_m}\right)}{dt} \tag{2.37}$$

An identical procedure can be followed with the rotor voltage equation (2.4), which is reproduced here for convenience.

$$\begin{bmatrix} v_{ar} \\ v_{br} \\ v_{cr} \end{bmatrix} = R_r \begin{bmatrix} i_{ar} \\ i_{br} \\ i_{cr} \end{bmatrix} + \frac{d}{dt} \begin{bmatrix} \psi_{ar} \\ \psi_{br} \\ \psi_{cr} \end{bmatrix} \tag{2.38}$$

The corresponding voltage, current, and flux linkage complex space vectors are defined by (2.39), (2.40), and (2.41), where superscript r stands for a rotor reference frame:

$$\vec{v}_r^r = \frac{2}{3}\left(v_{ar} + a \cdot v_{br} + a^2 \cdot v_{cr}\right) \tag{2.39}$$

$$\vec{i}_r^r = \frac{2}{3}\left(i_{ar} + a \cdot i_{br} + a^2 \cdot i_{cr}\right) \tag{2.40}$$

$$\vec{\psi}_r^r = \frac{2}{3}\left(\psi_{ar} + a \cdot \psi_{br} + a^2 \cdot \psi_{cr}\right) \tag{2.41}$$

Applying the transformations described by (2.24)–(2.31) to the rotor circuit, the following expression of the rotor flux space vector is obtained:

$$\vec{\psi}_r^r = \left(\frac{3}{2}L_h + L_{\sigma r}\right)\vec{i}_r^r + \frac{3}{2}L_h\vec{i}_s^s e^{-j\theta_m} \tag{2.42}$$

Using the definition of the mutual inductance in (2.32), and defining the rotor inductance as (2.43), the rotor flux can be written as (2.44).

$$L_r = L_m + L_{\sigma r} \tag{2.43}$$

$$\vec{\psi}_r^r = L_r\vec{i}_r^r + L_m\vec{i}_s^s e^{-j\theta_m} \tag{2.44}$$

The space vector voltage equation of the rotor circuit (2.45) is obtained from (2.4) and (2.6):

$$\vec{v}_r^r = R_r\vec{\psi}_r^r + \frac{d\vec{\psi}_r^r}{dt} \tag{2.45}$$

which combined with (2.44) results in:

$$\vec{v}_r^r = R_r\vec{i}_r^r + L_r\frac{d\vec{i}_r^r}{dt} + L_m\frac{d\left(\vec{i}_s^s e^{-j\theta_m}\right)}{dt} \tag{2.46}$$

In squirrel cage induction machines there is no voltage supply to the rotor circuit, the rotor voltage equation being therefore:

$$0 = R_r\vec{i}_r^r + L_r\frac{d\vec{i}_r^r}{dt} + L_m\frac{d\left(\vec{i}_s^s e^{-j\theta_m}\right)}{dt} \tag{2.47}$$

2.2.4   Dynamic model in the stationary reference frame

Transforming the rotor voltage equation (2.47) to the stationary reference frame and using (2.20), (2.48) is obtained. This equation, combined with (2.36), provides the induction machine voltage equations in the stationary reference frame.

$$0 = R_r\vec{i}_r^s + \frac{d\vec{\psi}_r^s}{dt} - j\cdot\omega_m\cdot\vec{\psi}_r^s \tag{2.48}$$

Separating the stator voltage equation (2.36) and the rotor voltage equation (2.48) into their $\alpha\beta$ components, (2.49)–(2.50) and (2.51)–(2.52) are obtained respectively.

$$v_{as} = R_s i_{as} + \frac{d\psi_{as}}{dt} \tag{2.49}$$

$$v_{\beta s} = R_s i_{\beta s} + \frac{d\psi_{\beta s}}{dt} \tag{2.50}$$

$$0 = R_r i_{\alpha r} + \frac{d\psi_{\alpha r}}{dt} + \omega_m \psi_{\beta r} \tag{2.51}$$

$$0 = R_r i_{\beta r} + \frac{d\psi_{\beta r}}{dt} - \omega_m \psi_{\alpha r} \tag{2.52}$$

On the other hand, from (2.35) and (2.44), the corresponding stator and rotor flux space complex vectors in the stationary reference frame are given by (2.53)–(2.54).

$$\vec{\psi}_s^s = L_s \cdot \vec{i}_s^s + L_m \cdot \vec{i}_r^s \tag{2.53}$$

$$\vec{\psi}_r^s = L_m \cdot \vec{i}_s^s + L_s \cdot \vec{i}_r^s \tag{2.54}$$

Taking the real and imaginary components, (2.55)–(2.58) are obtained.

$$\psi_{\alpha s} = L_s i_{\alpha s} + L_m i_{\alpha r} \tag{2.55}$$

$$\psi_{\beta s} = L_s i_{\beta s} + L_m i_{\beta r} \tag{2.56}$$

$$\psi_{\alpha r} = L_m i_{\alpha s} + L_r i_{\alpha r} \tag{2.57}$$

$$\psi_{\beta r} = L_m i_{\beta s} + L_r i_{\beta r} \tag{2.58}$$

The set of equations above can be represented by two electric circuits for the α and β components. This is shown in Fig. 2-7. The stator and rotor circuits of the α and β components are linked by the magnetizing inductance L_m. It is seen that coupling between the $\alpha\beta$ circuits only occurs in the rotor part.

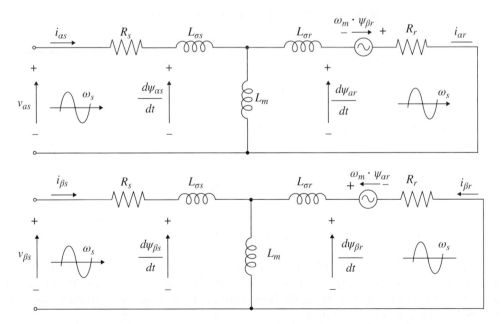

Fig. 2-7   *Electric equivalent circuit in the αβ (stationary) reference frame*

2.2.5   Dynamic models in a synchronous reference frame

The dynamic equations of the stator and rotor circuits of the induction machine in the stationary reference frame (2.36) and (2.48) can be transformed to a reference frame synchronous with the fundamental excitation by multiplying them by $e^{-j\theta_s}$. The resulting stator and rotor voltage equations are (2.59)–(2.60), superscript a denoting space vectors referred to a synchronously rotating reference frame.

$$\vec{v}_s^a = R_s \vec{i}_s^a + \frac{d\vec{\psi}_s^a}{dt} + j\omega_s \vec{\psi}_s^a \tag{2.59}$$

$$0 = R_s \vec{i}_s^a + \frac{d\vec{\psi}_r^a}{dt} + j(\omega_s - \omega_m)\vec{\psi}_r^a \tag{2.60}$$

Taking the real and imaginary components of the stator voltage equation yields:

$$v_{ds} = R_s i_{ds} + \frac{d\psi_{ds}}{dt} - \omega_s \psi_{qs} \tag{2.61}$$

$$v_{qs} = R_s i_{qs} + \frac{d\psi_{qs}}{dt} + \omega_s \psi_{ds} \tag{2.62}$$

Doing the same the rotor voltage equation:

$$0 = R_r i_{dr} + \frac{d\psi_{dr}}{dt} - (\omega_s - \omega_m)\psi_{qr} \tag{2.63}$$

$$0 = R_r i_{qr} + \frac{d\psi_{qr}}{dt} + (\omega_s - \omega_m)\psi_{dr} \tag{2.64}$$

Realizing the same transformation with the stator and rotor fluxes (2.51) and (2.52), the stator and rotor flux equations referenced to the synchronously rotating reference frame are obtained.

$$\vec{\psi}_s^a = L_s \cdot \vec{i}_s^a + L_m \cdot \vec{i}_r^a \tag{2.65}$$

$$\vec{\psi}_r^a = L_m \cdot \vec{i}_s^a + L_s \cdot \vec{i}_r^a \tag{2.66}$$

And decomposing into the real and imaginary parts:

$$\psi_{ds} = L_s i_{ds} + L_m i_{dr} \tag{2.67}$$

$$\psi_{qs} = L_s i_{qs} + L_m i_{qr} \tag{2.68}$$

$$\psi_{dr} = L_m i_{ds} + L_r i_{dr} \tag{2.69}$$

$$\psi_{qr} = L_m i_{qs} + L_r i_{qr} \tag{2.70}$$

Thus, the equivalent electric circuit of the induction machine, with space vectors referenced to the synchronous reference frame is shown in Fig. 2-8. It is similar to the equivalent electric circuit with space vectors referenced to the stationary reference frame. However, coupling terms between the d and q axis sub-circuits exist now both in the stator and in the rotor.

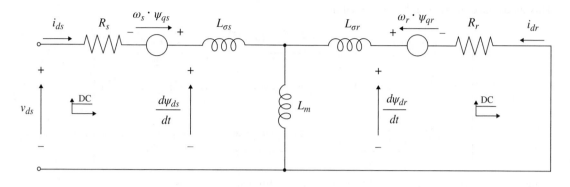

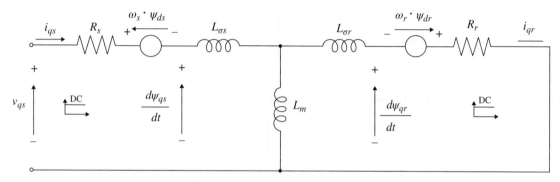

Fig. 2-8   dq *model electric equivalent circuit*

2.2.6   Torque and power equations

The torque equation of a three-phase induction machine can be represented as:

$$T_{em} = \begin{bmatrix} i_{as} & i_{bs} & i_{cs} \end{bmatrix} \cdot \frac{dL_{sr}}{d\theta_m} \cdot \begin{bmatrix} i_{ar} \\ i_{br} \\ i_{cr} \end{bmatrix} \qquad (2.71)$$

being the L_{sr} matrix inductance:

$$L_{sr} = L_h \begin{bmatrix} \cos\theta_m & \cos\left(\theta_m + \dfrac{2\pi}{3}\right) & \cos\left(\theta_m - \dfrac{2\pi}{3}\right) \\ \cos\left(\theta_m - \dfrac{2\pi}{3}\right) & \cos\theta_m & \cos\left(\theta_m + \dfrac{2\pi}{3}\right) \\ \cos\left(\theta_m + \dfrac{2\pi}{3}\right) & \cos\left(\theta_m - \dfrac{2\pi}{3}\right) & \cos\theta_m \end{bmatrix} \qquad (2.72)$$

This inductance matrix can be expressed as:

$$
L_{sr} = L_h
\begin{bmatrix}
\dfrac{e^{j\theta_m} + e^{-j\theta_m}}{2} & \dfrac{e^{j\left(\theta_m + \frac{2\pi}{3}\right)} + e^{-j\left(\theta_m + \frac{2\pi}{3}\right)}}{2} & \dfrac{e^{j\left(\theta_m - \frac{2\pi}{3}\right)} + e^{j\left(\theta_m - \frac{2\pi}{3}\right)}}{2} \\[4mm]
\dfrac{e^{j\left(\theta_m - \frac{2\pi}{3}\right)} + e^{-j\left(\theta_m - \frac{2\pi}{3}\right)}}{2} & \dfrac{e^{j\theta_m} + e^{-j\theta_m}}{2} & \dfrac{e^{j\left(\theta_m + \frac{2\pi}{3}\right)} + e^{-j\left(\theta_m + \frac{2\pi}{3}\right)}}{2} \\[4mm]
\dfrac{e^{j\left(\theta_m + \frac{2\pi}{3}\right)} + e^{-j\left(\theta_m + \frac{2\pi}{3}\right)}}{2} & \dfrac{e^{j\left(\theta_m - \frac{2\pi}{3}\right)} + e^{-j\left(\theta_m - \frac{2\pi}{3}\right)}}{2} & \dfrac{e^{j\theta_m} + e^{-j\theta_m}}{2}
\end{bmatrix}
\tag{2.73}
$$

Using (2.7), the derivative of the inductance matrix with respect to θ_m is:

$$
\frac{dL_{sr}}{d\theta_m} = j \cdot e^{j\theta_m} \cdot \frac{L_h}{2} \cdot
\begin{bmatrix}
1 & a & a^2 \\
a^2 & 1 & a \\
a & a^2 & 1
\end{bmatrix}
- j \cdot e^{-j\theta_m} \cdot \frac{L_h}{2} \cdot
\begin{bmatrix}
1 & a^2 & a \\
a & 1 & a^2 \\
a^2 & a & 1
\end{bmatrix}
\tag{2.74}
$$

the torque equations as a function of the three-phase variables being:

$$
\begin{aligned}
T_{em} = {} & j \cdot e^{j\theta_m} \cdot \frac{L_h}{2} \cdot [i_{as} \ \ i_{bs} \ \ i_{cs}] \cdot
\begin{bmatrix}
1 & a & a^2 \\
a^2 & 1 & a \\
a & a^2 & 1
\end{bmatrix}
\cdot
\begin{bmatrix}
i_{ar} \\
i_{br} \\
i_{cr}
\end{bmatrix} \\[4mm]
& - j \cdot e^{-j\theta_m} \cdot \frac{L_h}{2} \cdot [i_{as} \ \ i_{bs} \ \ i_{cs}] \cdot
\begin{bmatrix}
1 & a^2 & a \\
a & 1 & a^2 \\
a^2 & a & 1
\end{bmatrix}
\cdot
\begin{bmatrix}
i_{ar} \\
i_{br} \\
i_{cr}
\end{bmatrix}
\end{aligned}
\tag{2.75}
$$

which after rearranging terms results in:

$$
\begin{aligned}
T_{em} = {} & j \cdot e^{j\theta_m} \cdot \frac{L_h}{2} \cdot [i_{as} \ \ i_{bs} \ \ i_{cs}] \cdot
\begin{bmatrix}
1 \\
a^2 \\
a
\end{bmatrix}
\cdot [1 \ \ a \ \ a^2] \cdot
\begin{bmatrix}
i_{ar} \\
i_{br} \\
i_{cr}
\end{bmatrix} \\[4mm]
& - j \cdot e^{-j\theta_m} \cdot \frac{L_h}{2} \cdot [i_{as} \ \ i_{bs} \ \ i_{cs}] \cdot
\begin{bmatrix}
1 \\
a \\
a^2
\end{bmatrix}
\cdot [1 \ \ a^2 \ \ a] \cdot
\begin{bmatrix}
i_{ar} \\
i_{br} \\
i_{cr}
\end{bmatrix}
\end{aligned}
\tag{2.76}
$$

Replacing the phase currents by the corresponding space vectors, the following torque equation is obtained, superscript * stating for the complex conjugate of the corresponding space vector.

$$
T_{em} = \frac{9}{4} \cdot \frac{L_h}{2} \cdot \left(j \cdot \vec{i}_s^{\,*} \cdot e^{j\theta_m} \cdot \vec{i}_r - j \cdot \vec{i}_s \cdot e^{-j\theta_m} \cdot \vec{i}_r^{\,*} \right)
\tag{2.77}
$$

By analyzing the last equation, it is possible to distinguish the difference between two conjugated space vectors:

$$\vec{x} = \vec{i}_s \cdot e^{-j\theta_m} \cdot \vec{i}_r^* \tag{2.78}$$

$$\vec{x}^* = \vec{i}_s^* \cdot e^{j\theta_m} \vec{i}_r \tag{2.79}$$

These resulting space vectors are complex conjugated, therefore we have:

$$\vec{x} - \vec{x}^* = (a + j \cdot b) - (a - j \cdot b) = 2 \cdot j \cdot b \tag{2.80}$$

This means that the subtraction of a space vector and its complex conjugate results in two times the imaginary part. Therefore, by using this last result, the torque expression yields:

$$T_{em} = -\frac{9}{4} \cdot \frac{L_h}{2} \cdot j \cdot (2j) \cdot \text{Im}\{\vec{x}\} \tag{2.81}$$

which is equal to:

$$T_{em} = \frac{3}{2} \cdot \left(\frac{3}{2} \cdot L_h\right) \cdot \text{Im}\{\vec{x}\} \tag{2.82}$$

Or equivalently:

$$T_{em} = \frac{3}{2} \cdot L_m \cdot \text{Im}\left\{\vec{i}_s \cdot e^{-j\theta_m} \cdot \vec{i}_r^*\right\} \tag{2.83}$$

It can be noted that the torque is equal to a constant $\left(\frac{3}{2} \cdot L_m\right)$, multiplied by the imaginary part of a product of current space vectors.

Combining the stator and rotor current and flux complex space vectors, several alternative expressions of the electromagnetic torque produced by the machine can be obtained (2.84)–(2.89):

$$T_{em} = \frac{3}{2} \cdot p \cdot \text{Im}\left\{\psi_{\alpha\beta r} \cdot i_{\alpha\beta r}^*\right\} = \frac{3}{2} \cdot p \cdot \left(\psi_{\beta r} \cdot i_{\alpha r} - \psi_{\alpha r} \cdot i_{\beta r}\right) \tag{2.84}$$

$$T_{em} = \frac{3}{2} \cdot p \cdot \frac{L_m}{L_s} \cdot \text{Im}\left\{\vec{\psi}_s^{s*} \vec{i}_r^s\right\} \tag{2.85}$$

$$T_{em} = \frac{3}{2} \cdot p \cdot \text{Im}\left\{\vec{\psi}_s^{s*} \vec{i}_s^s\right\} \tag{2.86}$$

$$T_{em} = \frac{3}{2} \cdot \frac{L_m}{L_r} \cdot p \cdot \text{Im}\left\{\vec{\psi}_r^{s*} \vec{i}_s^s\right\} \tag{2.87}$$

$$T_{em} = \frac{3}{2} \cdot \frac{L_m}{\sigma \cdot L_r \cdot L_s} \cdot p \cdot \text{Im}\left\{\vec{\psi}_r^{s*} \vec{\psi}_s^s\right\} \tag{2.88}$$

$$T_{em} = \frac{3}{2} \cdot L_m \cdot p \cdot \text{Im}\left\{\vec{i}_r^{s*} \vec{i}_s^s\right\} \tag{2.89}$$

the leakage coefficient in (2.88) being defined as:

$$\sigma = 1 - \frac{L_m^2}{L_s L_r} \tag{2.90}$$

The torque equation as a function of the rotor flux and the stator current in the synchronous reference frame (2.91) can be obtained directly from (2.87):

$$T_{em} = \frac{3}{2}\frac{L_m}{L_r}p\,\mathrm{Im}\left\{\vec{\psi}_r^{a*}\cdot\vec{i}_s^{a}\right\} = \frac{3}{2}\frac{L_m}{L_r}p\left(\psi_{dr}\cdot i_{qs} - \psi_{qr}\cdot i_{ds}\right) \tag{2.91}$$

The apparent power is given by (2.92). The term 3/2 is needed to maintain the power between *abc* and the $\alpha\beta$ systems invariant.

$$S = \frac{3}{2}\left(\vec{v}_s^{\,s}\,\vec{i}_s^{\,s*}\right) \tag{2.92}$$

Developing (2.92), (2.93) is obtained, the active power (2.94) and reactive power (2.95) being the real and imaginary components respectively.

$$S = \frac{3}{2}\left(\left(v_\alpha + j\cdot v_\beta\right)\cdot\left(i_\alpha - j\cdot i_\beta\right)\right) = \frac{3}{2}\left(\left(v_\alpha i_\alpha + i_\beta v_\beta\right) + j\left(v_\beta i_\alpha - v_\alpha i_\beta\right)\right) \tag{2.93}$$

$$P = \frac{3}{2}\left(v_\alpha i_\alpha + i_\beta v_\beta\right) \tag{2.94}$$

$$Q = \frac{3}{2}\left(v_\beta i_\alpha - v_\alpha i_\beta\right) \tag{2.95}$$

2.3 Rotor flux oriented vector control

Referring the voltage, current, and flux complex space vectors to a reference frame which rotates synchronously with the fundamental excitation has been shown to convert quantities which are AC in steady state into DC. While the speed of rotation of such a synchronous reference frame is well defined, there are in principle infinite implementations, depending upon the initial angle used for the integration in (2.12). Implementation in which the synchronous reference frame is aligned with the fundamental wave of one of the fluxes in the machine (rotor, stator, or airgap) is advantageous for analysis and control purposes [2]. Consequently, three different types of vector control schemes can be considered, namely:

- rotor flux oriented vector control;
- stator flux oriented vector control;
- airgap flux oriented vector control.

In rotor flux oriented control, also called field-oriented control, all the variables are referred to a reference frame aligned with the rotor flux. By doing this, the rotor flux and the torque produced by the machine can be

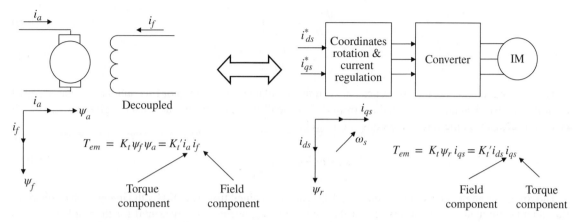

Fig. 2-9 *Separately excited DC motor versus vector controlled induction motor*

controlled separately through the d and q axis components of the stator current, eventually resulting in simpler control structures. Rotor flux oriented vector control follows therefore a philosophy similar to the control of separately excited DC machines. This equivalence is graphically shown in Fig. 2-9. Referring the stator currents to a reference frame aligned with the rotor flux allows an independent control of flux and torque, similarly as for the DC machine, with i_{ds} controlling the flux, i_{qs} being used to control the torque.

Conversely, decoupling between the d and q axis currents for the control of flux and torque is lost when stator or airgap fluxes are used as the reference frame. Owing to this, rotor flux oriented control has become the most popular choice. Only this strategy is covered in this book.

Initially, two different implementations were developed using rotor flux oriented vector control:

- direct vector control (feedback method), first introduced by Blaschke [3];
- indirect vector control (feedforward method), first introduced by Hasse [4].

2.3.1 Fundamentals of rotor flux oriented control

The fundamentals of rotor flux oriented control can be deduced from the analysis of the rotor voltage equations (2.63)–(2.64) and the rotor flux equations (2.69)–(2.70), in the synchronous reference frame. Since the d axis is defined to be aligned with the rotor flux, then the q axis component of the rotor flux is zero, (2.96)–(2.99) being then obtained.

$$\frac{d\psi_{dr}}{dt} + R_r i_{dr} = 0 \tag{2.96}$$

$$R_r i_{qr} - (\omega_s - \omega_m)\psi_{dr} = 0 \tag{2.97}$$

$$\psi_{dr} = L_m i_{ds} + L_r i_{dr} \tag{2.98}$$

$$L_m i_{qs} + L_r i_{qr} = 0 \tag{2.99}$$

Combining (2.96) and (2.98), results in the rotor flux dynamic equation (2.100) being obtained. The rotor flux is seen to depend only on the d axis component of the stator current, the dynamics being those of a first-order system governed by the rotor time constant τ_r (2.101).

$$\tau_r \frac{d\psi_{dr}}{dt} + \psi_{dr} = L_m i_{ds} \tag{2.100}$$

$$\tau_r = \frac{L_r}{R_r} \tag{2.101}$$

The slip frequency—i.e. the difference between the angular speed of the rotor flux and the angular speed of the rotor (in electrical units)—is obtained from (2.97) and (2.99). The slip frequency ωr is given by (2.102); it is seen to depend on the rotor time constant as well.

$$\omega_r = \omega_s - \omega_m = \tau_r \frac{L_m}{\psi_{dr}} i_{qs} \tag{2.102}$$

The rotor flux angle relative to the stator (2.103) can be obtained from (2.102) by the integration of the rotor flux angular frequency. It is observed that the rotor angle θ_m needs to be measured for the implementation of (2.103).

$$\theta_s = \int \omega_s dt = \int \left(\omega_m + \tau_r \frac{L_m}{\psi_{dr}} i_{qs} \right) dt = \theta_m + \int \tau_r \frac{L_m}{\psi_{dr}} i_{qs} dt \tag{2.103}$$

Fig. 2-10 schematically shows the implementation of (2.100) and (2.103) to obtain the rotor flux magnitude and phase angle. A time derivative has been replaced by the Laplace transform in this figure.

The torque equation in the rotor flux coordinates can be obtained particularizing (2.91) for the case of $\psi_{qr} = 0$.

$$T_{em} = \frac{3}{2} \frac{L_m}{L_r} p \psi_{dr} \cdot i_{qs} \tag{2.104}$$

The q axis component of the rotor current in the rotor flux reference frame is given by (2.105), which is obtained from (2.99).

$$i_{qr} = -\frac{L_m}{L_r} \cdot i_{qs} \tag{2.105}$$

On the other hand, the d axis component of the rotor current is obtained from (2.96), and is seen to be zero if the rotor flux magnitude remains constant.

Fig. 2-11 shows the rotor flux and stator current space vectors in a synchronous reference frame when the d axis is aligned with the rotor flux.

Fig. 2-12 and Fig. 2-13 illustrate the transformations of the stator currents needed to realize field oriented control of an induction machine. The transient shown in the figure corresponds to a speed reversal in which the

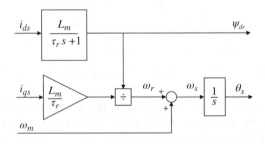

Fig. 2-10   *Block diagram showing the rotor flux angle and magnitude dynamic model*

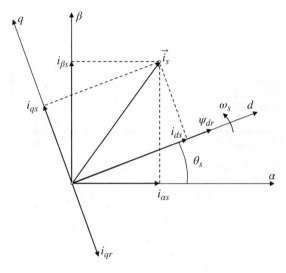

Fig. 2-11   *Stator and rotor current vector components in the stationary and rotor flux reference frames*

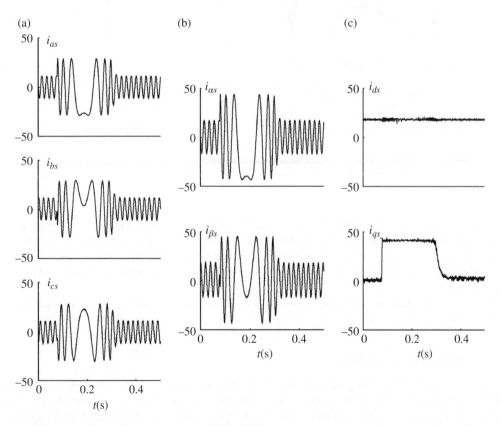

Fig. 2-12   *(a) Phase currents, (b) components of the resulting stator current vector in the stationary reference frame ($i_{\alpha s}$, $i_{\beta s}$), and (c) in the synchronous reference frame (i_{ds}, i_{qs})*

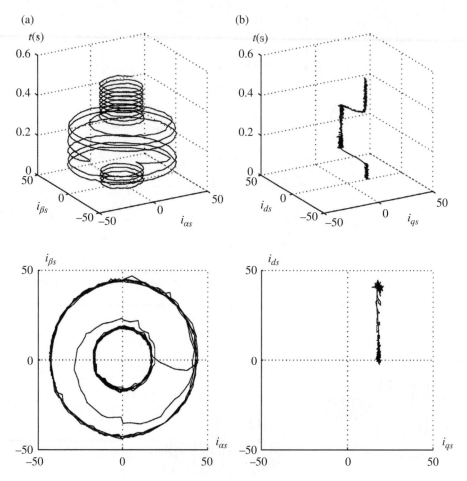

Fig. 2-13   *Stator current complex vector (a) in the stationary reference frame and (b) in the synchronous reference frame*

machine operates with rated rotor flux (the speed is not shown in the figure). Fig. 2-12 shows the phase currents, as well as the components of the stator current vector in the stationary and synchronous reference frame, obtained after applying (2.13) and (2.18) respectively. It can be observed from Fig. 2-12(c) that the d axis current, and therefore the rotor flux, is maintained constant. The torque is therefore proportional to the q axis current, as indicated by (2.104).

Fig. 2-13 shows the stator current complex vector both in the stationary and synchronous reference frames. In the upper subplot, the time is explicitly shown in the z axis. In the lower subplot the time is not shown.

2.3.2   The stator voltage equation

It has been shown in the preceding sections that the rotor flux and torque produced by the machine can be independently controlled through the d and q axis components of the stator current respectively. However, most of electronic power converters used in electric drives operate as a voltage source (i.e. they apply a voltage to the machine), the resulting current being a function of the machine parameters and of its operating point.

Combining (2.59), (2.65), and (2.66), it is possible to express the stator voltage as a function of the stator currents and the rotor flux in a rotor flux synchronous reference frame:

$$v_{ds} = R_s i_{ds} + \sigma L_s \frac{di_{ds}}{dt} - \omega_s \sigma L_s i_{qs} + \frac{L_m}{L_r}\frac{d\psi_{dr}}{dt} \tag{2.106}$$

$$v_{qs} = R_s i_{qs} + \sigma L_s \frac{di_{qs}}{dt} + \omega_s L_s i_{ds} - \omega_s \frac{L_m}{R_r}\frac{d\psi_{dr}}{dt} \tag{2.107}$$

Defining the stator transient resistance (2.108), these equations can be rearranged as (2.109)–(2.110):

$$R_s' = R_s + R_r \frac{L_m^2}{L_r^2} \tag{2.108}$$

$$v_{ds} = R_s' i_{ds} + \sigma L_s \frac{di_{ds}}{dt} - \omega_s \sigma L_s i_{qs} - R_r \frac{L_m}{L_r^2}\psi_{dr} \tag{2.109}$$

$$v_{qs} = R_s' i_{qs} + \sigma L_s \frac{di_{qs}}{dt} + \omega_s \sigma L_s i_{ds} + \omega_m \frac{L_m}{L_r}\psi_{dr} \tag{2.110}$$

These equations can be further simplified. The cross-coupling terms in the right side of (2.109) and (2.110), namely (2.111)–(2.112), are due to the coordinate transformation of the stator equivalent RL circuit to the synchronous reference frame.

$$v_{ccd} = -\omega_s \sigma L_s i_{qs} \tag{2.111}$$

$$v_{ccq} = \omega_s \sigma L_s i_{ds} \tag{2.112}$$

On the other hand, the rotor flux dependent terms in the right side of (2.109) and (2.110), namely (2.113)–(2.114) account for the effect of the rotor flux on the stator voltage (i.e. the back-emf).

$$e_d = -R_r \frac{L_m}{L_r^2}\psi_{dr} \tag{2.113}$$

$$e_q = \omega_m \frac{L_m}{L_r}\psi_{dr} \tag{2.114}$$

Combining (2.111)–(2.112) and (2.113)–(2.114) into the equivalent terms c_{ds} and c_{qs} (2.115)_(2.116), the stator voltage equations can be written now as (2.117)–(2.118).

$$c_{ds} = e_d + v_{ccd} = -\omega_s \sigma L_s i_{qs} - R_r \frac{L_m}{L_r^2}\psi_{dr} \tag{2.115}$$

$$c_{qs} = e_q + v_{ccq} = \omega_s \sigma L_s i_{ds} + \omega_m \frac{L_m}{L_r}\psi_{dr} \tag{2.116}$$

$$v_{ds} = R_s' i_{ds} + \sigma L_s \frac{di_{ds}}{dt} + c_{ds} \tag{2.117}$$

$$v_{qs} = R_s' i_{qs} + \sigma L_s \frac{di_{qs}}{dt} + c_{qs} \tag{2.118}$$

It can be observed from (2.117)–(2.118) that if the terms c_{ds} and c_{qs} are neglected the relationship between the d and q components of the stator voltage and current corresponds to an *RL* load (i.e. a first-order system). Accurate, high bandwidth control of the current is therefore possible with proper current regulator designs. The terms c_{ds} and c_{ds} can be seen therefore as disturbances, which can be either explicitly decoupled or compensated by the current regulators. It is noted in this regard that both c_{ds} and c_{qs} consist of two different parts that have very different characteristics: cross-coupling terms and back-emf terms.

As already mentioned, the cross-coupling voltages v_{ccd} and v_{ccq} in (2.111) and (2.112) result from the transformation of the stator winding corresponding transient impedance to the synchronous reference frame. These voltages vary proportional to the d and q axis components of the stator current, and to the fundamental frequency. The stator current vector in current-regulated drives (especially the q axis component) can change very quickly (from zero to its rated value in ms or even less). This means that the voltages in v_{ccd} and v_{ccq} can change very quickly as well, especially v_{ccd}. These effects will be more relevant at high fundamental excitation frequencies, owing to the presence of ω_s in the equations. It is concluded that v_{ccd} and v_{ccq} can produce large disturbances with very fast dynamics, which can therefore have significant adverse effects on the current dynamics when the drive operates at high speeds.

On the other hand, the induced voltages due to the back-emf e_d and e_q, (2.113) and (2.114), account for the effect of the rotor flux on the stator voltage. Both terms are proportional to the rotor flux, e_q also being proportional to the rotor speed. Since the rotor flux in vector controlled machines is kept constant, or changes relatively slowly (e.g. in the field weakening region), back-emf voltage will be characterized by slow dynamics. It is also noted that the rotor speed in electric drives change relatively slowly compared to the stator current dynamics. Consequently, e_d and e_q can be considered disturbances with relatively slow dynamics. Because of this, their effects are easier to compensate.

2.3.3   Synchronous current regulators

The voltage commands for v_{ds} and v_{qs} required to obtain the desired currents i_{ds} and i_{qs} could in principle be obtained in a feedforward manner using (2.109) and (2.110). However, the sensitivity of these equations to machine parameters as well as the need to calculate the currents' derivatives present in the right hand of these equations make this solution highly inadvisable. Instead, feedback-based solutions using current regulators are preferred, owing to their simplicity and robustness.

Current regulation of AC drives presents two distinguishing characteristics:

- Though all the three-phase currents need to be controlled simultaneously, only two independent variables exist, as the phase currents add to zero. Any strategy trying to control the three-phase current using three independent current regulators is therefore intrinsically incorrect.
- Contrary to most systems, the electrical variables in AC machines are sinusoidal in steady state, with frequencies that can go from zero up to several hundred Hz or even higher. Because of this, the use of an integral action in the controller is no longer capable of providing zero steady-state errors, assuming the currents are controlled in the stationary reference frame.

A variety of current regulator designs have been proposed to realize the current control in AC drives, but a thorough discussion of this topic is beyond the scope of this chapter. Among all the proposed solutions, PI regulators implemented in a synchronous reference frame are widely accepted as the standard solution for current regulation in vector controlled AC drives. The resulting current regulator design is called the synchronous PI current regulator [1], [5], [6].

Controlling the d and q axis currents in a synchronous reference using a PI regulator provides an appropriate solution to the two aforementioned problems. Since in the synchronous reference frame the AC variables

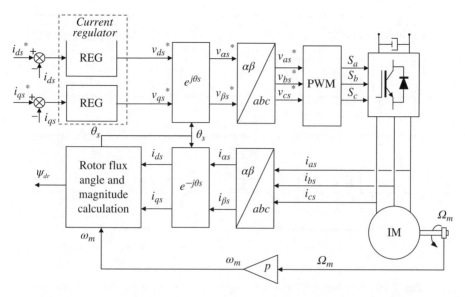

Fig. 2-14 *Block diagram of a current-regulated induction motor drive. It can be noted that the method may be implemented measuring only two stator currents, as all three stator currents add up to zero*

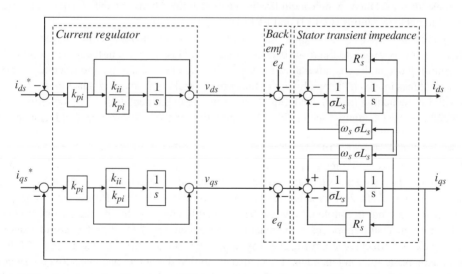

Fig. 2-15 *Synchronous PI current regulator, without back-emf and cross-coupling decoupling*

become DC in steady state, the integral action of the PI design guarantees zero steady-state error. Also, only two independent regulators are used (one for the d and one for the q axis), which is consistent with the fact that only two independent currents exist.

Fig. 2-14 shows the schematic representation of a current-regulated AC drive using synchronous PI current regulators.

Fig. 2-15 shows a more detailed representation of the current regulator structure. The electronic power converter and the coordinate transformations have been eliminated for the sake of simplicity, all the variables

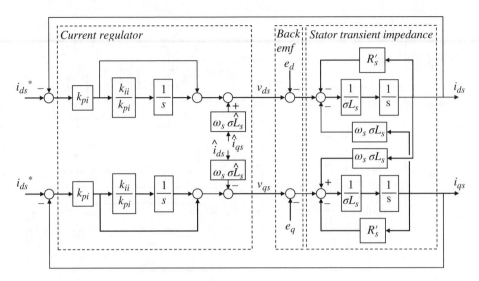

Fig. 2-16   *Synchronous PI current regulator with cross-coupling decoupling*

being shown in the synchronous reference frame. The induction machine in this figure is modeled by means of the stator transient impedance, which includes the cross-coupling between the *d* and *q* axes ((2.111) and (2.112)) and the back-emf terms ((2.113) and (2.114)).

As already mentioned, the cross-coupling terms can have a severe adverse effect on the current regulator performance, especially owing to the disturbances produced by fast changes in the *q* axis current. Because of this, cross-coupling decoupling is advisable, especially in machines which need to operate at very high frequencies, or machines in which the bandwidth of the current regulator is low (e.g. machines fed from inverters operating with low switching frequencies). Fig. 2-16 shows the implementation of the synchronous PI current regulator including cross-coupling decoupling. An estimation of the stator transient inductance is needed in this case.

The design of the synchronous PI current regulator can be further refined by including decoupling of the back-emf, the corresponding implementation being shown in Fig. 2-17. It should be noted, however, that, in general, decoupling the back-emf is less critical than cross-coupling decoupling. This is due to the fact that the dynamics of the back-emf are slow by nature, as they are governed by the rotor flux and the rotor speed. Owing to this, even if they are not decoupled, they can be acceptably compensated by the current regulator.

Fig. 2-18 shows the resulting model of the machine and current regulator if perfect compensation of cross-coupling and back-emf effects is achieved. These dynamics are seen to correspond to a simple RL load, control of the *d* and *q* axis being fully independent. It can also be observed that the machine show the same dynamics for the *d* and *q* axis, meaning that the same set of gains can be used in both current regulators. It should be noted that the implementation shown in Fig. 2-18 requires knowledge of several machines' parameters, which can raise parameter sensitivity concerns.

2.3.3.1   *Current regulator tuning using pole-zero cancellation*

Tuning of the synchronous PI current regulators can be addressed using well-known tools for linear systems. If ideal decoupling of the effects due to cross-coupling and the back-emf is assumed, the dynamic behavior of the *d* and *q* axis currents is given by (2.119).

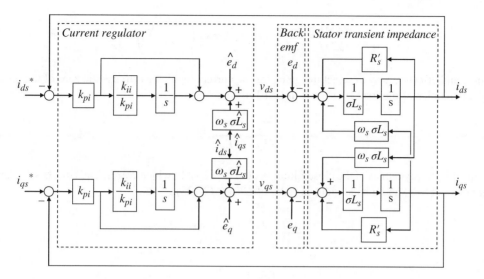

Fig. 2-17 *Synchronous PI current regulator with cross-coupling and back-emf decoupling*

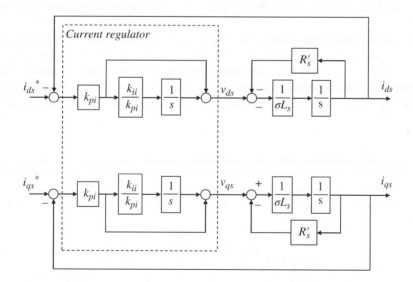

Fig. 2-18 *Equivalent model with ideal cross-coupling and back-emf decoupling*

$$\frac{i_{ds}(s)}{v_{ds}(s)} = \frac{i_{qs}(s)}{v_{qs}(s)} = \frac{1}{\sigma L_s s + R_s'} \tag{2.119}$$

On the other hand, the transfer function of the synchronous PI current regulators in the synchronous reference frame are given by (2.120), with ε_{ds} and ε_{qs} being the error between the corresponding current commands and the actual current for the d and q axis respectively.

$$\frac{v_{ds}^*(s)}{\varepsilon_{ds}(s)} = \frac{v_{qs}^*(s)}{\varepsilon_{qs}(s)} = k_{pi} + \frac{k_{ii}}{s} = k_{pi}\left(\frac{s + \frac{k_{ii}}{k_{pi}}}{s}\right) \tag{2.120}$$

A simple and effective way to select the gains for the current regulator is to use the zero of the regulator to cancel the dynamics of the load, as shown in (2.121).

$$\frac{k_{ii}}{k_{pi}} = \frac{R_s'}{\sigma L_s s} \tag{2.121}$$

By doing this, the dynamics of the closed-loop system are (2.122), which corresponds to a first-order system with a gain equal to 1, and a bandwidth in rad/s of *CRBW* (2.123).

$$\frac{i_{ds}(s)}{i_{ds}^*(s)} = \frac{i_{qs}(s)}{i_{qs}^*(s)} = \frac{k_{pi}}{\sigma L_s s + k_{pi}} \tag{2.122}$$

$$CRBW = \frac{k_{pi}}{\sigma L_s} \tag{2.123}$$

Therefore, knowing the desired bandwidth of the current regulator *CRBW*, the proportional gain of the current regulator k_{pi} is obtained using (2.123), the integral gain needed to obtain pole-zero cancellation k_{pi} being given by (2.121).

It is deduced from the preceding discussion that implementation of pole-zero cancellation requires knowledge of the stator transient impedance and resistance. These parameters can be estimated relatively easy and with adequate accuracy. It is noted in this regard that even if the pole-zero cancellation described by (2.121) is not ideal, this does not normally produce a significant deterioration of the current regulator response and does not affect its stability, as the system pole is located in the stable half of the *s* plane, and relatively far from the origin.

2.3.3.2   Current regulator tuning using a second-order equivalent system

Alternatively to the use of pole-zero cancellation, the dynamics of the current control loop can be approximated by a second-order system. Consequently, by deriving the closed-loop transfer function, from the block diagrams we get the equation in *d* coordinate:

$$R_s' \cdot i_{ds} + \sigma L_s \frac{di_{ds}}{dt} = \left(k_{pi} + \int k_{ii}dt\right) \cdot (i_{ds}^* - i_{ds}) \tag{2.124}$$

Applying the Laplace transform:

$$R_s' \cdot i_{ds} + s(\sigma L_s)i_{ds} + \left(k_{pi} + \frac{k_{ii}}{s}\right)i_{ds} = \left(k_{pi} + \frac{k_{ii}}{s}\right)i_{ds}^* \tag{2.125}$$

Resulting in the following second-order transfer function:

$$\frac{i_{ds}(s)}{i_{ds}^*(s)} = \frac{sk_{pi} + k_{ii}}{s^2(\sigma L_s) + s(R_s' + k_{pi}) + k_{ii}} \tag{2.126}$$

Identical results are obtained for the q axis current:

$$\frac{i_{qs}(s)}{i_{qs}^*(s)} = \frac{sk_{pi} + k_{ii}}{s^2(\sigma L_s) + s(R_s' + k_{pi}) + k_{ii}} \tag{2.127}$$

The closed-loop dynamics of the current loops can be adjusted by any known classic method, by appropriately selecting the k_{pi} and k_{ii} values of the PI regulators. In general, both loops are equally tuned. The closed-loop dynamics are determined by a second-order denominator (two poles that can be real or imaginary, depending on the k_{pi} and k_{ii} choice) and a first-order numerator (one zero that can also be selected). The closed-loop dynamics (location of the closed-loop poles) are often chosen to be faster than those of the open-loop system. Then, the sampling time where the controller is going to be implemented must be accordingly chosen. In addition, always the current closed-loop dynamics are tuned to be significantly faster than the flux and speed loops dynamics.

Tuning example   It is possible to use many different methods provided by classic control theories to tune the PI regulators of the closed current control loops. This subsection shows a simple method, in which the closed-loop denominators of equations (2.126) and (2.127) are matched to a second-order denominator:

$$s^2(\sigma L_s) + s(R_s' + k_{pi}) + k_{ii} \equiv s^2 + 2\xi\omega_n s + \omega_n^2 \tag{2.128}$$

ω_n being the natural frequency and ξ the damping ratio. Thus, by developing the denominator of the closed-loop system, we obtain:

$$s^2 + s\left(\frac{R_s' + k_{pi}}{\sigma L_s}\right) + \frac{k_{ii}}{\sigma L_s} \equiv s^2 + 2\xi\omega_n s + \omega_n^2 \tag{2.129}$$

which means that the required k_{pi} and k_{ii} values are:

$$k_{ii} = \sigma L_s \omega_n^2 \tag{2.130}$$

$$k_{pi} = \sigma L_s 2\xi\omega_n - R_s' \tag{2.131}$$

Taking, for instance, double real poles, the damping ratio is $\xi = 1$, which yields:

$$\xi = 1 \rightarrow (s + \omega_n)^2 \tag{2.132}$$

Therefore:

$$s^2 + s\left(\frac{R_s' + k_{pi}}{\sigma L_s}\right) + \frac{k_{ii}}{\sigma L_s} \equiv (s + \omega_n)^2 \tag{2.133}$$

The poles of the closed-loop system (ω_n) can be chosen by the designer to be x times faster than the open-loop poles $\dfrac{\sigma L_s}{R_s'}$.

$$\xi = 1, \quad \omega_n = \dfrac{x}{\dfrac{\sigma L_s}{R_s{'}}} \tag{2.134}$$

Obviously, the switching frequency of the converter must be faster than the chosen dynamics.

2.3.3.3   *Digital implementation of current regulators*

Linear, continuous models using Laplace transform have been used in the preceding discussion on the design and tuning of synchronous current regulators. However, this type of regulator is always implemented in some type of digital signal processor (microcontroller, DSP, etc.), meaning that the current regulator transfer function (2.120) needs to be transformed to the discrete domain.

A common approach for the design of digital controllers for continuous time systems is to first develop a continuous controller, and then use a continuous-to-discrete approximation to obtain the discrete equivalent controller. For the particular case of current regulators for electric drives, this approximation can be adequate when the maximum operating frequency of the drive and the current regulator bandwidth are below approximately 5–10% of the sampling frequency. While many electric drives meet this restriction, there are applications in which this does not occur. Examples of this can include large machines fed from inverters operating with low switching frequencies; machines operating at very high speeds; and machines required to provide very fast torque response, tuned therefore to have very high bandwidth regulators.

Adverse effects intrinsic to discretization when the sampling frequency is not sufficiently high compared to the fundamental frequency, and/or to the intended current regulator bandwidth, include cross-coupling among d and q axis currents, oscillatory behavior, and large settling times of the current control loop, and instability concerns can eventually arise.

To overcome, or at least partially compensate for, these problems, the delays intrinsic to digital controllers can be considered during the design process of the current regulator. One way to achieve this is by realizing the design in the discrete domain. The continuous model of the machine is replaced in this case by a discrete model using z-transfer functions. The delays due to the sampling process and PWM can then be considered, the resulting model representing more precisely the actual dynamics of the controlled system. Significant improvements in the performance of the current regulator can be achieved using this methodology compared to the conventional design method [7], [8].

2.3.4   Rotor flux estimation

Rotor flux oriented control of induction machines requires knowledge of the rotor flux magnitude and phase angle. Though the use of rotor flux sensors was investigated in the past, such sensors are never used in practice nowadays, owing to cost and reliability issues. Therefore, the rotor flux needs to be estimated from measurable variables. These include in principle the stator voltages and currents and the rotor speed/position.

A number of topologies have been proposed for the estimation of the rotor flux, which are derived from the machine equations. Two of the most widely used methods are now discussed: the current model and voltage model.

2.3.4.1   *Rotor flux estimation using the current model*

The rotor flux can be estimated from the measured stator current and rotor angle using (2.96), (2.102), and (2.103), the equations of the resulting observer being (2.135), (2.136), and (2.137) respectively. Estimated variables and parameters are indicated with "^". This method is often referred to as a current model-based flux observer.

$$\hat{\tau}_r \frac{d\hat{\psi}_{dr}}{dt} + \hat{\psi}_{dr} = \hat{L}_m \hat{i}_{ds} \tag{2.135}$$

$$\hat{\omega}_s = \hat{\tau}_r \frac{\hat{L}_m}{\hat{\psi}_{dr}} \hat{i}_{qs} + \omega_m \tag{2.136}$$

$$\hat{\theta}_s = \int \hat{\omega}_s dt = \theta_m + \int \hat{\tau}_r \frac{\hat{L}_m}{\hat{\psi}_{dr}} \hat{i}_{qs} dt \tag{2.137}$$

It is noted that, though the stator current space vector in the stationary reference frame is a measured variable, its transformation to the rotor flux reference frame requires the use of the estimated rotor flux angle. Therefore, the stator current space vector in the synchronous reference frame is an estimated variable. The block diagram showing the estimation of the rotor flux is shown in Fig. 2-19. The correspondence with the rotor flux dynamic model shown in Fig. 2-10 is evident.

Regardless of its simplicity, the current model-based flux observer presents some drawbacks. On the one hand, it suffers from a certain parameter sensitivity, as the model depends on the rotor time constant and on the magnetizing inductance. The rotor time constant will be especially sensitive to variations on the rotor resistance with temperature. Owing to this, implementation of an on-line rotor resistance adaptation method can be advisable. Also, inductances can be affected by saturation, which might need to be considered too.

A second drawback of this method is that it requires knowledge of the rotor position. The use of a position sensor is therefore required in principle. While many electrical drives include such sensors both for the implementation of vector control as well as for motion control purposes, the use of a position sensor is unadvisable in some applications. The use of rotor flux estimation methods which do not require measurement of the rotor position would be needed in this case.

It should be noted that the rotor flux observed in Fig. 2-19 could also be implemented in the stationary reference frame. However, both implementations can be shown to have the same parameter sensitivity, the computational requirements also being similar.

It should also be noted that the rotor flux observed in Fig. 2-19 can be implemented using the commanded d and q axis currents instead of the actual values. The resulting implementation is shown in Fig. 2-20, being often referred to as an indirect vector control [9]. One advantage of this implementation, compared to the one shown in Fig. 2-19, is that the current commands are cleaner signals than the actual currents, as they do not include switching harmonics. Owing to this, the noise content of the estimated rotor flux angle would be reduced.

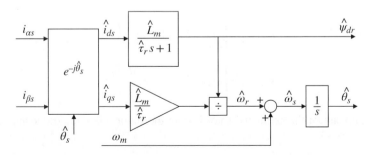

Fig. 2-19 *Rotor flux angle and magnitude estimation using the current model*

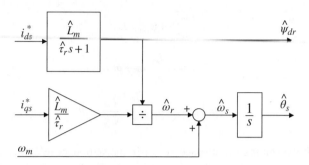

Fig. 2-20   *Indirect vector control using the current model*

2.3.4.2 *Rotor flux estimation using the voltage model*

It is possible to estimate the rotor flux using the stator voltage equations in the stationary reference frame. Using (2.49) and (2.50), the stator flux in the stationary reference frame can be estimated as:

$$\hat{\psi}_{\alpha s} = \int \left(v_{\alpha s} - \hat{R}_s \cdot i_{\alpha s} \right) dt \tag{2.138}$$

$$\hat{\psi}_{\beta s} = \int \left(v_{\beta s} - \hat{R}_s \cdot i_{\beta s} \right) dt \tag{2.139}$$

It is observed that estimation of the stator flux requires measurement of the stator voltage and current vectors, as well as knowledge of the stator resistance.

Using the known relationships among the stator and rotor fluxes and currents (2.67)–(2.70), the following equations linking the rotor fluxes with the stator fluxes and currents are obtained:

$$\hat{\psi}_{\alpha s} = \hat{\sigma}\hat{L}_s i_{\alpha s} + \frac{\hat{L}_m}{\hat{L}_r}\hat{\psi}_{\alpha r} \tag{2.140}$$

$$\hat{\psi}_{\beta s} = \hat{\sigma}\hat{L}_s i_{\beta s} + \frac{\hat{L}_m}{\hat{L}_r}\hat{\psi}_{\beta r} \tag{2.141}$$

Combining (2.138)–(2.139) and (2.140)–(2.141), the α and β components of the rotor flux in the stationary reference frame are obtained:

$$\hat{\psi}_{\alpha r} = \frac{\hat{L}_r}{\hat{L}_m}\int \left(v_{\alpha s} - \hat{R}_s i_{\alpha s} \right) dt - \frac{\hat{\sigma}\hat{L}_s\hat{L}_r}{\hat{L}_m} i_{\alpha s} \tag{2.142}$$

$$\hat{\psi}_{\beta r} = \frac{\hat{L}_r}{\hat{L}_m}\int \left(v_{\beta s} - \hat{R}_s i_{\beta s} \right) dt - \frac{\hat{\sigma}\hat{L}_s\hat{L}_r}{\hat{L}_m} i_{\beta s} \tag{2.143}$$

The amplitude and phase of the estimated rotor flux in the stator reference frame are given by:

$$\left| \vec{\psi}_r \right| = \sqrt{\psi_{\alpha r}^2 + \psi_{\beta r}^2} \tag{2.144}$$

$$\theta_s = a\tan\left(\frac{\psi_{\beta r}}{\psi_{ar}}\right) \tag{2.145}$$

Alternatively to the scalar notation, complex vector notation can be used for the analysis of the voltage model. Equations (2.138) and (2.139) can be represented in complex vector form as (2.146). If the machine is assumed to operate at a constant frequency, the time derivative can then be replaced by $j\omega_s$, (2.147) being obtained, from which the stator flux can be estimated as (2.148). The complex vector equation relating the stator and rotor fluxes and the stator current vectors (2.149) is obtained from (2.140) and (2.141). Combining (2.148) and (2.149), the estimation of the rotor flux in complex vector form using the measured stator voltage and current complex vectors is obtained. This is graphically shown in Fig. 2-21.

$$\vec{v}_s - \hat{R}_s \vec{i}_s = \frac{d\hat{\vec{\psi}}_s}{dt} \tag{2.146}$$

$$\vec{v}_s - \hat{R}_s \vec{i}_s = j\omega_s \hat{\vec{\psi}}_s \tag{2.147}$$

$$\hat{\vec{\psi}}_s = \frac{\vec{v}_s - \hat{R}_s \vec{i}_s}{j\omega_s} \tag{2.148}$$

$$\hat{\vec{\psi}}_r = \frac{\hat{L}_r}{\hat{L}_m}\hat{\vec{\psi}}_s - \frac{\hat{\sigma}\hat{L}_s\hat{L}_r}{\hat{L}_m}\vec{i}_s \tag{2.149}$$

The rotor flux angle and amplitude estimation block uses the measured voltages and currents in $\alpha\beta$ coordinates as depicted in Fig. 2-22, the detailed block diagram being shown in Fig. 2-23. A remarkable characteristic of the voltage model-based flux observer is that the rotor angle is not needed. Consequently, it is a natural

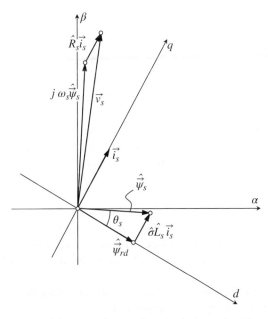

Fig. 2-21 *Graphical representation of the rotor flux angle calculation using the voltage model. The machine is assumed to operate at constant frequency*

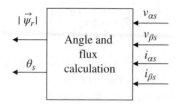

Fig. 2-22   *Schematic representation of the rotor flux estimation using the voltage model*

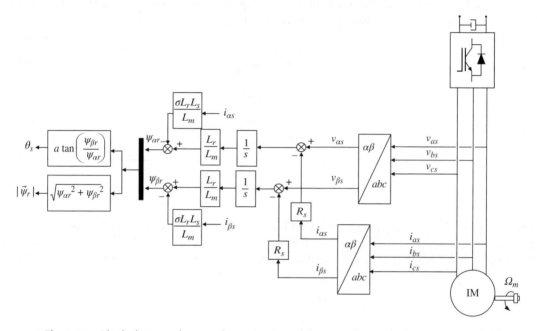

Fig. 2-23   *Block diagram showing the estimation of the rotor flux using the voltage model*

candidate for the implementation of field-oriented control in drives which do not use a position sensor (i.e. sensorless control).

It must be remarked regarding the implementation of the voltage model-based rotor flux observer that the stator voltages are not normally measured in industrial electric drives but derived from the switching states imposed on the inverter and the measured DC bus voltage. While cheaper and simpler, this solution places several problems at low or zero fundamental frequency:

- Effects due to the non-ideal behavior of the inverter (dead time, voltage drop in the power switches, etc.) will produce a mismatch between the commanded voltages used for the rotor flux estimation and the fundamental wave of the actual voltage applied to the machine. The resulting errors will become more relevant as the fundamental frequency decreases, as the stator voltage magnitude is proportional to the fundamental frequency.
- Integration needed to estimate the stator flux (see (2.146)) becomes especially critical at very low or zero frequency owing to offsets in the signals (integrator drift).
- Errors in the estimated machine parameters can also have a severe adverse impact on the accuracy of the estimation. Especially critical is the stator resistance R_s, which varies with temperature.

To avoid the drift of the pure integrators in Fig. 2-23, they are often replaced by a first-order system (2.150), with a cut-off frequency f_{co} in the range of 1–4Hz.

$$\frac{1}{s} \succ \frac{1}{s + 2\pi f_{co}} \tag{2.150}$$

At frequencies higher than the cut-off frequency, the low-pass filter behaves the same as the integrator. However, at frequencies near and below the cut-off frequency, including zero frequency, the deterioration of the rotor flux estimation makes this flux observer design inoperable.

2.3.4.3   *Parameter sensitivity and closed-loop flux observers*

As already mentioned, the current and voltage model-based flux observers discussed in the preceding section are sensitive to errors in their parameters and in their input signals. The current model-based flux observer is primarily sensitive to the rotor resistance and the magnetizing inductance, especially when operated near to rated slip [10]. The estimation error affects both the magnitude and the phase angle of the estimated rotor flux. Conversely, the estimation of the rotor flux is insensitive to errors in the estimated leakage inductance [10]. The voltage model-based flux observer is completely insensitive to errors in the estimated rotor resistance. However, it is highly sensitive to errors in the estimated stator resistance, especially at low speeds [10]. Since the commanded voltage is normally used instead of the actual voltage, mismatch between the commanded and the actual voltage is the other relevant source of error, especially at low speeds. Careful compensation of the sources of distortion of the output voltage (inverter dead time, voltage drop in the semiconductors, etc.) is therefore needed to extend the lower frequency limit of operation of the observer [11].

Both the current and the voltage model-based rotor flux observers previously discussed are open-loop systems (i.e. do not incorporate any feedback mechanism). To improve the performance of open-loop estimators, numerous closed-loop rotor flux observers have been proposed. Frequently, these closed-loop systems estimate the stator current, the error between the estimated and the measured stator current being used to drive the rotor flux to the actual rotor flux [10].

Independent of the flux observer structure, the performance of all model-based flux observers will depend on the accuracy in the estimation of the machine parameters that they use. To guarantee adequate performance for all operating conditions, numerous methods to estimate the motor parameters can be found in the literature [12]. This can be done before the operation of the motor, as a part of the commissioning process. Such methods are normally referred to as off-line. One limitation of off-line methods is that the actual value of the motor parameters can significantly change during the normal operation of the machine. Variations of the resistance are mainly due to changes in the temperature, being therefore slow. Changes in the inductances are mainly due to saturation, and can be therefore very rapid. To compensate for these variations, several on-line estimation methods intended to estimate machine parameters dynamically and without interfering with the normal operation of the drive have been proposed [12].

2.4   Torque capability of the induction machine

The torque capability of the induction machine is limited by different factors. At low speeds, the current is limited by thermal issues. Also the rotor flux (and the other fluxes) is limited by saturation. As the speed increases, voltage limits place further restrictions, owing to the induced back-emf. At speeds above rated speed, operation at rated flux is not possible. Even if the stator current vector magnitude can still be maintained, the rotor flux needs then to be decreased. This mode of operation is referred to as field weakening.

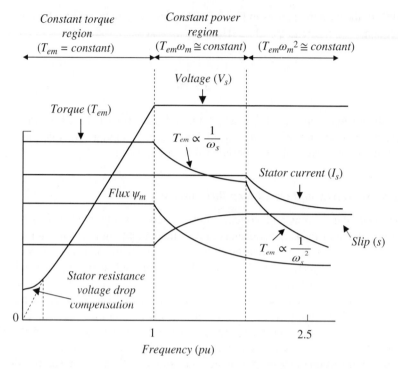

Fig. 2-24 *Operating regions of the induction machine considering voltage and current limits*

Fig. 2-24 shows the torque and rotor flux, as well as magnitude of the stator current and voltage vectors as a function of the fundamental frequency. Three different regions of operation are observed:

2.4.1 Constant torque region

- Maximum torque can be produced. Machine and power converter operate at their current limits, but below their voltage limits.
- In this mode of operation, the flux is normally kept constant and equal to its rated valued, the torque being controlled through i_{qs}.

2.4.2 Flux-weakening region I (constant power region)

- Machine and inverter reach their voltage limit. Reduction of the rotor flux is needed to operate beyond this speed. This is done by reducing the d axis current.
- Reduction of the rotor flux results in a decrease of the torque production capability. However, the reduction of the d axis current allows an increase of the q axis current to fully utilize the current capability of machine and inverter, partially compensating the torque reduction due to the decreased rotor flux. In this region, the machine operates therefore with reduced flux (reduced d axis current) but with rated voltage and current.

2.4.3 Flux-weakening region II (constant $T_{em}\omega_m{}^2$)

- As the speed increases, the current limit cannot be reached anymore, even if the rotor flux is reduced. Consequently, both d and q axis stator currents need to be limited. In this case, the machine operates with rated voltage but with reduced rotor flux and stator current.

The three different regions of operation are discussed in more detail in the next section.

2.5 Rotor flux selection

As already mentioned, when the machine operates below its rated speed, the rotor flux is often maintained constant and equal to its rated value. This provides excellent dynamic response, as the rotor is fully magnetized. The machine is therefore able to provide its maximum torque almost instantaneously (limited by the current control bandwidth) if required by the application. However, owing to the voltage constraints imposed by the inverter, at speeds higher than rated the rotor flux must be weakened. It is noted that this is independent of the control strategy (vector control, DTC, DSC, etc.).

Fig. 2-25 shows a generic implementation of the rotor flux reference selection block. The rotor flux limit has been shown to be a function of the machine speed. Simpler field-weakening methods therefore use the rotor speed to obtain the rotor flux command. Either the actual rotor speed or the commanded rotor speed could be used, the first option being shown in the figure. The accuracy of the rotor flux command can be enhanced if the voltage limits are taken into consideration (i.e. at high speeds the rotor flux command can be dynamically adapted to fully utilize the voltage available from the inverter, which depends on the DC bus voltage V_{bus}). In this case the inputs to the field-weakening block can be augmented with the stator voltage command and the stator voltage limit, $V_{s\ limit}$. This is schematically indicated by the dashed arrows in Fig. 2-25.

Strategies for the selection of the rotor flux are described in the following subsections.

2.5.1 Rotor flux reference selection below rated speed

It has already been shown that the rotor flux can be precisely controlled by means of the d axis current in the rotor flux reference frame. Some considerations can be made regarding the selection of the rotor flux reference.

Below rated speed, the rotor flux reference is often maintained constant and equal to its rated value. By doing this, the torque produced by the machine is only a function of the q axis current, which can be changed very quickly thanks to the high-bandwidth current regulator. This strategy therefore provides excellent dynamic response. At rated speed, the voltage limit is reached, the control entering in the field-weakening region. In this region, the flux reference decreases with the speed, to match the back-emf with the available voltage. This enables the operation at speed higher than rated, but with a reduction of the maximum torque available.

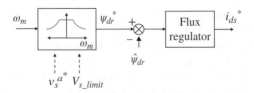

Fig. 2-25 *Rotor flux reference generation and the subsequent rotor flux regulator*

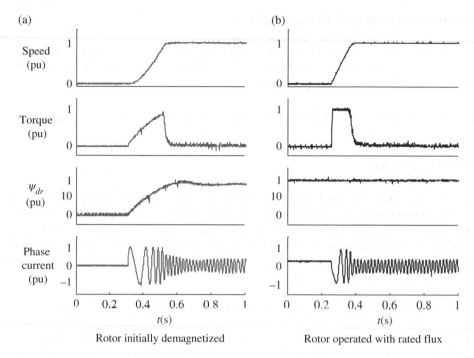

Fig. 2-26   *Dynamic response of a vector controlled induction machine during a start-up process from zero to rated speed, for the case of (a) rotor initially demagnetized and (b) rotor flux maintained equal to its rated value*

Although maintaining the rotor flux equal to its rated value at low speeds is adequate to be able to obtain good dynamic response, this is at the price of increased losses when the machine operates with light loads for long periods, as the rotor is unnecessarily magnetized in this case.

Alternatively, the machine can be operated with reduced rotor flux even below rated speed, if it is not fully loaded. With this strategy, the stator losses can be decreased thanks to the reduction of the d axis current. The reduction of the rotor flux (and of the d axis current) needs to be compensated for by an increase of the q axis current to maintain the produced torque constant, whilst the magnitude of the stator current vector is reduced. However, this reduction is at the price of a reduction of the torque dynamic response. Effectively, if the machine operates with reduced flux and the torque needs to be increased, the rotor must first be energized. The dynamics of this process are controlled by the rotor time constant, as shown by (2.135), which are significantly slower than those of the stator current control loop.

Fig. 2-26 shows the acceleration process of a speed controlled induction machine from zero to its rated speed, using the two strategies already mentioned. In the case shown in the Fig. 2-26(a), the rotor flux is initially equal to zero. It is noted that the phase current is therefore zero too; consequently, no losses exist when the machine does not produce torque. When a speed step is commanded, the rotor flux needs to be built up, the d axis current being used for this purpose (not shown in the figure). The slow dynamics building the rotor flux up results in a poor dynamic response in the torque produced by the machine, even though the q axis current was produced in the range of ms (not shown in the figure). This eventually affects to the dynamic response of the speed control loop, increasing the settling time. Conversely, in the case shown in Fig. 2-26(b), the rotor flux is maintained at its rated value. This results in losses even when the machine is not producing torque. However, when the speed step is commanded, the torque is seen to build up in a few ms (torque shown in the figure mirrors the q axis current), therefore enabling a faster response of the speed control loop.

2.5.2   Accurate criteria for flux reference generation

The method for flux reference generation explained in this section can be found in [1]. It is reproduced here for convenience, owing to its wide acceptance among electric drive designers.

First of all, the voltage and current constraints must be taken into consideration. The voltage constraints depend on the converter and the modulation. The maximum voltage provided by the converter when it operates in the linear region, and therefore the maximum available voltage that can be applied to the stator windings, is given by (2.153).

$$V_{s\_limit} = \frac{V_{bus}}{\sqrt{3}} \eta \qquad (2.151)$$

It is assumed that PWM with triplen harmonic injection or space vector modulation (SVM) is used, with V_{bus} being the DC bus voltage of the power electronic converter, η: voltage efficiency of the converter (0.9–1), and V_{s_limit}: maximum phase voltage (peak).

Thus, the stator voltage limit is:

$$v_{ds}^2 + v_{qs}^2 = v_{\alpha s}^2 + v_{\beta s}^2 \leq V_{s\_limit}^2 \qquad (2.152)$$

On the other hand, the d–q axis stator current should comply with the current limit, I_{s_limit}, which is usually established by the thermal limit of the inverter or the AC machine itself.

$$i_{ds}^2 + i_{qs}^2 = i_{\alpha s}^2 + i_{\beta s}^2 \leq I_{s\_limit}^2 \qquad (2.153)$$

Performing the study at a synchronously rotating reference frame, the stator voltage can be modeled by means of the previously obtained equations (rotor flux orientation):

$$v_{ds} = R_s i_{ds} + \sigma L_s \frac{di_{ds}}{dt} - \omega_s \sigma L_s i_{qs} + \frac{L_m}{L_r} \frac{d\psi_{dr}}{dt} \qquad (2.154)$$

$$v_{qs} = R_s i_{qs} + \sigma L_s \frac{di_{qs}}{dt} + \omega_s \sigma L_s i_{ds} + \omega_s \frac{L_m}{L_r} \psi_{dr} \qquad (2.155)$$

Assuming steady-state conditions and neglecting the voltage drop in the stator resistances, these equations are simplified to:

$$v_{ds} \cong -\omega_s \sigma L_s i_{qs} \qquad (2.156)$$

$$v_{qs} \cong \omega_s \sigma L_s i_{ds} + \omega_s \frac{L_m}{L_r} \psi_{dr} \qquad (2.157)$$

Again, emphasizing that we are only considering the steady-state operation of the machine, the rotor flux is equal to:

$$\psi_{dr} = L_m i_{ds} \qquad (2.158)$$

Resulting in a more compact and simplified q voltage component:

$$v_{qs} \cong \omega_s L_s i_{ds} \qquad (2.159)$$

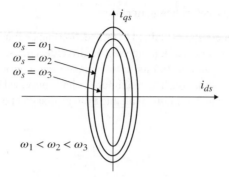

Fig. 2-27   *Ellipses at different ω$_s$ due to the voltage constraint*

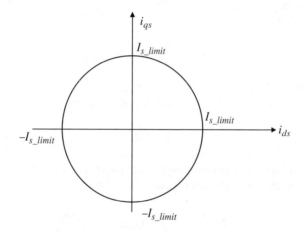

Fig. 2-28   *Circle due to the current limit*

With these simplified voltage equations in steady state, it is possible to calculate the module of the stator voltage by substituting (2.156) and (2.159) in (2.152):

$$\left(\omega_s \sigma L_s i_{qs}\right)^2 + \left(\omega_s L_s i_{ds}\right)^2 \leq V_{s\_limit}^2 \tag{2.160}$$

This inequality corresponds to the area enclosed by an ellipse in the i_{ds}, i_{qs} plane, as illustrated in Fig. 2-27. Thus, the bigger ω_s it is, the smaller becomes the area of the ellipse.

On the other hand, the current constraint in the i_{ds}, i_{qs} plane corresponds to a circle, also depicted as in Fig. 2-28.

The torque can be represented as a function of the d and q axis stator currents in a rotor flux reference frame:

$$T_{em} = \frac{3}{2} p \frac{L_m^2}{L_r} i_{ds} i_{qs} \tag{2.161}$$

Constant torque values in the i_{ds}, i_{qs} plane follow the trajectory of a hyperbola, as illustrated in Fig. 2-29.

The feasible operating region (the shaded area) corresponds to the cross-section between the ellipse and the circle. Thus, in this presented example of Fig. 2-29, if the torque curve represents the maximum torque value,

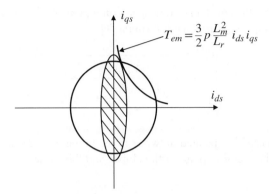

Fig. 2-29 *Constant torque characteristic, voltage, and current constraints and the permitted operation region (shaded area)*

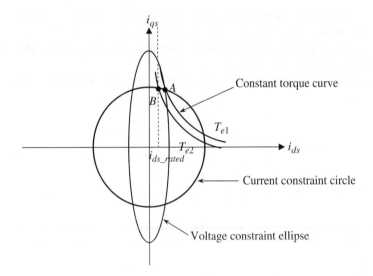

Fig. 2-30 *Voltage and current constraints and the limitation of the rated* i$_{ds}$ *value*

only one point can satisfy the voltage and current limits. However, this point must be checked with the data given by the manufacturer of the machine. If the rated flux (rated i_{ds} because $\psi_{dr} = \psi_{dr\_rated} = L_m i_{ds} = L_m i_{ds\_rated}$) corresponds to a value smaller than the corresponding value at point A in Fig. 2-30, the maximum available torque of this machine will be the hyperbola which crosses point B in Fig. 2-30. Note that going above the rated i_{ds} value means excessively saturating the machine, increasing the losses, and deteriorating the ideal performance of the machine in general. Consequently, during constant torque region, the i_{ds} value is kept constant to:

$$i_{ds} = i_{ds\_rated} \tag{2.162}$$

while the maximum available i_{qs} value is:

$$i_{qs} = \sqrt{I^2_{s\_limit} - i^2_{ds\_rated}} \tag{2.163}$$

The angular frequency where the constant torque operation region ends ω_b is defined as the base frequency and can be calculated according to the following formula [1]:

$$\omega_b = \sqrt{\dfrac{(V_{s\_limit})^2}{\psi_{dr\_rated}^2 \dfrac{L_s^2-(\sigma L_s)^2}{L_m^2} + (\sigma L_s I_{s\_limit})^2}} \tag{2.164}$$

Field weakening operation can be separated in two regions. In region 1, the d axis current to maximize the torque can be derived from the crossing point of the ellipse and the circle:

$$i_{ds} = \sqrt{\dfrac{\left(\dfrac{V_{s\_limit}}{\omega_s}\right)^2 - (\sigma L_s I_{s\_limit})^2}{L_s^2-(\sigma L_s)^2}} \tag{2.165}$$

Note that in region 1 the d axis current reference is always smaller than the rated d axis current, since the ellipse becomes smaller and smaller. This fact is graphically represented in Fig. 2-31. Note also that the torque that can be provided in this region is below the rated, owing to the voltage (ellipse) limitation. However, the maximum available current can still be reached.

Consequently, the rotor flux reference can be generated according to the following expression:

$$\psi_{dr} = L_m \sqrt{\dfrac{\left(\dfrac{V_{s\_limit}}{\omega_s}\right)^2 - (\sigma L_s I_{s\_limit})^2}{L_s^2-(\sigma L_s)^2}} \tag{2.166}$$

while the remaining current for i_{qs} is:

$$i_{qs} = \sqrt{I_{s\,limit}^2 - (i_{ds})^2} \tag{2.167}$$

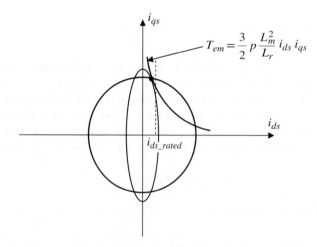

Fig. 2-31   *Voltage and current constraints in flux-weakening region 1*

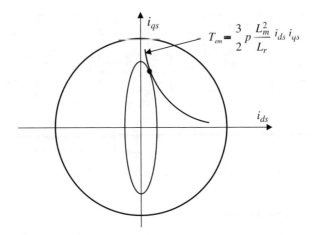

Fig. 2-32 *Voltage and current constraints in flux-weakening region 2*

Finally, if the speed further increases above a certain limit, the ellipse will be inside the circle, as depicted in Fig. 2-32. In this situation, the torque is only limited by the voltage constraint, its not being possible to reach the maximum current limit. This corresponds to region 2.

The frequency, ω_1, where the flux-weakening region 2 starts, can be derived as from the fact that at that frequency the circle meets the ellipse at a single point:

$$\omega_1 = \sqrt{\frac{L_s^2 + (\sigma L_s)^2}{2(L_s \sigma L_s)^2} \times \frac{V_{s\_limit}}{I_{s\_limit}}} \tag{2.168}$$

The reference points to maximize the torque at minimum current can be derived at the meeting points of torque and voltage curves:

$$i_{ds} = \frac{V_{s\,limit}}{\sqrt{2}\omega_s L_s} \tag{2.169}$$

Hence, the rotor flux reference yields:

$$\psi_{dr} = \frac{L_m V_{s\_limit}}{\sqrt{2}\omega_s L_s} \tag{2.170}$$

The q axis current can be calculated as:

$$i_{qs} = \frac{V_{s\_limit}}{\sqrt{2}\omega_s \sigma L_s} \tag{2.171}$$

As mentioned before, in this flux-weakening region 2, owing to the large values of the speed, the current limit cannot be reached because of a prevalence of the voltage limit.

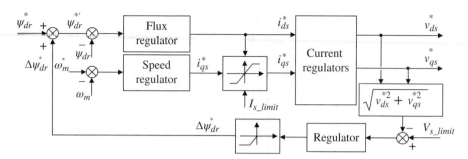

Fig. 2-33   *Flux weakening with closed-loop control*

2.5.3   Feedback based field weakening

Alternatively to feed-forward-based flux-weakening methods, a feedback-based strategy can be used. This method does not require the use of a machine model or pre-calculated look-up tables. Instead, the voltage command provided by the current regulator is used to dynamically adapt the flux reference to the machine speed. As illustrated in Fig. 2-33, the input to the flux-weakening regulator is the difference between the output voltage of the current regulator and the maximum available voltage, V_{s_limit}, which depends on the DC bus voltage, V_{bus}, and the PWM strategy ((2.151)). When the machine operates at frequencies below the rated frequency, the stator voltage reference set by the current regulators in steady state will be below the voltage limit V_{s_limit}. In this case, there is no need to decrease the rotor flux level, the incremental rotor flux $\Delta\psi_{dr}$ (i.e. variation of the flux reference with respect to its rated value) commanded by the flux-weakening control loop is zero. Conversely, when the frequency of the machine is equal to or larger than the rated frequency, the magnitude of the voltage reference set by the current regulators will be larger than V_{s_limit}, meaning that there is not enough voltage available in the inverter to compensate for the back-emf. This produces a negative error, which results in a decrease of the flux reference (i.e. a negative incremental rotor flux $\Delta\psi_{dr}$). The flux-weakening control loop therefore decreases the flux level until the commanded voltage matches the available voltage V_{s_limit}. It will be noted that, since the flux weakening produces a decrease of the current i_{ds}, the limit of the torque current, i_{qs}, can be increased to operate with maximum current, I_{s_limit}. A coordinated current reference limitation is implemented for this purpose, as shown in Fig. 2-33.

Tuning the flux-weakening regulator is not straightforward. Its bandwidth should be significantly lower than the current regulator bandwidth to prevent interference. However, and because of this, in a scenario of fast speed variations, the performance of this loop may present an oscillatory behavior.

2.6   Outer control loops

Though the current regulators and flux estimation blocks are the core of vector controlled induction machine drives, many applications may require the use of outer control loops to fulfill specific application requirements. These outer control loops are typically connected in cascade with the current control loops. A general block diagram is shown in Fig. 2-34. Two outer control loops for the speed and rotor flux provide the q and d axis current commands respectively. Though not shown in the figure, a position control loop can also be included, normally in cascade with the speed control loop. PI regulators are typically used for the speed and flux control, while a P controller is used for position control.

Design and tuning of speed and rotor flux control loops are discussed in this section. Position control is not discussed further in this book, as it has limited interest for the applications considered.

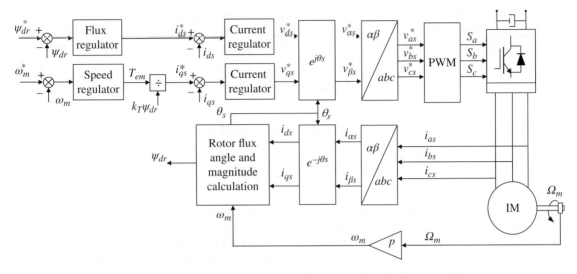

Fig. 2-34 *Field oriented control including speed and rotor flux control loops*

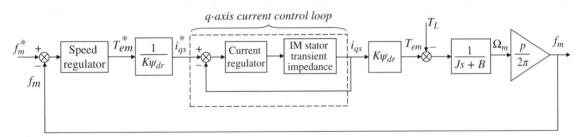

Fig. 2-35 *Simplified block diagram of the speed control loop. The controlled variable is the electrical speed in Hz* f_m, Ω_m *being the mechanical speed in rad/s*

2.6.1 Speed control

In applications requiring speed control, an outer speed control loop is added. The speed control loop provides the torque (or alternatively the q axis current) to an inner loop including current regulation and field-oriented control. This is schematically represented in Fig. 2-35. The term $1/K\psi_{dr}$ is included to make the speed-torque loop dynamics independent of the rotor flux magnitude. The speed regulator is typically PI type.

In high-performance electric drives, the current loop is tuned to have a bandwidth much faster than the speed control loop. The current control dynamics can therefore be safely neglected. Consequently, the actual torque can be assumed to be nearly equal to the commanded torque (2.172):

$$T_{em}^* \approx T_{em} = K\psi_{dr} i_{qs} \tag{2.172}$$

The overall mechanical equation will depend on the mechanical characteristics of the machine and load. The machine can be considered as nearly a pure mechanical, invariant inertia, friction being almost negligible. However, the mechanical characteristics of the load strongly depend on the application. Consequently, selection of the parameters for the speed regulator will depend on the mechanical load characteristic; no general tuning procedure can be given. As an example, tuning of the speed controller for the case of a load including mechanical inertia and a viscous friction, as well as the load torque (2.173), is briefly discussed here.

$$T_{em} = T_L + J\frac{d\Omega_m}{dt} + B\Omega_m \tag{2.173}$$

The transfer function that links the electrical speed with the torque produced by the machine is:

$$\frac{f_m(s)}{T_{em}(s)} = \frac{p}{2\pi} \cdot \left(\frac{1}{Js + B}\right) \tag{2.174}$$

The dynamics of the speed control loop are normally selected to be significantly slower than those of the current control loop. This restriction comes first from the physical nature of the mechanical systems, whose time constant is significantly larger than that of the electrical subsystem. In addition, proper operation of cascaded control systems, as that shown in Fig. 2-34, requires that the inner (current) control loop behaves significantly faster than the outer (speed) control loop, as this enables independent design and tuning of both control loops.

Being the transfer function of the PI speed controller (2.175), the transfer function linking the actual speed with the speed command in electrical units is given by (2.176), while the transfer function linking the speed with the disturbance torque is (2.177).

$$PI_f = k_{pf} + \frac{k_{if}}{s} \cdot \tag{2.175}$$

$$\frac{f_m(s)}{f_m^*(s)} = \frac{k_{pf} \cdot s + k_{if}}{\frac{2\pi}{p}J \cdot s^2 + \left(\frac{2\pi}{p}B \cdot + k_{pf}\right) \cdot s + k_{if}} \tag{2.176}$$

$$\frac{f_m(s)}{T_L(s)} = \frac{s}{\frac{2\pi}{p}J \cdot s^2 + \left(\frac{2\pi}{p}B \cdot + k_{pf}\right) \cdot s + k_{if}} \tag{2.177}$$

The system dynamics are seen to correspond to those of a second-order system. Similar to that discussed for the tuning of the synchronous PI current regulator, zero-pole cancellation can be used to reduce the dynamics to those of a first-order system. To achieve this, the relationship between the mechanical parameters and the speed PI controller gains is given by (2.178). The proportional gain of the speed controller k_{pf} is then selected to achieve the desired bandwidth.

$$\frac{k_{if}}{k_{pf}} = \frac{B}{J} \tag{2.178}$$

Alternatively, the natural frequency and damping coefficient for the second order can be specified. The speed controller gains k_{pf} and k_{if} are obtained as for the current controller case, being given by (2.179)–(2.180):

$$k_{if} = \frac{2\pi J}{p}\omega_n^2 \tag{2.179}$$

$$k_{pf} = \frac{2\pi J}{p} 2\xi\omega_n \tag{2.180}$$

There are applications in which the mechanical friction is negligible compared to the mechanical inertia, the mechanical transfer function being in this case (2.181).

$$\frac{f_m(s)}{T_{em}(s)} = \frac{p}{2\pi} \cdot \frac{1}{Js} \tag{2.181}$$

These types of systems are often prone to show oscillatory behavior, owing to the double integrator in the closed-loop system, coming from the mechanical load and from the PI speed regulator respectively. Zero-pole cancellation cannot be used in this case, as the load pole is located at the origin. A fictitious friction can be used in this case, which is often called active damping. This is schematically shown in Fig. 2-36.

When seen from the speed regulator, the fictitious friction B' is connected in parallel with the actual friction B (if this exists), meaning that B' can be used to adjust the overall friction to the desired value. Some consideration needs to be made in this regard:

- It is noted that B' physically does not consume torque (or power). This means that the torque commanded by the speed regulator $T_{em}^{*'}$ can be larger than the maximum torque that can be provided by the machine. Of course, the torque command once the torque consumed by the fictitious friction is subtracted, T_{em}^* has to be limited to the maximum torque that can be provided by the machine. This needs to be taken into account when implementing the torque limitation at the speed regulator output.
- It is observed from Fig. 2-36 that when seen from the speed regulator the fictitious friction B' is connected in parallel to the actual friction. However, when seen from the load torque, B' is in the feedback path, being connected in parallel with k_{pf}. Consequently, actual and fictitious frictions have different behaviors regarding load disturbance rejection.
- The fictitious damping acts through the q axis current control loop, contrary to the actual friction, which is mechanically coupled. Therefore, in order for the fictitious damping to be effective, the dynamic response or the q axis current regulator has to be much faster than the speed control loop dynamics. It should be noted, however, that this assumption is realistic in drives using high bandwidth current regulators.

It should be added that a large number of design and tuning methodologies for speed controllers have been proposed to improve command tracking and disturbance rejection capabilities, including predictive control, sliding mode control, robust control, neural networks, and so on. Further improvements—like feedforward and load torque observers—can also be used. If the mechanical characteristics of the load can change

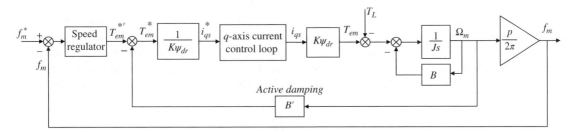

Fig. 2-36 *Speed control including active damping*

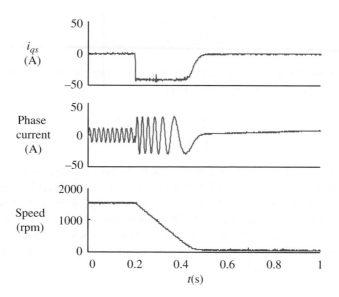

Fig. 2-37   *Dynamic response of the speed control loop*

significantly during the normal operation of the drive, the speed regulator might need to be adaptive to guarantee adequate performance under all operating conditions. On-line estimation of the load parameters is needed in this case.

Fig. 2-37 shows the typical response of the speed control loop of a vector controlled drive during a breaking process from the rated speed (1500rpm) to 0.5rpm. It is observed that for most of the transient the q axis current is saturated at its rated value. This minimizes the settling time, the resulting transient of the speed being characterized by a constant acceleration for the case of inertial loads. It can also be observed in Fig. 2-37 that if correctly tuned the speed control loop combines a fast transient with a perfectly damped response.

2.6.2   Rotor flux control loop

When a rotor flux control loop is to be implemented, the block diagram in Fig. 2-38(a) can be used. It is assumed that the rotor flux is measured. However, and contrary to the current and speed control loops, the rotor flux is never measured in practice but only ever estimated. This means that the accuracy of the rotor flux control is subject to the accuracy of the rotor flux observer. The block diagram in Fig. 2-38(b) shows the implementation of rotor flux control when the rotor flux is estimated using a current model. Obviously, the performance of the rotor flux control loop will be conditioned by the accuracy of the rotor flux observer.

For the tuning of the rotor flux regulator, it is safe to assume that the current loop bandwidth is tuned to be much faster than the rotor flux loop (as done before for the speed control loop). The dynamics of the current control loop can therefore be safely neglected. With this assumption, the equation linking the d axis current and the rotor flux is (2.182). Assuming a PI regulator with gains $k_{p\psi}$ and $k_{i\psi}$ is used, the closed-loop transfer function of the rotor flux control loop (2.183) is obtained. It should be noted that perfect rotor flux estimation has been assumed.

$$\frac{\psi_{dr}}{i_{ds}} = \frac{L_m}{\tau_r s + 1}. \qquad (2.182)$$

(a)

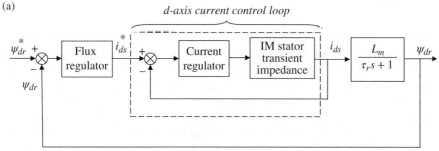

Rotor flux control loop assumed that the rotor flux can be measured

(b)

Rotor flux control loop including rotor flux estimation using a current model

Fig. 2-38   *Ideal and actual rotor flux control loops*

$$\frac{\psi_{dr}}{\psi_{dr}^{*}} = \frac{k_{p\psi}s + k_{i\psi}}{s^{2}\dfrac{\tau_{r}}{L_{m}} + s\left(\dfrac{1}{L_{m}} + k_{p\psi}\right) + k_{i\psi}} \tag{2.183}$$

It is observed from (2.183) that the dynamics of the rotor flux control loops corresponds to a second-order system. Therefore, the same methodology described for the current and speed control loops can be used for the rotor flux control loop. The controller gains can be selected to realize zero-pole cancellation (2.184), the resulting closed system being now reduced to a first-order system, its bandwidth being given by the selection of $k_{p\psi}$.

$$\frac{k_{i\psi}}{k_{p\psi}} = \frac{1}{\tau_{r}} \tag{2.184}$$

Alternatively, the gains can be selected to obtain the desired response for the second-order system in (2.183). Being ω_{n} and ξ, the desired natural frequency and damping factor of the closed-loop system, the rotor flux controller gains would be obtained as (2.185) and (2.186) respectively.

$$k_{i\psi} = \frac{\tau_{r}}{L_{m}}\omega_{n}^{2} \tag{2.185}$$

$$k_{p\psi} = \frac{\tau_{r}}{L_{m}}2\xi\omega_{n} - \frac{1}{L_{m}} \tag{2.186}$$

2.7   Sensorless control

Control of induction motor drives without speed/position sensors has been an intensive line of research for more than two decades. The methods developed to achieve this goal are often referred to as sensorless control. Elimination of the speed/position sensor will reduce the hardware needed to realize the control (not only the sensor but also the cabling and electronics needed to capture the sensor signals), which is advantageous in terms of the cost, size, reliability, and maintenance requirements of the electric drive.

Sensorless control methods for induction machines that rely on the fundamental excitation—often referred to as model-based methods—have been shown to provide high-performance control in the medium- and high-speed ranges, and are offered in standard electric drives by many manufacturers. However, as the speed decreases, the performance of these methods degrades, eventually failing at very low and zero speed. To overcome these limitations, sensorless methods that track the position of saliencies (asymmetries) in the rotor by using some type of high frequency signal excitation have been proposed. This signal is superimposed onto the fundamental excitation used for torque production.

While model-based sensorless methods can be considered a mature technology, several challenges persist for the case of saliency-tracking-based methods, for which the number of industrial experiences is still limited.

In this section, some of the most widely used methods for sensorless speed control of electric drives using induction machines are presented and discussed. This chapter focuses on model-based methods. Discussion of saliency-tracking-based methods, able to operate at very low speed and in position control, is presented in Chapter 3, as such capabilities are normally more important for permanent magnet motor drives than for induction machine drives. The particularities for the use of saliency-tracking-based methods with induction machines is also covered in Chapter 3. It is important to bear in mind, however, that this section accounts only for a small portion of the methods that can be found in the specialized literature and industry patents [13].

2.7.1   Sensorless control of induction machines using model-based methods

Model-based methods make use of the equations that define the electromagnetic behavior of the machine, and which are reproduced here in a complex vector form for convenience.

$$\vec{v}_s^s = R_s \vec{i}_s^s + \frac{d\vec{\psi}_s^s}{dt} \tag{2.187}$$

$$0 = R_r \cdot \vec{i}_r^s + \frac{d\vec{\psi}_r^s}{dt} - j \cdot \omega_m \cdot \vec{\psi}_r^s \tag{2.188}$$

$$\vec{\psi}_s^s = L_s \cdot \vec{i}_s^s + L_m \cdot \vec{i}_r^s \tag{2.189}$$

$$\vec{\psi}_r^s = L_m \cdot \vec{i}_s^s + L_s \cdot \vec{i}_r^s \tag{2.190}$$

These equations can be rewritten in a state space form:

$$\frac{d\vec{i}_s^s}{dt} = \frac{1}{L_{\sigma s}} \left(-R_s' \vec{i}_s^s + \frac{L_m}{L_r} \left(\frac{1}{\tau_r} - j\omega_m \right) \vec{\psi}_r^s \right) + \frac{1}{L_{\sigma s}} \vec{v}_s^s \tag{2.191}$$

$$\frac{d\vec{\psi}_r^s}{dt} = \frac{L_m}{\tau_r} \vec{i}_s^s - \left(\frac{1}{\tau_r} - j\omega_m \right) \vec{\psi}_r^s \tag{2.192}$$

It can be noted from (2.188), or alternatively (2.191)–(2.192), that the behavior of the machine is affected by the rotor speed, meaning that it should be possible to estimate the rotor speed by solving these equations. While this is true in principle, there are several practical aspects that need to be considered for the use of these equations, as they will strongly condition the results that can be expected.

- Equations (2.187)–(2.190) and (2.191)–(2.192) are seen to be a function of the machine inductances and resistances, both in the stator and rotor sides. Consequently, inaccuracies in the estimated parameters used in the model will directly impact the accuracy of the estimated speed.
- The input to the model in (2.191)–(2.192) is the stator voltage, the states being the stator current and the rotor flux. The stator current is measured with acceptable accuracy in practice. The stator voltage is practically never measured, as this would imply the use of voltage sensors, which are relatively expensive, as well as the implementation of precise acquisition and/or filtering strategies to get rid of the switching harmonics due to the PWM. Instead, the stator voltage is estimated from the voltage commands sent to the inverter and the measured DC bus voltage. However, errors between the commanded and actual voltages will exist in practice, mainly because of the non-ideal behavior of the inverter (dead-time, voltage drop in the power devices, etc.) [11]. These errors will be especially relevant at low speeds. Furthermore, the rotor flux is never measured, but also needs to be estimated, which can further increase the parameter sensitivity of the speed estimation.
- Though not explicitly shown in (2.191)–(2.192), this model uses the stator flux as an internal variable. Estimation of the stator flux from the stator voltage and current (2.187) involves an open-loop integration. This is not viable in practice, owing to DC offsets always being present, which eventually produces a signal drift. To overcome this problem, the pure integrator is replaced by a first-order system. However, this produces a mismatch between the physical system and the model, which gets worse as the frequency decreases, and which eventually makes the model unviable at very low or zero speeds.

2.7.1.1 Speed estimation using model reference adaptive systems (MRAS)

A widely studied method to estimate the rotor velocity is the use of a model reference adaptive system (MRAS) [14]. A potential implementation is shown in Fig. 2-39. The voltage model (2.193)–(2.194) is used as the reference model, while the current model (2.195), which depends on the rotor speed, is used as the adaptive system, with "^" indicating estimated variables/parameters.

$$\hat{\vec{\psi}}_s^s = \frac{\vec{v}_s^s - \hat{R}_s \vec{i}_s^s}{s} \tag{2.193}$$

$$\hat{\vec{\psi}}_{rV}^s = \frac{L_r}{L_m}\left(\hat{\vec{\psi}}_s^s - \vec{i}_s^s \hat{L}_{\sigma s}\right) \tag{2.194}$$

$$\frac{\hat{\vec{\psi}}_{rC}^s}{\vec{i}_s^s} = \frac{\hat{L}_m}{\hat{\tau}_r}\left(\frac{1}{s + \frac{1}{\hat{\tau}_r} - j\hat{\omega}_m}\right) \tag{2.195}$$

The error signal needed to adapt the estimated rotor speed can be obtained as the cross product between the rotor flux estimates provided by the voltage and current models (2.194)–(2.195), which can be written as (2.197).

$$\varepsilon = \hat{\psi}_{qrC}^s \hat{\psi}_{drV}^s - \hat{\psi}_{drC}^s \hat{\psi}_{qrV}^s \tag{2.196}$$

$$\varepsilon = \hat{\psi}_{rC}^s \hat{\psi}_{rV}^s \sin\left(\hat{\theta}_{rfC} - \hat{\theta}_{rfV}\right) \approx \hat{\psi}_{rC}^s \hat{\psi}_{rV}^s \left(\hat{\theta}_{rfC} - \hat{\theta}_{rfV}\right) \tag{2.197}$$

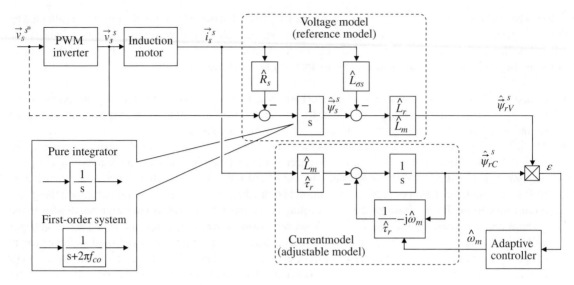

Fig. 2-39   *Rotor speed estimation using an MRAS*

MRAS methods suffer from the same parameter sensitivity as any other method that relies on the back-emf. The voltage model in Fig. 2-39 will be highly sensitive to the accuracy in the estimate of the stator resistance. This is especially critical because of the pure integrator needed to estimate the stator flux. To alleviate the problems of the pure integrator infinite gain at DC, it is often replaced by a first-order system (see Fig. 2-39). However, as already mentioned, the selection of the filter bandwidth f_{co} intrinsically limits the lowest frequency at which the model can operate. As for the current model, it is especially sensitive to the rotor time constant. To overcome these problems, the development of methods to dynamically estimate some of the machine parameters has been a very active area of research [11].

2.7.1.2   *Speed estimation using an adaptive flux observer*

A different approach for the estimation of the rotor speed is the use of an adaptive flux observer, which combines rotor speed and rotor flux estimation, with parameter adaptation. An example of this can be found in [15]. The state space model of the induction machine in matrix form is obtained by combining (2.193) and (2.194).

$$\frac{d}{dt}\begin{bmatrix} \vec{i}_s^{\,s} \\ \vec{\psi}_r^{\,s} \end{bmatrix} = \begin{bmatrix} A_{11} & A_{12} \\ A_{21} & A_{22} \end{bmatrix} \begin{bmatrix} \vec{i}_s^{\,s} \\ \vec{\psi}_r^{\,s} \end{bmatrix} + \begin{bmatrix} B_1 \\ 0 \end{bmatrix} \vec{v}_s^{\,s} \tag{2.198}$$

$$\frac{d}{dt}\vec{x} = A\,\vec{x} + B\,\vec{v}_s^{\,s} \tag{2.199}$$

$$\vec{i}_s^{\,s} = C\,\vec{x} \tag{2.200}$$

where

$$A_{11} = -\frac{R_s'}{L_{\sigma s}}; \quad A_{12} = \frac{L_m}{L_{\sigma s}L_r}\left(\frac{1}{\tau_r} - j\omega_m\right);$$

$$A_{21} = \frac{L_m}{\tau_r}; \quad A_{22} = -\frac{1}{\tau_r} + j\omega_m; \quad B_1 = \frac{1}{L_{\sigma s}}; \quad C^T = [1 \quad 0]$$

The full order observer which estimates the stator current and the rotor flux is then written as (2.201). The error signal formed by the estimated and measured stator current is used to feedback the observer through a gain matrix G, which is selected to make the observer stable and with adequate dynamics [15]. The corresponding block diagram is shown in Fig. 2-40.

$$\frac{d}{dt}\hat{x} = \hat{A}\hat{x} + \hat{B}v_{qds}^s + G\left(\hat{\vec{i}}_s - \vec{i}_s\right) \tag{2.201}$$

The inputs to the adaptive controller in Fig. 2-40 are the error in the estimated current (2.202), as well as the full order observer estates (i.e. rotor flux and estimated stator current). The adaptive controller estimates the stator resistance, rotor time constant, and rotor speed using the control laws defined by (2.203), (2.204), and (2.205) respectively.

$$\vec{e}_{qds}^s = \vec{i}_s - \hat{\vec{i}}_s \tag{2.202}$$

$$\frac{d\hat{R}_s}{dt} = -\lambda 1 \left(e_{ids}^s \hat{i}_{ds}^s + e_{iqs}^s \hat{i}_{qs}^s\right) \tag{2.203}$$

$$\frac{d(1/\hat{\tau}_r)}{dt} = \frac{\lambda_2}{L_r}\lambda_2\left(e_{ids}^s\left(\psi_{dr}^s - L_m\hat{i}_{ds}^s\right) + e_{iqs}^s\left(\psi_{qr}^s - L_m\hat{i}_{qs}^s\right)\right) \tag{2.204}$$

$$\hat{\omega}_m = K_p\left(e_{ids}^s\psi_{qr}^s - e_{iqs}^s\psi_{dr}^s\right) + K_i\int\left(e_{ids}^s\psi_{qr}^s - e_{iqs}^s\psi_{dr}^s\right)dt \tag{2.205}$$

Comparative analysis between MRAS solutions and full order observers, including tuning methodology, can be found in [16].

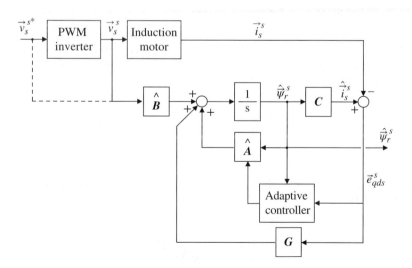

Fig. 2-40   *Rotor speed estimation using a full order observer*

2.7.2   Sensorless control using saliency-tracking-based methods

As already mentioned, sensorless control techniques that rely on the fundamental excitation have been shown to be capable of providing high performance control in the medium- to high-speed range. However, as the speed decreases, the performance of these methods decrease, eventually failing in the very low-speed range and/or for DC excitation. To overcome this limitation, sensorless control methods based on tracking the position of saliencies (asymmetries) in the rotor of the machine have been proposed. These techniques measure the response of the machine when a high-frequency excitation is applied via the inverter. Potentially, they have the capability of providing accurate, high-bandwidth, position, speed, and/or flux estimates in the low-speed range, including zero speed.

Though all saliency-tracking-based sensorless methods respond to the same physical principles, their practical implementation, as well as their performance, can strongly depend on the machine design. Machines which are inherently asymmetric (e.g. interior permanent magnet synchronous machines and synchronous reluctance machines) are in principle natural candidates for the use of these methods. However, it is also possible to extend the concept to machines which are symmetric in principle, like the induction machine.

For the sake of maintaining consistency with Chapters 2 and 3, saliency-tracking-based sensorless methods are discussed in the context of the synchronous machines in Chapter 3. However, a section specific to the application of these methods to induction machines is included in Chapter 3.

2.8   Steady-state equations and limits of operation of the induction machine

This section describes how it is possible to obtain the steady-state values of the magnitudes of the induction machine by using the dynamic model. First of all, a procedure for obtaining the maximum capability curves of magnitudes of the induction machine is presented, after which a procedure for obtaining steady-state values in general is presented.

2.8.1   Calculation of the maximum capability curves

First, it is necessary to evaluate the steady state of the model differential equations presented in previous sections. Thus, Table 2-1 shows the steps that must be followed to obtain the maximum available magnitudes of an induction machine in function of ω_s. Input data are P_{rated}, V_{rated}, n_{rated}, f_{rated}, and electric parameters of the induction machine. The rated torque is calculated as:

$$T_{em\,rated} = \frac{P_{rated}}{\omega_{m\,rated}/p} \tag{2.206}$$

Then, the rated flux is obtained by substituting (2.102) into torque equation (2.104) yielding a relation between the torque and flux as:

$$T_{em} = \frac{3}{2}p\frac{\omega_r}{R_r}\psi_{dr}^{\,2} \tag{2.207}$$

Table 2-1 Maximum capability curves for the induction machine

	Given data: P_{rated}, V_{rated}, n_{rated}, f_{rated}, and electric parameters of the induction machine (motor mode operation)			
1. Torque and flux	$\omega_{s\,rated} = 2\pi f_{rated}$ $\omega_{m\,rated} = n_{rated}(2\pi/60)p$	$\omega_{r\,rated} = \omega_{s\,rated} - \omega_{m\,rated}$	$T_{em\,rated} = \dfrac{P_{rated}}{\omega_{m\,rated}/p}$	$\psi_{dr\_rated} = \sqrt{\dfrac{T_{em\,rated}}{\frac{3}{2}p\dfrac{\omega_{r\_rated}}{R_r}}}$
2. Stator currents	$i_{ds\,rated} = \dfrac{\psi_{dr\_rated}}{L_m}$	$i_{qs\,rated} = \dfrac{\psi_{dr\_rated}\,\omega_{r\,rated}}{L_m R_r/L_r}$	$I_{s\,rated} = \sqrt{i_{ds\,rated}^2 + i_{qs\,rated}^2}$	$\theta_{i_s} = \alpha\tan\left(\dfrac{i_{qs}}{i_{ds}}\right)$
3. Limit frequencies	$\omega_b = \sqrt{\dfrac{(V_{s\,rated})^2}{\psi_{dr\_rated}^2\,\dfrac{L_s^2-(\sigma L_s)^2}{L_m^2}+(\sigma L_s I_{s\,rated})^2}}$		$\omega_1 = \sqrt{\dfrac{L_s^2+(\sigma L_s)^2}{2(L_s\sigma L_s)^2}}\times\dfrac{V_{s\_rated}}{I_{s\_rated}}$	

Constant torque region ($0<\omega_s<\omega_b$)

1. Torque and flux	$\psi_{dr} = \psi_{dr\_rated}$		$T_{em} = T_{em\,rated}$	
2. Stator currents	$i_{ds} = i_{ds\,rated}$	$i_{qs} = i_{qs\,rated}$	$I_{s\,rated} = \sqrt{i_{ds\,rated}^2 + i_{qs\,rated}^2}$	$\theta_{i_s} = \alpha\tan\left(\dfrac{i_{qs}}{i_{ds}}\right)$
3. Stator voltages	$v_{ds} = R_s i_{ds\,rated} - \omega_s\sigma L_s i_{qs\,rated}$	$v_{qs} = R_s i_{qs\,rated} + \omega_s\sigma L_s i_{ds\,rated} + \omega_s\dfrac{L_m}{L_r}\psi_{dr\_rated}$	$V_{s\,rated} = \sqrt{v_{ds}^2 + v_{qs}^2}$	$\theta_{v_s} = \alpha\tan\left(\dfrac{v_{qs}}{v_{ds}}\right)$
4. Frequencies and powers	$\omega_r = \dfrac{i_{qs\,rated}(L_m R_r/L_r)}{\psi_{dr\_rated}}$	$\omega_m = \omega_s - \omega_r$	$P_s = \dfrac{3}{2}\left(v_{ds}i_{ds}+v_{qs}i_{qs}\right)$	$P = \dfrac{T_{em}}{\omega_m/p}$
5. Efficiency, phase shift, and apparent power	$\eta = \dfrac{P}{P_s}$	$\varphi = \theta_{v_s} - \theta_{i_s}$		$S_s = \dfrac{P_s}{\cos(\varphi)}$

(continued overleaf)

Table 2-1 (continued)

Flux-weakening region I (variable $\omega_b < \omega_s < \omega_1$)

1. Stator currents	$i_{ds} = \sqrt{\dfrac{\left(\dfrac{V_{s\,rated}}{\omega_s}\right)^2 - (\sigma L_s I_{s\,rated})^2}{L_s^2 - (\sigma L_s)^2}}$	$i_{qs} = \sqrt{I_s^2 - i_{ds}^2}$	$I_s = \sqrt{i_{ds}^2 + i_{qs}^2} \equiv I_{s\,rated}$	$\theta_{i_s} = a\tan(i_{qs}/i_{ds})$
2. Torque and flux	$\psi_{dr} = L_m i_{ds}$		$T_{em} = \dfrac{3}{2}p\,\dfrac{L_m^2}{L_r}i_{ds}i_{qs}$	
3. Stator voltages	$v_{ds} = R_s i_{ds} - \omega_s\sigma L_s i_{qs}$	$v_{qs} = R_s i_{qs} + \omega_s\sigma L_s i_{ds} + \omega_s\dfrac{L_m}{L_r}\psi_{dr}$	$V_s = \sqrt{v_{ds}^2 + v_{qs}^2} \equiv V_{s\,rated}$	$\theta_{v_s} = a\tan\left(\dfrac{v_{qs}}{v_{ds}}\right)$
4. Frequencies and powers	$\omega_r = \dfrac{i_{qs}(L_m R_r/L_r)}{|\vec{\psi}_r|}$	$\omega_m = \omega_s - \omega_r$	$P_s = \dfrac{3}{2}(v_{ds}i_{ds} + v_{qs}i_{qs})$	$P = \dfrac{T_{em}}{\omega_m/F}$
5. Efficiency, phase shift, and apparent power	$\eta = \dfrac{P}{P_s}$	$\varphi = \theta_{v_s} - \theta_{i_s}$		$S_s = \dfrac{P_s}{\cos(\varphi)}$

Flux-weakening region II (variable $\omega_s > \omega_1$)

1. Stator currents	$i_{ds} = \dfrac{V_{s\,rated}}{\sqrt{2}\omega_s L_s}$	$i_{qs} = \dfrac{V_{s\,rated}}{\sqrt{2}\omega_s L_s}$	$I_s = \sqrt{i_{ds}^2 + i_{qs}^2} < I_{s\,rated}$	$\theta_{i_s} = a\tan\left(\dfrac{i_{qs}}{i_{ds}}\right)$
2. Torque and flux	$\psi_{dr} = L_m i_{ds}$		$T_{em} = \dfrac{3}{2}p\,\dfrac{L_m^2}{L_r}i_{ds}i_{qs}$	
3. Stator voltages	$v_{ds} = R_s i_{ds} - \omega_s\sigma L_s i_{qs}$	$v_{qs} = R_s i_{qs} + \omega_s\sigma L_s i_{ds} + \omega_s\dfrac{L_m}{L_r}\psi_{dr}$	$V_s = \sqrt{v_{ds}^2 + v_{qs}^2} \equiv V_{s\,rated}$	$\theta_{v_s} = a\tan\left(\dfrac{v_{qs}}{v_{ds}}\right)$
4. Frequencies and powers	$\omega_r = \dfrac{i_{qs}(L_m R_r/L_r)}{|\vec{\psi}_r|}$	$\omega_m = \omega_s - \omega_r$	$P_s = \dfrac{3}{2}(v_{ds}i_{ds} + v_{qs}i_{qs})$	$P = \dfrac{T_{em}}{\omega_m/F}$
5. Efficiency, phase shift, and apparent power	$\eta = \dfrac{P}{P_s}$	$\varphi = \theta_{v_s} - \theta_{i_s}$		$S_s = \dfrac{P_s}{\cos(\varphi)}$

Knowing the rated torque, the rated slip $\omega_{r\ rated}$ is given by:

$$\omega_{r\ rated} = \omega_{s\ rated} - \omega_{m\ rated} \tag{2.208}$$

The rated flux is obtained as:

$$\psi_{dr-rated} = \sqrt{\frac{T_{em\ rated}}{\frac{3}{2}p\dfrac{\omega_{r\ rated}}{R_r}}} \tag{2.209}$$

Then, once the rated flux is obtained, the rated current is obtained as:

$$i_{ds\ rated} = \frac{\psi_{dr\_rated}}{L_m} \quad i_{qs\ rated} = \frac{\psi_{dr\_rated}\omega_{r\ rated}}{L_m R_r / L_r} \tag{2.210}$$

It must be remarked that these magnitudes should be checked with data provided by the manufacturer, in an attempt to fix any possible inconsistency. Then, as shown in Table 2-1, the rest of the magnitudes at maximum capability (maximum torque, current, etc.) can be calculated step by step, at the three operation regions of the machine.

Fig. 2-41 shows an example of the maximum capability curves obtained for an induction machine of data: 160kW, 400V$_{LL}$, 1487rpm, and 50Hz, obtained from [17]. The electrical parameters of this machine are: $R_s = 1.405\Omega$, $R_r = 1.395\Omega$, $L_{\sigma s} = L_{\sigma r} = 5.84$mH, $L_m = 172.2$mH. By evaluating the procedure, Fig. 2-41 shows the most representative curves of the machine at three operating regions, in function of ω_s. By careful analysis, the reader can check the performance and behavior of every magnitude. It can also be seen that the rotor flux magnitude is obtained, which can be used as the flux-weakening control strategy by means of the pre-calculated look-up table mentioned in Section 2.5.2. The table can be indexed by ω_s, as illustrated in Fig. 2-42. Conversely, if ω_m is to be used as an index, ω_r must be subtracted from the table.

2.8.2 Calculation of the steady-state operation

The steady-state operation of the induction machine is determined by the specific torque load placed at the shaft at any given moment. Thus, this section shows how the most interesting magnitudes of the machine at steady state can be derived, by using the dynamic model seen in previous sections. Accordingly, the procedure to derive the steady-state magnitudes is described by means of Table 2-2. The input data are supposed to be the load placed at the shaft (i.e. the pair ω_m-T_{load}; see Chapter 1). Thus, the pair torque and speed at the shaft, together with the rotor flux reference commanded by the vector control, determine the steady-state operation of the machine. This means that the rest of the magnitudes are derived from these three initial values: torque, speed, and rotor flux.

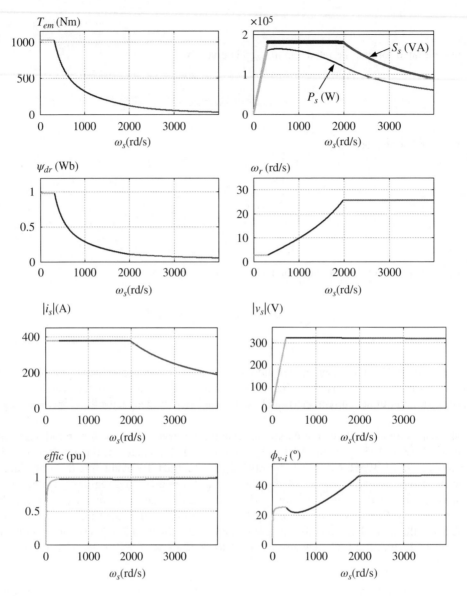

Fig. 2-41 *Maximum capability curves of a 160kW induction machine of 400V$_{LL}$, 1487rpm, 50Hz*

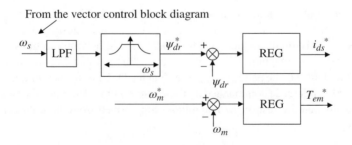

Fig. 2-42 *Pre-calculated look-up table*

Table 2-2 Procedure to obtain the magnitudes of an induction machine at steady state

	Given data: Operation pair array ω_m-T_{load}			
Constant torque region ($0 < \omega_m < \omega_b$)				
1. Torque and flux	$\psi_{dr} = \psi_{dr\_rated}$		$T_{em} = T_{load}$	
2. Stator currents	$i_{ds} = \dfrac{\psi_{dr\_rated}}{L_m}$	$i_{qs} = \dfrac{T_{load}}{\dfrac{3}{2}p\dfrac{L_m^{\,2}}{L_r}i_{ds}}$	$I_s = \sqrt{i_{ds}^{\,2} + i_{qs}^{\,2}}$	$\theta_{is} = a\tan\left(\dfrac{i_{qs}}{i_{ds}}\right)$
3. Frequencies		$\omega_r = \dfrac{i_{qs}\left(L_m R_r / L_r\right)}{\psi_{dr\_rated}}$	$\omega_s = \omega_m + \omega_r$	
4. Stator voltages		$v_{ds} = R_s i_{ds} - \omega_s \sigma L_s i_{qs}$	$v_{qs} = R_s i_{qs} + \omega_s \sigma L_s i_{ds} + \omega_s \dfrac{L_m}{L_r}\psi_{dr\_rated}$	$\theta_{vs} = a\tan\left(\dfrac{v_{qs}}{v_{ds}}\right)$
			$V_s = \sqrt{v_{ds}^{\,2} + v_{qs}^{\,2}}$	
5. Powers		$P_s = \dfrac{3}{2}\left(v_{ds}i_{ds} + v_{qs}i_{qs}\right)$	$P = \dfrac{T_{em}}{\omega_m/p}$	
			$S_s = \dfrac{P_s}{\cos(\varphi)}$	
6. Efficiency, phase shift, and apparent power		$\eta = \dfrac{P}{P_s}$		
		$\varphi = \theta_{vs} - \theta_{is}$		
Flux-weakening region I (variable $\omega_b < \omega_m < \omega_1$)				
1. Torque and flux	$\psi_{dr} = L_m \sqrt{\dfrac{\left(\dfrac{V_{s\,rated}}{\omega_m}\right)^2 - \left(\sigma L_s I_{s\,rated}\right)^{2^*}}{L_s^2 - \left(\sigma L_s\right)^2}}$		$T_{em} = T_{load}$	
2. Stator currents	$i_{ds} = \sqrt{\dfrac{\left(\dfrac{V_{s\,rated}}{\omega_m}\right)^2 - \left(\sigma L_s I_{s\,rated}\right)^{2^*}}{L_s^2 - \left(\sigma L_s\right)^2}}$	$i_{qs} = \dfrac{T_{load}}{\dfrac{3}{2}p\dfrac{L_m^{\,2}}{L_r}i_{ds}}$	$I_s = \sqrt{i_{ds}^{\,2} + i_{qs}^{\,2}}$	$\theta_{is} = a\tan\left(i_{qs}/i_{ds}\right)$

(continued overleaf)

Table 2-2 (continued)

3. Frequencies	$\omega_r = \dfrac{i_{qs}(L_m R_r/L_r)}{\psi_{dr}}$		$\omega_s = \omega_m + \omega_r$
4. Stator voltages	$v_{ds} = R_s i_{ds} - \omega_s \sigma L_s i_{qs}$	$v_{qs} = R_s i_{qs} + \omega_s \sigma L_s i_{ds} + \omega_s \dfrac{L_m}{L_r}\psi_{dr}$	$V_s = \sqrt{v_{ds}^2 + v_{qs}^2}$ $\theta_{v_s} = a\tan\left(\dfrac{v_{qs}}{v_{ds}}\right)$
5. Powers	$P_s = \dfrac{3}{2}\left(v_{ds}i_{ds} + v_{qs}i_{qs}\right)$		$P = \dfrac{T_{em}}{\omega_m/p}$
6. Efficiency, phase shift, and apparent power	$\eta = \dfrac{P}{P_s}$	$\varphi = \theta_{v_s} - \theta_{i_s}$	$S_s = \dfrac{P_s}{\cos(\varphi)}$
Flux-weakening region II (variable $\omega_m > \omega_1$)			
1. Torque and flux	$\psi_{dr} = L_m \dfrac{V_{s\,rated}^{*}}{\sqrt{2}\,\omega_m L_s}$		$T_{em} = T_{load}$
2. Stator currents	$i_{ds} = \dfrac{V_{s\,rated}^{*}}{\sqrt{2}\,\omega_m L_s}$	$i_{qs} = \dfrac{T_{load}}{\dfrac{3}{2}p\dfrac{L_m^2}{L_r}i_{ds}}$	$I_s = \sqrt{i_{ds}^2 + i_{qs}^2}$ $\theta_{i_s} = a\tan(i_{qs}/i_{ds})$
3. Frequencies	$\omega_r = \dfrac{i_{qs}(L_m R_r/L_r)}{\psi_{dr}}$		$\omega_s = \omega_m + \omega_r$
4. Stator voltages	$v_{ds} = R_s i_{ds} - \omega_s \sigma L_s i_{qs}$	$v_{qs} = R_s i_{qs} + \omega_s \sigma L_s i_{ds} + \omega_s \dfrac{L_m}{L_r}\psi_{dr}$	$V_s = \sqrt{v_{ds}^2 + v_{qs}^2}$ $\theta_{v_s} = a\tan\left(\dfrac{v_{qs}}{v_{ds}}\right)$
5. Powers	$P_s = \dfrac{3}{2}\left(v_{ds}i_{ds} + v_{qs}i_{qs}\right)$		$P = \dfrac{T_{em}}{\omega_m/p}$
6. Efficiency, phase shift, and apparent power	$\eta = \dfrac{P}{P_s}$	$\varphi = \theta_{v_s} - \theta_{i_s}$	$S_s = \dfrac{P_s}{\cos(\varphi)}$

^{a} ω_m is used instead of ω_s for easier calculation.

Consequently, first of all from the known speed, it is necessary to derive in which operation region the machine is working: constant torque, flux-weakening region I, or flux-weakening region II, by means of (2.164) and (2.168). Note that ω_m is used in the equations instead of ω_s for easier derivation. Once the regions are known, first electromagnetic torque and flux are derived, as described in Table 2-2. Then, the stator current *dq* components are calculated. After that, ω_r and ω_s are calculated. Then stator voltages, and finally calculating powers, efficiencies, and phase shift between the stator voltage and current. It will be noted that the procedure itself is in some way similar to the procedure presented in Table 2-1 to calculate the maximum capability curves of the induction machine.

Therefore, Fig. 2-43 gives one example of steady-state magnitudes evaluation, using the procedure described in Table 2-2, for the same machine used in the previous subsection: 160kW, 400V$_{LL}$, 1487rpm, and 50Hz, obtained from [17]. The load torque applied to the shaft is supposed to be given by the law: $T_{load} = 0.0025\omega_m^2$, which means that the torque evolves quadratic to the speed. Note that in this case the *x* axis (i.e. the magnitudes) represented in function of ω_m (instead of ω_s as done when deriving the maximum capability curves; however, note that ω_r takes very small values). Only the speed region contained in the constant torque region and flux-weakening region I are shown. The reader will observe the evolution and behavior of all the magnitudes of this specific example. As already mentioned, the torque presents a quadratic evolution, while the active power presents a cubic evolution in function of the speed. With regards to the apparent power, it increases at two different rhythms: one being the rated speed and the other when flux weakening. For this specific example, something similar also happens with ω_r, which evolves as determined by i_{qs} and rotor flux evolutions. It will be noticed that during all the operation points ω_r takes relatively small values. With the stator current, the increase is similar to the rest of the variables. Until the rated speed, the increase is less pronounced than when flux weakening, basically because when flux weakening the i_{qs} must be increased much more pronouncedly since the i_{ds} is decreased, owing to the flux weakening and the torque (which continues to increase). With regards to the stator flux voltage amplitude required at motor input terminals, it is seen that, until the rated value (at rated speed) is achieved, the voltage is increased in a similar way as a ramp. However, once the maximum voltage value is reached, this voltage is approximately maintained as constant. In a different way, the efficiency of the machine is increased approximately until the rated speed, to later decrease above that speed where the maximum has been achieved. Finally, with regards to the phase shift between the voltage and current, it is seen that almost descends during all range of speeds, while the decrease is much more pronounced at flux-weakening region.

On the other hand, Fig. 2-44 illustrates in a similar way the behavior of the most interesting magnitudes of the machine, when the torque load is kept constant during the entire speed range. The reader can observe the main differences in the behavior of the magnitudes by comparing Fig. 2-43 with Fig. 2-44.

On the other hand, as described in Chapter 7, often in electric machines (induction and synchronous), efficiency maps are obtained and employed for optimized designs. These maps are obtained by computing the efficiency of the motor at every possible torque-speed point. This analysis shows that the machines in general present approximately constant efficiency regions. This analysis allows one to perform optimized designs of electric drives, mechanical systems and energy sources, taking into account the global efficiency and particular efficiencies of the different parts of the system (e.g. in hybrid vehicles: these would be the electric motor, diesel engine, gear boxes, etc.).

Finally Table 2-3 provides examples of parameters for different power ratings.

Here, some common tendencies can be noted, for instance the stator and rotor resistances and leakage inductances decrease with power increase. This results in a larger stator current ripple. It will be noted, however, that the fundamental current amplitude also increases, owing to the decrease of the magnetizing inductance. Finally, both the stator transient time constant and the rotor time constant are seen to increase with power, the second being always larger.

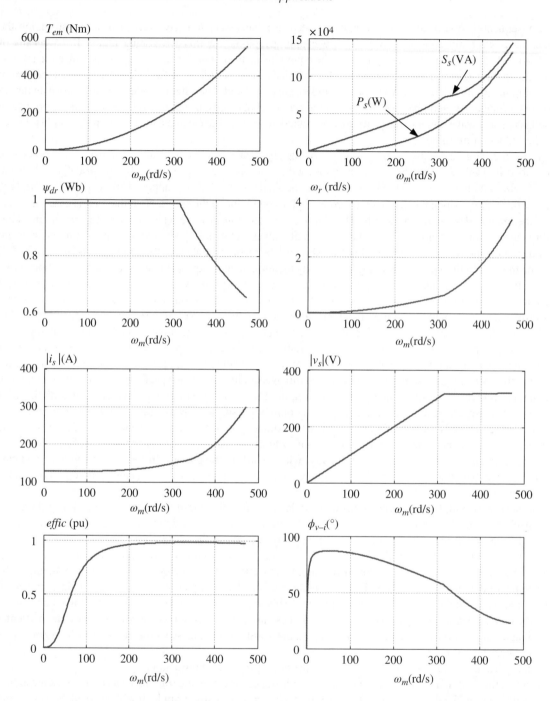

Fig. 2-43   *Steady-state curves of a 160kW induction machine of 400V$_{LL}$, 1487rpm, 50Hz, with a quadratic torque-speed load* (T$_{load}$ = 0.0025ω_m^2)

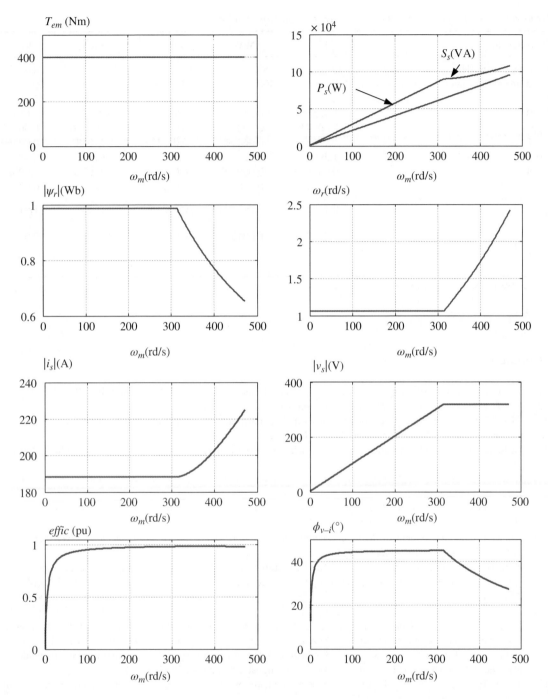

Fig. 2-44 *Steady-state curves of a 160kW induction machine of 400V$_{LL}$, 1487rpm, 50Hz, with constant torque load*

Table 2-3   *Electric parameters of different power induction machines*

Characteristic	4kW	15kW	37kW	110kW	160kW
Rated frequency (Hz)	50	50	50	50	50
Rated line to line stator voltage (V_{RMS})	400	400	400	400	400
p	2	2	2	2	2
$R_s(\Omega)$	1.405	0.2147	0.08233	0.02155	0.0137
$L_{\sigma s}(H)$	0.005839	0.000991	0.000724	0.000226	0.000152
$L_m(H)$	0.1722	0.06419	0.02711	0.01038	0.00769
$R_r(\Omega)$	1.395	0.2205	0.0503	0.01231	0.007728
$L_{\sigma r}(H)$	0.005839	0.000991	0.000724	0.000226	0.000152
σ	0.0645	0.03017	0.05134	0.04216	0.03838
$\dfrac{L_r}{R_r}$(sec)	0.127	0.2956	0.55335	0.86157	1.01475
$\dfrac{\sigma L_s}{R_s + R_r \dfrac{L_m^2}{L_r^2}}$(sec)	0.00423	0.00458	0.01098	0.01341	0.01418

(Source: Data from MATLAB-Simulink using preset models [17])

References

[1]   Sul S-K. *Control of Electric Machine Drive Systems*. Hoboken, NJ: John Wiley & Sons, Inc, 2011.

[2]   Vas P. *Vector Control of AC Machines*. London: Clarendon Press, 1990.

[3]   Blaschke F. The principle of field orientation as applied to the new transvector closed loop control system for rotating field machines. *Siemens Rev May* 1972; **34**: 217–220.

[4]   Hasse K. Zur Dynamik drehzahlgeregelter Antriebe mit stromrichtergespeisten Asynchron-Kurzsehlublaufermasehinen. *Techn Hochsch Diss* 1969.

[5]   Rowan TR, Kerkman RL. A new synchronous current regulator and an analysis of current-regulated PWM inverters. *IEEE Trans Ind Appl* July/August 1986; **22**: 678–690.

[6]   Briz F, Degner MW, Lorenz RD. Dynamic analysis of current regulators for AC motors using complex vectors. *IEEE Trans Ind Appl* November/December 1999; **35**(6): 1424, 1432.

[7]   Kim H, Degner MW, Guerrero JM, et al. Discrete-time current regulator design for AC machine drives. *IEEE Trans Ind Appl* July/August 2010; **46**(4): 1425–1435.

[8]   Yim J-S, Sul S-K, Bae B-H, et al. Modified current control schemes for high-performance permanent-magnet AC drives with low sampling to operating frequency ratio. *IEEE Trans Ind Appl* March/April 2009; **45**(2): 763–771.

[9]   Novotny DW, Lipo TA. *Vector Control and Dynamics of AC Drives*. New York: Oxford University Press, 1996.

[10]   Jansen PL, Lorenz RD. A physically insightful approach to the design and accuracy assessment of flux observers for field oriented induction machine drives. *IEEE Trans Ind Appl* January/February 1994; **30**(1): 101, 110.

[11]   Holtz J, Quan J. Sensorless vector control of induction motors at very low speed using a nonlinear inverter model and parameter identification. *IEEE Trans Ind Appl* July/August 2002; **38**(4): 1087, 1095.

[12]   Toliyat HA, Levi E, Raina M. A review of RFO induction motor parameter estimation techniques. *IEEE Trans Energy Convers* June 2003; **18**(2):271, 283.

[13]   Holtz J. Sensorless control of induction machines: With or without signal injection? *IEEE Trans Ind Electron* February 2006; **53**(1): 7, 30.

[14] Schauder C. Adaptive speed identification for vector control of induction motors without rotational transducers. *IEEE Trans Ind Appl* September/October 1992; **28**(5): 1054, 1061.

[15] Kubota K, Matsuse K. Speed sensorless field-oriented control of induction motor with rotor resistance adaptation. *IEEE Trans Ind Appl* September/October 1994; **30**(5): 1219, 1224.

[16] Ohyama K, Asher GM, Sumner M. Comparative analysis of experimental performance and stability of sensorless induction motor drives. *IEEE Trans Ind Electron* February 2006; **53**(1): 178, 186.

[17] Induction machine POWER library of Matlab-Simulink.

3

Control of synchronous machines

Fernando Briz and Gonzalo Abad

3.1 Introduction

Synchronous AC machines are a type of machine in which, in steady-state conditions, the rotor speed in electrical units is equal to the frequency of the stator current. Generally speaking, synchronous machines provide higher performance compared to induction motors, but at an increased cost and reduced robustness, mainly owing to a more elaborate rotor construction. Synchronous machines have been used for decades and have played a major role in high-power applications. However, it has been during the last two decades, with the introduction of rare-earths magnet materials, that the permanent magnet synchronous machines (PMSMs) have emerged as a key element for the development of high-performance electric drives. PMSMs are excellent candidates for applications requiring high torque and power densities, as well as high efficiency. They can be found in multiple sectors, with powers ranging from fractional horsepower up to the MW, and in multiple applications including hybrid and electric vehicles, trains, large wind-turbines, servo-drives, and household appliances. Owing to this, although other types of synchronous machines will also be covered, this chapter is mainly devoted to the PMSM. Thus this chapter presents a similar structure to the previous chapter. First of all, the model of the machine is analyzed; based on these models, vector control is addressed. After that, sensorless control of PMSM is discussed. Finally, a short section is dedicated to steady-state analysis.

3.2 Types of synchronous machines

Synchronous machines have a stator design which responds to the same principles and is therefore similar to that of induction machines. The rotor of the synchronous machine is, however, significantly different and can be of various types, depending on the design and materials. The rotor always moves in sync with the magnetic

Power Electronics and Electric Drives for Traction Applications, First Edition. Edited by Gonzalo Abad.
© 2017 John Wiley & Sons, Ltd. Published 2017 by John Wiley & Sons, Ltd.

field produced by the stator windings, the synchronously rotating d–q reference frame being aligned with the rotor.

Attending to the rotor construction, several types of synchronous machines can be found. In all cases the rotor core is made of laminated iron.

- **Wound rotor synchronous machine:** the rotor of this type of machine has a winding which is excited with a DC current, the electromagnetic torque resulting from the interaction of the resulting field and the flux produced in the stator windings [1]. The DC current is normally supplied to the rotor by means of brushes and slip rings. However, implementations in which the energy is transferred to the rotor by means of auxiliary windings are also popular. This rotor design is often found in very high-power machines. Fig. 3-1 shows a simplified illustration of a wound field synchronous machine with one pole pair.
- **Permanent magnet synchronous machine (PMSM):** The rotor flux is created by magnets located in the rotor. There is a large variety of rotor designs, attending to the shape, size, and location of the rotor magnets [2], [3]. Fig. 3-2 shows two examples. In the example in Fig. 3-2(a), the magnets are located on the rotor surface, the resulting design being called a surface permanent magnet synchronous machine (SPMSM). In the example in Fig. 3-2(b), the magnets are buried in the rotor, the resulting design being called an interior permanent magnet synchronous machine (IPMSM).

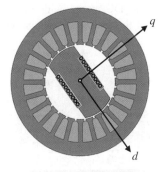

Fig. 3-1   *Simplified representation of a wound rotor synchronous machine*

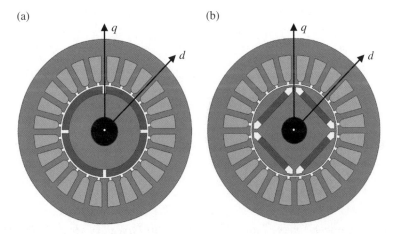

Fig. 3-2   *Schematic representation of a four-pole (a) SPMSM and (b) IPMSM*

Modern PMSMs often use rare-earth magnets with high magnetic field strength (e.g. NdFeB). This results in machines of a smaller size, and consequently of higher power/torque densities and lower inertia, compared to the induction machine. The fact that no energy has to be transferred to the rotor also results in an increase in efficiency. However, the presence of the magnets in the rotor adversely impacts cost and robustness.

- **Synchronous reluctance machine (SynRM):** These types of machines have a stator which is identical to the stator of other three-phase machines, like induction machines or PMSMs. However, the rotor does not incorporate any mechanism to produce flux. Instead, the rotor is designed to be salient, therefore producing reluctance torque. An example of synchronous reluctance machine design is shown in Fig. 3-3.

  SynRMs perform worse than permanent magnet machines in terms of torque density and power factor. However, they are advantageous in terms of cost and robustness, owing to the absence of magnets in the rotor. On the other hand, SynRMs are more efficient than induction motors, as the losses in the rotor are very small, compared to the 20–30% typically found in induction motors. This is due to the fact that there are no conductors in the rotor, and that it rotates synchronously with the stator fundamental excitation. The stator losses tend to be similar for both types of motors.

- **Brushless DC motor (BLDC):** The stator of this type of machine uses concentrated windings, in contrast to the distributed windings more often used in the previous synchronous machine designs. This results in a trapezoidal back-emf, vs. the sinusoidal back-emf of other designs. Normally the magnets of the rotor are mounted on the rotor surface. An electronic converter is used to provide a six-step current to the stator windings, whose interaction with the rotor magnet results (ideally) in a constant torque. Fig. 3-4 shows a simplified representation of a brushless DC motor, which in essence is similar to the surface PMSM, as well as a schematic representation of the back-emf voltage and current waveforms.

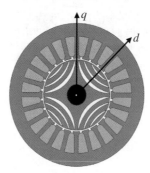

Fig. 3-3   *Schematic representation of a SynRM*

(a)   (b)

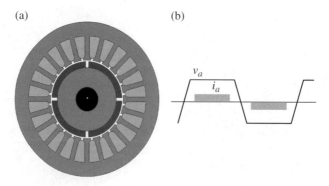

Fig. 3-4   *(a) Simplified representation of a brushless DC motor, (b) Typical voltage and current waveforms of one of the phases*

Compared to their AC permanent magnet machines counterpart, BLDC machines have a simpler design and construction, their control also being simpler in principle. This results in a cheaper electric drive, which is often found in low-power applications, with power densities than can be higher than for sinusoidal back-emf PM machines. However, they present lower performance in terms of torque smoothness.

3.3   Modeling of synchronous machines

Dynamic models of three-phase synchronous machines are presented in this section. These models are used later in this chapter when we come on to discuss the development of control strategies. The three-phase model is developed first, from which the complex vector model is later obtained. Similar to the induction machine case, the use of a *d–q* reference frame that rotates synchronously with the rotor will be shown to provide simple and insightful dynamic models, eventually enabling the development of high-performance control methods.

3.3.1   Dynamic models of synchronous machines using three-phase variables

The stator of three-phase synchronous machines shares major characteristics of the induction machine case studied in the previous chapter. The stator voltage equations of a three-phase synchronous machine are therefore identical to those of the induction machine. They are reproduced here for convenience:

$$
\begin{bmatrix} v_{as} \\ v_{bs} \\ v_{cs} \end{bmatrix} = R_s \begin{bmatrix} i_{as} \\ i_{bs} \\ i_{cs} \end{bmatrix} + \frac{d}{dt} \begin{bmatrix} \psi_{as} \\ \psi_{bs} \\ \psi_{cs} \end{bmatrix}
\tag{3.1}
$$

The stator fluxes are:

$$
\begin{bmatrix} \psi_{as} \\ \psi_{bs} \\ \psi_{cs} \end{bmatrix} = \begin{bmatrix} L_{aa} & L_{ab} & L_{ac} \\ L_{ba} & L_{bb} & L_{bc} \\ L_{ca} & L_{ca} & L_{cc} \end{bmatrix} \begin{bmatrix} i_{as} \\ i_{bs} \\ i_{cs} \end{bmatrix} + \begin{bmatrix} \psi_{am} \\ \psi_{bm} \\ \psi_{cm} \end{bmatrix}
\tag{3.2}
$$

In the case of a non-salient synchronous machine, the self-inductances for each phase (3.3), as well as the mutual inductances (3.4), are constant and equal to each other, as they do not depend on rotor position.

$$
\text{Self-inductances: } L_{aa} = L_{bb} = L_{cc} = L_{\sigma s} = \text{constant}
\tag{3.3}
$$

$$
\text{Mutual inductances: } L_{ab} = L_{ba} = L_{bc} = L_{cb} = L_{ac} = L_{ca} = L_m = \text{constant}
\tag{3.4}
$$

The fundamental wave of the fluxes produced in the rotor, either by the magnets or by the rotor circuit, can be modeled as:

$$
\begin{bmatrix} \psi_{apm} \\ \psi_{bpm} \\ \psi_{cpm} \end{bmatrix} = \begin{bmatrix} \psi_{pm} \cdot \cos\theta_m \\ \psi_{pm} \cdot \cos(\theta_m - 2\pi/3) \\ \psi_{pm} \cdot \cos(\theta_m + 2\pi/3) \end{bmatrix}
\tag{3.5}
$$

with θ_m being the rotor position relative to the stator windings in electrical units. The derivative of the fluxes with respect to time is therefore given by:

$$\frac{d}{dt}\begin{bmatrix} \psi_{apm} \\ \psi_{bpm} \\ \psi_{cpm} \end{bmatrix} = \begin{bmatrix} -\psi_{pm} \cdot \sin\theta_m \cdot \dfrac{d\theta_m}{dt} \\ -\psi_{pm} \cdot \sin(\theta_m - 2\pi/3) \cdot \dfrac{d\theta_m}{dt} \\ -\psi_{pm} \cdot \sin(\theta_m + 2\pi/3) \cdot \dfrac{d\theta_m}{dt} \end{bmatrix} \tag{3.6}$$

which combined with (3.2) and in the case of a constant rotor speed of ω_m yields:

$$\frac{d}{dt}\begin{bmatrix} \psi_{as} \\ \psi_{bs} \\ \psi_{cs} \end{bmatrix} = \begin{bmatrix} L_{\sigma s} & -L_m & -L_m \\ -L_m & L_{\sigma s} & -L_m \\ -L_m & -L_m & L_{\sigma s} \end{bmatrix} \cdot \frac{d}{dt}\begin{bmatrix} i_{as} \\ i_{bs} \\ i_{cs} \end{bmatrix} - \omega_m \cdot \psi_{pm} \begin{bmatrix} \sin\theta_m \\ \sin(\theta_m - 2\pi/3) \\ \sin(\theta_m + 2\pi/3) \end{bmatrix} \tag{3.7}$$

Substituting this last expression into the voltage equation (3.1) yields:

$$\begin{bmatrix} v_{as} \\ v_{bs} \\ v_{cs} \end{bmatrix} = R_s \cdot \begin{bmatrix} i_{as} \\ i_{bs} \\ i_{cs} \end{bmatrix} + \begin{bmatrix} L_{\sigma s} & -L_m & -L_m \\ -L_m & L_{\sigma s} & -L_m \\ -L_m & -L_m & L_{\sigma s} \end{bmatrix} \cdot \frac{d}{dt}\begin{bmatrix} i_{as} \\ i_{bs} \\ i_{cs} \end{bmatrix} - \omega_m \cdot \psi_{pm} \begin{bmatrix} \sin\theta_m \\ \sin(\theta_m - 2\pi/3) \\ \sin(\theta_m + 2\pi/3) \end{bmatrix} \tag{3.8}$$

Considering that all three stator currents add up to zero (i.e. $i_{as} + i_{bs} + i_{cs} = 0$), the stator fluxes can be represented as:

$$\begin{bmatrix} \psi_{as} \\ \psi_{bs} \\ \psi_{cs} \end{bmatrix} = \begin{bmatrix} L_{\sigma s} i_{as} - L_m(i_{bs} + i_{cs}) \\ L_{\sigma s} i_{bs} - L_m(i_{as} + i_{cs}) \\ L_{\sigma s} i_{cs} - L_m(i_{bs} + i_{as}) \end{bmatrix} = \begin{bmatrix} (L_{\sigma s} + L_m)i_{as} \\ (L_{\sigma s} + L_m)i_{bs} \\ (L_{\sigma s} + L_m)i_{cs} \end{bmatrix} = \begin{bmatrix} L_s i_{as} \\ L_s i_{bs} \\ L_s i_{cs} \end{bmatrix} \tag{3.9}$$

Combining the previous equations, the following expression of the stator-voltage equation for a symmetric synchronous machine is finally obtained:

$$\begin{bmatrix} v_{as} \\ v_{bs} \\ v_{cs} \end{bmatrix} = R_s \cdot \begin{bmatrix} i_{as} \\ i_{bs} \\ i_{cs} \end{bmatrix} + \begin{bmatrix} L_s & 0 & 0 \\ 0 & L_s & 0 \\ 0 & 0 & L_s \end{bmatrix} \cdot \frac{d}{dt}\begin{bmatrix} i_{as} \\ i_{bs} \\ i_{cs} \end{bmatrix} - \omega_m \cdot \psi_{pm} \begin{bmatrix} \sin\theta_m \\ \sin(\theta_m - 2\pi/3) \\ \sin(\theta_m + 2\pi/3) \end{bmatrix} \tag{3.10}$$

3.3.2   Dynamic model of synchronous machines in the stationary reference frame using complex space vectors

Using the space vector theory, the three-phase stator voltage equations can be transformed into an equivalent $\alpha\beta$ system. Applying the definition of the space complex vectors (2.6) seen in the previous chapter, the complex vector representation of the stator voltage equation (3.1) is obtained.

$$\vec{v}_s^{\,s} = R_s \cdot \vec{i}_s^{\,s} + \frac{d\vec{\psi}_s^{\,s}}{dt} \tag{3.11}$$

This equation can be separated into its real and imaginary ($\alpha\beta$) components:

$$v_{\alpha s} = R_s i_{\alpha s} + \frac{d\psi_{\alpha s}}{dt} \tag{3.12}$$

$$v_{\beta s} = R_s i_{\beta s} + \frac{d\psi_{\beta s}}{dt} \tag{3.13}$$

The stator fluxes, in the case of a non-salient machine, are given by:

$$\psi_{\alpha s} = \psi_{pm} \cos \theta_m + L_s i_{\alpha s} \tag{3.14}$$

$$\psi_{\beta s} = \psi_{pm} \sin \theta_m + L_s i_{\beta s} \tag{3.15}$$

Substituting the stator flux expressions (3.14) and (3.15) into the voltage equations (3.12) and (3.13), the following expressions of the stator voltages are obtained:

$$v_{\alpha s} = R_s i_{\alpha s} + \frac{d}{dt}\left(\psi_{pm} \cos \theta_m + L_s i_{\alpha s}\right) \tag{3.16}$$

$$v_{\beta s} = R_s i_{\beta s} + \frac{d}{dt}\left(\psi_{pm} \sin \theta_m + L_s i_{\beta s}\right) \tag{3.17}$$

3.3.3 Dynamic model of synchronous machines in the synchronous reference frame

Similar to the induction motor, the dynamic model of synchronous machines can be transformed to a rotating d–q reference frame with the d axis aligned with the rotor flux. It will be noted that for synchronous machines the rotor flux angle coincides with the rotor angle. This transformation provides important benefits both for analysis and control purposes. Equations in the d–q reference frame are obtained by multiplying (3.11) by $e^{-j\theta_m}$, the resulting voltage equations after some algebraic transformations being:

$$v_{ds} = R_s i_{ds} - \omega_m \psi_{qs} + \frac{d\psi_{ds}}{dt} \tag{3.18}$$

$$v_{qs} = R_s i_{qs} + \omega_m \psi_{ds} + \frac{d\psi_{qs}}{dt} \tag{3.19}$$

The stator flux equations are given by:

$$\psi_{ds} = i_{ds} L_{ds} + \psi_{pm} \tag{3.20}$$

$$\psi_{qs} = i_{qs} L_{qs} \tag{3.21}$$

Fig. 3-5 shows the equivalent circuit of the synchronous machine obtained from (3.18) to (3.21).

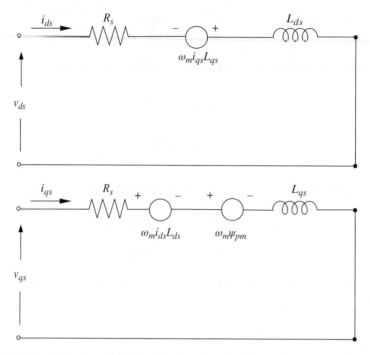

Fig. 3-5   *Equivalent circuit for the d and q axis of a synchronous machine*

Combining the flux and voltage equations, the relationship between the stator voltages and currents in the synchronous reference frame are obtained.

$$v_{ds} = R_s i_{ds} + L_{ds}\frac{di_{ds}}{dt} - \omega_m i_{qs} L_{qs} \tag{3.22}$$

$$v_{qs} = R_s i_{qs} + L_{qs}\frac{di_{qs}}{dt} + \omega_m \psi_{pm} + \omega_m i_{ds} L_{ds} \tag{3.23}$$

It should be noted that the term corresponding to the rotor flux only exists in the q axis circuit.

3.4   Torque equation for synchronous machines

The general expression of the torque produced by a synchronous machine is given by (3.24), which is equivalent to (2.87) in Chapter 2 for the induction machine.

$$T_{em} = \frac{3}{2}p\left(\psi_{ds}i_{qs} - \psi_{qs}i_{ds}\right) \tag{3.24}$$

The torque equation can be rewritten as a function of the stator inductances and magnet field as:

$$T_{em} = \frac{3}{2}p\left(\psi_{pm}i_{qs} + \left(L_{ds} - L_{qs}\right)i_{ds}i_{qs}\right) \tag{3.25}$$

The torque is seen to consist of two terms: the electromagnetic torque, which is produced by the interaction between the rotor magnet and the stator current, and the reluctance torque, which is due to the different inductances in the d and q axis. Depending on the design, synchronous machines can produce electromagnetic torque, reluctance torque, or both. The torque production mechanisms for the SPMSM (Fig. 3-2(a)), IPMSM (Fig. 3-2(b)), and synchronous reluctance machine (Fig. 3-3) are discussed in the following subsections.

3.4.1   Surface permanent magnet synchronous machine (non-salient machines)

Non-salient machines have the same inductance in the d and q axis, that is

$$L_s = L_{ds} = L_{qs} \tag{3.26}$$

This is the case in surface permanent magnet motors (Fig. 3-2(a)) because the magnet permeability is normally close to that of air. Consequently, the torque equation in (3.25) simplifies to:

$$T_{em} = \frac{3}{2} p \psi_{pm} i_{qs} \tag{3.27}$$

It can be observed from this expression that the electromagnetic torque is proportional to the q axis current i_{qs} and independent of the d axis current i_{ds}. This is shown graphically in Fig. 3-6. For this type of machine, the d axis current is normally set to zero. However, injection of a negative d axis current can be necessary at high speeds to weaken the magnet flux linkage, to operate with limited stator voltage. This is known as a

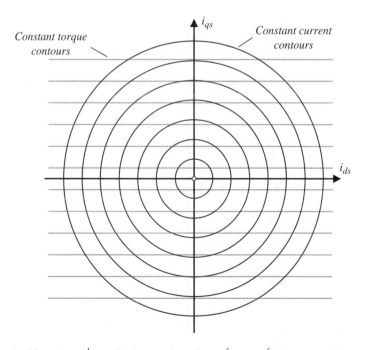

Fig. 3-6   *Constant torque and constant current contours for a surface permanent magnet machine*

field-weakening operation. Consequently, control strategies for SPMSM use the q axis component of the stator current to control torque, and the d axis current to weaken the flux if needed.

3.4.2   Interior permanent magnet synchronous machine (salient machines with magnets)

In salient machines which also include magnets in the rotor, both terms in the right hand of (3.25) contribute to the overall torque. The electromagnetic torque (3.28) results from the interaction of the magnet with the component of the stator current on the q axis. The reluctance torque (3.29) results from the different values of the inductance for the d and q axis, being proportional to the level of asymmetry.

$$\text{Electromagnetic torque: } \frac{3}{2}p\psi_{pm}i_{qs} \tag{3.28}$$

$$\text{Reluctance torque: } \frac{3}{2}p\left(L_{ds}-L_{qs}\right)i_{ds}i_{qs} \tag{3.29}$$

An example of this type of machine is the IPMSM (see Fig. 3-2(b)). The d axis shows a large airgap, due to the magnet. Conversely, the rotor in the q axis is mostly iron. Consequently, the permeability in the d axis direction is significantly smaller than the permeability in the q axis direction (i.e. $L_{ds}<L_{qs}$). It is readily seen from (3.29) that in this case both components of the stator current, i_{ds} and i_{qs}, contribute to the overall torque. This means that for a given torque, if no further constraints are considered, there are in principle infinite sets of d and q axis current values that satisfy the torque equation. This is graphically shown in Fig. 3-7. The use of this degree of freedom to select the stator current can be used to optimize some performance index, maximum torque per ampere (MTPA) being one of the most popular strategies. MTPA methods are targeted to minimize

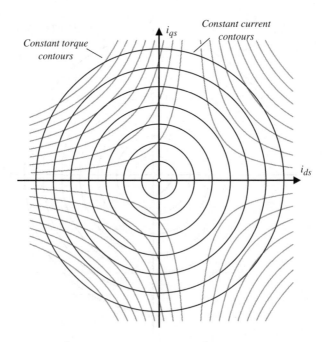

Fig. 3-7   *Constant torque and constant current contours for an interior permanent magnet machine*

the stator current required to produce a certain amount of torque, with the objective of minimizing the stator losses, as well as the power converter rating. MTPA methods are discussed later in this chapter.

3.4.3   Synchronous reluctance machines (salient machines without magnets)

Synchronous reluctance machines (SynRM; Fig. 3-3) are a type of salient synchronous machine which do not use any form of excitation, and neither do they include magnets in the rotor. The only mechanism to produce torque is therefore the reluctance torque (3.29). Consequently, the rotor is designed to provide a large L_{ds} vs. L_{qs} ratio. It should be said that for the sake of consistency with Fig. 3-2(b), in the design shown in Fig. 3-3, the d axis is defined to be aligned with the minimum permeability direction (minimum inductance), the q axis being aligned with the maximum permeability direction, i.e. $L_{ds} < L_{qs}$. However, a different definition of the d and q axis, in which the d axis corresponds to the maximum inductance direction and the q axis to the minimum inductance direction, can also be found in the literature.

Fig. 3-8 shows the constant torque contours for SynRM. It will be observed from this figure that, as in the case of the interior permanent magnet machines, there are infinite sets of d and q axis current values that can be used to produce a certain amount of torque. However, contrary to the interior permanent magnet case, the torque contours are ideally symmetrical with respect to i_{ds} and i_{qs}. Symmetry among quadrants therefore exists. It should be noted, however, that this is without considering saturation and iron losses [4].

3.4.4   Maximum torque per ampere (MTPA) in interior permanent magnet machines

As already mentioned, salient synchronous machines that include a magnet in the rotor offer two different mechanisms to produce torque: electromagnetic torque and reluctance torque. This provides an extra degree of freedom in the selection of the stator current needed to produce a certain amount of torque. This can be used

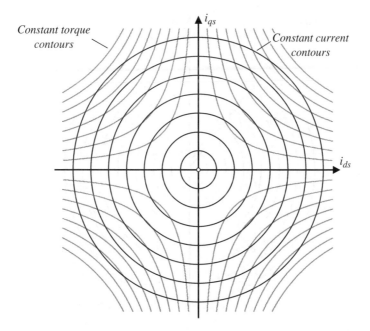

Fig. 3-8   *Constant torque and constant current contours for a synchronous reluctance machine*

to reduce the losses of the machine and power converter, therefore improving the overall efficiency of the drive.

MTPA strategies maximize the torque produced by interior permanent magnet machines for a given amount of stator current. The purpose of these methods is therefore to select i_{ds} and i_{qs} to produce a certain torque (3.25), such that the stator current vector magnitude (3.30) is minimized.

$$I_s = \sqrt{i_{qs}^2 + i_{ds}^2} \qquad (3.30)$$

Equations (3.31) and (3.32) express the d and q axis components of the stator current as a function of the stator current magnitude and phase angle (see Fig. 3-9).

$$i_{qs} = I_s \cos(\varphi) \qquad (3.31)$$

$$i_{ds} = -I_s \sin(\varphi) \qquad (3.32)$$

Using these expressions, it is possible to rewrite the torque equation (3.25) as:

$$T_{em} = \frac{3}{2} p \left(\psi_{pm} I_s \cos(\varphi) + \left(L_{ds} - L_{qs} \right) I_s^2 \cos(\varphi) \sin(\varphi) \right) \qquad (3.33)$$

For a certain magnitude of the stator current, the condition for MTPA is then given by:

$$\frac{dT_{em}}{d\varphi} = 0 \qquad (3.34)$$

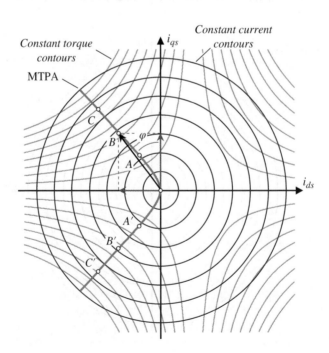

Fig. 3-9   *Constant torque contours and MTPA trajectories for positive and negative torque values in the d–q plane*

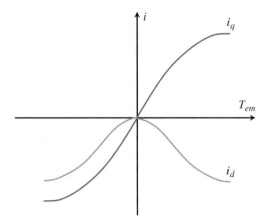

Fig. 3-10 i_{ds} *and* i_{qs} *as a function of the torque using MTPA*

Solving this equation, and combining it with (3.33), the relationship between the q axis current and torque fulfilling the MTPA condition is obtained (3.35), the required d axis current being (3.36).

$$T_{em} = \frac{3}{2}P\left(\frac{\psi_{pm}}{2}i_{qs} + (L_{ds} - L_{qs})\sqrt{\frac{\psi_{pm}^2}{4(L_{qs} - L_{ds})}i_{qs}^2 + i_{qs}^4}\right) \tag{3.35}$$

$$i_{ds} = \frac{\psi_{pm}}{2(L_{qs} - L_{ds})} - \sqrt{\frac{\psi_{pm}^2}{4(L_{qs} - L_{ds})} + i_{qs}^2} \tag{3.36}$$

Fig. 3-9 shows the constant torque curves, and the sets of values for i_{qs} and i_{ds} (A, B, C, ... for increasing, positive torques and A', B', C', ... for increasing negative torques) that satisfy the MTPA condition. Fig. 3-10 shows the resulting i_{qs} and i_{ds} currents as a function of the torque.

Use of (3.35) and (3.36) to obtain the required d and q axis currents to provide the desired torque might not be advisable in a practical implementation of the MTPA strategy, owing to the computational requirements to solve these equations is real time. Alternatively, (3.35) and (3.36) can be approximated by polynomial functions of the type shown in (3.37). Given the desired torque, the necessary q axis current is obtained using (3.37), which can be easily processed in real time by modern digital signal processors used in electric drives.

$$i_{qs} = k_0 + k_1 T_{em} + k_2 T_{em}^2 + k_3 T_{em}^3 + k_4 T_{em}^4 + k_5 T_{em}^5 \tag{3.37}$$

Alternatively, (3.35) and (3.36) can be solved off-line and the results stored in look-up tables. The stored values, combined with an interpolation process, are used later during the normal operation of the drive.

3.5   Vector control of permanent magnet synchronous machines

The same principles discussed in Chapter 2 for the vector control of the induction motor apply to the vector control of PMSMs. One simplification in the case of synchronous machines comes from the fact that the rotor flux (created either by magnets or by a wound rotor) is aligned with the rotor position. This means that an encoder, or other type of position sensor attached to the rotor, will directly provide the angle needed for the transformation to the synchronous d–q reference frame. Therefore, no flux observers are needed in this case, simplifying the control and reducing the parameter sensitivity compared to the induction motor case.

3.5.1   Vector control of non-salient synchronous machines

The control principles in this case are identical to those of the induction machine case. The q axis current is used to control the torque, the d axis current being used to control the stator flux. The stator current vector relative to the rotor angle, and the resulting decomposition into its d and q axis components, can be seen in Fig. 3-11. The resulting control block diagram is shown in Fig. 3-12. The torque command is converted into a q axis current command. The d axis current command is made in principle equal to zero, as the d axis

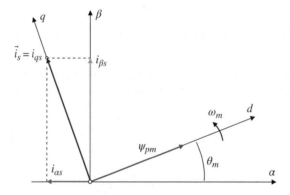

Fig. 3-11   *Stator current vector components for a non-salient machine in the stationary and rotor reference frames. The d axis current is made equal to zero*

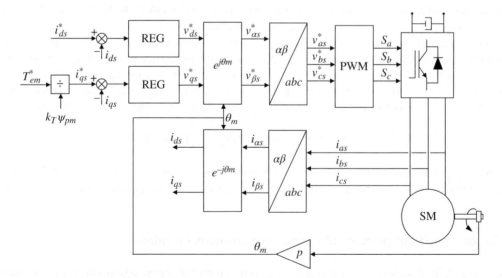

Fig. 3-12   *Schematic representation of the vector control of a non-salient synchronous machine. Back-emf decoupling and cross-coupling decoupling are not implemented*

current does not contribute to torque production. However, at high speeds, negative values of the d axis current can be needed to reduce the stator flux. This is discussed in Section 3.6. Current regulators, typically synchronous PI type, are used to guarantee that the actual currents precisely follow the current commands. Further discussion on the current regulator design can be found in this section.

3.5.2 Vector control of salient synchronous machines

Vector controlled salient synchronous machines follow the same philosophy as non-salient synchronous machines. However, to optimize the current consumption and take advantage of the reluctance torque in (3.25), the i_{ds} current reference is normally set to a value different from zero, with the purpose of minimizing the current needed to produce the desired torque. The stator current vector relative to the rotor angle and the resulting decomposition into its d and q axis components can be seen in Fig. 3-13.

Salient PMSMs typically have $L_{ds} < L_{qs}$, meaning that negative values of i_{ds} produce positive reluctance torque (motoring operation assumed). The resulting control block diagram is depicted in Fig. 3-14. The only difference with respect to the non-salient machine (see Fig. 3-12) is the MTPA block (i.e. both i_{ds} and i_{qs} current commands are now provided to comply with the required torque).

3.5.3 Synchronous current regulators

Fig. 3-15 shows the block diagram of a current-regulated PMSM using synchronous PI current regulators. Cross-coupling decoupling and back-emf decoupling are not implemented in the figure. All the discussion in Chapter 2 on the design and tuning of the current control loop, as well as on cross-coupling decoupling and back-emf decoupling, is applicable to the synchronous machine case.

One difference compared to the induction machine case is that the d and q axis inductances can be different in the case of IPMSM. This means that the gains of the current regulators for both axes might be different. Another potential difference compared to the induction machine case is the dynamics of the back-emf and their effect on the current regulator performance. As discussed in Chapter 2, the dynamics of the back-emf are related to the rotor speed, and therefore depend on the dynamics of the mechanical system. This is typically much slower than the dynamics of the current control loop. Therefore, even if back-emf

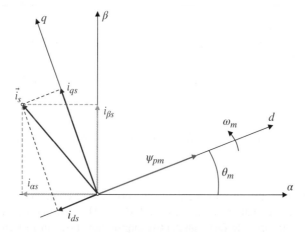

Fig. 3-13   *Stator current vector components for a salient machine in the stationary and rotor reference frames, using both electromagnetic torque and reluctance torque*

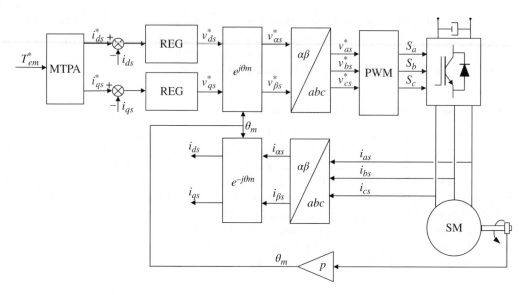

Fig. 3-14 *Schematic representation of the vector control of a salient synchronous machine. Back-emf decoupling and cross-coupling decoupling are not implemented*

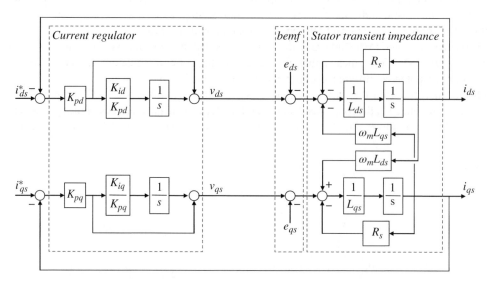

Fig. 3-15 *Synchronous PI current regulator. Back-emf decoupling and cross-coupling decoupling are not implemented*

decoupling is not implemented, the current regulator can effectively compensate the resulting disturbance. However, this assumption is debatable in the case of PMSM. High torque density PMSM with reduced mechanical inertia designed to provide fast transient response can show an fast changing back-emf, which is difficult to compensate for by the current regulator. Back-emf decoupling, or otherwise tuning of the current regulators to provide very high bandwidths (several hundred Hz or even kHz), might be mandatory in this case.

3.6 Operation under voltage and current constraints

This section analyses the operating limits of synchronous machines, taking into consideration the voltage and current constraints imposed both by the machine and by the power converter. These limits of operation need to be considered for the proper selection of the current references commanded to the current regulators.

3.6.1 Current and voltage limits

The current limit (3.38) mainly comes from thermal restrictions both in the machine and power converter. The risk of demagnetization due to an excessive negative d axis current in the machine might also need to be considered.

$$i_{ds}^2 + i_{qs}^2 \leq I_{s\_limit}^2 \tag{3.38}$$

On the other hand, the voltage limit (3.39) comes both from the design of the machine (mainly due to the isolation) and from the maximum voltage that can be supplied by the power converter. Of course, both limits should coincide in a properly designed electric drive.

$$v_{ds}^2 + v_{qs}^2 \leq V_{s\_limit}^2 \tag{3.39}$$

At low speeds, the power converter feeding the machine has enough voltage to impose the desired currents. This means that the limitation imposed by (3.39) has no effect. Only the current limit needs to be considered in this case. As the speed increases, the back-emf will approach the maximum voltage available from the inverter. The resulting modes of operation, as well as the strategies to adopt in this case, will depend on the machine's design. Both non-salient and salient machines are analyzed in the following subsections.

3.6.2 Stator voltage equation at high speeds: Field weakening

Voltage constraints given by (3.39) become relevant at speeds near or above rated speed. To analyze the effects and remedies due to voltage constraints at high speeds, some simplifications can be introduced in the machine model. At high speeds, the contribution of the voltage drop in the stator resistances to the overall stator voltage equations (3.22) and (3.23) can be neglected. If steady-state operation is also assumed, the stator voltage equations in the case of an IPMSM are simplified to:

$$v_{ds} \cong -\omega_m i_{qs} L_{qs} \tag{3.40}$$

$$v_{qs} \cong \omega_m \psi_{pm} + \omega_m i_{ds} L_{ds} = \omega_m \left(\psi_{pm} + i_{ds} L_{ds} \right) \tag{3.41}$$

Taking into account (3.26), the stator voltage equations in the case of an SPMSM are obtained.

$$v_{ds} \cong -\omega_m i_{qs} L_s \tag{3.42}$$

$$v_{qs} \cong \omega_m \psi_{pm} + \omega_m i_{ds} L_s = \omega_m \left(\psi_{pm} + i_{ds} L_s \right) \tag{3.43}$$

It is seen from these equations that the voltage induced in the stator winding due to the magnet flux linkage occurs in the q axis voltage, being proportional to the rotor speed. At rated speed, the induced voltage due to the back-emf will match the voltage available from the inverter. Because of this, the speed cannot be further increased in principle, as no voltage is left to produce q axis current, and therefore to produce torque. However, it is possible in this case to inject a negative d axis current (i.e. $i_{ds} < 0$) to reduce the d axis stator flux, and consequently v_{qs}.

Equations (3.40) and (3.41) are graphically represented in Fig. 3-16 and Fig. 3-17 when the machine operates above rated speed, in the case of $i_{ds} = 0$ and $i_{ds} < 0$ respectively. In the first case, the resulting stator voltage is seen to be larger than the stator voltage limit. Consequently, this operating condition cannot be achieved in practice. Injection of negative d axis current shown in Fig. 3-17 reduces the stator flux and consequently the stator voltage, which now matches the stator voltage limit.

It should be noted, however, that the use of d axis current for field weakening implies a reduction of the available q axis current, as otherwise the current limit of the machine and inverter would be exceeded. Consequently, the torque production capability of the machine is reduced.

The effect of injecting d axis current, and the subsequent reduction of the available q axis current, will depend on the machine design. In the case of the SPMSM, reduction of the q axis current will directly impact the torque production capability, since the torque is proportional to the q axis current. Conversely, in the case of IPMSM, the decrease of the electromagnetic torque caused by the reduction of the q axis current can be partially compensated for by the increase of the reluctance torque caused by the increase of the d axis current.

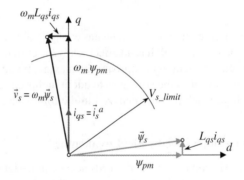

Fig. 3-16   *Space vector diagram of a permanent magnet machine above rated speed with* $i_{ds} = 0$

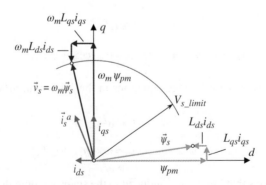

Fig. 3-17   *Space vector diagram of a permanent magnet machine above rated speed with* $i_{ds} < 0$

3.6.3 Control of non-salient machines under voltage constraints

Combining (3.40) and (3.41) with the stator voltage constraint (3.39), the following expression is obtained:

$$\left(i_{ds} + \frac{\psi_{pm}}{L_s}\right)^2 + i_{qs}^2 \le \left(\frac{V_{s\_limit}}{L_s\omega_m}\right)^2 \tag{3.44}$$

This equation corresponds to a circle whose center, C, and radius, R, are (3.45) and (3.46) respectively.

$$R = \frac{V_{s\_limit}}{L_{qs}\omega_m} \tag{3.45}$$

$$C = I_{s\_sc} = -\frac{\psi_{pm}}{L_s} \tag{3.46}$$

The center is seen to be a function of design parameters. It corresponds to the case of a stator voltage equal to zero, which is identical to the case of an infinite speed. It coincides therefore with the short-circuit current I_{s_sc}. The radius, R, depends on the rotational speed. The larger the speed, the smaller the radius. Fig. 3-18 shows the graphical representation of the constant voltage circles for different speeds, as well as the current limit and constant torque contours. Since the torque depends only on the q axis current, the constant torque contours are in this case horizontal lines. The maximum torque is obtained when $i_{qs} = I_{s\_limit}$, and corresponds to the constant torque region in Fig. 3-18.

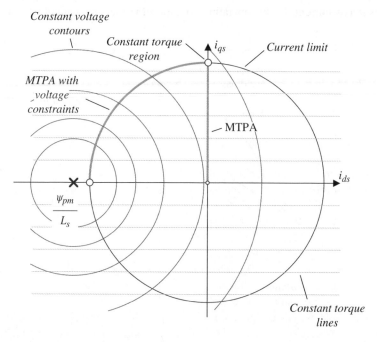

Fig. 3-18 *Constant voltage, current, and torque contours in the case of an SPMSM*

Below the base speed, the available voltage is enough to counteract the back-emf induced by the rotation of the rotor magnets. The base speed at which the back-emf matches the available voltage is given by:

$$\omega_{base} = \frac{V_{s\_limit}}{\sqrt{\psi_{pm}^2 + (L_s \cdot I_{s\_limit})^2}} \tag{3.47}$$

For rotor speeds below the base speed, only the q axis current is injected, meaning that all the stator current produces torque. Therefore, the machine provides MTPA. If the speed increases beyond the base speed, the back-emf will surpass the stator voltage limit. Injection of the d axis current is needed in this case to weaken the stator flux. Thus, the q axis current has to be reduced to hold the current limit equation (3.38), therefore decreasing the torque production capability. The maximum speed at which the machine can rotate is given by (3.48), and occurs when the d axis current equals the rated current. It should be noted that this is of little interest in practice, as the machine cannot produce any torque.

$$\omega_{max} = \frac{V_{s\_limit}}{\psi_{pm} - L_s \cdot I_{s\_limit}} \tag{3.48}$$

Fig. 3-19 shows the regions of operation of the SPMSM as a function of the rotor speed. For a given rotor speed, $\omega_{base} < \omega_m < \omega_{max}$, the intersection between the voltage and current limits (shadowed area) corresponds to the feasible region in which the machine can operate.

Fig. 3-20 shows the limits of operation of the SPMSM as a function of the speed. Below the base speed ω_{base}, the machine can provide its nominal torque. Once the base speed is reached, the d axis current is used to reduce stator flux linkage to match the available voltage. The subsequent reduction of the q axis current produces a reduction of the torque production capability. The maximum speed, ω_{max}, is reached when all the stator current is d axis current. In this operating condition the machine cannot produce any torque.

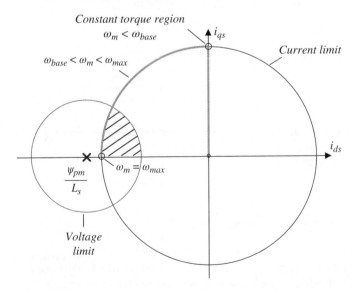

Fig. 3-19   *Regions of operation in the d–q plane of the SPMSM*

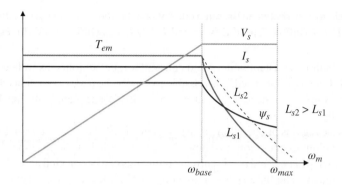

Fig. 3-20 *Maximum capability curves of the SPMSM*

It should also be noted that in all the preceding discussion it is assumed that the rated current is smaller than the short-circuit current ($I_{s\_limit} < I_{s\_sc}$), that is that the center of the constant voltage circles lies outside the current limit circle. It is also possible, depending on the machine design, that the center of the constant voltage circles lies within the current limit circle ($I_{s\_limit} > I_{s\_sc}$). In this case, the voltage and current limits would not impose any limit to the maximum rotor speed. The machine could therefore operate (ideally) at an infinite speed. However, this case is unusual for SPMSMs. This is discussed in the following section for IPMSM.

3.6.4   Control of salient machines under voltage constraints

The voltage limit equation in the case of salient machines, assuming a steady-state operation, is directly obtained from (3.42)–(3.43) and (3.39) [5].

$$\left(L_{ds}\omega_m\right)^2 \left(i_{ds} + \frac{\psi_{pm}}{L_{ds}}\right)^2 + \left(L_{qs}\omega_m\right)^2 i_{qs}^2 \leq V_{s\_limit}^2 \tag{3.49}$$

This equation corresponds to an ellipse of center C. The lengths of the minor and major axis of the ellipse are R_{minor} and R_{major} respectively. As for the SPMSM case, the center of the ellipse is equal to the short-circuit current I_{s_sc}. This is identical to the case when the machine rotates at an infinite speed [5].

$$C = I_{s\_sc} = -\frac{\psi_{pm}}{L_{ds}} \tag{3.50}$$

$$R_{minor} = \frac{V_{s\_limit}}{L_{qs}\omega_m} \tag{3.51}$$

$$R_{major} = \frac{V_{s\_limit}}{L_{ds}\omega_m} \tag{3.52}$$

The center of the ellipse, C, (and consequently the short-circuit current) depends on the machine design. The length of the axis of the ellipse depends on the operating point. For a given rated voltage, the higher the speed, the smaller the axis lengths.

Depending on the relationship between the rated current and the short-circuit current, two different cases can be considered: $I_{s\_sc} < I_{s\_limit}$ (infinite speed drives) and $I_{s\_sc} > I_{s\_limit}$ (finite speed drives):

- $I_{s\_sc} < I_{s\_limit}$: The resulting constant voltage and constant torque contours, as well as the current limit, are shown in Fig. 3-21. For any rotor speed, it is possible to supply the d axis current needed to compensate the magnet flux without surpassing the voltage and current limits. Consequently, the machine can operate at an infinite speed (ideally).

  Fig. 3-22 shows the regions of operation in the d–q plane. It can be observed from Fig. 3-21 and Fig. 3-22 that three different regions of operation exist now [5]. For speeds below the base speed, the voltage limits do not need to be considered. The d and q axis current commands can be selected according to the MTPA strategy discussed previously. The base speed is given by (3.53), the current i_{ds1} being (3.54) and i_{qs1} being obtained from i_{ds1} and (3.38) [5].

$$\omega_{base} = \frac{V_{s\_limit}}{\sqrt{\left(\psi_{pm}^2 + L_{ds}i_{ds1}\right)^2 + \left(L_{qs}i_{qs1}\right)^2}} \tag{3.53}$$

$$i_{ds1} = \frac{\psi_{pm} - \sqrt{\psi_{pm}^2 + 8I_{s\_limit}^2\left(L_{qs} - L_{ds}\right)^2}}{4\left(L_{qs} - L_{ds}\right)}, i_{qs1} = \pm\sqrt{I_{s\_limit}^2 - i_{ds1}^2} \tag{3.54}$$

For rotor speeds above the base speed, the MTPA strategy needs to be modified, because of the restriction imposed by the voltage limits. A larger (negative) d axis current is needed to reduce the stator flux. This implies a reduction of the q axis current caused by the current constraints, and consequently of the torque production capability. The intersection between the constant voltage ellipse and the current limit (shadowed area in Fig. 3-22) corresponds to the feasible region in which the machine can operate.

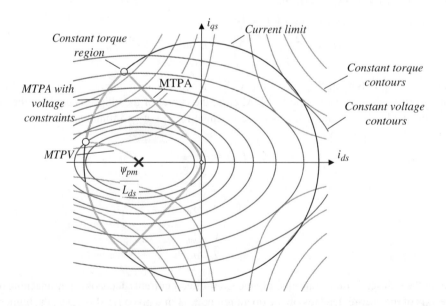

Fig. 3-21   *Constant voltage, current, and torque contours in the case of an IPMSM with infinite rotor speed*

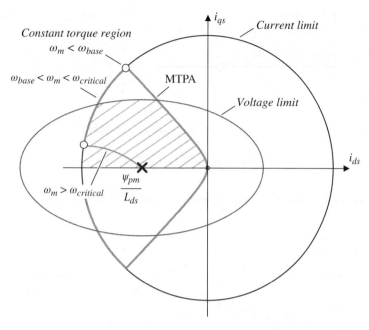

Fig. 3-22   *Regions of operation in the d–q plane of the IPMSM*

At a certain critical rotor speed (3.55), it is not possible to supply the maximum current because of the reduction of the voltage ellipse. The machine enters then a second region of field weakening, where it operates with a reduced current. The torque production capability is further reduced in this region [5].

$$\omega_{critial} = \frac{V_{s\_limit}}{\sqrt{(L_{ds}I_{s\_limit})^2 - \psi_{pm}^2}} \tag{3.55}$$

Fig. 3-23 shows the limits of operation of the IPMSM as a function of the speed. Below the base speed ω_{base}, the machine can provide its nominal torque, being operated using MTPA. For rotor speeds $\omega_{base} < \omega_m < \omega_{critical}$, the machine operates with a constant current, but the torque production capability is decreased, owing to the need for an increased d axis current. At speeds above the critical value, the reduction of the current further decreases the torque production capability.

- $I_{s\_sc} > I_{s\_limit}$: The resulting constant voltage and constant torque contours, as well as the current limit, are shown in Fig. 3-24. In this case, the voltage limit imposes a physical constraint on the maximum rotor speed. Contrary the case when $I_{s\_sc} < I_{s\_limit}$, only one field-weakening region exists now. The base speed (i.e. the maximum speed at which the machine can operate with the MTPA strategy) is given by (3.53). At speeds above the base speed, the machine enters into the field-weakening region; the discussion in the preceding subsection for Fig. 3-21, Fig. 3-22, and Fig. 3-23 applies to this case too. However, now there is a maximum speed which cannot be surpassed, as it is not possible to produce enough d axis current to compensate for the magnet flux linkage. The maximum speed (3.56) can be obtained from the stator voltage equations (3.40) and (3.41). The limits of operation of the IPMSM as a function of the speed in this case are similar to those shown in Fig. 3-20.

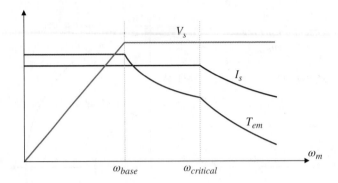

Fig. 3-23 *Maximum capability curves of the IPMSM infinite speed drive*

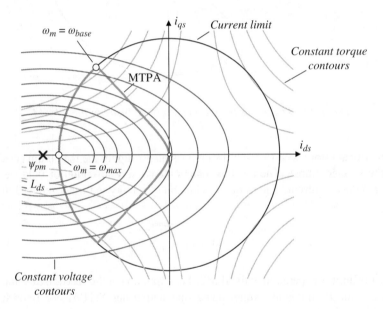

Fig. 3-24 *Constant voltage, current, and torque contours in the case of an IPMSM with finite rotor speed*

$$\omega_{max} = \frac{V_{s\_limit}}{\psi_{pm} - L_{ds}I_{s\_limit}} \tag{3.56}$$

3.6.5 Synchronous reluctance machines

Fig. 3-25 shows the constant torque, current, and voltage contours for the SynRM. It will be observed that both torque and voltage contours are now centered at the origin of the *d–q* plane. This is due to the fact that no magnetization exists now in the rotor. It is straightforward to particularize the preceding discussion for the IPMSM to the SynRM case. As can be observed in Fig. 3-25, the machine can operate (ideally) at infinite

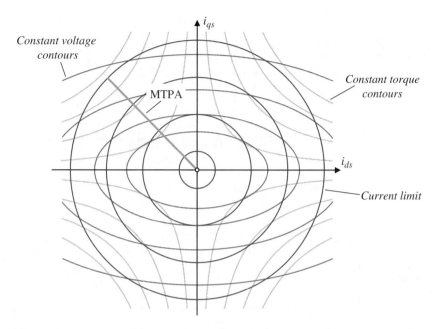

Fig. 3-25   *Constant voltage, current, and torque contours in the case of a SynRM*

speed, since the center of the voltage ellipsis is located at the origin, and consequently within the current limit circle. The machine will show two different modes of operation. For rotor speeds below the base speed, the machine can provide its rated current, and consequently the rated torque. For rotor speeds above the base speed, the current, and consequently the torque, have to be decreased. If saturation and rotor losses are not considered, it is readily seen from Fig. 3-25 that MTPA is achieved by making $i_{ds} = i_{qs}$. However, this assumption can be unrealistic, especially in the case of machines with a high saliency ratio, owing to the large asymmetries in the magnetic circuit of the d and q axis. This has boosted the development of advanced MTPA algorithms specific to SynRM that take into consideration saturation and iron losses [4].

3.6.6   Feedback-based flux weakening of vector controlled synchronous machines

The flux-weakening method presented in this section is based on a closed-loop control strategy, and is equivalent to the feedback-based flux-weakening method described in Chapter 2 for induction machines. The method can be used both with non-salient and salient synchronous machines. The corresponding block diagram is shown in Fig. 3-26 [5]. For a given torque reference, the current reference calculator block provides the i_{ds} and i_{qs} current references. This can be done using an MTPA strategy in salient machines, or making $i_{ds} = 0$ in non-salient machines. The voltage reference amplitude is compared with the rated voltage, i_{ds} and i_{qs} references in the flux-weakening region being adjusted according to the available voltage. If the stator voltage reference amplitude increases above the rated value, the regulator produces a negative i_{ds} current variation (Δi_{ds}), which is added to the i_{ds} current reference provided by the current reference calculator block. This decreases the overall i_{ds} current, enabling the speed to increase further without exceeding the voltage limit. On the contrary, if field weakening is not required (constant torque region), the voltage regulator imposes zero i_{ds} current variation ($\Delta i_{ds} = 0$), since the voltage reference is lower than the stator voltage

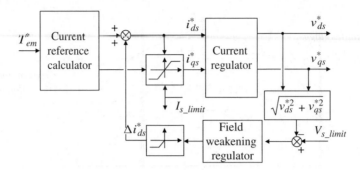

Fig. 3-26   *Feedback-based flux weakening*

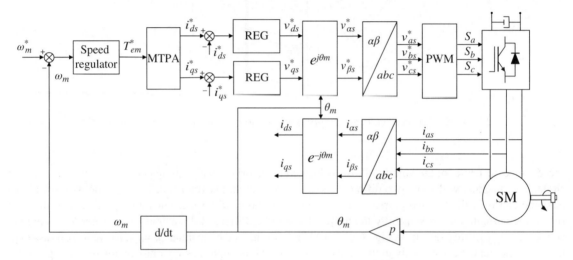

Fig. 3-27   *Schematic representation of the vector control of a salient synchronous machine, including a speed control loop*

limit. It should be noted that, if i_{ds} is different from zero, i_{qs} must be decreased accordingly, as the current limit cannot be exceeded.

The bandwidth of the flux-weakening regulator should be low enough compared to the current regulator bandwidth to prevent interferences with the current control loop. However, this is not problematic in principle, as the dynamics of the flux-weakening control loop are mainly related to the rotor speed, being therefore much slower than the dynamics of the stator current [5].

3.7   Speed control

All the discussion on the design and tuning of the speed control loop in the case of induction machines is applicable here. Fig. 3-27 shows the schematic representation of the speed control loop. A speed regulator, typically the PI type, provides the required torque. A salient machine is assumed, the corresponding d and q axis current commands being obtained from the commanded torque through the MTPA block. In the case of an SPMSM, the speed controller directly provides the q axis current command, the d axis being used for field weakening if needed.

3.8   Sensorless control

Control of the torque produced by synchronous machines requires precise knowledge of the rotor position. Rotor position and/or speed are also needed if motion control is required by the application. Consequently, electric drives using synchronous machines normally include a position sensor, which is attached to the motor shaft. As already discussed for the induction machine case, elimination of this sensor will reduce the hardware needed to realize the control, being therefore advantageous in terms of cost, size, reliability, and maintenance requirements of the electric drive.

Sensorless control of PM machines responds to the same general principles as induction machines. However, several relevant differences exist:

- Sensorless control of induction machines requires the estimation of both the rotor flux angle (for field oriented control purposes) and the rotor angle (for motion control purposes). Both angles coincide in the case of permanent magnet machines.
- Control of permanent magnet machines requires knowledge of the absolute position of the rotor, as the rotor flux is not induced from the stator circuit, but inherent to the rotor construction. This implies the need to detect the rotor magnet polarity.
- Some synchronous machines are asymmetric by design (e.g. interior permanent magnet machines or SynRMs). Such asymmetry comes from the fact that these machines are designed to produce reluctance torque. Asymmetric machine designs are natural candidates for the use of saliency-tracking-based sensorless methods.

Sensorless control methods developed for permanent magnet machines are typically separated into model-based methods and saliency-tracking-based methods. Those in the first category use the fundamental model of the machine, and are suitable for medium-high speeds. However, they cannot operate at very low or zero speed. Those making up the second category use saliencies which are intrinsic to the machine design. These methods require the use of some form of high-frequency excitation, being intended to operate at very low and zero speed, or in position control.

The use of saliency-tracking-based methods at high speeds is unadvisable, for several reasons. First, they need a certain voltage to produce a high-frequency excitation, which is superimposed on the fundamental voltage used for torque production. This is not problematic at low speeds, where the stator voltage is significantly smaller than the voltage available in the inverter. However, such a voltage margin might not exist at high speeds. Second, at high fundamental frequencies, the required spectral separation between the fundamental excitation and the high-frequency excitation needed to realize the signal processing of the high-frequency signals can be compromised. This will result in a deterioration of the sensorless control performance. It is concluded from the previous discussion that model-based methods and saliency-tracking-based methods should be combined to cover the whole speed range of the machine.

3.8.1   Permanent magnet synchronous machine model

It is useful for the analysis of rotor position estimation methods for PMSMs to transform the stator voltage equations (3.22)–(3.23) to the stationary reference frame (3.57). In the case of IPMSM, d and q axis inductances are different (i.e. $L_{ds} \neq L_{qs}$), the average and differential inductances ΣL_s and ΔL_s being defined by (3.58).

$$\begin{bmatrix} v_{as} \\ v_{\beta s} \end{bmatrix} = \left(R_s + \frac{d}{dt} \begin{bmatrix} \Sigma L_s + \Delta L_s \cos(2\theta_m) & \Delta L_s \sin(2\theta_m) \\ \Delta L_s \sin(2\theta_m) & \Sigma L_s - \Delta L_s \cos(2\theta_m) \end{bmatrix} \right) \begin{bmatrix} i_{as} \\ i_{\beta s} \end{bmatrix} + \omega_m \Psi_{pm} \begin{bmatrix} -\sin(\theta_m) \\ \cos(\theta_m) \end{bmatrix} \tag{3.57}$$

$$\Sigma L_s = \frac{L_{qs} + L_{ds}}{2} \; ; \; \Delta L_s = \frac{L_{qs} - L_{ds}}{2} \tag{3.58}$$

Separating the terms in the right-hand side of the equation which do not depend on the rotor position from those which change with the rotor position, (3.59) is obtained.

$$\begin{bmatrix} v_{as} \\ v_{\beta s} \end{bmatrix} = \left(R_s + \frac{d}{dt} \Sigma L_s \right) \begin{bmatrix} i_{as} \\ i_{\beta s} \end{bmatrix} + \frac{d}{dt} \Delta L_s \begin{bmatrix} \cos(2\theta_m) & \sin(2\theta_m) \\ \sin(2\theta_m) & \cos(2\theta_m) \end{bmatrix} \begin{bmatrix} i_{as} \\ i_{\beta s} \end{bmatrix} + \omega_m \Psi_{pm} \begin{bmatrix} -\sin(\theta_m) \\ \cos(\theta_m) \end{bmatrix} \tag{3.59}$$

In the case of SPMSM, the d and q axis inductances are the same and equal to the average inductance (i. e. $\Sigma L_s = L_d = L_{qs}$), therefore $\Delta L_s = 0$. In this case (3.59) simplifies to (3.60).

$$\begin{bmatrix} v_{as} \\ v_{\beta s} \end{bmatrix} = \left(R_s + \frac{d}{dt} \Sigma L_s \right) \begin{bmatrix} i_{as} \\ i_{\beta s} \end{bmatrix} + \omega_m \Psi_{pm} \begin{bmatrix} -\sin(\theta_m) \\ \cos(\theta_m) \end{bmatrix} \tag{3.60}$$

It is observed from (3.59) and (3.60) that the right-hand side of both equations depends on the rotor position. Consequently, it should be possible to estimate the rotor positon from this model, assuming that the stator voltage and current are measured or can be estimated, and that the machine parameters are known. Methods to do this are presented in the next subsections.

3.8.1.1   *The extended back-emf*

Obtaining the rotor position for an SPMSM using (3.60) is in principle straightforward. In the case of IPMSM, however, it is not; as will be observed from (3.59), there are two terms in the right-hand side of the stator voltage equation which depend on the rotor position. This makes estimation of the rotor position in IPMSM more challenging in terms of computational difficulty and parameter sensitivity compared to the SPMSM case. A convenient way to address this issue is the use of the extended back-emf. The idea behind this concept is to transform the asymmetric IPMSM into a symmetric machine (i.e. a machine with identical d and q axis inductances). This implies the definition of a fictitious (extended) back-emf, which accounts for all the rotor position dependent terms in the stator voltage equation. This can be done in two different manners. In the implementation given by (3.61)–(3.62), the inductance of the equivalent symmetric machine is L_{ds} [6], while in the implementation given by (3.63)–(3.64), the inductance of the equivalent symmetric machine is L_{qs} [7].

$$\begin{bmatrix} v_{as} \\ v_{\beta s} \end{bmatrix} = \begin{bmatrix} R_s & -2\omega_m \Delta L_s \\ 2\omega_m \Delta L_s & R_s \end{bmatrix} \begin{bmatrix} i_{as} \\ i_{\beta s} \end{bmatrix} + \frac{d}{dt} \begin{bmatrix} L_{ds} & 0 \\ 0 & L_{ds} \end{bmatrix} \begin{bmatrix} i_{as} \\ i_{\beta s} \end{bmatrix} + \begin{bmatrix} e_{as} \\ e_{\beta s} \end{bmatrix} \tag{3.61}$$

$$\begin{bmatrix} e_{as} \\ e_{\beta s} \end{bmatrix} = \left\{ (L_{ds} - L_{qs}) \left(\omega_m i_{ds} - \frac{d i_{qs}}{dt} \right) + \omega_m \Psi_{pm} \right\} \begin{bmatrix} -\sin(\theta_m) \\ \cos(\theta_m) \end{bmatrix} \tag{3.62}$$

$$\begin{bmatrix} v_{as} \\ v_{\beta s} \end{bmatrix} = \begin{bmatrix} R & 0 \\ 0 & R \end{bmatrix} \begin{bmatrix} i_{as} \\ i_{\beta s} \end{bmatrix} + \frac{d}{dt} \begin{bmatrix} L_{qs} & 0 \\ 0 & L_{qs} \end{bmatrix} \begin{bmatrix} i_{as} \\ i_{\beta s} \end{bmatrix} + \begin{bmatrix} e_{as} \\ e_{\beta s} \end{bmatrix} \tag{3.63}$$

$$\begin{bmatrix} e_{\alpha s} \\ e_{\beta s} \end{bmatrix} = \frac{d}{dt} \psi_{ext} \begin{bmatrix} \cos(\theta_m) \\ \sin(\theta_m) \end{bmatrix} \tag{3.64}$$

$$\psi_{ext} = \psi_{pm} + (L_{ds} - L_{qs}) i_{ds} \tag{3.65}$$

Implementation using (3.61)–(3.62) or (3.63)–(3.65) have similarities in terms of parameter sensitivity and computational requirements. One remarkable difference is the need to calculate the derivative of i_{qs}^e on the right-hand side of (3.62), as this can be problematic in the event of fast changes of the q axis current or noisy signals.

Independent of which implementation is chosen, it can be observed that the use of the extended back-emf makes the inductance matrix independent of the rotor position, the resulting models being therefore equivalent to (3.60). This enables the use of the same signal processing for rotor position estimation both with SPMSM and IPMSM.

3.8.2 Model-based sensorless control of PMSM

Model-based sensorless methods estimate the rotor position from the stator voltage equation of the machine (3.60). As discussed in the preceding section, the use of the extended back-emf transforms the asymmetric model of the IPMSM (3.57) into an equivalent symmetric model. Consequently, the following discussion applies both to SPMSM and IPMSM.

For the following analysis, it is useful to rearrange (3.60) using complex vector notation, (3.66) being obtained in this case. The time derivative has been replaced by the Laplace operator s.

$$\vec{v}_s^s = R_s \vec{i}_s^s + sL_s \vec{i}_s^s + j\omega_m \psi_{pm} e^{j\theta_m} \tag{3.66}$$

Fig. 3-28 shows graphically the stator voltage equation of the machine (3.66). The projection of the complex vector quantities in (3.66) on the $\alpha\beta$ stationary reference frame would provide the corresponding scalar representation. The known (measurable) variables in Fig. 3-28 are the stator voltage and current, $\vec{v}_s^s$ and $\vec{i}_s^s$, the target variable to obtain being the rotor angle θ_m.

3.8.2.1 Open-loop methods

There are two different approaches to estimate the rotor angle from the stator voltage equation: (1) using the back emf and (2) using the stator flux.

The back-emf $\vec{e}_s^s$ (3.67) is directly obtained from the stator voltage and current using (3.66); knowledge of the stator inductance and resistance is also needed.

$$\vec{e}_s^s = j\omega_m \psi_{pm} e^{j\theta_m} = \vec{v}_s^s - R_s \vec{i}_s^s - sL_s \vec{i}_s^s \tag{3.67}$$

Fig. 3-29 shows the implementation of (3.67). Discussion on the signal processing used to estimate the speed and the position ("speed and position estimation" block in Fig. 3-29) can be found in Section 3.8.4.3.

While simple, this strategy has several drawbacks:

- Implementation of (3.67) requires differentiation of the stator current, which is difficult when the signals are contaminated with noise. Low-pass filters (LPFs) can be used to eliminate the noise, but at the cost of a degradation of the dynamic response.

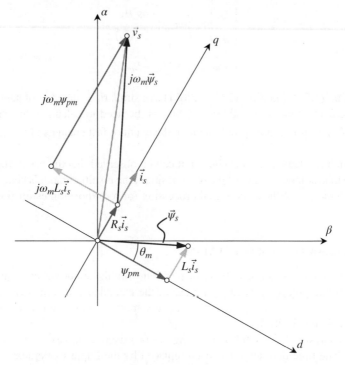

Fig. 3-28   *Complex vector representation of the stator voltage equation*

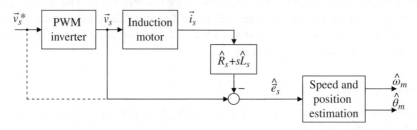

Fig. 3-29   *Rotor speed estimation using the back-emf. Details on the implementation of the "speed and position estimation" block can be found in Section 3.8.4.3*

- It is observed from (3.67) that the back-emf is proportional to the rotor speed. As the rotor speed reduces, the magnitude of $\vec{e}_s^s$ will reduce too, the signal-to-noise relationship getting worse. Consequently, this method cannot be used at very low or zero speed, not being capable therefore to provide position control.
- As discussed in Chapter 2 for the induction motor, the stator voltage is practically never measured in practice. Instead, the stator voltage is estimated from the voltage commands sent to the inverter and the measured DC bus voltage. Errors between the commanded and actual voltages always exist, mainly because of the non-ideal behavior of the inverter (dead-time, voltage drop in the power devices, etc.), these errors being more relevant at low speeds.

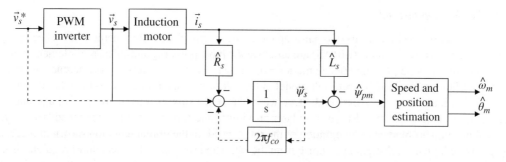

Fig. 3-30 *Rotor speed estimation using the stator flux. By inserting the block with dashed line, the pure integrator is transformed into an LPF of bandwidth in Hz of* f_{co}

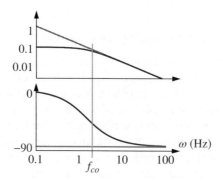

Fig. 3-31 *Bode diagram of the pure integrator and the LPF used to estimate the stator flux*

As an alternative to the back-emf, the stator flux can be used for the estimation of the rotor angle, the corresponding model being (3.68)–(3.69). Fig. 3-30 shows the implementation of this method. Two major differences are observed when comparing Fig. 3-29 and Fig. 3-30.

$$\vec{\psi}\,_s^s = \frac{\vec{v}\,_s^s - R_s \vec{i}\,_s^s}{s} \tag{3.68}$$

$$\psi_{pm} e^{j\theta_m} = \vec{\psi}\,_s^s - L_s \vec{i}\,_s^s \tag{3.69}$$

- Differentiation of the stator current is not needed if the stator flux is used, which is advantageous.
- Contrary to the back-emf, the magnitude of the stator flux is independent from the frequency. Consequently, this option would be capable in principle of providing the same performance at any speed, and therefore to operate at zero speed and in position control. Unfortunately, this is not true in practice. It will be observed from (3.68) and Fig. 3-30 that obtaining the stator flux from the stator voltage and current complex vectors requires a pure, open-loop integration. However, this is not viable in practice. Minor errors in the measured/estimated signals or in the estimations of the machine parameters will produce a drift of the integrator output. This drift will be more relevant when the machine operates at very low or zero speed, as the gain of the integrator at DC is infinite. To avoid the use of the pure integrator, a first-order system can be used instead (dashed line in the integrator feedback in Fig. 3-30) [8]. Fig. 3-31 compares the response of the pure integrator and that of the first-order system. While both systems perform similarly at frequencies starting around one decade above the corner frequency f_{co}, discrepancies are readily observable at low frequencies, which makes the method inviable at low and zero speeds.

3.8.2.2   *Closed-loop methods*

To overcome some of the limitations of open-loop methods, several closed-loop methods have been proposed. Model reference adaptive systems (MRAS) are amongst the most popular solutions; an implementation of this type is shown in Fig. 3-32 [9]. The voltage model estimating the stator flux acts as the reference model, as it is independent of the rotor speed. The current model in the rotor reference frame acts as the adaptive model. An error signal is obtained as the cross-product between the stator flux estimated by the voltage and current models. This error signal drives a PI type regulator providing an estimation of the rotor speed. The estimation of the rotor position is obtained by integration of the rotor speed. In the implementation shown in Fig. 3-32, an LPF is used to eliminate high-frequency noise from the estimated rotor speed. A second PI regulator is used to feed back the integrator of the voltage model, avoiding in this way the problems due to the pure integration mentioned above.

3.8.3   Sensorless control using saliency-tracking-based methods

The sensorless control techniques that rely on fundamental excitation (model-based methods) have been shown to be capable of providing high-performance control in the medium- to high-speed range. However, as the speed decreases, the performance of these methods decreases, eventually failing in the very low-speed range and/or for DC excitation. To overcome this limitation, sensorless control methods based on tracking the position of saliencies (asymmetries) in the rotor of the machine have been proposed. These techniques measure the response of the machine when a high-frequency excitation is applied via the inverter. Potentially, they have the capability of providing accurate, high-bandwidth position and speed estimation in the low-speed range, including zero speed. The IPMSM will be used to analyze the principles of these methods, owing to the asymmetric nature of this machine design. Extension of the proposed concepts to the case of SynRM is immediate. On the other hand, though SPMSMs are not normally asymmetric by design, saturation of the iron caused by the magnet often produce an asymmetry, which is associated with the rotor position, and can therefore be used for sensorless control. Use of saliency tracking methods with induction machines is also briefly discussed in this section.

It is noted that model-based methods and saliency-tracking-based methods are complementary and can therefore be combined to provide sensorless control over the full speed range, with the second covering the low-speed range, including zero speed, and the first being used at high speeds. Transition between both types of methods typically occurs in the range of 10–15% of rated speed.

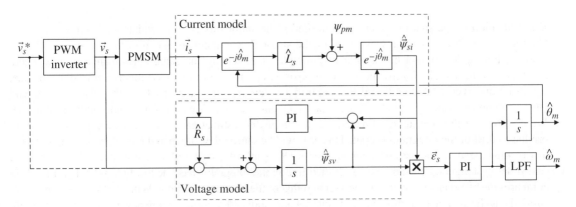

Fig. 3-32   *Rotor speed and position estimation using an MRAS*

3.8.3.1   *High-frequency model of the interior permanent magnet synchronous machine*

The high-frequency models used for the analysis of saliency-tracking-based methods can be derived from the corresponding fundamental frequency model. The model of the IPMSM in matrix form in a reference frame which is synchronous with the rotor is obtained from (3.22)–(3.23):

$$
\begin{bmatrix} v_{ds} \\ v_{qs} \end{bmatrix} = \begin{bmatrix} R_s & -\omega_m L_{qs} \\ \omega_m L_{ds} & R_s \end{bmatrix} \begin{bmatrix} i_{ds} \\ i_{qs} \end{bmatrix} + \begin{bmatrix} L_{ds} & 0 \\ 0 & L_{qs} \end{bmatrix} \frac{d}{dt} \begin{bmatrix} i_{ds} \\ i_{qs} \end{bmatrix} + \begin{bmatrix} 0 \\ \omega_m \psi_{pm} \end{bmatrix}
\tag{3.70}
$$

The fact that some form of high-frequency signal injection will be used to track the rotor position allows the resistive terms to be eliminated from this equation. In addition, the frequency of the injected signal can be assumed to be much higher than that of the rotor speed ω_m, meaning the terms in (3.70) that depend on the rotor frequency can be safely neglected compared to the terms that depend on the current derivative. Based on the previous assumptions, the model in (3.70) can be simplified to (3.71), where the subscript *hf* in the variables indicates that the corresponding voltages and currents are high-frequency components, which are different (superimposed) from the fundamental voltages and currents responsible for torque production.

$$
\begin{bmatrix} v_{ds\_hf} \\ v_{qs\_hf} \end{bmatrix} = \begin{bmatrix} L_{ds} & 0 \\ 0 & L_{qs} \end{bmatrix} \frac{d}{dt} \begin{bmatrix} i_{ds\_hf} \\ i_{qs\_hf} \end{bmatrix}
\tag{3.71}
$$

By transforming (3.71) to the stationary reference frame, and using the definition of the average and differential inductances ΣL_s and ΔL_s, (3.58), (3.72) is obtained.

$$
\begin{bmatrix} v_{\alpha s\_hf} \\ v_{\beta s\_hf} \end{bmatrix} = \begin{bmatrix} \Sigma L_s + \Delta L_s \cos(2\theta_m) & -\Delta L_s \sin(2\theta_m) \\ -\Delta L_s \sin(2\theta_m) & \Sigma L_s - \Delta L_s \cos(2\theta_m) \end{bmatrix} \frac{d}{dt} \begin{bmatrix} i_{\alpha s\_hf} \\ i_{\beta s\_hf} \end{bmatrix}
\tag{3.72}
$$

According to (3.72), it is possible to estimate the rotor position from the response of the machine to a high-frequency excitation signal. It should be noted that all the discussion above does not introduce any particular restriction to the voltages feeding the machine, other than being "high frequency", meaning that these equations are valid for all high-frequency excitation methods.

3.8.3.2   *Forms of high-frequency excitation and signal measurement*

A relative large number of different implementations of saliency-tracking-based sensorless methods have been proposed [10], major differences among these are: (1) the type of high-frequency excitation, (2) the type and number of signals measured, and (3) the signal processing used to estimate the rotor position. These methods share the same physical principles. However, there are a number of implementation issues, including the non-ideal behavior of the inverter, parasitic effects in the cables and machine windings, and so on, that can have significant adverse effects on the overall performance and can vary significantly from method to method. Fig. 3-33 summarizes the main high-frequency sensorless control methods that have been proposed, based on the type of high-frequency signal excitation (columns) and the signals measured (rows). Boxes in the array indicate the combinations of signal injection and signal measurement that have been studied and reported. They are briefly discussed below. Further details can be found in [10].

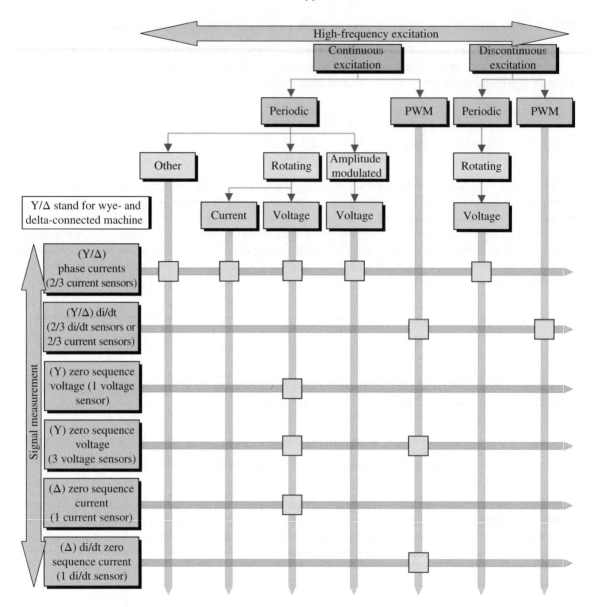

Fig. 3-33   *Classification of high-frequency signal-injection-based sensorless methods*

- **Continuous vs. discontinuous excitation:** If the high-frequency signal used to estimate the rotor position is always present along with the fundamental excitation, it is referred to as continuous excitation. Discontinuous excitation methods inject the high-frequency signal periodically, either because they require discontinuing the regular operation of the inverter or because they have to reduce its adverse effects on the normal operation of the machine. A drawback of these methods is that they do not provide a continuous rotor position estimate.

- **Periodic vs. PWM excitation:** Periodic injection methods typically inject a periodic high-frequency voltage (usually in the range of several hundred Hz up to a few kHz) superimposed on the fundamental

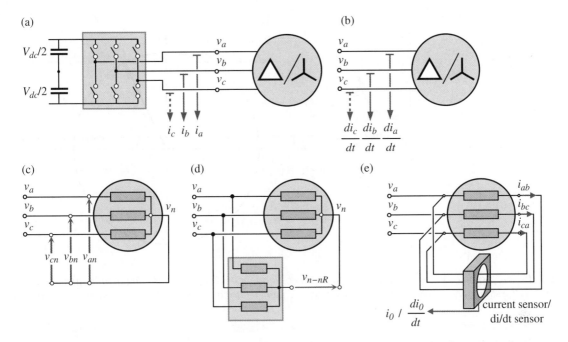

Fig. 3-34 *Signal measurement: (a) phase currents using two/three sensors, (b) phase currents derivative using two/three sensors, (c) and (d) zero sequence voltage using three voltage sensors and a single sensor with an auxiliary resistor network (wye-connected machines), and (e) zero sequence current/zero sequence current derivative (delta-connected machines)*

excitation. The rotor position information is encoded in the magnitude/phase of the resulting high-frequency currents. Use of the machine response to particular states of the inverter has also been proposed. These methods commonly require modified forms of PWM.

- **Signal measurement:** The number and type of signals that can be measured and processed to obtain the rotor position vary from method to method, with more than one option existing for each form of high-frequency excitation (see Fig. 3-33). Fig. 3-34 shows the configuration of the sensors for each case. Most industrial drives include phase current sensors and often a DC bus voltage sensor. Therefore, sensorless methods that rely on only these signals can be considered "no cost" from a hardware perspective. On the contrary, methods that require additional signals and associated hardware (sensors, cabling, A/D channels, etc.) replace a position/speed sensor by a different type of sensor, which obviously limits the intended benefits of sensorless control.

3.8.4 Position estimation using rotating high-frequency voltage injection and the negative sequence current

Sensorless control using rotating voltage injection is discussed in this subsection. This method injects a high-frequency voltage vector superposed onto the fundamental voltage, and estimates the rotor position from the measured current. Since the current sensors are already present, no additional hardware is required for its implementation.

(a)   (b)

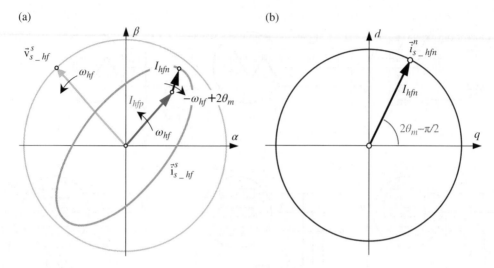

Fig. 3-35   *(a) Complex vector representation of the high-frequency voltage and resulting high-frequency current, shown in the stationary reference frame, (b) Negative sequence current in the negative sequence reference frame*

When a rotating, high-frequency voltage (3.73) is applied to the interior permanent magnet machine, it interacts with the saliencies in the inductances matrix. Substituting (3.73) in (3.72), the resulting high-frequency current (3.74) is obtained, where V_{hf} and ω_{hf} are the magnitude and frequency of the applied high-frequency voltage.

$$\vec{v}^{\,s}_{s\_hf} = \begin{bmatrix} v_{\alpha s\_hf} \\ v_{\beta s\_hf} \end{bmatrix} = V_{hf} \begin{bmatrix} \cos(\omega_{hf}t) \\ -\sin(\omega_{hf}t) \end{bmatrix} = V_{hf}e^{j\omega_{hf}t} \tag{3.73}$$

$$\vec{i}^{\,s}_{s\_hf} = I_{hfp} \begin{bmatrix} \sin(\omega_{hf}t) \\ \cos(\omega_{hf}t) \end{bmatrix} - I_{hfn} \begin{bmatrix} \sin(2\theta_m - \omega_{hf}t) \\ \cos(2\theta_m - \omega_{hf}t) \end{bmatrix} \tag{3.74}$$

Writing the high-frequency current in complex vector form (3.75), it will be seen that it consists of a positive sequence component and a negative sequence component, of magnitude I_{hfp} (3.76) and I_{hfn} (3.77) respectively. The high-frequency current is schematically shown in Fig. 3-35. The positive sequence does not contain any rotor position information. On the contrary, the phase angle of the negative sequence component is modulated by the rotor angle, meaning that this component can be used to estimate the rotor position.

$$\vec{i}^{\,s}_{s\_hf} = \vec{i}^{\,s}_{s\_phf} + \vec{i}^{\,s}_{s\_nhf} = -jI_{hfp}e^{j\omega_{hf}t} - jI_{hfn}e^{j(-\omega_{hf}t + 2\theta_m)} \tag{3.75}$$

$$I_{hfp} = \frac{V_{hf}}{\omega_{hf}} \frac{\Sigma L_s}{\Sigma L_s^2 - \Delta L_s^2} \tag{3.76}$$

$$I_{hfn} = \frac{V_{hf}}{\omega_{hf}} \frac{\Delta L_s}{\Sigma L_s^2 - \Delta L_s^2} \tag{3.77}$$

3.8.4.1   *Negative sequence current isolation*

Equation (3.75) only includes the current that resulted from the high-frequency voltage. In normal operation, the measured stator current will also include the fundamental current used for torque production. Isolating the negative sequence current and preventing interference between the negative sequence current and the fundamental current is of great importance. This can be achieved using the scheme shown in Fig. 3-36. Band stop filters (BSFs) are used to eliminate the high frequency current from the feedback signal used for current regulation, preventing reaction of the current regulator to the high frequency current. Similarly, BSFs which eliminate the fundamental current and the positive sequence high frequency current are used to isolate the negative sequence current. Bandwidths in the range of several Hz to some tens of Hz are typically used for these filters.

It is advantageous for further processing to refer the negative sequence current to a reference frame rotating at $-\omega_{hf}$ (see Fig. 3-36), (3.78) being obtained after the coordinate rotation. Superscript n in (3.78) stands for the negative sequence reference frame. The resulting (ideal) negative sequence current after the described signal processing is shown in Fig. 3-35(b). The rotor angle is obtained from its phase angle.

$$\vec{i}^{\,n}_{s\_hfn} = \vec{i}^{\,s}_{s\_hfn} e^{j\omega_{hf}t} = -jI_{hfn}e^{j2\theta_m} \tag{3.78}$$

One interesting fact which is deduced from (3.78) is that the negative sequence high-frequency current rotates at twice the rotor speed (note the 2 multiplying θ_m in the complex exponent). This means that for each rotation of 180° electrical degrees of the rotor the negative sequence current rotates 360° degrees. In other words, there is an uncertainty of 180 electrical degrees in the estimation of the rotor angle in electrical units from the negative sequence current. This is because the rotor asymmetry is not affected by the magnet polarity. A consequence of this is that saliency-tracking methods have necessarily to be combined with a magnet polarity detection mechanism. This is discussed in a further section.

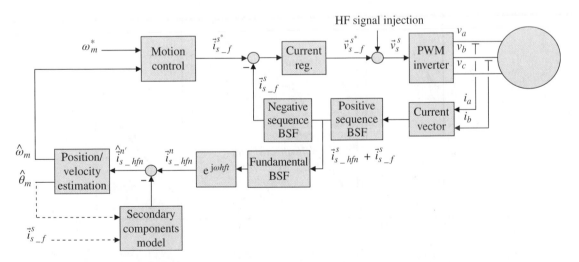

Fig. 3-36   *Schematic representation of the signal processing of the negative sequence high-frequency current to estimate the rotor speed and position. BSF stands for band stop filter*

3.8.4.2   *Secondary saliencies decoupling*

All the modeling and discussion presented so far has assumed that the saliency in the machine consists of a single, sinusoidally varying, spatial harmonic. However, secondary saliencies, mainly owing to saturation, and non-sinusoidal variation of the desired saliency always exist. Secondary saliencies cause additional components in the negative sequence current, typically resulting in unacceptable levels of estimation error, and very often in instability. Some form of compensation is therefore required. The methods that have been developed to address this issue often use the implementation shown at the bottom of Fig. 3-36 (Secondary components model). An estimation of the secondary components is first measured and stored (e.g. in look-up tables) for different operating conditions, as part of a commissioning process. The stored information is later accessed during the normal sensorless operation of the drive [10].

Fig. 3-37 shows the estimated position using the negative sequence current, with and without secondary saliencies decoupling. It will be observed that saturation-induced harmonics produce inadmissible errors in the estimated position. It will also be seen that decoupling these harmonics using the mechanism shown in Fig. 3-36 reduces the error to reasonable limits, making sensorless control feasible.

3.8.4.3   *Rotor position and speed estimation*

Once the component of the negative current containing the rotor position information (3.78) has been isolated, it can be processed in different ways to obtain the desired rotor angle. It should be noted that all the discussion presented so far has focused on the estimation of the rotor position $\hat{\theta}_m$; the rotor speed $\hat{\omega}_m$ is also often needed for control purposes.

The use of a *tan*$^{-1}$ function is very likely the most intuitive solution to obtain the phase angle from the current vector in (3.78), as it is conceptually simple and provides instantaneous response (see Fig. 3-38(a)). However,

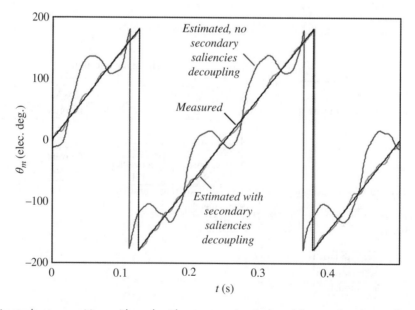

Fig. 3-37   *Estimated rotor position with and without saturation-induced harmonics decoupling. The machine operated at a constant speed of 4Hz, 50% of rated load*

this solution has some drawbacks. Digital implementation of tan^{-1} function can be computationally more expensive compared to other solutions. Also, it only provides $\hat{\theta}_m$. If $\hat{\omega}_m$ is also needed, discrete differentiation of the estimated rotor position is required, which is problematic in practice because of the noise present in the signals. To alleviate the problems with noise, the output of the tan^{-1} function is often low-pass filtered (the *LPF* block in the figure), but this will limit the estimation bandwidth.

A different approach for obtaining the speed and position is the use of tracking filter/observers. These methods use a vector cross-product to generate an error signal that drives a controller from which the rotor position $\hat{\theta}_m$ is obtained. An example of a tracking observer is shown in Fig. 3-38(b). It will be observed that the rotor speed $\hat{\omega}_m$ is also obtained as an internal variable without the need for realizing a differentiation of the estimated rotor position.

Fig. 3-39 shows an example of sensorless control of an IPMSM operating in position control using the implementation shown in Fig. 3-36 and Fig. 3-38.

(a)   (b)

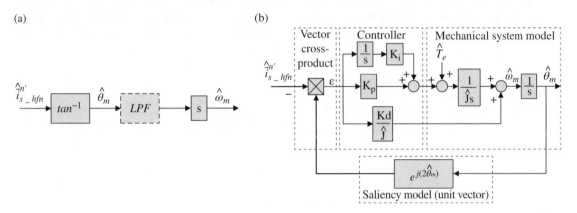

Fig. 3-38 *Rotor position estimation using (a) a* tan^{-1} *function and (b) a tracking observer implemented in the negative synchronous reference frame*

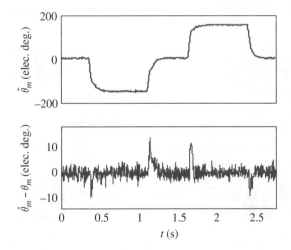

Fig. 3-39 *Sensorless position control of an IPMSM. Estimated position (top) and estimation error (bottom) with the machine operated full load*

3.8.5   Magnet polarity detection

Saliency-tracking-based methods previously discussed detect the rotor position with a periodicity of 180 electrical degrees, as they cannot distinguish the magnet polarity. However, detection of the magnet polarity is mandatory to control the torque produced by the machine. This is required not only to enable a smooth start-up of the drive but also to guarantee that the rotor will rotate in the desired direction when the q axis current is applied. Detection of the magnet polarity can be done by measuring the effects of iron saturation when a d axis current is injected. This is shown in Fig. 3-40. If a positive d axis current is injected, the resulting flux linkage adds to the magnet, increasing the iron saturation and decreasing the inductance. Conversely, a negative d axis current will reduce the iron saturation, increasing the inductance. Both cases are shown schematically in Fig. 3-41(a) and (b) respectively. It should be noted that the injection of the d axis current will not produce torque, preventing any unwanted movement of the rotor.

Several methods for magnet polarity detection using this principle have been proposed, including methods which operate in the time domain as well as methods which operate in the frequency domain. An implementation in the time domain is shown in Fig. 3-42. Two successive positive and negative voltage pulses (according to the initial assumption of the magnet polarity) are injected in the d axis direction. If the assumed polarity

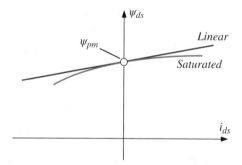

Fig. 3-40   *Variation of the flux linkage with the d axis current*

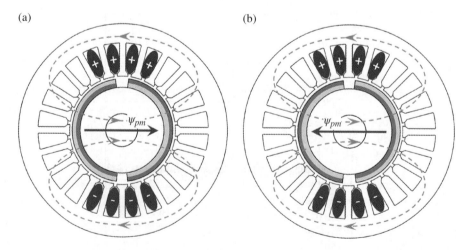

Fig. 3-41   *Injection of d axis current: (a) in the direction of the rotor magnetization and (b) opposite to the rotor magnetization*

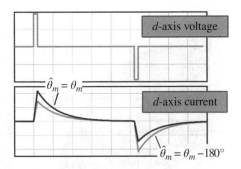

Fig. 3-42 *Top: successive positive-negative voltage pulses in the* d *axis. Bottom: resulting* d *axis current in the case of a correct magnet polarity estimation* ($\hat{\theta}_m = \theta_m$) *and an incorrect magnet polarity estimation* ($\hat{\theta}_m = \theta_m - 180°$)

is correct, the positive pulse will produce a positive d axis current, further saturating the iron (lower d axis inductance). On the contrary, the negative d axis current resulting from the negative d axis voltage pulse will reduce the iron saturation (larger d axis inductance). The peak value of the d axis current resulting from the positive d axis voltage pulse will be therefore larger than the peak value of the current resulting from the negative pulse. Thus, a larger positive peak value compared to the negative peak value of the d axis current will indicate that the initial magnet polarity assumption was correct. Conversely, a smaller positive peak value compared to the negative peak value will indicate that the initial magnet polarity assumption was incorrect (i.e. the initial rotor angle needs to be corrected by 180°).

3.8.6   Use of saliency-tracking methods with induction machines

Though induction machines are designed to be symmetric, there are asymmetries, either due to construction issues or created on purpose, that can be used for sensorless control using the saliency-tracking methods described in Section 3.8.3 [10].

A first mechanism that produces asymmetries in the rotor is the combined effect of the stator and rotor slots. This effect can occur in principle in standard induction machines, meaning that no special design or modification would be needed for their use in a sensorless drive. However, not all machine designs are adequate for this purpose. Semi-open or open rotor slot designs are needed. Also, the relationship between the pole number, p, number of stator slots, S, and rotor slots, R, has to meet the criteria shown in (3.79) in order for the slotting saliency to couple with the stator windings. An example of this rotor design is shown in Fig. 3-43(a).

$$n \cdot p = |R - S| \quad n = 1, 2, 4, 5, \ldots \tag{3.79}$$

The high-frequency signal current is given by (3.80)–(3.81) in this case. It will be observed that these expressions are similar to (3.75) and (3.78), the relationship between the rotor angle and the negative sequence current angle being now R/p. Consequently, all the signal processing described in the preceding sections in the case of the PMSM is directly applicable to the induction machine case.

$$\vec{i}^{\,s}_{s\_hf} = -jI_{hfp}e^{j\omega_{hf}t} - jI_{hfn}e^{j\left(-\omega_{hf}t \pm \frac{R}{p}\theta_m\right)} \tag{3.80}$$

$$\vec{i}^{\,n}_{s\_hfn} = -jI_{hfn}e^{\pm j\frac{R}{p}\theta_m} \tag{3.81}$$

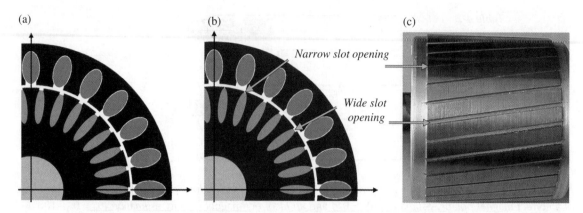

Fig. 3-43 *Induction machines with saliencies. Schematic representation of an induction machine (a) with semi-open rotor slots (constant slot opening), (b) with semi-open rotor slots with a modulation of the rotor slot width with a period of 90 mechanical degrees (180° electrical), and (c) image of the rotor of an induction motor with a modulation of the rotor slots' width*

Alternatively to the use of the rotor-stator slotting effect, methods for deliberately creating rotor-position-dependent saliencies have also been proposed (e.g. by modulating the rotor slot widths). An example of this is shown in Fig. 3-43(b) and (c). These modifications complicate the design and/or manufacturing process of the machine but are viable for high-volume production.

3.9   Numerical calculation of the steady-state of synchronous machines

This section, as with induction machines in Chapter 2, utilizes the dynamic model of the synchronous machine seen in previous sections to derive the steady-state behavior of its most representative magnitudes. The section is divided into two parts. Derivation of the maximum capability curves is first addressed, equations for steady-state operation being derived later.

3.9.1   Calculation of the maximum capability curves

This section briefly describes how the maximum capability curves of the synchronous machines can be calculated. In a similar way as we did with the induction machines, this capability curves calculation procedure is summarized in a table. In this case, two different tables are utilized—one for non-salient and another one for salient machines. First of all, Table 3-1 illustrates the procedure to calculate the maximum capability curves of a non-salient synchronous machine. This first (simpler) case takes the supposed input data and shows all the required equations to calculate the most representative expressions of the machine at both constant torque and flux-weakening regions. Note that for the torque and stator current calculations at flux-weakening regions, there is not an explicit equation used. Instead, the numerical closed-loop calculation procedure of Fig. 3-44 is utilized. It finds the pair of currents i_{ds} and i_{qs} which maximize the torque at every single rotational speed, by means of a numerical swept calculation approach.

Table 3-1 Maximum capability curves derivation procedure for non-salient synchronous machines

	Given data: P_{rated}, V_{rated}, n_{rated}, p, and electric parameters of the machine (motor mode of operation)			
1. Torque and current	$T_{em\ rated} = \dfrac{P_{rated}}{\omega_{m\ rated}/p}$			$i_{qs\ rated} = \dfrac{T_{em\ rated}}{\frac{3}{2}p\psi_{pm}} = I_{s\_rated}$
2. Limit Frequencies	$\omega_{m\ rated} = n_{rated}(2\pi/60)p$	$\omega_{base} = \dfrac{V_{s\_rated}}{\sqrt{\psi_{pm}^2 + (L_s \cdot I_{s\_rated})^2}}$		$\omega_{max} = \dfrac{V_{s\_rated}}{\psi_{pm} - L_s \cdot I_{s\_rated}}$
Constant torque region ($0 < \omega_m < \omega_{base}$)				
1. Torque	$T_{em} = T_{em\ rated}$			
2. Stator currents	$i_{ds} = 0$	$i_{qs} = \dfrac{T_{em\ rated}}{\frac{3}{2}p\psi_{pm}}$	$I_{s\ rated} = \sqrt{i_{ds}^2 + i_{q\ srated}^2}$	$\theta_{i_s} = a\tan\left(\dfrac{i_{qs}}{i_{ds}}\right)$
3. Stator voltages	$v_{ds} = -\omega_m i_{qs}L_{qs}$	$v_{qs} = R_s i_{qs} + \omega_m \psi_{pm}$	$V_s = \sqrt{v_{ds}^2 + v_{qs}^2}$	$\theta_{v_s} = a\tan\left(\dfrac{v_{qs}}{v_{ds}}\right)$
4. Stator flux	$\psi_{ds} = \psi_{pm}$	$\psi_{qs} = i_{qs}L_{qs}$	$\psi_s = \sqrt{\psi_{ds}^2 + \psi_{qs}^2}$	$\theta_{\psi_s} = a\tan\left(\dfrac{\psi_{qs}}{\psi_{ds}}\right)$
5. Powers	$P_s = \dfrac{3}{2}\left(v_{ds}i_{ds} + v_{qs}i_{qs}\right)$		$P = \dfrac{T_{em}}{\omega_m/p}$	
6. Efficiency, phase shift, and apparent power	$\eta = \dfrac{P}{P_s}$	$\phi = \theta_{v_s} - \theta_{i_s}$	$S_s = \dfrac{P_s}{\cos(\phi)}$	
Flux-weakening region (variable $\omega_{base} < \omega_m < \omega_{max}$)				
1. Stator currents and torque	Procedure of Fig. 3-44			
2. Stator voltages	$v_{ds} = R_s i_{ds} - \omega_m i_{qs}L_{qs}$	$v_{qs} = R_s i_{qs} + \omega_m \psi_{pm} + \omega_m i_{ds}L_{ds}$	$V_s = \sqrt{v_{ds}^2 + v_{qs}^2}$	$\theta_{v_s} = a\tan\left(\dfrac{v_{qs}}{v_{ds}}\right)$
3. Stator flux	$\psi_{ds} = i_{ds}L_{ds} + \psi_{pm}$	$\psi_{qs} = i_{qs}L_{qs}$	$\psi_s = \sqrt{\psi_{ds}^2 + \psi_{qs}^2}$	$\theta_{\psi_s} = a\tan\left(\dfrac{\psi_{qs}}{\psi_{ds}}\right)$
4. Powers	$P_s = \dfrac{3}{2}\left(v_{ds}i_{ds} + v_{qs}i_{qs}\right)$		$P = \dfrac{T_{em}}{\omega_m/p}$	
5. Efficiency, phase shift, and apparent power	$\eta = \dfrac{P}{P_s}$	$\phi = \theta_{v_s} - \theta_{i_s}$	$S_s = \dfrac{P_s}{\cos(\phi)}$	

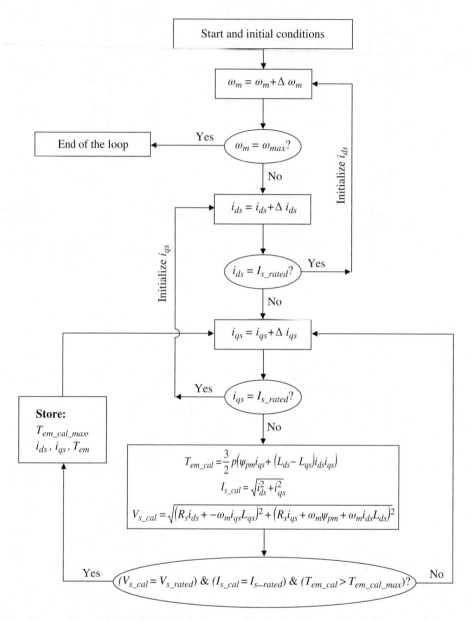

Fig. 3-44   *Procedure to calculate the maximum torque available and* i_{ds}–i_{qs} *pair, at flux-weakening region*

On the other hand, in a similar way for salient synchronous machines, since an MTPA must be performed, the deviation becomes slightly more complex. Table 3-2 illustrates the procedure, for which in this case the method of Fig. 3-44 for calculating the pair of i_{ds} and i_{qs} at rated torque is employed at constant torque region.

Table 3-2 Maximum capability curves derivation procedure for salient synchronous machines (finite speed drive)

	Given data: P_{rated}, V_{rated}, n_{rated}, p, and electric parameters of the machine (motor mode of operation)			
1. Torque and speed	$\omega_{m\,rated} = n_{rated}(2\pi/60)p$			$T_{em\,rated} = \dfrac{P_{rated}}{\omega_{m\,rated}/p}$
2. Limit Frequencies	$\omega_{base} = \dfrac{V_{s\_rated}}{\sqrt{\left(\psi_{pm}^2 + L_{ds}\cdot i_{ds1}\right)^2 + \left(L_{qs}\cdot i_{qs1}\right)^2}}$			$\omega_{max} = \dfrac{V_{s\_rated}}{\psi_{pm} - L_{ds}\cdot I_{s\_rated}}$
Constant torque region ($0 < \omega_m < \omega_{base}$)				
1. Torque	$T_{em} = T_{em\,rated}$			
2. Stator currents	Procedure of Fig. 3-44 (for rated torque)		$I_{s\_limit} = \sqrt{i_{ds}^2 + i_{qs}^2}$	$\theta_{i_s} = a\tan\left(\dfrac{i_{qs}}{i_{ds}}\right)$
3. Stator voltages	$v_{ds} = R_s i_{ds} - \omega_m i_{qs} L_{qs}$	$v_{qs} = R_s i_{qs} + \omega_m \psi_{pm} + \omega_m i_{ds} L_{ds}$	$V_s = \sqrt{v_{ds}^2 + v_{qs}^2}$	$\theta_{v_s} = a\tan\left(\dfrac{v_{qs}}{v_{ds}}\right)$
4. Stator flux	$\psi_{ds} = i_{ds}L_{ds} + \psi_{pm}$	$\psi_{qs} = i_{qs}L_{qs}$	$\psi_s = \sqrt{\psi_{ds}^2 + \psi_{qs}^2}$	$\theta_{\psi_s} = a\tan\left(\dfrac{\psi_{qs}}{\psi_{ds}}\right)$
5. Powers	$P_s = \dfrac{3}{2}\left(v_{ds}i_{ds} + v_{qs}i_{qs}\right)$		$P = \dfrac{T_{em}}{\omega_m/p}$	
			$S_s = \dfrac{P_s}{\cos(\phi)}$	
6. Efficiency, phase shift, and apparent power	$\eta = \dfrac{P}{P_s}$	$\phi = \theta_{v_s} - \theta_{i_s}$		
Flux-weakening region (variable $\omega_{base} < \omega_m < \omega_{max}$)				
1. Stator currents and torque	Procedure of Fig. 3-44			
2. Stator voltages	$v_{ds} = R_s i_{ds} - \omega_m i_{qs} L_{qs}$	$v_{qs} = R_s i_{qs} + \omega_m \psi_{pm} + \omega_m i_{ds} L_{ds}$	$V_s = \sqrt{v_{ds}^2 + v_{qs}^2}$	$\theta_{v_s} = a\tan\left(\dfrac{v_{qs}}{v_{ds}}\right)$
3. Stator flux	$\psi_{ds} = i_{ds}L_{ds} + \psi_{pm}$	$\psi_{qs} = i_{qs}L_{qs}$	$\psi_s = \sqrt{\psi_{ds}^2 + \psi_{qs}^2}$	$\theta_{\psi_s} = a\tan\left(\dfrac{\psi_{qs}}{\psi_{ds}}\right)$
4. Powers	$P_s = \dfrac{3}{2}\left(v_{ds}i_{ds} + v_{qs}i_{qs}\right)$		$P = \dfrac{T_{em}}{\omega_m/p}$	
			$S_s = \dfrac{P_s}{\cos(\phi)}$	
5. Efficiency, phase shift, and apparent power	$\eta = \dfrac{P}{P_s}$	$\phi = \theta_{v_s} - \theta_{i_s}$		

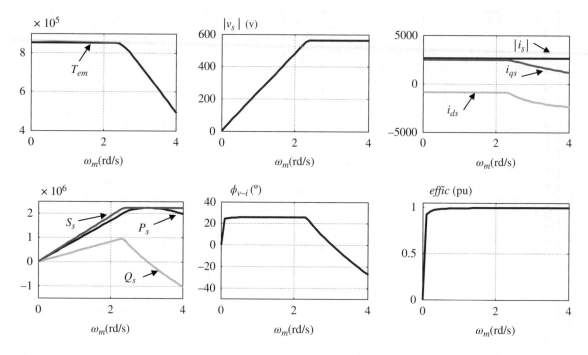

Fig. 3-45 *Maximum capability curves of a 2MW, 690V$_{LL}$, 1867.76A$_{RMS}$, 22.5rpm, and 30 pole pairs, salient synchronous machine (Source: Machine parameters based on [11])*

Thus, as shown in Fig. 3-45, the maximum capability curves of a 2 MW salient synchronous machine are presented. The main electric characteristics of the machine are 2MW, 690V$_{LL}$, 1867.76A$_{RMS}$, 22.5rpm, and 30 pole pairs. On the other hand, the main characteristic parameters of this machine are: $L_{ds} = 1.21$mH, $L_{qs} = 2.31$mH, $R_s = 0.73$mΩ, and $\psi_{pm} = 4.696$Wb$_{RMS}$.

3.9.2   Calculation of the steady-state operation

This section shows how the steady-state operation of synchronous machines can be calculated, when the load applied to the shaft is known. The procedure is represented in Table 3-3. It will be seen that the input is the load torque applied to the shaft, for each shaft's speed. The stator currents are obtained according to the MTPA criteria. Given the stator currents, the rest of the machine's variables can be calculated as described in Table 3-3. Note that this procedure is very similar to that followed in previous sections for deriving the maximum capability curves.

Therefore, for the machine utilized in previous sections (2MW, 690V$_{LL}$, 1867.76A$_{RMS}$, 22.5rpm, and 30 pole pairs salient synchronous motor), a load torque according to the following formula is applied:

$$T_{load} = \left(14\omega_m^2 + 1\right) \cdot 10000 \tag{3.82}$$

The most representative magnitudes are depicted in Fig. 3-46.

Finally, Table 3-4 provides some electric parameters of different salient synchronous machines [12].

Table 3-3 Steady-state operation of a salient synchronous machine

	Given data: Operation pair array ω_m-T_{load}		
Constant torque region ($0 < \omega_m < \omega_{base}$)			
1. Torque	$T_{em} = T_{em\,load}$		
2. Stator currents	Procedure of Fig. 3-44 (at load torque and not considering the voltage constraint)	$I_s = \sqrt{i_{ds}^{\,2} + i_{qs}^{\,2}}$	$\theta_{i_s} = a\tan\left(\dfrac{i_{qs}}{i_{ds}}\right)$
3. Stator voltages	$v_{ds} = R_s i_{ds} - \omega_m i_{qs} L_{qs}$	$v_{qs} = R_s i_{qs} + \omega_m \psi_{pm} + \omega_m i_{ds} L_{ds}$	$V_s = \sqrt{v_{ds}^{\,2} + v_{qs}^{\,2}}$ $\theta_{v_s} = a\tan\left(\dfrac{v_{qs}}{v_{ds}}\right)$
4. Stator flux	$\psi_{ds} = i_{ds} L_{ds} + \psi_{pm}$	$\psi_{qs} = i_{qs} L_{qs}$	$\psi_s = \sqrt{\psi_{ds}^{\,2} + \psi_{qs}^{\,2}}$ $\theta_{\psi_s} = a\tan\left(\dfrac{\psi_{qs}}{\psi_{ds}}\right)$
5. Powers	$P_s = \dfrac{3}{2}\left(v_{ds} i_{ds} + v_{qs} i_{qs}\right)$		$P = \dfrac{T_{em}}{\omega_m/p}$
6. Efficiency, phase shift, and apparent power	$\eta = \dfrac{P}{P_s}$	$\phi = \theta_{v_s} - \theta_{i_s}$	$S_s = \dfrac{P_s}{\cos(\phi)}$
Flux-weakening region (variable $\omega_{base} < \omega_m < \omega_{max}$)			
1. Stator currents and torque	Procedure of Fig. 3-44		
2. Stator voltages	$v_{ds} = R_s i_{ds} - \omega_m i_{qs} L_{qs}$	$v_{qs} = R_s i_{qs} + \omega_m \psi_{pm} + \omega_m i_{ds} L_{ds}$	$V_s = \sqrt{v_{ds}^{\,2} + v_{qs}^{\,2}}$ $\theta_{v_s} = a\tan\left(\dfrac{v_{qs}}{v_{ds}}\right)$
3. Stator flux	$\psi_{ds} = i_{ds} L_{ds} + \psi_{pm}$	$\psi_{qs} = i_{qs} L_{qs}$	$\psi_s = \sqrt{\psi_{ds}^{\,2} + \psi_{qs}^{\,2}}$ $\theta_{\psi_s} = a\tan\left(\dfrac{\psi_{qs}}{\psi_{ds}}\right)$
4. Powers	$P_s = \dfrac{3}{2}\left(v_{ds} i_{ds} + v_{qs} i_{qs}\right)$		$P = \dfrac{T_{em}}{\omega_m/p}$
5. Efficiency, phase shift, and apparent power	$\eta = \dfrac{P}{P_s}$	$\phi = \theta_{v_s} - \theta_{i_s}$	$S_s = \dfrac{P_s}{\cos(\phi)}$

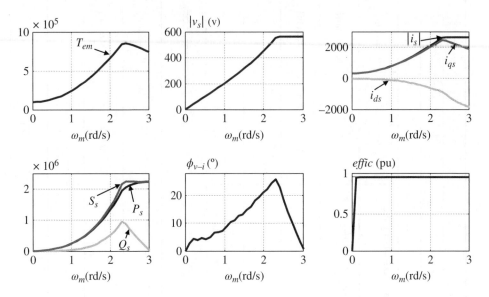

Fig. 3-46 *Curves of a 2MW, 690V_{LL}, 22.5rpm salient synchronous machine, with load torque* $T_{load} = (14\omega_m^2 + 1) \cdot 10000$ *until base speed, then at flux-weakening region, the maximum torque available of the machine is provided (Source: Machine parameters based on [11])*

Table 3-4 *Electric parameters of different synchronous machines (Source: Data from MATLAB-Simulink using preset models [12])*

Characteristic	7.14Nm	26.13Nm	67.27Nm
Rated speed (rev/m)	5000	3000	1700
Rated frequency (Hz)	333.3	200	113.3
Rated line-to-line stator voltage (V_{RMS})	400	400	400
p	4	4	4
ψ_{pm} (Wb)$_{peak}$	0.06784	0.1119	0.192
R_s (Ω)	0.24	0.11	0.085
L_{ds} (mH)	0.9642	0.9215	0.9025
L_{qs} (mH)	1.066	1.018	0.9975
$\frac{L_{ds}}{R_s}$ (msec)	4.01	8.36	10.61
$\frac{L_{qs}}{R_s}$ (msec)	4.44	9.25	11.73

References

[1] Novotny DW, Lipo TA. *Vector Control and Dynamics of AC Drives.* New York: Oxford University Press, 1996.

[2] Krishnan R. *Permanent Magnet Synchronous and Brushless DC Motor Drives.* Boca Raton, FL: CRC Press, 2009.

[3] Hendershot JR, Miller TJE. *Design of Brushless Permanent Magnet Motors.* New York: Oxford University Press, 1994.

[4] Xu L, Xu X, Lipo TA, Novotny DW. Vector control of a synchronous reluctance motor including saturation and iron loss. *IEEE Trans Ind Appl* September/October 1991; **27**(5): 977, 985.

[5]   Sul S-K. *Control of Electric Machine Drive Systems*. New York: John Wiley & Sons, Inc, 2011.

[6]   Chen Z, Tomita M, Ichikawa S, et al. Sensorless control of interior permanent magnet synchronous motor by estimation of an extended electromotive force. *Industry Applications Conference* 2000; **3**: 1814, 1819.

[7]   Zhao Y, Zhang Z, Qiao W, Wu L. An extended flux model-based rotor position estimator for sensorless control of salient-pole permanent-magnet synchronous machines. *IEEE Trans Power Electron* 2015; **30**(8): 4412–4422.

[8]   Shen JX, Zhu ZQ, Howe D. Improved speed estimation in sensorless PM brushless AC drives. *IEEE Trans Ind Appl* July/August 2002; **38**(4): 1072, 1080.

[9]   Andreescu GD. Position and speed sensorless control of PMSM drives based on adaptive observer. *Proceedings EPE '99*. Lausanne, Switzerland, September 1999.

[10] Briz F, Degner MW. Rotor position estimation: A review of high-frequency methods. *IEEE Ind Electron Mag* June 2011; **5**(2): 24–36.

[11] Wu B, Lang Y, Zargari N, Kouro S, *Power Conversion and Control of Wind Energy Systems*. New York: John Wiley & Sons, Inc, 2011.

[12] SIMPOWER Library of Matlab-Simulink.

4

Control of grid-connected converters

Aritz Milicua and Gonzalo Abad

4.1   Introduction

In this chapter the control of AC grid-connected converters is studied. More specifically, regenerative, also known as reversible or back-to-back, converters connected to the AC grid are studied. As was described in Chapter 1, these are controlled converters typically constructed by arrangements of insulated-gate bipolar transistors (IGBTs) and diodes, enabling one to bi-directionally control the exchange of power with the grid. According to the applications that will be later studied in this book, regenerative converters can be in some cases employed in ship propulsion in a three-phase converter version and can also be employed in a single-phase configuration in railway traction. A schematic representation of the two possible scenarios is depicted in Fig. 4-1.

Basically, the usage of reversible converters in these two applications against passive or diode front ends can make sense when regenerative power produced by the application can be delivered to the ship's electric grid or to the catenary. As is described in forthcoming chapters, this regenerative possibility can be useful in certain circumstances. Therefore, owing to this, the necessity of studying grid-connected converters arises in this chapter.

On the other hand, as can be seen in Fig. 4-1, in general in railway traction applications the converter is connected to the grid or catenary through a single-phase transformer that provides an adequate voltage to the converter. However, in ship propulsion, although the connection can be performed through a transformer, more often the transformer is eliminated in order to save space and reduce weight and costs. In addition, a filter between the grid and the converter is necessary to obtain a good quality of current exchange with the grid. There are different types of possible filters, as will be seen later this chapter. It must be remarked that if a transformer is located between the grid and the converter, as typically is

Power Electronics and Electric Drives for Traction Applications, First Edition. Edited by Gonzalo Abad.
© 2017 John Wiley & Sons, Ltd. Published 2017 by John Wiley & Sons, Ltd.

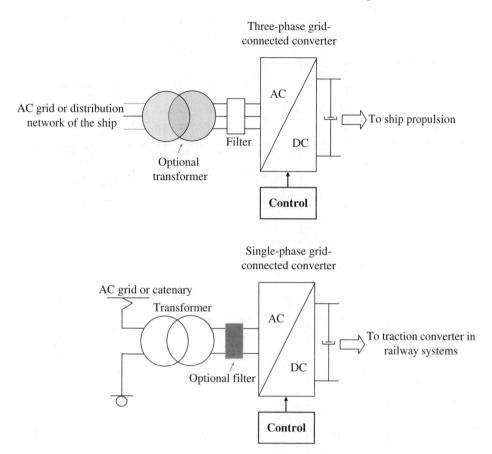

Fig. 4-1 *Grid-connected converter configurations*

the case in railway applications, its leakage inductance can actuate as a filter therefore minimizing or even eliminating the necessity of an additional filter.

Thus, in summary, basically this chapter first studies modeling and certain aspects of three-phase grid-connected converters. The converter itself is introduced and then the control of these types of system is described and analyzed. The controls described are divided into two parts: one is devoted to the control strategy when the system works with balanced voltages and currents exchanged with the grid. The other is oriented to the control when the system works with unbalanced voltages and/or currents. Finally, the single-phase connected converter's modeling and control are described.

4.2   Three-phase grid-connected converter model

When we refer to a three-phase converter connected to the grid in this book, it is supposed to be as the power electric circuit that is presented in Fig. 4-2. It is an active front end (AFE) composed by the bi-directional power electronic converter, the grid side filter, and the grid three-phase voltages. The converter is modeled by bi-directional switches composed by IGBTs and diodes in antiparallel, as depicted in Fig. 4-2.

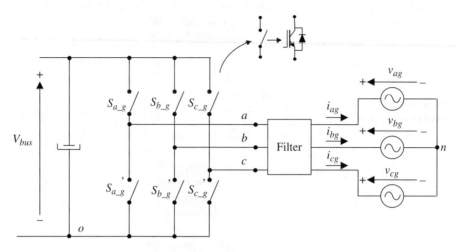

Fig. 4-2   *Grid-connected converter power circuit*

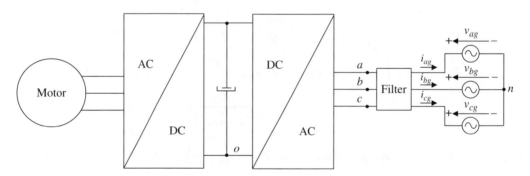

Fig. 4-3   *Back-to-back converter supplying a motor, with a grid-connected converter*

It converts voltage and currents from DC to AC, while the exchange of power can be in both directions from AC to DC (rectifier mode) and from DC to AC (inverter mode). A classic two-level converter has been considered, but any other multilevel converter topology could be considered. In addition, this chapter considers an idealized behavior of IGBTs and diodes, and does not take into account phenomena such as voltage drop in the semiconductors or switching times.

On the other hand, the grid side filter is composed of at least three inductances, one inductance per phase. This type of filter is called an L filter. Then, it is also possible to build structures with inductance and capacitance arrangements (LC filter), or even inductance–capacitance–inductance arrangements (LCL filter). The grid voltage is often supplied by a transformer in many applications. In this section, it is assumed that a three-phase balanced voltage is provided, while the grid side impedance of the transformer has been neglected.

At the DC side of the converter, it is possible to connect another three-phase converter, forming, for instance, a three-phase back-to-back power electronic converter, as illustrated in Fig. 4-3. Alternatively in some other applications, it is also possible to leave the DC-link without connection of any other element, resulting in a reactive power compensator device for instance.

4.2.1   Converter model

The output voltages of the converter are synthesized by commanding the appropriate switching orders of the IGBTs: S_{a_g}, S_{b_g}, S_{c_g}. The output voltages referred to the negative point of the DC bus can be calculated according to the following general expression:

$$v_{jo} = V_{bus} \cdot S_{j\_g}$$
$$\text{with} \quad S_{j\_g} \in \{0, 1\} \quad \text{and} \quad j = a, b, c \tag{4.1}$$

Thus, by different combinations of S_{a_g}, S_{b_g}, and S_{c_g}, it is possible to create AC output voltages with a fundamental component of different amplitude and frequency. On the other hand, it is also useful to derive the output AC voltages referred to the neutral point of the grid side voltages (n). For that purpose, by checking the single-phase AC equivalent circuit of the system depicted in Fig. 4-4, the following relation between voltages can be obtained:

$$v_{jn} = v_{jo} - v_{no}$$
$$\text{with} \quad j = a, b, c \tag{4.2}$$

From this last expression, v_{no} voltage is needed for its resolution. First, it is necessary to consider the following equation:

$$v_{an} + v_{bn} + v_{cn} = 0 \tag{4.3}$$

Combining these last two expressions, we obtain:

$$v_{no} = \frac{1}{3} \cdot (v_{ao} + v_{bo} + v_{co}) \tag{4.4}$$

which is a useful result, if we substitute it for (4.2):

$$v_{an} = \frac{2}{3} \cdot v_{ao} - \frac{1}{3} \cdot (v_{bo} + v_{co}) \tag{4.5}$$

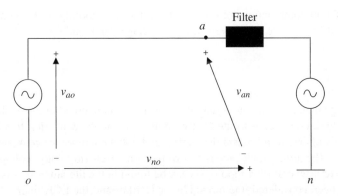

Fig. 4-4 *Simplified equivalent single-phase grid circuit (a phase) (Source: [1]. Reproduced with permission of John Wiley & Sons)*

Table 4-1   *Different output voltage combinations of 2L-VSC*

S_{a_g}	S_{b_g}	S_{c_g}	v_{ao}	v_{bo}	v_{co}	v_{an}	v_{bn}	v_{cn}
0	0	0	0	0	0	0	0	0
0	0	1	0	0	V_{bus}	$-\dfrac{V_{bus}}{3}$	$-\dfrac{V_{bus}}{3}$	$2\cdot\dfrac{V_{bus}}{3}$
0	1	0	0	V_{bus}	0	$-\dfrac{V_{bus}}{3}$	$2\cdot\dfrac{V_{bus}}{3}$	$-\dfrac{V_{bus}}{3}$
0	1	1	0	V_{bus}	V_{bus}	$-2\cdot\dfrac{V_{bus}}{3}$	$\dfrac{V_{bus}}{3}$	$\dfrac{V_{bus}}{3}$
1	0	0	V_{bus}	0	0	$2\cdot\dfrac{V_{bus}}{3}$	$-\dfrac{V_{bus}}{3}$	$-\dfrac{V_{bus}}{3}$
1	0	1	V_{bus}	0	V_{bus}	$\dfrac{V_{bus}}{3}$	$-2\cdot\dfrac{V_{bus}}{3}$	$\dfrac{V_{bus}}{3}$
1	1	0	V_{bus}	V_{bus}	0	$\dfrac{V_{bus}}{3}$	$\dfrac{V_{bus}}{3}$	$-2\cdot\dfrac{V_{bus}}{3}$
1	1	1	V_{bus}	V_{bus}	V_{bus}	0	0	0

$$v_{bn} = \frac{2}{3}\cdot v_{bo} - \frac{1}{3}\cdot\left(v_{ao} + v_{co}\right) \tag{4.6}$$

$$v_{cn} = \frac{2}{3}\cdot v_{co} - \frac{1}{3}\cdot\left(v_{bo} + v_{ao}\right) \tag{4.7}$$

or simpler, directly from the order commands:

$$v_{an} = \frac{V_{bus}}{3}\cdot\left(2\cdot S_{a\_g} - S_{b\_g} - S_{c\_g}\right) \tag{4.8}$$

$$v_{bn} = \frac{V_{bus}}{3}\cdot\left(2\cdot S_{b\_g} - S_{a\_g} - S_{c\_g}\right) \tag{4.9}$$

$$v_{cn} = \frac{V_{bus}}{3}\cdot\left(2\cdot S_{c\_g} - S_{a\_g} - S_{b\_g}\right) \tag{4.10}$$

There are eight different combinations of output AC voltages, according to the eight permitted switching states of S_{a_g}, S_{b_g}, and S_{c_g}. Table 4-1 serves all these voltage combinations.

4.2.2   Filter model

Once the most important modeling voltage equations of the power converter have been derived, the next step is to consider the grid side filter, which is located between the grid and the converter. It is important to highlight that the objective of this filter is to minimize the current and voltage harmonics generated by the switching of the converter. Owing to the different characteristics of the converters (topology, voltage and current ratings, switching frequency, etc.), several filter topologies can be found in the literature. In this section, two different filter topologies have been considered: the inductive, or L, filter and the LCL filter.

From the system power circuit presented in Fig. 4-2, the three-phase system can be modeled as three independent, but equivalent, single-phase systems, as depicted in Fig. 4-5. The voltages created by the power

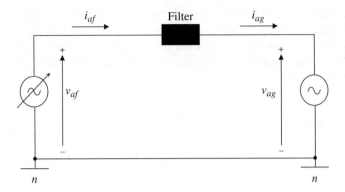

Fig. 4-5 *Simplified equivalent single-phase grid circuit (a phase)*

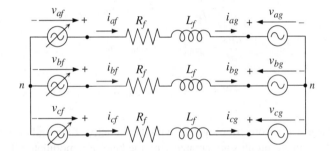

Fig. 4-6 *Simplified representation of the three-phase grid system with an L filter*

electronic converter have been replaced by variable AC voltage sources. These output AC voltages of the converter, referred to the neutral point n, as well as the output current of the converter, are named with sub-index f.

The first filter to be analyzed corresponds to the inductive filter, resulting in the simplest modeling approach from the used filter point of view. The three-phase equivalent circuit of such a filter is represented in Fig. 4-6. It is important to remember that the output AC voltages and currents of the converter are named with the sub-index f. Finally, the inductance (L_f) is accompanied by an unavoidable resistance (R_f) that is also considered in the model.

Being:

- L_f: Inductance of the filter (H)
- R_f: Resistive part of the filter (Ω)
- v_{af}, v_{bf}, v_{cf}: Output voltages of the converter referred to the neutral point of the load n (V)
- i_{af}, i_{bf}, i_{cf}: Output currents of the converter (A)
- v_{ag}, v_{bg}, v_{cg}: Grid voltages (V), with ω electric angular speed in (rad/sec)
- i_{ag}, i_{bg}, i_{cg}: Currents flowing through the grid side of the filter (A).

Obviously in this filter topology, both the grid side currents (i_g) and converter output currents (i_f) are identical. Therefore, the following mathematical expressions can be obtained using both of the currents. However, in this section, the grid side current (i_g) has been considered in the mathematical expressions.

Thus, the electric differential equations of the converter connected to the grid by means of an L filter can be derived from Fig. 4-6 as:

$$v_{af} = R_f \cdot i_{ag} + L_f \cdot \frac{di_{ag}}{dt} + v_{ag} \tag{4.11}$$

$$v_{bf} = R_f \cdot i_{bg} + L_f \cdot \frac{di_{bg}}{dt} + v_{bg} \tag{4.12}$$

$$v_{cf} = R_f \cdot i_{cg} + L_f \cdot \frac{di_{cg}}{dt} + v_{cg} \tag{4.13}$$

Consequently, for modeling purposes, it is necessary to isolate the first derivative of the currents:

$$\frac{di_{ag}}{dt} = \frac{1}{L_f} \cdot \left(v_{af} - R_f \cdot i_{ag} - v_{ag}\right) \tag{4.14}$$

$$\frac{di_{bg}}{dt} = \frac{1}{L_f} \cdot \left(v_{bf} - R_f \cdot i_{bg} - v_{bg}\right) \tag{4.15}$$

$$\frac{di_{cg}}{dt} = \frac{1}{L_f} \cdot \left(v_{cf} - R_f \cdot i_{cg} - v_{cg}\right) \tag{4.16}$$

Thus, the model of the grid side converter together with the inductive filter and the grid itself is represented in Fig. 4-7. The converter voltage outputs are created with the help of a modulator. The grid ideal sinusoidal voltages are generated at constant amplitude and frequency. Then the currents exchanged with the grid are calculated taking into consideration the filter, according to expressions (4.14), (4.15), and (4.16).

The second type of filter considered in this section is the LCL filter. In this case the filter is composed of two inductances (L_f and L_g), a capacitor (C), and a damping resistance (R) per phase. It is important to highlight that the damping resistances are used in order to avoid possible resonances between the inductances and the capacitors. The three-phase equivalent circuit of this filter is represented in Fig. 4-8. Notice that as in the L filter the parasitic resistances of all the inductances have been considered (R_f and R_g). In addition, the whole system has been considered to be connected in a wye configuration.

Being:

- L_f: Inductance of the converter side of the filter (H)
- R_f: Resistive part of the converter side of the filter (Ω)
- L_g: Inductance of the grid side of the filter (H)
- R_g: Resistive part of the grid side of the filter (Ω)
- C: Capacitance of the filter (F)
- R: Damping resistance (Ω)
- v_{af}, v_{bf}, v_{cf}: Output voltages of the converter referred to the neutral point of the load n (V)
- i_{af}, i_{bf}, i_{cf}: Output currents of the converter (A)
- v_{ag}, v_{bg}, v_{cg}: Grid voltages (V), with ω electric angular speed in (rad/sec)
- i_{ag}, i_{bg}, i_{cg}: Currents flowing through the grid side of the filter (A).

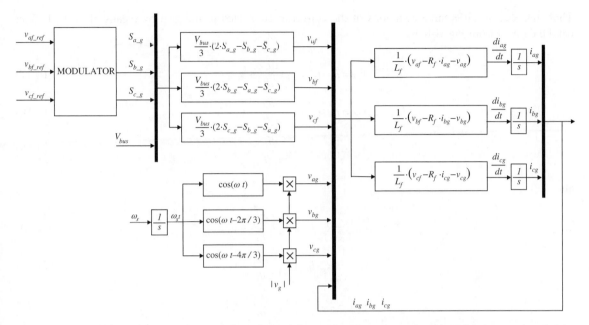

Fig. 4-7 *Simplified converter, filter, and grid model (Source: [1]. Reproduced with permission of John Wiley & Sons)*

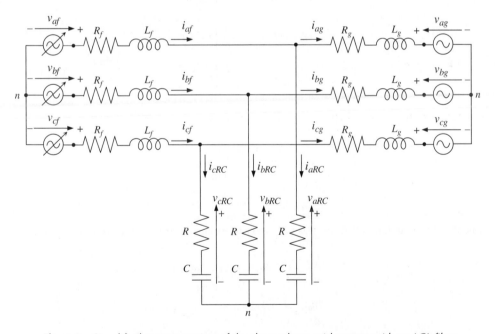

Fig. 4-8 *Simplified representation of the three-phase grid system with an LCL filter*

Thus, the electric differential equations of the converter connected to the grid by means of an LCL filter can be derived from Fig. 4-8 as:

$$v_{af} = R_f \cdot i_{af} + L_f \cdot \frac{di_{af}}{dt} + v_{aRC} \tag{4.17}$$

$$v_{bf} = R_f \cdot i_{bf} + L_f \cdot \frac{di_{bf}}{dt} + v_{bRC} \tag{4.18}$$

$$v_{cf} = R_f \cdot i_{cf} + L_f \cdot \frac{di_{cf}}{dt} + v_{cRC} \tag{4.19}$$

and:

$$v_{aRC} = R_g \cdot i_{ag} + L_g \cdot \frac{di_{ag}}{dt} + v_{ag} \tag{4.20}$$

$$v_{bRC} = R_g \cdot i_{bg} + L_g \cdot \frac{di_{bg}}{dt} + v_{bg} \tag{4.21}$$

$$v_{cRC} = R_g \cdot i_{cg} + L_g \cdot \frac{di_{cg}}{dt} + v_{cg} \tag{4.22}$$

The voltage drop across the filtering capacitors can also be expressed as a function of the currents:

$$v_{aRC} = R \cdot i_{aRC} + \frac{1}{C} \int i_{aRC} dt \tag{4.23}$$

$$v_{bRC} = R \cdot i_{bRC} + \frac{1}{C} \int i_{bRC} dt \tag{4.24}$$

$$v_{cRC} = R \cdot i_{cRC} + \frac{1}{C} \int i_{cRC} dt \tag{4.25}$$

On the other hand, the relation between the different currents can be developed as:

$$i_{aRC} = i_{af} - i_{ag} \tag{4.26}$$

$$i_{bRC} = i_{bf} - i_{bg} \tag{4.27}$$

$$i_{cRC} = i_{cf} - i_{cg} \tag{4.28}$$

Consequently, for modeling purposes, it is necessary to isolate the first derivative of the converter currents:

$$\frac{di_{af}}{dt} = \frac{1}{L_f} \cdot \left(v_{af} - R_f \cdot i_{af} - v_{aRC} \right) \tag{4.29}$$

$$\frac{di_{bf}}{dt} = \frac{1}{L_f} \cdot \left(v_{bf} - R_f \cdot i_{bf} - v_{bRC} \right) \tag{4.30}$$

$$\frac{di_{cf}}{dt} = \frac{1}{L_f} \cdot \left(v_{cf} - R_f \cdot i_{cf} - v_{cRC} \right) \tag{4.31}$$

And the first derivative of the grid side currents:

$$\frac{di_{ag}}{dt} = \frac{1}{L_g} \cdot \left(v_{aRC} - R_g \cdot i_{ag} - v_{ag} \right) \tag{4.32}$$

$$\frac{di_{bg}}{dt} = \frac{1}{L_g} \cdot \left(v_{bRC} - R_g \cdot i_{bg} - v_{bg} \right) \tag{4.33}$$

$$\frac{di_{cg}}{dt} = \frac{1}{L_g} \cdot \left(v_{cRC} - R_g \cdot i_{cg} - v_{cg} \right) \tag{4.34}$$

Thus, the model of the grid side converter together with the LCL filter and the grid itself is represented in Fig. 4-9. The converter voltage outputs are created with the help of a modulator. The ideal grid sinusoidal voltages are generated at constant amplitude and frequency. Then the converter side and grid side currents are calculated according to (4.29)–(4.34).

4.2.3 DC-link model

The DC side of the power electronic converter is commonly called the DC-link. It stores a certain amount of energy in the capacitor (or association of capacitors). Fig. 4-10 illustrates a possible model for the DC-link. A resistor in parallel with the capacitors bank has been introduced. This R_{bus} is typically of very high resistive value.

Thus, in order to derive the model of the DC-link, the DC bus voltage must be calculated. This voltage is dependent on the current through the capacitor:

$$V_{bus} = \frac{1}{C_{bus}} \cdot \int i_c \cdot dt \tag{4.35}$$

The current through the capacitor can be found as:

$$i_c = i_{r\_dc} - i_{g\_dc} - i_{res} \tag{4.36}$$

Being:

- i_{res}: The current through the resistance (A)
- i_{g_dc}: The DC current flowing from the DC-link to the grid (A)
- i_{r_dc}: The DC current flowing from the left part to the DC-link (A).

On the other hand, the DC current can be calculated as follows from the output AC currents of the converter:

$$i_{g\_dc} = S_{a\_g} \cdot i_{af} + S_{b\_g} \cdot i_{bf} + S_{c\_g} \cdot i_{cf} \tag{4.37}$$

while the current through the resistance:

$$i_{res} = \frac{V_{bus}}{R_{bus}} \tag{4.38}$$

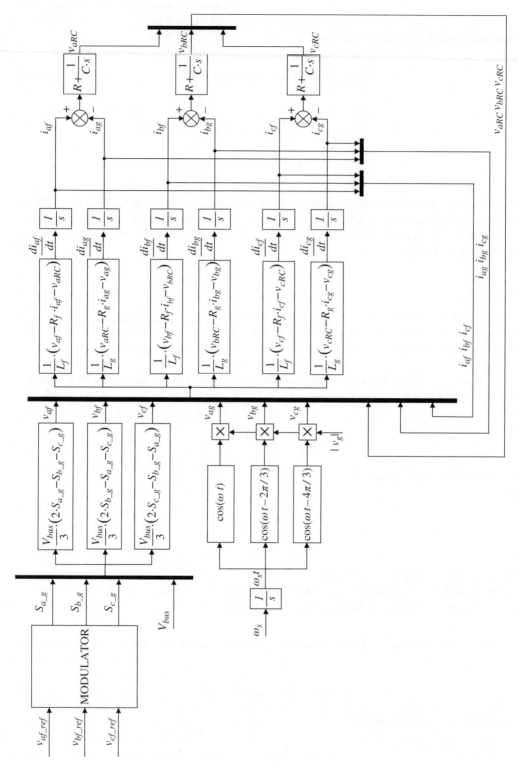

Fig. 4-9 *Simplified converter, LCL filter, and grid model*

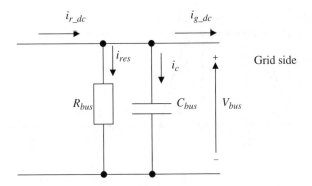

Fig. 4-10 *DC-link system (Source: [1]. Reproduced with permission of John Wiley & Sons)*

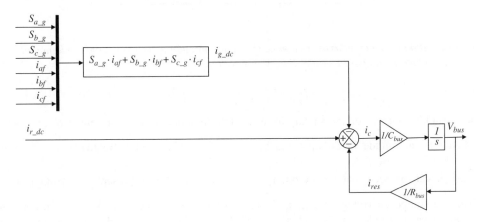

Fig. 4-11 *DC-link model*

Consequently, with all these equations, the model of the DC-link system yields as illustrated in Fig. 4-11. Note that the current i_{r_dc} depends on what it is connected to the left-hand side of the DC-link. If another two-level power electronic converter were to be connected, this current could be calculated in an equivalent way to (4.37).

4.2.4 Sinusoidal PWM with third harmonic injection

This subsection describes one possible modulation solution, for the three-phase power electronic converter. The sinusoidal PWM technique allows different modifications in order to achieve several specific improvements. In this case, an increase of the maximum achievable output voltage amplitude is studied. Hence, by simply adding a third harmonic signal to each of the reference signals, it is possible to obtain a significant amplitude increase at the output voltage without loss of quality, as represented in Fig. 4-12.

It is remarkable that the reference signal resulting from the addition of the third (V_3) and first harmonic (V_1) is smaller in amplitude than the first harmonic. At the output, the obtained amplitude of the first harmonic is equal to the amplitude of the first harmonic reference. Note also that the third harmonic does not appear at the output voltage.

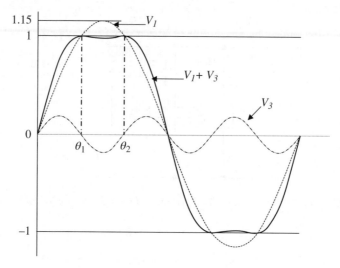

Fig. 4-12   *Third harmonic injection to the reference signal, for amplitude increase of the fundamental component (Source: [1]. Reproduced with permission of John Wiley & Sons)*

Thus, the resultant output voltage signals by using this modulation are represented in Fig. 4-13. It has been employed an amplitude modulation index, $m_a = \dfrac{|V_1^*|}{|v_{tri}|} = 0.75$, and a frequency modulation index, $m_f = \dfrac{f_{tri}}{f_{ref}} = 20$. It can be seen that the output v_{ao}, v_{bo}, and v_{co} voltages present two voltage levels, V_{bus} and 0. Then, the composed voltage v_{ab} presents three voltage levels (V_{bus}, 0, $-V_{bus}$), while the output v_{an} voltage presents five voltage levels ($2V_{bus}/3$, $V_{bus}/3$, 0, $-V_{bus}/3$, $-2V_{bus}/3$). Note that the maximum achievable fundamental component of the v_{an} voltage is given at the linear modulation region:

$$\text{Linear modulation region: } 0 \le m_a \le 1.15 \tag{4.39}$$

On the other hand, the third harmonic injection signal can be easily obtained by means of the following approximated expression:

$$V_3 = -\frac{\max\{v_a^*, v_b^*, v_c^*\} + \min\{v_a^*, v_b^*, v_c^*\}}{2} \tag{4.40}$$

The block diagram for implementation is depicted in Fig. 4-14(a). With this procedure, not only a third harmonic is injected but some other harmonics are also injected, since the resulting injecting signal presents a triangular shape, as depicted in Fig. 4-14(b).

In Fig. 4-15 the spectrum of the converter's output voltage, v_{af}, is depicted, with $m_f = 20$, $m_a = 0.9$, and with third harmonic injection. It is possible to see that apart from the fundamental component harmonic there are several groups of harmonics around the switching frequency and its multiples that, although not desired, will always be present. Thanks to the filter effect, these harmonics will not appear at the current with such a high magnitude, since they will be attenuated.

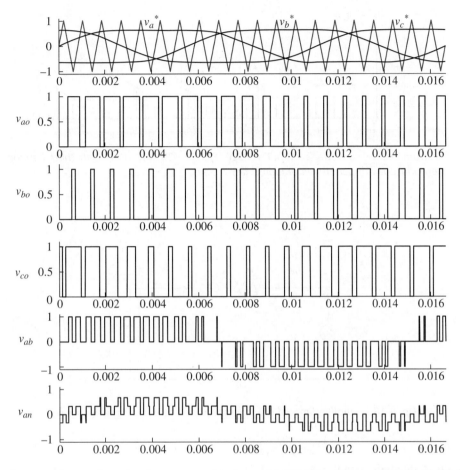

Fig. 4-13   *Output voltages of two-level converter with sinusoidal PWM, m$_f$ = 20, m$_a$ = 0.75, with third harmonic injection*

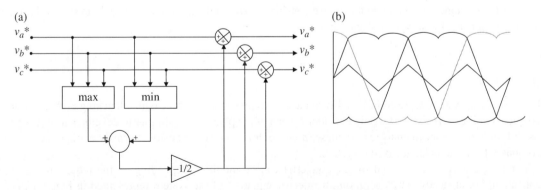

Fig. 4-14   *(a) Block diagram for simplified third harmonic injection, (b) Resultant signals (Source: [1]. Reproduced with permission of John Wiley & Sons)*

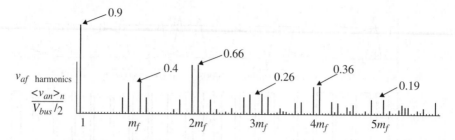

Fig. 4-15   *Spectrum of the converter's output voltage* v_{af}, $m_f = 20$, $m_a = 0.9$, *with third harmonic injection (Source: [1]. Reproduced with permission of John Wiley & Sons)*

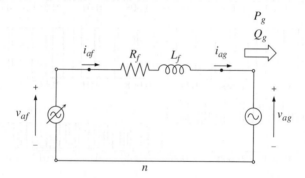

Fig. 4-16   *Simplified model of single-phase grid side system with an L filter*

4.2.5   Steady-state model equations

In this subsection the steady-state model equations of the system are deduced considering the two analyzed filters: the L filter and the LCL filter. As in previous sections the voltages created by the power electronic converter are replaced by three equivalent and variable AC voltage sources: v_{af}, v_{bf}, and v_{cf}. On the other hand, the parasitic resistances of the inductances are considered: (R_f) and (R_g). Finally, the whole system is considered to be connected in a wye configuration.

4.2.5.1   L filter

The three-phase system depicted in Fig. 4-6 can also be represented by three equivalent single-phase systems, as depicted in Fig. 4-16. Thus, by only considering one of the phases, it is simpler to develop subsequent analysis. It is only necessary to analyze one phase (*a* phase, for instance) to be able to easily extrapolate the other two phases from the behavior of the system.

Therefore, the voltage imposed by the converter can be modified depending on the requirements of the application. So, at steady-state operation, the electric equation of the system represented in Fig. 4-16 is:

$$\underline{V}_{af} = \underline{V}_{ag} + \left(R_f + j \cdot L_f \cdot \omega\right) \cdot \underline{I}_{ag} \tag{4.41}$$

Phasor representation is used ($\underline{V}_{af}$, $\underline{V}_{ag}$, and $\underline{I}_{ag}$) for steady-state. The general phasor diagram is constructed as illustrated in Fig. 4-17(a) to (d).

4.2.5.2 *LCL filter*

The same procedure can be followed with the LCL filter, representing the three-phase system depicted in Fig. 4-8 by means of three equivalent single-phase systems. As in the previous analysis, only one-phase is considered (*a* phase, for instance), because it is easy to extrapolate the other two phases from the behavior of the system. The single-phase system is depicted in Fig. 4-18.

At steady-state operation, the electric equation of the system represented in Fig. 4-18 is:

$$\underline{V}_{af} = \underline{V}_{ag} + \left(R_g + j \cdot L_g \cdot \omega\right) \cdot \underline{I}_{ag} + \left(R_f + j \cdot L_f \cdot \omega\right) \cdot \underline{I}_{af} \tag{4.42}$$

Where the relation between the three currents can be defined as:

$$\underline{I}_{af} = \underline{I}_{aRC} + \underline{I}_{ag} \tag{4.43}$$

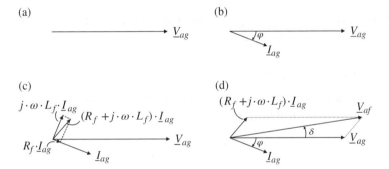

Fig. 4-17 *Phasor diagram construction of the grid side system with an L filter. (a) Grid voltage phasor, (b) grid voltage and current phasors, (c) voltage drops across the filter impedances, (d) converter output voltage phasor (Source: [1]. Reproduced with permission of John Wiley & Sons)*

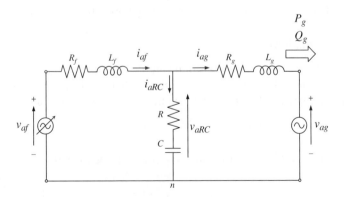

Fig. 4-18 *Simplified model of single-phase grid side system with an LCL filter*

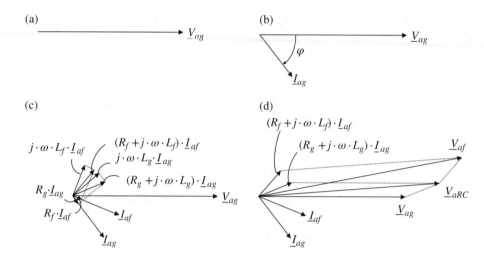

Fig. 4-19 *Phasor diagram construction of the grid side system with an LCL filter. (a) Grid voltage phasor, (b) grid voltage and current phasors, (c) voltage drops across the filter impedances, (d) converter output voltage phasor*

Phasor representation is used ($\underline{V}_{af}$, $\underline{V}_{ag}$, $\underline{I}_{aRC}$, $\underline{I}_{af}$, and $\underline{I}_{ag}$) for steady-state. The general phasor diagram is constructed as illustrated in Fig. 4-19(a) to (d).

4.2.5.3   *Powers transferred to the grid*

Regarding to the active and reactive powers transferred to the grid, they can be calculated using voltages and currents in the grid side terminals and using the same expressions for both filters.

$$P_g = \frac{3}{2} \cdot |\underline{V}_{ag}| \cdot |\underline{I}_{ag}| \cdot \cos\varphi \tag{4.44}$$

$$Q_g = \frac{3}{2} \cdot |\underline{V}_{ag}| \cdot |\underline{I}_{ag}| \cdot \sin\varphi \tag{4.45}$$

The power sign convention shown in Fig. 4-20 is adopted. When $P_g > 0$, the converter is delivering power to the grid. Conversely, when $P_g < 0$, the converter is receiving power from the grid.

4.2.6   **Dynamic model equations**

In Section 4.2.2, the model differential equations of the grid-connected converter are developed, considering two different filters: the L filter and the LCL filter. In this subsection the dynamic mathematical expressions are obtained in both stationary and synchronous reference frames for these two filters.

4.2.6.1   *L filter*

For the inductive filter (the L filter), the differential equations are defined as:

$$v_{af} = R_f \cdot i_{ag} + L_f \cdot \frac{di_{ag}}{dt} + v_{ag} \tag{4.46}$$

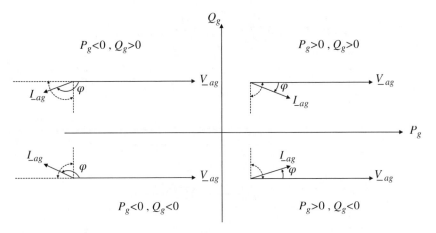

Fig. 4-20 *Convention of power signs (Source: [1]. Reproduced with permission of John Wiley & Sons)*

$$v_{bf} = R_f \cdot i_{bg} + L_f \cdot \frac{di_{bg}}{dt} + v_{bg} \tag{4.47}$$

$$v_{cf} = R_f \cdot i_{cg} + L_f \cdot \frac{di_{cg}}{dt} + v_{cg} \tag{4.48}$$

By applying the space vector theory to these *abc* modeling equations, it is possible to represent these electric equations in $\alpha\beta$ components. Thus, by multiplying (4.46) by $\frac{2}{3}$, then (4.47) by $\frac{2}{3}a$, and (4.48) by $\frac{2}{3}a^2$, being $a = e^{j\frac{2\pi}{3}}$, the addition of the resulting equations yields:

$$v_{\alpha f} = R_f \cdot i_{\alpha g} + L_f \cdot \frac{di_{\alpha g}}{dt} + v_{\alpha g} \tag{4.49}$$

$$v_{\beta f} = R_f \cdot i_{\beta g} + L_f \cdot \frac{di_{\beta g}}{dt} + v_{\beta g} \tag{4.50}$$

This expression can be also represented in only one compact equation, using the space vector notation:

$$\vec{v}_f^s = R_f \cdot \vec{i}_g^s + L_f \cdot \frac{d\vec{i}_g^s}{dt} + \vec{v}_g^s \tag{4.51}$$

Referring the system to a stationary $\alpha\beta$ reference frame, being the space vectors defined in the previous equation, we get:

$$\vec{v}_f^s = v_{\alpha f} + j \cdot v_{\beta f} \tag{4.52}$$

$$\vec{v}_g^s = v_{\alpha g} + j \cdot v_{\beta g} \tag{4.53}$$

$$\vec{i}_g^s = i_{\alpha g} + j \cdot i_{\beta g} \tag{4.54}$$

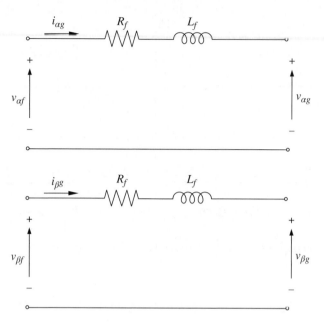

Fig. 4-21   *An αβ model of the grid side system in stationary coordinates using an L filter*

The electric circuit representation of the resulting equations is depicted in Fig. 4-21.

In a similar way, multiplying (4.51) by $e^{-j\theta}$, the *dq* expressions are also derived (referenced to a synchronously rotating reference frame):

$$\vec{v}_f^s \cdot e^{-j\theta} = R_f \cdot \vec{i}_g^s \cdot e^{-j\theta} + L_f \cdot \frac{d\vec{i}_g^s}{dt} \cdot e^{-j\theta} + \vec{v}_g^s \cdot e^{-j\theta} \tag{4.55}$$

resulting in:

$$\vec{v}_f^a = R_f \cdot \vec{i}_g^a + L_f \cdot \frac{d\vec{i}_g^a}{dt} + \vec{v}_g^a + j \cdot \omega_a \cdot L_f \cdot \vec{i}_g^a \tag{4.56}$$

Note that, being $\theta = \omega_a \cdot t$, the angular position of the rotatory reference frame:

$$\frac{d\vec{i}_g^s}{dt} \cdot e^{-j\theta} = \frac{d\left(\vec{i}_g^s \cdot e^{-j\theta}\right)}{dt} + j \cdot \omega_a \cdot \vec{i}_g^s \cdot e^{-j\theta} \tag{4.57}$$

with *dq* components:

$$\vec{v}_f^a = v_{df} + j \cdot v_{qf} \tag{4.58}$$

$$\vec{v}_g^a = v_{dg} + j \cdot v_{qg} \tag{4.59}$$

$$\vec{i}_g^a = i_{dg} + j \cdot i_{qg} \tag{4.60}$$

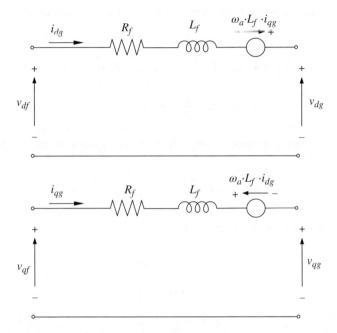

Fig. 4-22 *A dq model of the grid side system in stationary coordinates using an L filter*

Therefore, by decomposing into *dq* components, the basic equations for vector orientation are obtained:

$$v_{df} = R_f \cdot i_{dg} + L_f \cdot \frac{di_{dg}}{dt} + v_{dg} - \omega_a \cdot L_f \cdot i_{qg} \tag{4.61}$$

$$v_{qf} = R_f \cdot i_{qg} + L_f \cdot \frac{di_{qg}}{dt} + v_{qg} + \omega_a \cdot L_f \cdot i_{dg} \tag{4.62}$$

The schematic representation of the equivalent electric circuit is depicted in Fig. 4-22.

4.2.6.2 LCL filter

The same analysis can be carried out for the LCL filter, where the differential equations are defined as:

$$v_{af} = R_f \cdot i_{af} + L_f \cdot \frac{di_{af}}{dt} + R_g \cdot i_{ag} + L_g \cdot \frac{di_{ag}}{dt} + v_{ag} \tag{4.63}$$

$$v_{bf} = R_f \cdot i_{bf} + L_f \cdot \frac{di_{bf}}{dt} + R_g \cdot i_{bg} + L_g \cdot \frac{di_{bg}}{dt} + v_{bg} \tag{4.64}$$

$$v_{cf} = R_f \cdot i_{cf} + L_f \cdot \frac{di_{cf}}{dt} + R_g \cdot i_{cg} + L_g \cdot \frac{di_{cg}}{dt} + v_{cg} \tag{4.65}$$

Applying the αβ transformation, the three-phase system can be expressed by means of two mathematical expressions:

$$v_{\alpha f} = R_f \cdot i_{\alpha f} + L_f \cdot \frac{di_{\alpha f}}{dt} + R_g \cdot i_{\alpha g} + L_g \cdot \frac{di_{\alpha g}}{dt} + v_{\alpha g} \tag{4.66}$$

$$v_{\beta f} = R_f \cdot i_{\beta f} + L_f \cdot \frac{di_{\beta f}}{dt} + R_g \cdot i_{\beta g} + L_g \cdot \frac{di_{\beta g}}{dt} + v_{\beta g} \tag{4.67}$$

This expression can also be represented in only one compact equation, using the space vector notation:

$$\vec{v}_f^s = R_f \cdot \vec{i}_f^s + L_f \cdot \frac{d\vec{i}_f^s}{dt} + R_g \cdot \vec{i}_g^s + L_g \cdot \frac{d\vec{i}_g^s}{dt} + \vec{v}_g^s \tag{4.68}$$

Referring the system to a stationary αβ reference frame, we get:

$$\vec{v}_f^s = v_{\alpha f} + j \cdot v_{\beta f} \tag{4.69}$$

$$\vec{v}_g^s = v_{\alpha g} + j \cdot v_{\beta g} \tag{4.70}$$

$$\vec{i}_f^s = i_{\alpha f} + j \cdot i_{\beta f} \tag{4.71}$$

$$\vec{i}_g^s = i_{\alpha g} + j \cdot i_{\beta g} \tag{4.72}$$

The electric circuit representation of the resulting equations is depicted in Fig. 4-23.

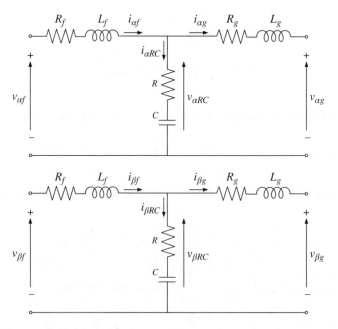

Fig. 4-23   *An αβ model of the grid side system in stationary coordinates using an LCL filter*

In a similar way, multiplying (4.68) by $e^{-j\theta}$, the *dq* expressions are also derived (referenced to a synchronously rotating reference frame):

$$\vec{v}_f^s \cdot e^{-j\theta} = R_f \cdot \vec{i}_f^s \cdot e^{-j\theta} + L_f \cdot \frac{d\vec{i}_f^s}{dt} \cdot e^{-j\theta} + R_g \cdot \vec{i}_g^s \cdot e^{-j\theta} + L_g \cdot \frac{d\vec{i}_g^s}{dt} \cdot e^{-j\theta} + \vec{v}_g^s \cdot e^{-j\theta} \tag{4.73}$$

resulting in:

$$\vec{v}_f^a = R_f \cdot \vec{i}_f^a + L_f \cdot \frac{d\vec{i}_f^a}{dt} + R_g \cdot \vec{i}_g^a + L_g \cdot \frac{d\vec{i}_g^a}{dt} + \vec{v}_g^a + j \cdot \omega_a \cdot L_f \cdot \vec{i}_f^a + j \cdot \omega_a \cdot L_g \cdot \vec{i}_g^a \tag{4.74}$$

Note that, being $\theta = \omega_a \cdot t$, the angular position of the rotatory reference frame:

$$\frac{d\vec{i}_f^s}{dt} \cdot e^{-j\theta} = \frac{d\left(\vec{i}_f^s \cdot e^{-j\theta}\right)}{dt} + j \cdot \omega_a \cdot \vec{i}_f^s \cdot e^{-j\theta} \tag{4.75}$$

$$\frac{d\vec{i}_g^s}{dt} \cdot e^{-j\theta} = \frac{d\left(\vec{i}_g^s \cdot e^{-j\theta}\right)}{dt} + j \cdot \omega_a \cdot \vec{i}_g^s \cdot e^{-j\theta} \tag{4.76}$$

with *dq* components:

$$\vec{v}_f^a = v_{df} + j \cdot v_{qf} \tag{4.77}$$

$$\vec{v}_g^a = v_{dg} + j \cdot v_{qg} \tag{4.78}$$

$$\vec{i}_f^a = i_{df} + j \cdot i_{qf} \tag{4.79}$$

$$\vec{i}_g^a = i_{dg} + j \cdot i_{qg} \tag{4.80}$$

Therefore, by decomposing into *dq* components, the basic equations for vector orientation are obtained:

$$v_{df} = R_f \cdot i_{df} + L_f \cdot \frac{di_{df}}{dt} + R_g \cdot i_{dg} + L_g \cdot \frac{di_{dg}}{dt} + v_{dg} - \omega_a \cdot L_f \cdot i_{qf} - \omega_a \cdot L_g \cdot i_{qg} \tag{4.81}$$

$$v_{qf} = R_f \cdot i_{qf} + L_f \cdot \frac{di_{qf}}{dt} + R_g \cdot i_{qg} + L_g \cdot \frac{di_{qg}}{dt} + v_{qg} + \omega_a \cdot L_f \cdot i_{df} + \omega_a \cdot L_g \cdot i_{dg} \tag{4.82}$$

The schematic representation of the equivalent electric circuit is depicted in Fig. 4-24.

4.2.7   LCL filter analysis and design

This section briefly analyzes the LCL filter of a grid-connected converter. It must be remarked that in order to cope with the current quality grid codes or standards, such as the IEEE 519-1992, it is typically necessary to use

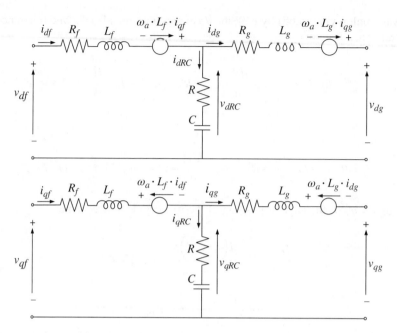

Fig. 4-24 *A dq model of the grid side system in stationary coordinates using an LCL filter*

an LCL filter. This type of filter often allows the current ripple or current harmonics exchanged with the grid to be reduced more effectively and efficiently than does the simpler L filter configuration.

Hence, first of all it is useful to mathematically derive the input/output and voltage/currents relations of the LCL filter. From the basic voltage and current equations of the filter and rearranging them into a matrix form, it is possible to derive the following:

$$\begin{bmatrix} I_f(s) \\ I_g(s) \end{bmatrix} = \begin{bmatrix} G_{11}(s) & G_{12}(s) \\ G_{21}(s) & G_{22}(s) \end{bmatrix} \cdot \begin{bmatrix} V_f(s) \\ V_g(s) \end{bmatrix}$$

(4.83)

being I_f, I_g, V_f, and V_g, the Laplace transformation of the converter current, grid current, converter voltage, and grid voltage respective of phases *abc*. The transfer functions G_{11}, G_{12}, G_{21}, and G_{22} after applying the Laplace transformation are defined in Table 4-2. Note that the equivalent electric circuit of the filter considered at the analysis neglects the parasitic resistance of the inductances, as depicted in Fig. 4-25.

Therefore, it is possible to see how the currents at both sides of the filter depend on the voltages also at both sides of the filter and on the filter parameters. By paying attention to the transfer functions G_{11}, G_{12}, G_{21}, and G_{22}, one can see that they share an equal denominator. By neglecting the effect of the damping resistance R, from the roots of the denominator it is possible to find the following expression for the resonance frequency of the filter in Hz:

$$f_{resonance} \approx \frac{1}{2\pi} \sqrt{\frac{(L_f + L_g)}{L_f \cdot L_g \cdot C}}$$

(4.84)

Table 4-2 *LCL filter transfer functions*

Name	Related variables	Transfer function
$G_{11}(s)$	$\dfrac{I_f(s)}{V_f(s)}$	$\dfrac{L_g \cdot C \cdot s^2 + R \cdot C \cdot s + 1}{L_f \cdot L_g \cdot C \cdot s^3 + R \cdot C \cdot \left(L_f + L_g\right) \cdot s^2 + \left(L_f + L_g\right) \cdot s}$
$G_{12}(s)$	$\dfrac{I_f(s)}{V_g(s)}$	$\dfrac{-\left(R \cdot C \cdot s + 1\right)}{L_f \cdot L_g \cdot C \cdot s^3 + R \cdot C \cdot \left(L_f + L_g\right) \cdot s^2 + \left(L_f + L_g\right) \cdot s}$
$G_{21}(s)$	$\dfrac{I_g(s)}{V_f(s)}$	$\dfrac{R \cdot C \cdot s + 1}{L_f \cdot L_g \cdot C \cdot s^3 + R \cdot C \cdot \left(L_f + L_g\right) \cdot s^2 + \left(L_f + L_g\right) \cdot s}$
$G_{22}(s)$	$\dfrac{I_g(s)}{V_g(s)}$	$\dfrac{-\left(L_f \cdot C \cdot s^2 + R \cdot C \cdot s + 1\right)}{L_f \cdot L_g \cdot C \cdot s^3 + R \cdot C \cdot \left(L_f + L_g\right) \cdot s^2 + \left(L_f + L_g\right) \cdot s}$

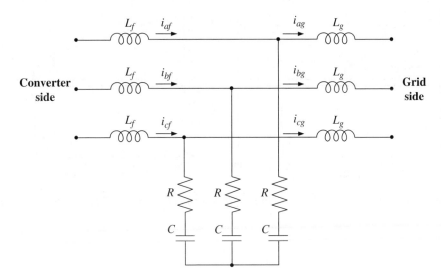

Fig. 4-25 *Equivalent electric circuit of the filter considered at the analysis neglecting the parasitic resistance of the inductances*

The Bode diagrams of the four transfer functions (G_{11}, G_{12}, G_{21}, and G_{22}) are depicted in Fig. 4-26(a) to (d) for a filter with parameters: $L_f = 60$uH, $L_g = 40$uH, $C = 500$uF, and $R = 20$mΩ. The resonance frequency is situated approximately at 1.45kHz. In these Bode diagrams it is possible to see how attenuated the grid side current harmonics are, in function of the harmonics generated by the converter. In general the switching frequency of the converter is located at the region of –40dB/decade of the filter, to the right side of the resonance frequency. Thus, for instance in this specific example, if the switching frequency of the two-level converter is 4kHz, the family of harmonics generated at the converter voltage, $V_f(s)$, owing to the switching frequency, will be attenuated approximately 23dB at the grid side current, $I_g(s)$, as can be seen in Fig. 4-26(c). On the contrary, if, for instance, the switching frequency of the converter was 1.5kHz, the chosen filter would not effectively attenuate the first family of harmonics at the grid side currents, since they would be amplified instead of being reduced.

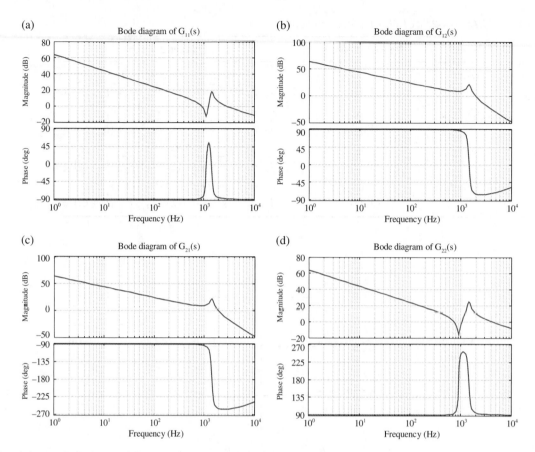

Fig. 4-26   *Bode diagrams of the* G_{11}, G_{12}, G_{21}, *and* G_{22} *transfer functions of an LCL filter, with values of parameters:* $L_f = 60uH$, $L_g = 40uH$, $C = 500uF$, *and* $R = 20m\Omega$

This detail can be observed in Fig. 4-26(c). Consequently, the grid side converter would not be able to operate properly.

In addition, in Fig. 4-27(a) to (d), Bode diagrams of transfer function G_{21} for the same filter are depicted when the values of the parameters of the filter are modified. Thus in Fig. 4-27(a) the grid side inductance (L_g) is set to values: 80uH, 40uH, and 20uH and $L_f = 60uH$, $C = 500uF$, and $R = 20m\Omega$ are left to their original values. Then, in Fig. 4-27(b) the converter side inductance (L_f) is set to values: 120uH, 60uH, and 30uH, and $L_g = 40uH$, $C = 500uF$, and $R = 20m\Omega$ are left to their original values. After that, in Fig. 4-27(c) the capacitance is set to values: 1000uF, 500uF, and 250uF, and $L_f = 60uH$, $L_g = 40uH$, and $R = 20m\Omega$ are left to their original values. In this scenario, it is possible to intuitively see how the resonance frequency moves according to the filter parameters. Then, if the switching frequency of the converter is located at a higher frequency than the resonance frequency, the harmonics produced by the switching frequency can be attenuated. It can be seen that, in general, the bigger we choose values for L_f, L_g, and C, lower frequency the resonance will be located and therefore a larger attenuation will be applied to the harmonics generated by the switching frequency of the converter. However, as will be seen later in the chapter, in general there are some other factors which

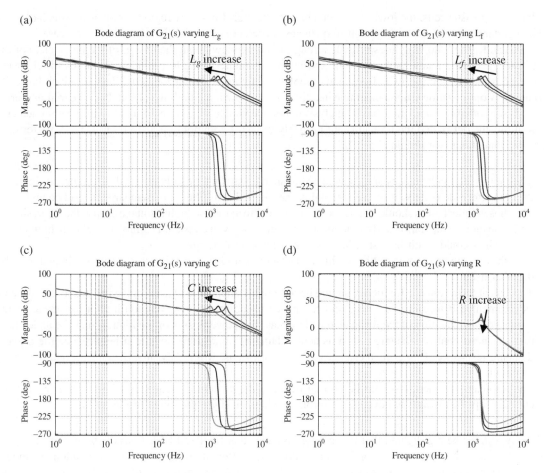

Fig. 4-27 *Bode diagrams of transfer function* G_{21}, *with values of original LCL filter:* $L_f = 60uH$, $L_g = 40uH$, $C = 500uF$, *and* $R = 20m\Omega$. *(a)* L_g: *80uH, 40uH, and 20uH, and* $L_f = 60uH$, $C = 500uF$, *and* $R = 20m\Omega$, *(b)* L_f: *120uH, 60uH, and 30uH, and* $L_g = 40uH$, $C = 500uF$, *and* $R = 20m\Omega$, *(c) C: 1000uF, 500uF, and 250uF, and* $L_f = 60uH$, $L_g = 40uH$, *and* $R = 20m\Omega$, *(d) R: 40mΩ, 20mΩ, and 10mΩ, and* $L_f = 60uH$, $L_g = 40uH$, *and* $C = 500uF$

recommend not bringing the resonance frequency too far to the left. One of them, for instance, is not to introduce the resonance frequency within the bandwidth of the current loops of the converter in order to avoid dangerous instabilities. Thus if, for instance, the application requires a current control bandwidth of 500Hz, the resonance frequency of the LCL filter should be of a greater value.

Finally, in Fig. 4-27(d) the damping resistance is set to values: 40mΩ, 20mΩ, and 10mΩ, and the rest $L_f = 60uH$, $L_g = 40uH$, and $C = 500uF$ are left to their original values. Increasing damping resistance helps to attenuate the gain at the resonance frequency, basically mitigating instabilities and amplified resonances when the nonlinearities of the converter or even of the control creates harmonics at that specific frequency. A minimum damping resistance in general is always necessary to avoid a dangerous resonance with a very large amplification ratio; however, conversely, the larger the resistance we choose, the larger the power losses we generate at the filter. Therefore, an effective balance must be carefully achieved. In addition, note also that the larger the resistance is, the lower the attenuation at the high-frequency region is obtained. This fact is not very important, since at those very high frequencies sufficient attenuation is guaranteed in all cases.

To conclude, the following lines provide a rough intuitive view about the design of the parameters of the LCL filter. There are many methods in the specialized literature trying to provide procedures for choosing optimized parameters of the LCL filter. This problem is a complex multi-objective task in which basically we can apply optimization criteria or simple iterative trial-and-error methods to solve it. Iterative trial-and-error methods supported by simulation-based analysis to deduce whether the ripple or quality of the obtained currents fulfills the demanded requirements are probably the most commonly practiced ones. Thus, performing iterative simulations, we should check [2]:

- Choose the filter inductances and capacitance in such a way that they obtain the desired current ripples at the converter side (safety of converter) and at the grid side (fulfilment of standards or grid codes).
- The chosen inductances should not be too high, in order to avoid too high a voltage drop at the filter, since it limits the range of operation and exchangeable power for a given constant DC-link. In addition, high inductance values could result in costly, large-volume inductance designs.
- Also the capacitance should not be too high, in order to avoid an excessive reactive power interchange by the filter itself.
- The gain at the resonance frequency should be reasonably attenuated by a damping resistance, for instance as commented before, in order to avoid resonance instabilities. However, too high a resistance which brings good attenuation of the resonance will increase the power losses of the filter. It must be added here that advanced control strategies, such as active damping control methods, can help reduce the passive damping resistance. However, these active control methods, while they do exist, are beyond the scope of this chapter.
- The resonance frequency should be chosen between the switching frequency of the converter and the current loops bandwidth achieved by control, as schematically represented in Fig. 4-28.

Finally, it must be highlighted that this chapter assumes that the fixed LCL filter configuration shown in Fig. 4-25 is used; however, other configurations do exist. Different combinations or dispositions of inductances, capacitances, and resistances can result in more-efficient and optimum filter configurations, in very specific circumstances or when there are certain restrictions, such as when a very low switching frequency is required for a medium-voltage application.

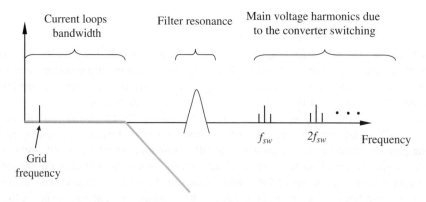

Fig. 4-28 *Schematic disposition of different key frequencies of the grid side converter*

4.3   Three-phase grid-connected converter control

The model of the three-phase grid-connected converter is described in previous sections. This section describes the control of that converter, namely the vector control strategy, as applied to the motors discussed in Chapters 2 and 3.

4.3.1   Alignment with the grid voltage space vector

The vector control strategy of the grid-connected converter is typically developed from the system modeled in dq components. Thus, in order to further simplify the model equations (4.61)–(4.62), typically, the space vector of the grid voltage $\vec{v}_g$ is aligned with the d axis of the dq rotating reference frame, as illustrated in Fig. 4-29. Thus, the resulting dq components of the grid voltage yield:

$$v_{dg} = |\vec{v}_g| \tag{4.85}$$

$$v_{qg} = 0 \tag{4.86}$$

Being the synchronous rotating angular speed of the reference frame equal to the angular speed of the grid voltages:

$$\theta = \omega_a \cdot t \Rightarrow \theta = \theta_g = \omega \cdot t \tag{4.87}$$

Therefore, (4.61) and (4.62) for the L filter are simplified to:

$$v_{df} = R_f \cdot i_{dg} + L_f \cdot \frac{di_{dg}}{dt} + v_{dg} - \omega \cdot L_f \cdot i_{qg} \tag{4.88}$$

$$v_{qf} = R_f \cdot i_{qg} + L_f \cdot \frac{di_{qg}}{dt} + \omega \cdot L_f \cdot i_{dg} \tag{4.89}$$

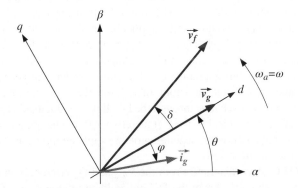

Fig. 4-29   *Alignment with the d axis of the grid voltage space vector*

And (4.81) and (4.82) for the LCL filter are simplified to:

$$v_{df} = R_f \cdot i_{df} + L_f \cdot \frac{di_{df}}{dt} + R_g \cdot i_{dg} + L_g \cdot \frac{di_{dg}}{dt} + v_{dg} - \omega \cdot L_f \cdot i_{qf} - \omega \cdot L_g \cdot i_{qg} \tag{4.90}$$

$$v_{qf} = R_f \cdot i_{qf} + L_f \cdot \frac{di_{qf}}{dt} + R_g \cdot i_{qg} + L_g \cdot \frac{di_{qg}}{dt} + \omega \cdot L_f \cdot i_{df} + \omega \cdot L_g \cdot i_{dg} \tag{4.91}$$

This grid voltage alignment not only slightly simplifies the voltage equations of the system but also reduces the active and reactive power computations. Thus, if the active and reactive powers exchanged with the grid are calculated:

$$P_g = \frac{3}{2}\mathrm{Re}\left\{\vec{v}_g \cdot \vec{i}_g^{\,*}\right\} = \frac{3}{2}\left(v_{dg} \cdot i_{dg} + v_{qg} \cdot i_{qg}\right) \tag{4.92}$$

$$Q_g = \frac{3}{2}\mathrm{Im}\left\{\vec{v}_g \cdot \vec{i}_g^{\,*}\right\} = \frac{3}{2}\left(v_{qg} \cdot i_{dg} - v_{dg} \cdot i_{qg}\right) \tag{4.93}$$

Considering relations (4.85) and (4.86), the power calculation can be simplified to:

$$P_g = \frac{3}{2} \cdot v_{dg} \cdot i_{dg} = \frac{3}{2} \cdot |\vec{v}_g| \cdot i_{dg} \tag{4.94}$$

$$Q_g = -\frac{3}{2} \cdot v_{dg} \cdot i_{qg} = -\frac{3}{2} \cdot |\vec{v}_g| \cdot i_{qg} \tag{4.95}$$

Note that the voltage terms of these last two expressions are constant under ideal conditions. Thus, decoupled expressions for the powers are obtained and it will be seen that the i_{dg} current is responsible for the P_g, while the i_{qg} current is responsible for Q_g.

4.3.2   Vector control strategy

The vector control strategy applied to the grid side connected converter is typically in charge of controlling the DC bus voltage of the DC-link (V_{bus}) and the reactive power exchange with the grid (Q_g), as depicted in Fig. 4-30. This control strategy is in charge of generating the pulses for the IGBTs of the converter. The vector control strategy, since it uses an alignment or orientation with the grid voltage space vector, is called a grid voltage oriented vector control (GVOVC). The vector control presented in this section is developed for a two-level converter; however, it would also be applicable to any other three-phase converter topology, for instance a multilevel converter topology. This converter must control the DC bus voltage of the converter (V_{bus})—whether the grid side converter presents another voltage source converter connected to the DC-link or even if nothing is connected. This will guarantee that the converter has the necessary DC-bus voltage for proper operation and to synthesize AC voltage output. In addition, thanks to the control of the DC bus voltage, this will ensure that the active power exchange through the DC-link is maintained properly. With regards to the reactive power reference value, it is typical to leave it at zero VAR, exchanging currents with a unity power factor at the grid voltage terminals. However, depending on the application, it can also be necessary to operate at reactive power values different from zero.

Fig. 4-31 depicts a GVOVC block diagram. From the V_{bus} and Q_g references, it creates the pulses for the controlled switches S_{a_g}, S_{b_g}, and S_{c_g}. From the DC bus voltage reference, its corresponding PI regulator generates the active power reference value (P_g^*) that later is transformed into a d component of grid current

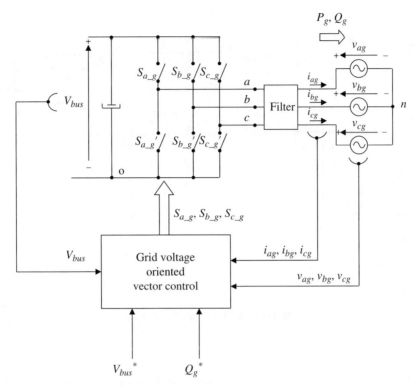

Fig. 4-30   *Grid-connected converter control*

reference (i_{dg}^*). On the other hand, form the reactive power reference value (Q_g^*), the q component of the grid current reference is generated (i_{qg}^*). The constant terms needed to transform from powers to dq current reference components are easily deduced from (4.94) and (4.95):

$$K_{Pg} = \frac{1}{\frac{3}{2} \cdot v_{dg}} \tag{4.96}$$

$$K_{Qg} = \frac{1}{-\frac{3}{2} \cdot v_{dg}} \tag{4.97}$$

Then, once the dq current reference components are obtained, they are passed by their corresponding PI regulators. Note that the cancelation of the coupling terms is also carried out at the output of the current PI regulators. The terms employed for the cancelation are deduced from (4.88) and (4.89) for the L filter:

$$e_{df} = -\omega \cdot L_f \cdot i_{qg} + v_{dg} \tag{4.98}$$

$$e_{qf} = \omega \cdot L_f \cdot i_{dg} \tag{4.99}$$

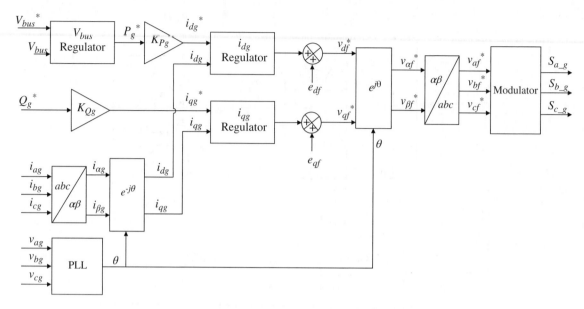

Fig. 4-31   *GVOVC block diagram*

On the other hand, the coupling terms for the LCL filter are deduced from (4.90) and (4.91):

$$e_{df} = -\omega \cdot L_f \cdot i_{qf} - \omega \cdot L_g \cdot i_{qg} + v_{dg} \tag{4.100}$$

$$e_{qf} = \omega \cdot L_f \cdot i_{df} + \omega \cdot L_g \cdot i_{dg} \tag{4.101}$$

In an ideal case, the proper cancelation of the coupling terms using the LCL filter would be achieved measuring both the converter side and the grid side currents (i_f and i_g), because both currents are involved in these coupling terms, as can be seen in (4.100) and (4.101). Assuming that both currents (i_f and i_g) have similar fundamental values, an approximation can be implemented minimizing the coupling terms for both cases where only the grid side current or the converter side current is measured. This approximation is derived from (4.100) and (4.101).

$$e_{df} \approx -\left(L_f + L_g\right)\cdot\omega\cdot i_{qf} + v_{dg} \approx -\left(L_f + L_g\right)\cdot\omega\cdot i_{qg} + v_{dg} \tag{4.102}$$

$$e_{qf} \approx \left(L_f + L_g\right)\cdot\omega\cdot i_{df} \approx \left(L_f + L_g\right)\cdot\omega\cdot i_{dg} \tag{4.103}$$

Then, by means of the rotational transformation and the Clarke transformation, the voltage references for the three-phase converter are generated. Finally, with the modulator, these voltage references are transformed into order commands for the IGBTs of the converter. The modulation strategy that can be employed could be, for instance, the one seen in Section 4.2.4, the sinusoidal PWM with third harmonic injection.

For voltage and currents coordinate transformations the angle of the grid voltage as depicted in Fig. 4-31 is needed: θ. In general, this angle is estimated by a phase-locked loop (PLL). Its closed-loop nature provides stability and perturbation rejections to the angle estimation.

4.3.3   Synchronization method

In the specialized literature, it is possible to find many synchronization methods analyzed. This section describes a simple but at the same time effective synchronization method: the PLL. Its simplified block diagram is illustrated in Fig. 4-32. The PLL seeks synchronization to a sinusoidally varying three-phase system, in this case the grid three-phase voltages. This PLL is synchronized by using the dq coordinates of the voltage. Thus, the d component of the grid voltage (v_{dg}) must be aligned with the d rotating reference frame, which means that the estimated θ must be modified until the q component of the voltage (v_{qg}) is zero. Thus, the abc voltages are taken and transformed into dq voltages by their own estimated angle. After that, the calculated q component of the voltage is passed through a PI regulator, modifying the estimated angular speed until the q component is made zero, which means that the synchronization process has been stabilized. This PLL must be continuously running with the control strategy, since the angle is not a constant magnitude but is modified according to the angular speed of the grid voltage. The dynamic of the synchronization process can be altered by tuning the PI controller.

4.3.4   Tuning of the current regulators

Since the control strategy is described in previous subsections, this subsection analyzes how the closed-loop system of the current loops behaves for both the L filter and the LCL filter.

4.3.4.1   L filter

From the voltage expressions (4.88) and (4.89), thanks to the cancelation of the coupling terms carried out (see Fig. 4-31), the current versus voltage equations of the electric model equations at open loop for the L filter yield:

$$v_{df} = R_f \cdot i_{dg} + L_f \cdot \frac{di_{dg}}{dt} \tag{4.104}$$

$$v_{qf} = R_f \cdot i_{qg} + L_f \cdot \frac{di_{qg}}{dt} \tag{4.105}$$

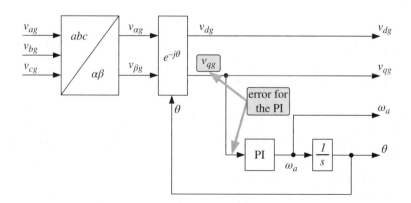

Fig. 4-32   *Classic PLL structure block diagram*

which means that if we apply the Laplace transformation the following transfer functions are derived:

$$\frac{i_{dg}(s)}{v_{df}(s)} = \frac{1}{L_f s + R_f} \tag{4.106}$$

$$\frac{i_{qg}(s)}{v_{qf}(s)} = \frac{1}{L_f s + R_f} \tag{4.107}$$

Thus, this fact can be represented in block diagrams, as illustrated in Fig. 4-33. With this approach, an idealized control strategy has been considered, where aspects such as delays of the converter and measurements, nonlinearities of the converter, harmonics presence, etc. have been neglected. Note also that the adopted regulator for the current control loops is an idealized PI regulator.

Thus, the current closed loops can be represented ideally according to the following transfer functions:

$$\frac{i_{ds}(s)}{i_{ds}^*(s)} = \frac{sk_p + k_i}{s^2(L_f) + s(R_f + k_p) + k_i} \tag{4.108}$$

$$\frac{i_{qs}(s)}{i_{qs}^*(s)} = \frac{sk_p + k_i}{s^2(L_f) + s(R_f + k_p) + k_i} \tag{4.109}$$

Thanks to these closed-loop current expressions, it is possible to tune the gains of the current PI regulators by following, for instance, the same procedure employed in Chapter 2. Hence, by equaling the denominators of these derived transfer functions, to the standard second-order denominator of classic control theories, we obtain:

$$s^2(L_f) + s(R_f + k_p) + k_i \equiv s^2 + 2\xi\omega_n s + \omega_n^2 \tag{4.110}$$

which means that the required k_p and k_i values are:

$$k_i = L_f \omega_n^2 \tag{4.111}$$

$$k_p = L_f 2\xi\omega_n - R_f \tag{4.112}$$

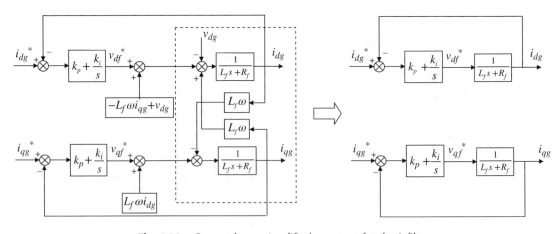

Fig. 4-33 *Current loops simplified structure for the L filter*

And ω_n and ξ values can be chosen by any of the pole placement methods of classic control theories.

4.3.4.2 LCL filter

The tuning of the current regulators in the case of an LCL filter is developed in a similar way. In this case the same approximation made in (4.102) and (4.103) can be implemented, assuming that both the converter side current and the grid side current (i_f and i_g) have similar fundamental values. Therefore, using the voltage equations (4.90) and (4.91) for the LCL filter, and owing to the cancelation of the coupling terms (see block diagram of Fig. 4-31), the converter voltage (v_f) can be expressed, in a simplified way, only as a function of the grid side current (i_g):

$$v_{df} \approx R_f \cdot i_{dg} + L_f \cdot \frac{di_{dg}}{dt} + R_g \cdot i_{dg} + L_g \cdot \frac{di_{dg}}{dt} \tag{4.113}$$

$$v_{qf} \approx R_f \cdot i_{qg} + L_f \cdot \frac{di_{qg}}{dt} + R_g \cdot i_{qg} + L_g \cdot \frac{di_{qg}}{dt} \tag{4.114}$$

Applying the Laplace transformation, the following transfer functions are obtained:

$$\frac{i_{dg}(s)}{v_{df}(s)} \approx \frac{1}{(L_f + L_g)s + (R_f + R_g)} \tag{4.115}$$

$$\frac{i_{qg}(s)}{v_{qf}(s)} \approx \frac{1}{(L_f + L_g)s + (R_f + R_g)} \tag{4.116}$$

As developed for the L filter, the closed loops can be represented by means of the following transfer functions:

$$\frac{i_{ds}(s)}{i_{ds}{}^*(s)} \approx \frac{sk_p + k_i}{s^2(L_f + L_g) + s(R_f + R_g + k_p) + k_i} \tag{4.117}$$

$$\frac{i_{qs}(s)}{i_{qs}{}^*(s)} \approx \frac{sk_p + k_i}{s^2(L_f + L_g) + s(R_f + R_g + k_p) + k_i} \tag{4.118}$$

Therefore, equaling the denominators of the transfer functions to the second-order standard denominator of classic control theories, we obtain:

$$s^2(L_f + L_g) + s(R_f + R_g + k_p) + k_i \equiv s^2 + 2\xi\omega_n s + \omega_n{}^2 \tag{4.119}$$

Finally, the required k_p and k_i values are obtained:

$$k_i = (L_f + L_g)\omega_n{}^2 \tag{4.120}$$

$$k_p = (L_f + L_g)2\xi\omega_n - (R_f + R_g) \tag{4.121}$$

As in the previous case, ω_n and ξ values can be chosen by any of the pole placement methods of classic control theories. As an illustrative example, Fig. 4-34 shows the Bode diagrams of d and q open-loop transfer

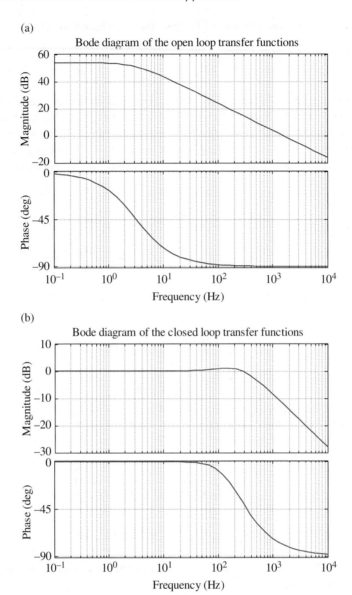

Fig. 4-34 *(a) Bode diagram of d and q open-loop transfer functions represented by (4.115) and (4.116), with filter values $L_c = 60uH$, $R_c = 1m\Omega$, $L_g = 40uH$, and $R_g = 1m\Omega$, (b) Bode diagram of d and q closed-loop current transfer functions represented by (4.117) and (4.118), with filter values $L_c = 60uH$, $R_c = 1m\Omega$, $L_g = 40uH$, and $R_g = 1m\Omega$ (C and R of LCL filter are neglected in this analysis)*

functions represented by (4.115) and (4.116) and closed-loop current transfer functions represented by (4.117) and (4.118), with filter values $L_c = 60uH$, $R_c = 1m\Omega$, $L_g = 40uH$, and $R_g = 1m\Omega$. Note that C and R of the LCL filter are neglected with this analysis. It is possible to observe at these Bode diagrams that by closing the loop the d and q currents are controlled increasing the bandwidth considerably. In this particular example, the $\xi = 1$

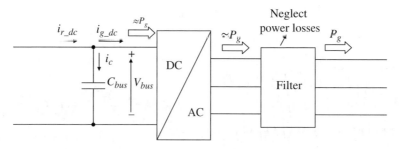

Fig. 4-35   *Assumption of conservation of the active power from DC to AC and to grid, in grid side converter*

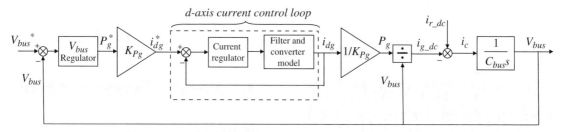

Fig. 4-36   *Simplified representation of DC bus voltage control loop*

and $\omega_n = 2\pi200$rad/sec has been chosen, which means that we obtain two equal real poles. For that purpose, according to (4.120) and (4.121), k_p and k_i gains results in 0.24 and 157.91 respectively.

On the other hand, with regards to the DC bus voltage loop, it must be remarked that this loop cannot be modeled with linear differential equations. However, it is possible to model it in a simplified way as the current control loops, by neglecting the power losses of the filter and assuming that the power delivered to the grid (P_g) is equal to the power provided at the DC side by the grid side converter, as depicted in Fig. 4-35.

In this way, assuming also that there is no power loss at the DC bus, the simplified representation of the DC bus voltage control loop results in the block diagram depicted in Fig. 4-36. It is possible to observe that the transformation from active power to DC current is nonlinear. We could, here, apply linearization techniques to this system to analyze the behavior of the system, but this would go far beyond the scope of this chapter.

4.3.5   Simulation-based example

This section examines an example representing the most representative magnitudes of a 15kW grid-connected converter (two-level converter), simulated with a Matlab-Simulink computer-based simulator. The converter is connected to a 380V grid. The grid side filter is a pure inductive L filter (with parasitic resistance). The simulation experiments show the steady-state and the transient performance of the grid side converter's magnitudes when it is controlled by the vector control presented in the previous sections.

Thus, Fig. 4-37(a) illustrates the grid voltage in pu where the converter system is connected. Fig. 4-37(b) shows the DC bus voltage performance when it is controlled by the converter itself. At time instant 1.5 sec, there is a reactive power demand variation from 0 to 0.4 in pu, which produces a small perturbation at the DC

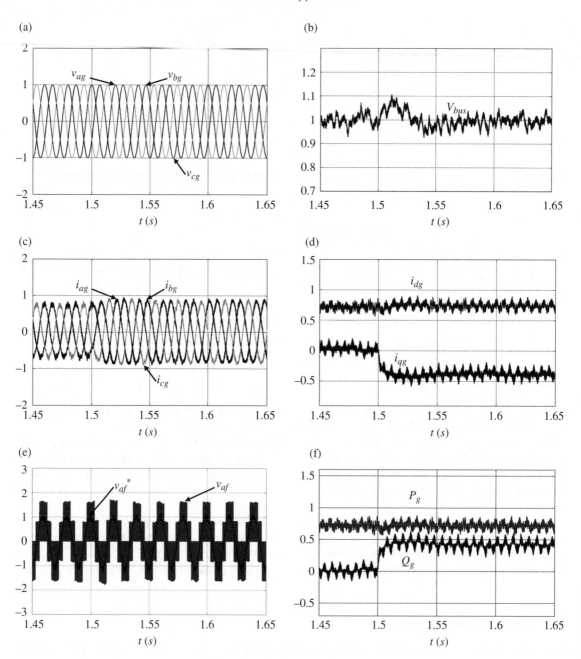

Fig. 4-37 *Most representative magnitudes of a grid-connected converter of 15 kW: (a) grid voltages in pu, (b) DC bus voltage in pu, (c) abc grid side converter currents in pu, (d) dq grid side converter currents in pu, (e) grid side converter output voltage in pu, and (f) power exchange with the grid in pu*

bus voltage. On the other hand, Fig. 4-37(c) shows the currents exchange with the grid. It will be noticed how after the reference variation of the reactive power the grid current's magnitude has been increased according to the power increase. In Fig. 4-37(d), the d and q components of the grid currents are presented. It will be seen that the reactive power variation produces a step variation of i_{qg}, and i_{dg} is not affected by the transient on the whole. Finally, Fig. 4-37(e) and (f) show the voltage generated by phase a of the converter and the active and reactive power exchange with the grid. It will be seen how the converter voltage behaves at steady state and how it is not affected transitorily by power variation.

4.4   Three-phase grid-connected converter control under unbalanced voltage conditions

This section deals with the control strategy that is required when the grid-connected converter is connected to a grid which presents unbalanced voltages. Under these exceptional unbalanced operating conditions, the control strategy analyzed in the previous subsection (i.e. the vector control strategy) must be especially adapted to this scenario. In general, the employed modified control strategy is specially designed to deal with the unbalance problem. Hence, this section describes the vector-control-based control strategy adapted to operate under unbalanced voltage conditions. Therefore, first of all a short overview of the unbalance theory developed to deal with unbalanced three-phase magnitudes is presented. Later, the control strategy itself is presented.

4.4.1   Unbalanced three-phase systems

A three-phase unbalanced voltage or current can take the shape of the waveforms presented in Fig. 4-38. Since Fortescue, in the early 20th century, developed his theory to represent unbalanced three-phase systems [3], most of the control strategies and theories dealing with unbalanced systems have utilized this theory. Thus, Fortescue demonstrated that a three-phase unbalanced system could be represented as the superposition of three balanced three-phase systems: one positive sequence system, one negative sequence system, and a zero sequence system, as illustrated in Fig. 4-39.

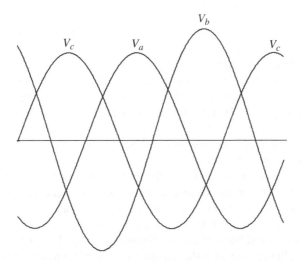

Fig. 4-38   *Three-phase unbalanced voltage*

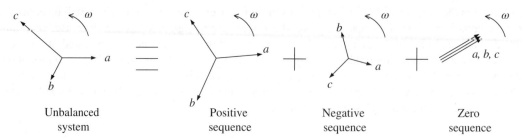

Fig. 4-39   *Decomposition of a three-phase unbalanced system, into positive, negative, and zero sequences*

As will be seen in Fig. 4-39, the positive sequence presents *abc* sequence of phases. The negative sequence presents *acb* sequence of phases, while the zero sequence presents three equal voltages not phase shifted.

The positive, negative, and zero sequences are related to the original *abc* three-phase system according to the following expression, in phasor representation:

$$
\begin{bmatrix} E_{a_0} \\ E_{a_1} \\ E_{a_2} \end{bmatrix} = \frac{1}{3} \begin{bmatrix} E_a + E_b + E_c \\ E_a + aE_b + a^2 E_c \\ E_a + a^2 E_b + aE_c \end{bmatrix}
\tag{4.122}
$$

being E_{a_0}, E_{a_1}, and E_{a_2} the phasors representing the *a* phases of the zero, positive and negative sequences. Then, E_a, E_b, and E_c are the phasors representing the *a*, *b*, and *c* phases of the unbalanced system, where $a = e^{j\frac{2\pi}{3}}$ is a mathematical operator. There is much mathematical analysis and development that accompanies this theory, but it is beyond the scope of this chapter.

On the other hand, by using the space vector representation given by the following expression:

$$
\vec{x}(t) = \frac{2}{3} \left(x_a(t) + a \cdot x_b(t) + a^2 \cdot x_c(t) \right)
\tag{4.123}
$$

It is possible to determine the space vector representation of each of the three sequence components. Thus, as illustrated in Fig. 4-40, a graphical representation of the space vector shows that the positive sequence space vector of a given three-phase unbalanced system rotates in an anticlockwise direction, while the negative sequence space vector rotates in a clockwise direction. In a general approach, both positive and negative sequences present different amplitude and initial phase shift angles, but the angular frequency for both space vectors is equal to the original ω. Hence, the trajectory described by both space vectors is circular as each of them represents a balanced system. Note that the zero sequence is not represented in this approach. The zero sequence will exist only if a path to the current is provided by a fourth neutral wire, which is something not very common in many power electronic applications.

Thus, the addition of both space vectors produces the total space vector of the unbalanced system, as illustrated in Fig. 4-41. Since both space vectors representing each of the sequences rotate in the opposite direction, neither the amplitude of the space vector nor the rotating speed will take constant values.

Consequently, the trajectory described by the space vector representing the unbalanced three-phase system during the whole period will be an ellipse, as illustrated in Fig. 4-42. The points where both positive and negative sequences are aligned determine the maximum and minimum amplitude of the total space vector. This means that these points would be the semi-major and the semi-minor axis of the ellipse. Accordingly, the

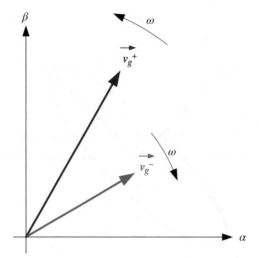

Fig. 4-40   *Space vectors of the positive and negative sequences of an unbalanced three-phase system*

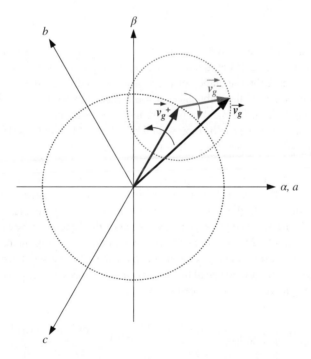

Fig. 4-41   *Addition of both positive and negative sequences, forming the total space vector* $\vec{v}_g$

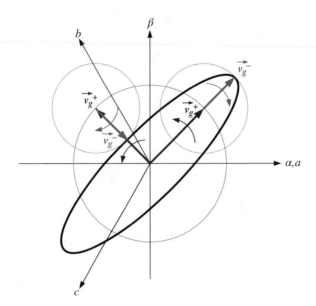

Fig. 4-42   *Ellipse trajectory described by the space vector representing the unbalanced system*

rotational speed of the total space vector will be maximum at the semi-minor axis and minimum at the semi-major axis. Depending on the nature of the unbalance of the three-phase system, the appearance of the ellipse will be different. For instance, if the unbalance of the system is small, the ellipse will present a shape close to the circumference. Conversely, if the unbalance of the system is very big, the ellipse will present a big difference between the longitude of the maximum and minimum semi-axis.

Therefore, as the unbalanced system can be represented by means of two balanced systems, a synchronously rotating reference frame can be generated for each of the sequences. This scenario is depicted in Fig. 4-43 for an unbalanced voltage.

Thus, the projections of both space vectors in their corresponding reference frames will be constant. However, the negative sequence will be projected in the positive sequence reference frame generating oscillating terms at twice the grid frequency. In this sense, the positive sequence will generate the same effect over the negative sequence reference frame. The projections over each of the synchronously rotating reference frames can be calculated according to (4.124) [4] for the steady-state operation, using the rotational speed of each of the reference frames (ω_a). Therefore, the *dq* components over the positive sequence reference frame will be obtained considering a positive rotational speed ($\omega_a = 2\pi f$) and the *dq* components over the negative sequence reference frame, using a negative rotational speed ($\omega_a = -2\pi f$).

$$\begin{bmatrix} v_{dg} \\ v_{qg} \end{bmatrix} = |V^+| \cdot \begin{bmatrix} \cos(\omega t - \omega_a t) \\ \sin(\omega t - \omega_a t) \end{bmatrix} + |V^-| \cdot \begin{bmatrix} \cos(-\omega t - \omega_a t) \\ \sin(-\omega t - \omega_a t) \end{bmatrix} \tag{4.124}$$

where ω is the rotational speed of the grid voltage space vector and ω_a is the rotational speed of the rotating reference frame. It is important to highlight that, although (4.124) is presented for the voltage, the same procedure can be followed to obtain the current projections over each of the rotating reference frames.

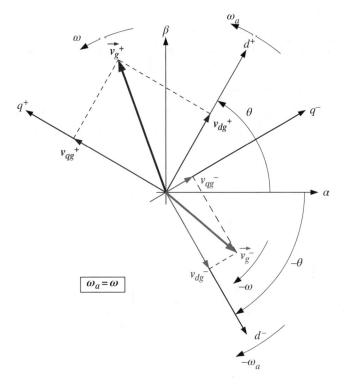

Fig. 4-43 *Double synchronous reference frame for an unbalanced voltage*

At the control stage, these oscillations can cause problems, so normally they are filtered. There are several options to filter the oscillations of the *dq* components, such as inserting low-pass filters or notch filters or employing the delayed signal cancelation (DSC) method. This DSC method is schematically represented in the block diagram depicted in Fig. 4-44. This technique adds to the original signal its own 180° delayed signal. If the frequency of the grid voltage is 50Hz, the delayed signal should be 10 msec delayed. In this way, under ideal conditions, the oscillations of the *dq* components can be canceled.

On the other hand, there also exists a property of unbalanced systems which determines that the perpendicular projections of the ellipse over each of the *abc* axes represent the maximum amplitude value of each sinusoidal *abc* magnitude (voltage or current) of the three-phase system. This fact is graphically illustrated in Fig. 4-45. Thus, knowing this, these peak amplitudes can be mathematically calculated according to the following procedure [5], [6]:

$$peak_a = |V_a| = \sqrt{B^2 \sin^2 \delta + A^2 \cos^2 \delta} \tag{4.125}$$

$$peak_b = |V_b| = \sqrt{B^2 \sin^2 (\delta - 2\pi/3) + A^2 \cos^2 (\delta - 2\pi/3)} \tag{4.126}$$

$$peak_c = |V_c| = \sqrt{B^2 \sin^2 (\delta + 2\pi/3) + A^2 \cos^2 (\delta + 2\pi/3)} \tag{4.127}$$

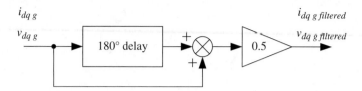

Fig. 4-44 *DSC method for filtering the oscillations of the dq components of an unbalanced three-phase system*

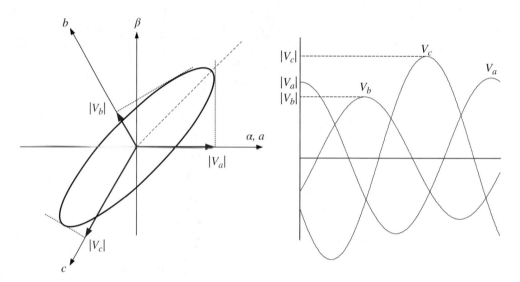

Fig. 4-45 *Perpendicular projections to the abc axis, results in the maximum amplitude of each sinusoidal abc magnitude*

being:

$$A = |\vec{v}_g^+| + |\vec{v}_g^-| \quad \text{and} \quad B = |\vec{v}_g^+| - |\vec{v}_g^-| \tag{4.128}$$

where:

$$|\vec{v}_g^+| = \sqrt{\left(v_{dg}^+\right)^2 + \left(v_{qg}^+\right)^2} \tag{4.129}$$

$$|\vec{v}_g^-| = \sqrt{\left(v_{dg}^-\right)^2 + \left(v_{qg}^-\right)^2} \tag{4.130}$$

and:

$$\delta = \frac{\alpha^+ + \alpha^-}{2} \tag{4.131}$$

with:

$$\alpha^+ = a\tan\left(\frac{v_{qg}^{\ +}}{v_{dg}^{\ +}}\right) \quad \text{and} \quad \alpha^- = a\tan\left(\frac{v_{qg}^{\ -}}{v_{dg}^{\ -}}\right) \tag{4.132}$$

being $v_{dg}^{\ +}$, $v_{qg}^{\ +}$, $v_{dg}^{\ -}$, and $v_{qg}^{\ -}$, d and q components of the positive and negative sequences as shown in Fig. 4-43. It is important to remember that although the example is developed for the voltage the peak values for the three current phases would be calculated following the same steps and using the d and q components of the positive and negative sequences of the current ($i_{dg}^{\ +}$, $i_{qg}^{\ +}$, $i_{dg}^{\ -}$ and $i_{qg}^{\ -}$).

On the other hand, it is also possible to calculate the total amplitude of the space vector describing the ellipse trajectory. Hence, the total space vector amplitude is the addition of the positive sequence space vector and the negative sequence space vector:

$$\vec{v}_g = \vec{v}_g^{\,s+} + \vec{v}_g^{\,s-} \tag{4.133}$$

Since the objective is only to demonstrate the existence of double-frequency oscillations in both the amplitude and the angle of the space vector, (4.134)–(4.136) are simplified for the particular case where the ellipse is in a horizontal position. Therefore, (4.133) can be rewritten as:

$$\vec{v}_g = \left(|\vec{v}_g^{\,+}| + |\vec{v}_g^{\,-}|\right)\cdot\cos(\omega t) + j\left(|\vec{v}_g^{\,+}| - |\vec{v}_g^{\,-}|\right)\cdot\sin(\omega t) \tag{4.134}$$

Therefore, the amplitude can be computed as:

$$|\vec{v}_g| = \sqrt{\left[\left(|\vec{v}_g^{\,+}| + |\vec{v}_g^{\,-}|\right)\cdot\cos(\omega t)\right]^2 + \left[\left(|\vec{v}_g^{\,+}| - |\vec{v}_g^{\,-}|\right)\cdot\sin(\omega t)\right]^2} \tag{4.135}$$

which results in:

$$|\vec{v}_g| = \sqrt{|\vec{v}_g^{\,+}|^2 + |\vec{v}_g^{\,-}|^2 + 2\cdot|\vec{v}_g^{\,+}|\cdot|\vec{v}_g^{\,-}|\cdot\cos(2\omega t)} \tag{4.136}$$

On the other hand, by means of (4.134), a mathematical expression can be obtained for the calculation of the angle (θ) of the space vector.

$$\theta = \arctan\left[\frac{|\vec{v}_g^{\,+}| - |\vec{v}_g^{\,-}|}{|\vec{v}_g^{\,+}| + |\vec{v}_g^{\,-}|}\cdot\frac{(1-\cos(2\omega t))}{\sin(2\omega t)}\right] \tag{4.137}$$

As will be noticed, the amplitude and the angle of the total space vector present a non-constant behavior, since they present an oscillatory term: ($2\omega t$). Consequently, Fig. 4-46 illustrates an example of the amplitude and rotating speed of the space vector representing an unbalanced system with $|\vec{v}_g^{\,+}| = 275\text{V}$ and $|\vec{v}_g^{\,-}| = 225\text{V}$.

Owing to these special characteristics of unbalanced voltages, the whole control strategy should be reconsidered in order to ensure the correct operation of the vector control. In this sense, one of the most important parts of the control is the grid synchronization block, since all these oscillations will complicate the correct synchronization of the control with the grid. Therefore, if the synchronous reference frame phase-locked loop

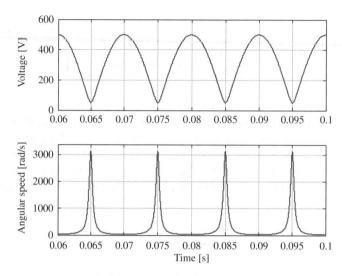

Fig. 4-46 *Amplitude and angular speed of the total voltage space vector representing an unbalanced three-phase system with* $|\vec{v}_g^+| = 275V$ *and* $|\vec{v}_g^-| = 225V$

(SRF-PLL) synchronization method, described in Section 4.3.3 for balanced systems, is used with unbalanced three-phase systems, the closed loop of this synchronization method will continue aligning the d axis of the rotating reference frame with the total voltage space vector. As a consequence, the estimated d component of the voltage (v_{dg}) will be equal to the total space vector and thus its amplitude and rotating speed will be identical to those presented in Fig. 4-46. The spectrum of these two waveforms is shown in Fig. 4-47.

Analyzing the spectrums of Fig. 4-47, two main conclusions can be obtained. On the one hand, it can be noticed that none of the harmonic components of the amplitude, Fig. 4-47(a), corresponds with either of the positive or negative sequence amplitudes. Hence, using an SRF-PLL in an unbalanced three-phase system, it will be impossible to separate the positive and negative sequences even filtering the estimated voltage amplitude. On the other hand, Fig. 4-47(b) shows a more distributed spectrum for the rotating speed (ω) but with a DC component equal to the rotating speed of the grid voltage ($\omega = 2 \cdot \pi \cdot f = 2 \cdot \pi \cdot 50 = 314.16 rad/s$). It suggests that it would be possible to estimate the angular speed (ω) and the grid frequency by means of the SRF-PLL. However, the Fig. 4-47(b) shows a very distributed spectrum and thus the existence of low-order harmonics requires high-order filters reducing the dynamic of the synchronization method.

As a consequence it can be concluded that the SRF-PLL is not an appropriate synchronization method for unbalanced grid voltages and therefore an alternative method is presented in the following sections.

4.4.2   Voltage equations of the grid-connected converter system

This subsection calculates the dq components of the entire grid-connected converter system, considering an L filter and an LCL filter, for both positive and negative sequence components.

4.4.2.1   L filter

From the space vector notation developed in previous subsections for the L filter, (4.51), the dq component equations of the positive sequence can be calculated by multiplying these mentioned expressions by $e^{-j\theta}$.

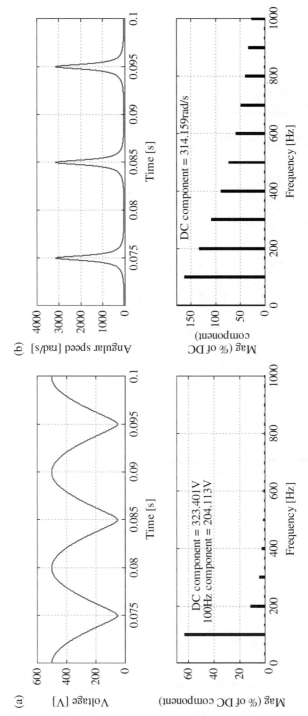

Fig. 4-47 *Oscillating amplitude and angular speed of the space vector representing an unbalanced system with $|\vec{v}_g^{\,+}| = 275V$ and $|\vec{v}_g^{\,-}| = 225V$, and their corresponding spectrums*

$$\vec{v}_f^{s+} \cdot e^{-j\theta} = R_f \cdot \vec{i}_g^{s+} \cdot e^{-j\theta} + L_f \cdot \frac{d\vec{i}_g^{s+}}{dt} \cdot e^{-j\theta} + \vec{v}_g^{s+} \cdot e^{-j\theta} \tag{4.138}$$

resulting in:

$$\vec{v}_f^{a+} = R_f \cdot \vec{i}_g^{a+} + L_f \cdot \frac{d\vec{i}_g^{a+}}{dt} + \vec{v}_g^{a+} + j \cdot \omega_a \cdot L_f \cdot \vec{i}_g^{a+} \tag{4.139}$$

Note that, being $\theta = \omega_a \cdot t$, the angular position of the rotatory reference frame:

$$\frac{d\vec{i}_g^{s+}}{dt} \cdot e^{-j\theta} = \frac{d\left(\vec{i}_g^{s+} \cdot e^{-j\theta}\right)}{dt} + j \cdot \omega_a \cdot \vec{i}_g^{s+} \cdot e^{-j\theta} \tag{4.140}$$

with *dq* components:

$$\vec{v}_f^{a+} = v_{df}^{\ +} + j \cdot v_{qf}^{\ +} \tag{4.141}$$

$$\vec{v}_g^{a+} = v_{dg}^{\ +} + j \cdot v_{qg}^{\ +} \tag{4.142}$$

$$\vec{i}_g^{a+} = i_{dg}^{\ +} + j \cdot i_{qg}^{\ +} \tag{4.143}$$

Therefore, by decomposing into *dq* components, the basic positive sequence equations for vector orientation are obtained:

$$v_{df}^{\ +} = R_f \cdot i_{dg}^{\ +} + L_f \cdot \frac{di_{dg}^{\ +}}{dt} + v_{dg}^{\ +} - \omega_a \cdot L_f \cdot i_{qg}^{\ +} \tag{4.144}$$

$$v_{qf}^{\ +} = R_f \cdot i_{qg}^{\ +} + L_f \cdot \frac{di_{qg}^{\ +}}{dt} + v_{qg}^{\ +} + \omega_a \cdot L_f \cdot i_{dg}^{\ +} \tag{4.145}$$

A similar procedure can be followed for the negative sequence, multiplying (4.51) by $e^{j\theta}$:

$$\vec{v}_f^{s-} \cdot e^{j\theta} = R_f \cdot \vec{i}_g^{s-} \cdot e^{j\theta} + L_f \cdot \frac{d\vec{i}_g^{s-}}{dt} \cdot e^{j\theta} + \vec{v}_g^{s-} \cdot e^{j\theta} \tag{4.146}$$

resulting in:

$$\vec{v}_f^{a-} = R_f \cdot \vec{i}_g^{a-} + L_f \cdot \frac{d\vec{i}_g^{a-}}{dt} + \vec{v}_g^{a-} - j \cdot \omega_a \cdot L_f \cdot \vec{i}_g^{a-} \tag{4.147}$$

Note that, being $\theta = \omega_a \cdot t$, the angular position of the rotatory reference frame:

$$\frac{d\vec{i}_g^{s-}}{dt} \cdot e^{j\theta} = \frac{d\left(\vec{i}_g^{s-} \cdot e^{j\theta}\right)}{dt} - j \cdot \omega_a \cdot \vec{i}_g^{s-} \cdot e^{j\theta} \tag{4.148}$$

with *dq* components:

$$\vec{v}_f^{a-} = v_{df}^- + j \cdot v_{qf}^- \tag{4.149}$$

$$\vec{v}_g^{a-} = v_{dg}^- + j \cdot v_{qg}^- \tag{4.150}$$

$$\vec{i}_g^{a-} = i_{dg}^- + j \cdot i_{qg}^- \tag{4.151}$$

Therefore, by decomposing into *dq* components, the basic negative sequence equations for vector orientation are obtained:

$$v_{df}^- = R_f \cdot i_{dg}^- + L_f \cdot \frac{di_{dg}^-}{dt} + v_{dg}^- + \omega_a \cdot L_f \cdot i_{qg}^- \tag{4.152}$$

$$v_{qf}^- = R_f \cdot i_{qg}^- + L_f \cdot \frac{di_{qg}^-}{dt} + v_{qg}^- - \omega_a \cdot L_f \cdot i_{dg}^- \tag{4.153}$$

4.4.2.2   *LCL filter*

On the other hand, considering an LCL filter and following the same procedure as carried out with the L filter, the *dq* expressions are obtained for both the positive and negative sequences. Multiplying (4.68) by $e^{-j\theta}$, the *dq* components of the positive sequence can be obtained:

$$\vec{v}_f^{s+} \cdot e^{-j\theta} = R_f \cdot \vec{i}_f^{s+} \cdot e^{-j\theta} + L_f \cdot \frac{d\vec{i}_f^{s+}}{dt} \cdot e^{-j\theta} + R_g \cdot \vec{i}_g^{s+} \cdot e^{-j\theta} + L_g \cdot \frac{d\vec{i}_g^{s+}}{dt} \cdot e^{-j\theta} + \vec{v}_g^{s+} \cdot e^{-j\theta} \tag{4.154}$$

resulting in:

$$\vec{v}_f^{a+} = R_f \cdot \vec{i}_f^{a+} + L_f \cdot \frac{d\vec{i}_f^{a+}}{dt} + R_g \cdot \vec{i}_g^{a+} + L_g \cdot \frac{d\vec{i}_g^{a+}}{dt} + \vec{v}_g^{a+} + j \cdot \omega_a \cdot L_f \cdot \vec{i}_f^{a+} + j \cdot \omega_a \cdot L_g \cdot \vec{i}_g^{a+} \tag{4.155}$$

Note that, being $\theta = \omega_a \cdot t$, the angular position of the rotatory reference frame:

$$\frac{d\vec{i}_f^{s+}}{dt} \cdot e^{-j\theta} = \frac{d\left(\vec{i}_f^{s+} \cdot e^{-j\theta}\right)}{dt} + j \cdot \omega_a \cdot \vec{i}_f^{s+} \cdot e^{-j\theta} \tag{4.156}$$

$$\frac{d\vec{i}_g^{s+}}{dt} \cdot e^{-j\theta} = \frac{d\left(\vec{i}_g^{s+} \cdot e^{-j\theta}\right)}{dt} + j \cdot \omega_a \cdot \vec{i}_g^{s+} \cdot e^{-j\theta} \tag{4.157}$$

with *dq* components:

$$\vec{v}_f^{a+} = v_{df}^+ + j \cdot v_{qf}^+ \tag{4.158}$$

$$\vec{v}_g^{a+} = v_{dg}^+ + j \cdot v_{qg}^+ \tag{4.159}$$

$$\vec{i}_f^{a+} = i_{df}^+ + j \cdot i_{qf}^+ \tag{4.160}$$

$$\vec{i}_g^{a+} = i_{dg}^+ + j \cdot i_{qg}^+ \tag{4.161}$$

Therefore, by decomposing into *dq* components, the basic equations for vector orientation are obtained:

$$v_{df}{}^+ = R_f \cdot i_{df}{}^+ + L_f \cdot \frac{di_{df}{}^+}{dt} + R_g \cdot i_{dg}{}^+ + L_g \cdot \frac{di_{dg}{}^+}{dt} + v_{dg}{}^+ - \omega_a \cdot L_f \cdot i_{qf}{}^+ - \omega_a \cdot L_g \cdot i_{qg}{}^+ \tag{4.162}$$

$$v_{qf}{}^+ = R_f \cdot i_{qf}{}^+ + L_f \cdot \frac{di_{qf}{}^+}{dt} + R_g \cdot i_{qg}{}^+ + L_g \cdot \frac{di_{qg}{}^+}{dt} + v_{qg}{}^+ + \omega_a \cdot L_f \cdot i_{df}{}^+ + \omega_a \cdot L_g \cdot i_{dg}{}^+ \tag{4.163}$$

Finally, a similar procedure can be followed to obtain the mathematical expressions for the *dq* components of the negative sequence, multiplying (4.68) by $e^{j\theta}$:

$$\vec{v}_f^{s-} \cdot e^{j\theta} = R_f \cdot \vec{i}_f^{s-} \cdot e^{j\theta} + L_f \cdot \frac{d\vec{i}_f^{s-}}{dt} \cdot e^{j\theta} + R_g \cdot \vec{i}_g^{s-} \cdot e^{j\theta} + L_g \cdot \frac{d\vec{i}_g^{s-}}{dt} \cdot e^{j\theta} + \vec{v}_g^{s-} \cdot e^{j\theta} \tag{4.164}$$

resulting in:

$$\vec{v}_f^{a-} = R_f \cdot \vec{i}_f^{a-} + L_f \cdot \frac{d\vec{i}_f^{a-}}{dt} + R_g \cdot \vec{i}_g^{a-} + L_g \cdot \frac{d\vec{i}_g^{a-}}{dt} + \vec{v}_g^{a-} - j \cdot \omega_a \cdot L_f \cdot \vec{i}_f^{a-} - j \cdot \omega_a \cdot L_g \cdot \vec{i}_g^{a-} \tag{4.165}$$

Note that, being $\theta = \omega_a \cdot t$, the angular position of the rotatory reference frame:

$$\frac{d\vec{i}_f^{s-}}{dt} \cdot e^{j\theta} = \frac{d\left(\vec{i}_f^{s-} \cdot e^{j\theta}\right)}{dt} - j \cdot \omega_a \cdot \vec{i}_f^{s-} \cdot e^{j\theta} \tag{4.166}$$

$$\frac{d\vec{i}_g^{s-}}{dt} \cdot e^{j\theta} = \frac{d\left(\vec{i}_g^{s-} \cdot e^{j\theta}\right)}{dt} - j \cdot \omega_a \cdot \vec{i}_g^{s-} \cdot e^{j\theta} \tag{4.167}$$

with *dq* components:

$$\vec{v}_f^{a-} = v_{df}{}^- + j \cdot v_{qf}{}^- \tag{4.168}$$

$$\vec{v}_g^{a-} = v_{dg}{}^- + j \cdot v_{qg}{}^- \tag{4.169}$$

$$\vec{i}_f^{a-} = i_{df}{}^- + j \cdot i_{qf}{}^- \tag{4.170}$$

$$\vec{i}_g^{a-} = i_{dg}{}^- + j \cdot i_{qg}{}^- \tag{4.171}$$

Therefore, by decomposing into *dq* components, the basic equations for vector orientation are obtained:

$$v_{df}{}^- = R_f \cdot i_{df}{}^- + L_f \cdot \frac{di_{df}{}^-}{dt} + R_g \cdot i_{dg}{}^- + L_g \cdot \frac{di_{dg}{}^-}{dt} + v_{dg}{}^- + \omega_a \cdot L_f \cdot i_{qf}{}^- + \omega_a \cdot L_g \cdot i_{qg}{}^- \tag{4.172}$$

$$v_{qf}{}^- = R_f \cdot i_{qf}{}^- + L_f \cdot \frac{di_{qf}{}^-}{dt} + R_g \cdot i_{qg}{}^- + L_g \cdot \frac{di_{qg}{}^-}{dt} + v_{qg}{}^- - \omega_a \cdot L_f \cdot i_{df}{}^- - \omega_a \cdot L_g \cdot i_{dg}{}^- \tag{4.173}$$

4.4.3   Power expressions

This subsection analyzes the powers that are obtained in a three-phase unbalanced system. Owing to the presence of unbalance in the voltages and currents, it arises with the presence of positive and negative sequences of both voltages and currents, yielding a more complex power expression than in a three-phase balanced system.

Hence, the original apparent power expression for a three-phase unbalanced system is equal to the power original expression of a three-phase balanced system:

$$S = \frac{3}{2} \cdot \vec{v}_g \cdot \vec{i}_g^* \tag{4.174}$$

Considering that both voltages and currents are unbalanced and substituting the positive and negative sequences, the apparent power expression results in:

$$S = \frac{3}{2} \cdot \left(\vec{v}_g^{a+} \cdot e^{j\omega t} + \vec{v}_g^{a-} \cdot e^{-j\omega t} \right) \cdot \left(\vec{i}_g^{a+} \cdot e^{j\omega t} + \vec{i}_g^{a-} \cdot e^{-j\omega t} \right)^* \tag{4.175}$$

Then, considering the *dq* components of both positive and negative sequences, the following equation is obtained:

$$S = \frac{3}{2} \left[\left(v_{dg}^+ + jv_{qg}^+ \right) \cdot e^{j\omega t} + \left(v_{dg}^- + jv_{qg}^- \right) \cdot e^{-j\omega t} \right] \cdot \left[\left(i_{dg}^+ - ji_{qg}^+ \right) \cdot e^{-j\omega t} + \left(i_{dg}^- - ji_{qg}^- \right) \cdot e^{j\omega t} \right] \tag{4.176}$$

Thus, multiplying all the terms of this last expression and separating into real and imaginary parts, we obtain the general active and reactive power expressions as:

$$P_g = \frac{3}{2} \cdot \left[P_0 + P_{C2} \cdot \cos(2\omega t) + P_{S2} \cdot \sin(2\omega t) \right] \tag{4.177}$$

$$Q_g = \frac{3}{2} \cdot \left[Q_0 + Q_{C2} \cdot \cos(2\omega t) + Q_{S2} \cdot \sin(2\omega t) \right] \tag{4.178}$$

being each P_0, P_{C2}, P_{S2}, Q_0, Q_{C2}, and Q_{S2} constant power terms equal to:

$$P_0 = \left(v_{dg}^+ i_{dg}^+ + v_{qg}^+ i_{qg}^+ + v_{dg}^- i_{dg}^- + v_{qg}^- i_{qg}^- \right) \tag{4.179}$$

$$P_{c2} = \left(v_{dg}^+ i_{dg}^- + v_{qg}^+ i_{qg}^- + v_{dg}^- i_{dg}^+ + v_{qg}^- i_{qg}^+ \right) \tag{4.180}$$

$$P_{s2} = \left(v_{qg}^- i_{dg}^+ - v_{dg}^- i_{qg}^+ - v_{qg}^+ i_{dg}^- + v_{dg}^+ i_{qg}^- \right) \tag{4.181}$$

$$Q_0 = \left(v_{qg}^+ i_{dg}^+ - v_{dg}^+ i_{qg}^+ + v_{qg}^- i_{dg}^- - v_{dg}^- i_{qg}^- \right) \tag{4.182}$$

$$Q_{c2} = \left(v_{qg}^+ i_{dg}^- - v_{dg}^+ i_{qg}^- + v_{qg}^- i_{dg}^+ - v_{dg}^- i_{qg}^+ \right) \tag{4.183}$$

$$Q_{s2} = \left(v_{dg}^+ i_{dg}^- + v_{qg}^+ i_{qg}^- - v_{dg}^- i_{dg}^+ - v_{qg}^- i_{qg}^+ \right) \tag{4.184}$$

Note that the total powers present an average constant value of P_0 and Q_0 and then they also oscillate because of the presence of terms $\cos(2\omega t)$ and $\sin(2\omega t)$, which means that the oscillations are at double frequency of the grid frequency ω.

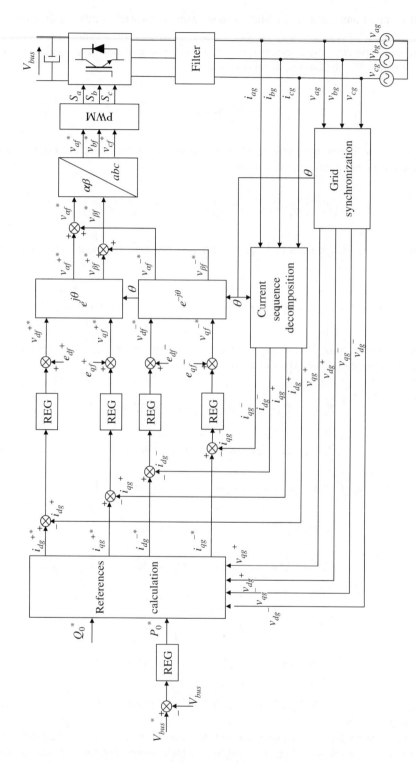

Fig. 4-48 Vector control strategy for an unbalanced three-phase grid system

On the other hand, it is quite typical to use a matrix expression to refer to all the positive and negative product of d and q components of voltages and currents:

$$
\begin{bmatrix} P_0 \\ P_{C2} \\ P_{S2} \\ Q_0 \\ Q_{C2} \\ Q_{S2} \end{bmatrix} = \frac{3}{2} \begin{bmatrix} v_{dg}^+ & v_{qg}^+ & v_{dg}^- & v_{qg}^- \\ v_{dg}^- & v_{qg}^- & v_{dg}^+ & v_{qg}^+ \\ v_{qg}^- & -v_{dg}^- & -v_{qg}^+ & v_{dg}^+ \\ v_{qg}^+ & -v_{dg}^+ & v_{qg}^- & -v_{dg}^- \\ v_{qg}^- & -v_{dg}^- & v_{qg}^+ & -v_{dg}^+ \\ -v_{dg}^- & -v_{qg}^- & v_{dg}^+ & v_{qg}^+ \end{bmatrix} \cdot \begin{bmatrix} i_{dg}^+ \\ i_{qg}^+ \\ i_{dg}^- \\ i_{qg}^- \end{bmatrix}
\tag{4.185}
$$

Thus, the last equations show how the P_g and Q_g expressions of a three-phase unbalanced system are oscillatory, in contrast to the powers of the three-phase balanced systems, which are constant.

4.4.4   Vector control under unbalanced conditions

The control strategy followed under an unbalanced voltage scenario is still based on the basic principles of the vector control strategy. The schematic block diagram is presented in Fig. 4-48. The control uses two current loops per sequence (i.e., two dq controllers for the positive sequence and two dq controllers for the negative sequence). Therefore, this control strategy is known in the specialized literature as dual vector control.

As in the vector control for balanced conditions, the coupling terms are canceled at the output of the current PI regulators. The cancelation terms for an inductive filter (i.e. an L filter) are obtained from (4.144)–(4.145) for the positive sequence and from (4.152)–(4.153) for the negative sequence.

$$
e_{df}^{\ +} = -\omega_a \cdot L_f \cdot i_{qg}^{\ +} + v_{dg}^{\ +}
\tag{4.186}
$$

$$
e_{qf}^{\ +} = \omega_a \cdot L_f \cdot i_{dg}^{\ +} + v_{qg}^{\ +}
\tag{4.187}
$$

$$
e_{df}^{\ -} = \omega_a \cdot L_f \cdot i_{qg}^{\ -} + v_{dg}^{\ -}
\tag{4.188}
$$

$$
e_{qf}^{\ -} = -\omega_a \cdot L_f \cdot i_{dg}^{\ -} + v_{qg}^{\ -}
\tag{4.189}
$$

On the other hand, the cancelation terms for the LCL filter are deduced from (4.162)–(4.163) for the positive sequence and from (4.172)–(4.173) for the negative sequence.

$$
e_{df}^{\ +} = -\omega_a \cdot L_f \cdot i_{qf}^{\ +} - \omega_a \cdot L_g \cdot i_{qg}^{\ +} + v_{dg}^{\ +}
\tag{4.190}
$$

$$
e_{qf}^{\ +} = \omega_a \cdot L_f \cdot i_{df}^{\ +} + \omega_a \cdot L_g \cdot i_{dg}^{\ +} + v_{qg}^{\ +}
\tag{4.191}
$$

$$
e_{df}^{\ -} = \omega_a \cdot L_f \cdot i_{qf}^{\ -} + \omega_a \cdot L_g \cdot i_{qg}^{\ -} + v_{dg}^{\ -}
\tag{4.192}
$$

$$
e_{qf}^{\ -} = -\omega_a \cdot L_f \cdot i_{df}^{\ -} - \omega_a \cdot L_g \cdot i_{dg}^{\ -} + v_{qg}^{\ -}
\tag{4.193}
$$

As explained in Section 4.3.2, an approximation can be implemented assuming that both the converter side current and the grid side current (i_f and i_g) have similar fundamental values. It is important to remember that this approximation is carried out because usually only one of these two currents (i_f and i_g) is measured. Therefore, the simplified cancelation terms are obtained for both the cases where only the grid side current or the converter side current is measured.

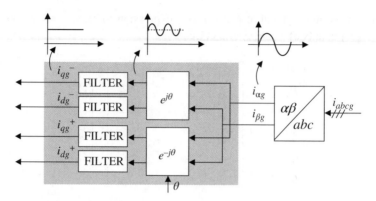

Fig. 4-49   *Current sequence decomposition block diagram*

$$e_{df}^{\ +} \approx -\left(L_f + L_g\right)\cdot\omega_a\cdot i_{qf}^{\ +} + v_{dg}^{\ +} \approx -\left(L_f + L_g\right)\cdot\omega_a\cdot i_{qg}^{\ +} + v_{dg}^{\ +} \tag{4.194}$$

$$e_{qf}^{\ +} \approx \left(L_f + L_g\right)\cdot\omega_a\cdot i_{df}^{\ +} + v_{qg}^{\ +} \approx \left(L_f + L_g\right)\cdot\omega_a\cdot i_{dg}^{\ +} + v_{qg}^{\ +} \tag{4.195}$$

$$e_{df}^{\ -} \approx \left(L_f + L_g\right)\cdot\omega_a\cdot i_{qf}^{\ -} + v_{dg}^{\ -} \approx \left(L_f + L_g\right)\cdot\omega_a\cdot i_{qg}^{\ -} + v_{dg}^{\ -} \tag{4.196}$$

$$e_{qf}^{\ -} \approx -\left(L_f + L_g\right)\cdot\omega_a\cdot i_{df}^{\ -} + v_{qg}^{\ -} \approx -\left(L_f + L_g\right)\cdot\omega_a\cdot i_{dg}^{\ -} + v_{qg}^{\ -} \tag{4.197}$$

The output of the current controllers creates the voltage references that, after canceling the coupling terms, are transformed into $\alpha\beta$ coordinates, by using the θ angle. Once transformed into positive and negative $\alpha\beta$ sequences, the positive and negative sequences are added, generating a total $\alpha\beta$ reference. After that, these are transformed into *abc* coordinates for the subsequent creation of the order commands for the IGBTs of the converter, by means of the PWM (for instance the one shown in Section 4.2.4).

On the other hand, for the current loop controllers and for the current reference generation, the positive and negative sequences of the voltages and currents must be calculated. This sequence calculation is performed in two different blocks, as depicted in Fig. 4-48. First of all the vector control is oriented to the grid voltage by means of the selected grid synchronization method. Although there are several grid synchronization techniques in the specialized literature for unbalanced three-phase systems, the one utilized in this chapter is analyzed in depth in the following section. This block calculates the *dq* components of the positive and negative sequences of the grid voltage and the angle (θ) for the different transformation blocks of the control algorithm. On the other hand, the *dq* components of the positive and negative sequences of the current are obtained according to the block diagram shown in Fig. 4-49 and using the angle (θ) provided by the grid synchronization block. Note that in the presence of unbalance currents the corresponding *dq* components include oscillatory terms at double the fundamental frequency (100Hz in a grid of 50Hz). To avoid these components, various filters are described in the specialized literature, such as low pass filters, notch filters tuned at the oscillatory frequency, or even DSC methods as described in Section 4.4.1. Note that the low-pass filter applied to the currents would significantly affect the bandwidth of the current loops, so maybe it is not the best option. Note also that for the cancelation of the coupling terms at the output of the regulators it is also necessary to filter the oscillations.

In addition, in general it is necessary to control the DC bus voltage with an extra voltage loop, as depicted in Fig. 4-48. This DC bus voltage regulator follows the same control philosophy as the one studied in Section 4.3.2, for three-phase balanced systems. The output of the DC bus voltage regulator is the reference for the active power exchanged with the grid (P_o^*). On the other hand, the other reference for the system is the mean value of the grid reactive power (Q_0^*). Thus, from these power references, the current positive and negative dq references are generated. For that purpose, the matrix equation (4.185) needs to be inverted:

$$
\begin{bmatrix} i_{dg}^{+*} \\ i_{qg}^{+*} \\ i_{dg}^{-*} \\ i_{qg}^{-*} \end{bmatrix} = \frac{2}{3} \begin{bmatrix} v_{dg}^{+} & v_{qg}^{+} & v_{dg}^{-} & v_{qg}^{-} \\ v_{dg}^{-} & v_{qg}^{-} & v_{dg}^{+} & v_{qg}^{+} \\ v_{qg}^{-} & -v_{dg}^{-} & -v_{qg}^{+} & v_{dg}^{+} \\ v_{qg}^{+} & -v_{dg}^{+} & v_{qg}^{-} & -v_{dg}^{-} \\ v_{qg}^{-} & -v_{dg}^{-} & v_{qg}^{+} & -v_{dg}^{+} \\ -v_{dg}^{-} & -v_{qg}^{-} & v_{dg}^{+} & v_{qg}^{+} \end{bmatrix}^{-1} \begin{bmatrix} P_0^* \\ P_{C2}^* \\ P_{S2}^* \\ Q_0^* \\ Q_{C2}^* \\ Q_{S2}^* \end{bmatrix} \tag{4.198}
$$

Note that for the calculation of the current references the dq positive and negative sequences of the voltages are necessary. However, for the inversion of the matrix, a squared matrix is preferable. For that purpose, the last two rows of the matrix equation (4.198) are omitted for the current reference calculation. In addition, it is possible to make zero the oscillatory terms of the active power $P_{c2} = 0$ and $P_{s2} = 0$. This fact is shown in the next expression:

$$
\begin{bmatrix} i_{dg}^{+*} \\ i_{qg}^{+*} \\ i_{dg}^{-*} \\ i_{qg}^{-*} \end{bmatrix} = \frac{2}{3} \begin{bmatrix} v_{dg}^{+} & v_{qg}^{+} & v_{dg}^{-} & v_{qg}^{-} \\ v_{dg}^{-} & v_{qg}^{-} & v_{dg}^{+} & v_{qg}^{+} \\ v_{qg}^{-} & -v_{dg}^{-} & -v_{qg}^{+} & v_{dg}^{+} \\ v_{qg}^{+} & -v_{dg}^{+} & v_{qg}^{-} & -v_{dg}^{-} \end{bmatrix}^{-1} \begin{bmatrix} P_0^* \\ 0 \\ 0 \\ Q_0^* \end{bmatrix} \tag{4.199}
$$

Thanks to this, the calculations are eased and the active power oscillations are eliminated, which is useful in order to minimize the oscillations of the DC bus voltage. (Note that if there are oscillations in the active power then there will be oscillations in the DC bus) Thus, from this the current references result in:

$$
i_{dg}^{+*} = \frac{2}{3} \left\{ \frac{v_{dg}^{+}}{\left| v_g^{+} \right|^2 - \left| v_g^{-} \right|^2} P_0^* + \frac{v_{qg}^{+}}{\left| v_g^{+} \right|^2 + \left| v_g^{-} \right|^2} Q_0^* \right\} \tag{4.200}
$$

$$
i_{qg}^{+*} = \frac{2}{3} \left\{ \frac{v_{qg}^{+}}{\left| v_g^{+} \right|^2 - \left| v_g^{-} \right|^2} P_0^* - \frac{v_{dg}^{+}}{\left| v_g^{+} \right|^2 + \left| v_g^{-} \right|^2} Q_0^* \right\} \tag{4.201}
$$

$$
i_{dg}^{-*} = \frac{2}{3} \left\{ -\frac{v_{dg}^{-}}{\left| v_g^{+} \right|^2 - \left| v_g^{-} \right|^2} P_0^* + \frac{v_{qg}^{-}}{\left| v_g^{+} \right|^2 + \left| v_g^{-} \right|^2} Q_0^* \right\} \tag{4.202}
$$

$$
i_{qg}^{-*} = \frac{2}{3} \left\{ -\frac{v_{qg}^{-}}{\left| v_g^{+} \right|^2 - \left| v_g^{-} \right|^2} P_0^* - \frac{v_{dg}^{-}}{\left| v_g^{+} \right|^2 + \left| v_g^{-} \right|^2} Q_0^* \right\} \tag{4.203}
$$

where:

$$\left|v_g^+\right|^2 = \left(v_{dg}^+\right)^2 + \left(v_{qg}^+\right)^2 \tag{4.204}$$

$$\left|v_g^-\right|^2 = \left(v_{dg}^-\right)^2 + \left(v_{qg}^-\right)^2 \tag{4.205}$$

These current references are in function of the average active and reactive power references and the positive and negative sequence *dq* voltages. Note that the reactive power value will present oscillations, since terms Q_{c2} and Q_{s2} are not made zero.

4.4.5 Synchronization method for unbalanced grid voltages

Finally, the angle θ for applying the rotational transformations of the vector control strategy (presented in Fig. 4-48) is indispensable. This angle can be obtained by any of the synchronization methods that can be found in the specialized literature. This section employs a DSOGI-PLL (dual second-order generalized integrator phase-locked loop) as the synchronization method, which is analyzed in depth in [4] and [7]. This method is based on the second-order generalized integrator (SOGI) whose transfer function is shown in the following expression:

$$C(s) = \frac{\omega_r s}{s^2 + \omega_r^2} \tag{4.206}$$

where ω_r is the resonance frequency.

The first step is to obtain the $\alpha\beta$ components of the grid voltage by means of the Clarke transformation. These two components are then transferred to two blocks where they are filtered and two other signals are generated in quadrature with the original $\alpha\beta$ components. These blocks are denominated SOGI-QSG (second-order generalized integrator quadrature signal generator) and its block diagram is depicted in Fig. 4-50.

The relations between the two outputs (x' and qx') and the input (x) are defined by means of the following transfer functions:

$$\frac{x'}{x}(s) = \frac{K\omega_r s}{s^2 + K\omega_r s + \omega_r^2} \tag{4.207}$$

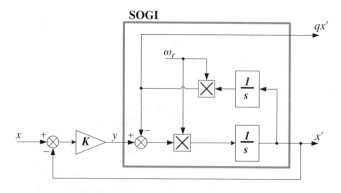

Fig. 4-50 *Block diagram of the SOGI-QSG*

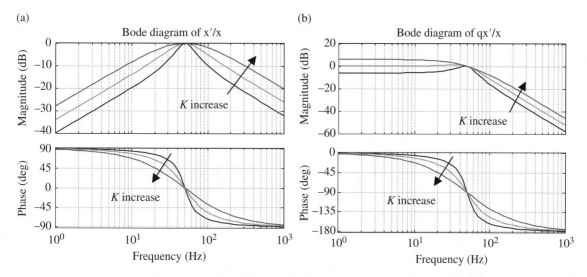

Fig. 4-51 *Bode diagrams of the SOGI-QSG with different values of the gain K: (a) Bode diagram of x'/x, (b) Bode diagram of qx'/x*

$$\frac{qx'}{x}(s) = \frac{K\omega_r^2}{s^2 + K\omega_r s + \omega_r^2} \tag{4.208}$$

The Bode diagrams of the transfer functions (4.207) and (4.208) are depicted in Fig. 4-51.

As can be noticed in the Bode diagrams of Fig. 4-51(a) and (b), both relations present a unitary gain at the resonance frequency. In case of the phase delays, x'/x Bode diagram shows an output in phase with the input at the resonance frequency, whereas the qx'/x relation generates a 90° phase delay between the output and the input at the same frequency. Therefore this resonance frequency must be adjusted to the grid frequency in order to maintain the unitary gains and the phase delays in the case of grid frequency variations.

As mentioned before, using the $\alpha\beta$ components of the voltage as inputs for the two SOGI-QSG blocks, four components of the voltage are calculated ($v_{\alpha g}'$, $v_{\beta g}'$, $qv_{\alpha g}'$, $qv_{\beta g}'$). By means of these four components it is possible to separate the positive and negative sequences of the grid voltage by applying the following relations:

$$\begin{bmatrix} v_{\alpha g}^{+} \\ v_{\beta g}^{+} \end{bmatrix} = \frac{1}{2} \begin{bmatrix} 1 & -q \\ q & 1 \end{bmatrix} \begin{bmatrix} v_{\alpha g} \\ v_{\beta g} \end{bmatrix} \tag{4.209}$$

$$\begin{bmatrix} v_{\alpha g}^{-} \\ v_{\beta g}^{-} \end{bmatrix} = \frac{1}{2} \begin{bmatrix} 1 & q \\ -q & 1 \end{bmatrix} \begin{bmatrix} v_{\alpha g} \\ v_{\beta g} \end{bmatrix} \tag{4.210}$$

being:

$$q = e^{-j\frac{\pi}{2}} \tag{4.211}$$

The block diagram representing the relations defined by (4.209) and (4.210) is depicted in Fig. 4-52, where the $\alpha\beta$ components of the positive and negative sequences are calculated.

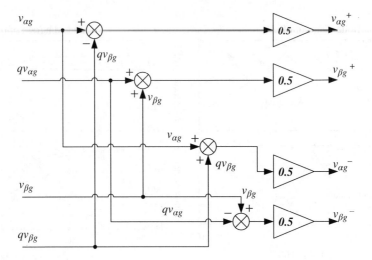

Fig. 4-52   *Block diagram of the positive and negative sequence calculator*

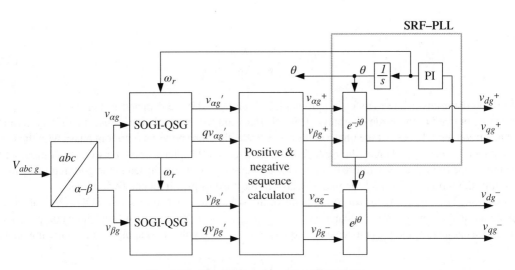

Fig. 4-53   *Block diagram of the DSOGI-PLL*

Once the positive and negative sequences of the grid voltage are calculated, the use of an algorithm to obtain the phase (θ) and the angular speed (ω_r) of the positive sequence is needed. This angular speed will be used in the SOGI-QSG blocks as described in Fig. 4-50. Owing to the proper separation of the positive and negative sequences, each of them is a balanced system and, thus, the calculation of the phase and angular speed can be achieved by means of any of the synchronization methods existing in the specialized literature for balanced systems. However, in this section the SRF-PLL (synchronous reference frame phase-locked loop) algorithm, described in Section 4.3.3 will be employed. This grid synchronization method, comprising two SOGI-QSG blocks together with the SRF-PLL is known in the specialized literature as DSOGI-PLL (dual second-order generalized integrator phase-locked loop), whose block diagram is shown in Fig. 4-53.

As will be noticed in the block diagram of Fig. 4-53, the angular speed obtained in the SRF-PLL is transferred to the SOGI-QSG blocks and the angle to all the rotational transformations existing in the vector control of Fig. 4-48.

4.4.6 Tuning of the current regulators

This subsection proposes a procedure for the tuning of the current regulators for dual vector control. Starting from the voltage equations for the positive sequence (expressions (4.144)–(4.145)), and for the negative sequence (expressions (4.152)–(4.153)), and owing to the cancelation of the coupling terms (expressions (4.186)–(4.189)), the current versus voltage equations can be deduced for the L filter:

$$v_{df}^{+} = R_f \cdot i_{dg}^{+} + L_f \cdot \frac{di_{dg}^{+}}{dt} \tag{4.212}$$

$$v_{qf}^{+} = R_f \cdot i_{qg}^{+} + L_f \cdot \frac{di_{qg}^{+}}{dt} \tag{4.213}$$

$$v_{df}^{-} = R_f \cdot i_{dg}^{-} + L_f \cdot \frac{di_{dg}^{-}}{dt} \tag{4.214}$$

$$v_{qf}^{-} = R_f \cdot i_{qg}^{-} + L_f \cdot \frac{di_{qg}^{-}}{dt} \tag{4.215}$$

The same relations can be obtained for the case where an LCL filter is used. In this situation an approximation is made assuming that both the converter side and grid side currents (i_f and i_g) present similar fundamental values. Therefore, using the voltage equations for the positive sequence (expressions (4.162)–(4.163)) and for the negative sequence (expressions (4.172)–(4.173)) and thanks again to the cancelation of the coupling terms (expressions (4.194)–(4.197)), the current versus voltage equations are obtained, in a simplified way, only as a function of the grid side current (i_g):

$$v_{df}^{+} \approx R_f \cdot i_{dg}^{+} + L_f \cdot \frac{di_{dg}^{+}}{dt} + R_g \cdot i_{dg}^{+} + L_g \cdot \frac{di_{dg}^{+}}{dt} \tag{4.216}$$

$$v_{qf}^{+} \approx R_f \cdot i_{qg}^{+} + L_f \cdot \frac{di_{qg}^{+}}{dt} + R_g \cdot i_{qg}^{+} + L_g \cdot \frac{di_{qg}^{+}}{dt} \tag{4.217}$$

$$v_{df}^{-} \approx R_f \cdot i_{dg}^{-} + L_f \cdot \frac{di_{dg}^{-}}{dt} + R_g \cdot i_{dg}^{-} + L_g \cdot \frac{di_{dg}^{-}}{dt} \tag{4.218}$$

$$v_{qf}^{-} \approx R_f \cdot i_{qg}^{-} + L_f \cdot \frac{di_{qg}^{-}}{dt} + R_g \cdot i_{qg}^{-} + L_g \cdot \frac{di_{qg}^{-}}{dt} \tag{4.219}$$

As can be observed, the expressions obtained have the same structure in both the positive and negative sequences for each of the filters. In addition, their structure is identical to those deduced in Section 4.3.4 for balanced scenarios. Therefore, the same tuning procedure presented in Section 4.3.4 can be followed to tune the four existing PI regulators in a dual vector control.

4.4.7 Simulation-based example

As we did in the balanced analysis in previous sections, this section shows an example representing the most representative magnitudes of a 10kW grid-connected converter (two-level converter), simulated with a Matlab-Simulink computer-based simulator. The converter is connected to a grid of 380V. The grid side filter is a pure inductive L filter (with parasitic resistance). The simulation experiments show the steady-state performance of the grid side converter's magnitudes when it is controlled by the vector control presented in the previous sections oriented to work at unbalanced situations.

Thus, Fig. 4-54(a) illustrates the grid voltage in pu where the converter system is connected. There is a voltage drop of equal magnitude at phase voltages v_{ag} and v_{bg}. Fig. 4-54(b) shows the DC bus voltage performance.

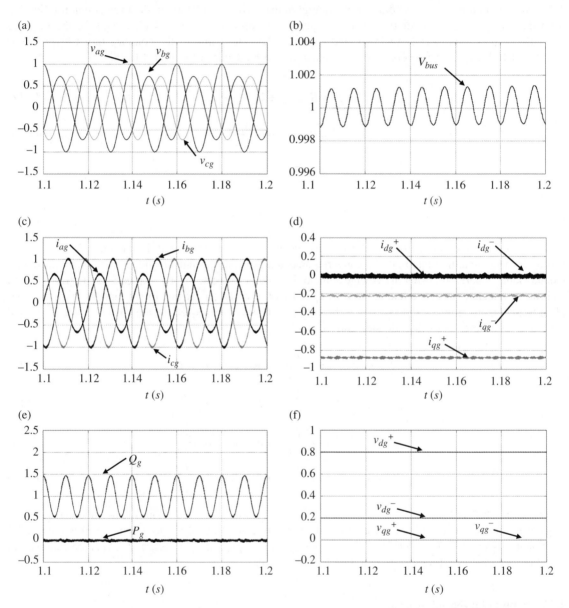

Fig. 4-54 *Most representative magnitudes of a grid-connected converter of 10kW: (a) grid abc voltages in pu, (b) DC bus voltage in pu, (c) abc grid side converter currents in pu, (d) dq grid side currents in pu (positive and negative sequences), (e) power exchange with the grid in pu, and (f) dq grid side voltages in pu (positive and negative sequences)*

It can be noticed that there are oscillations at double the grid frequency (at 100Hz), since the control strategy employed minimizes the active power oscillations at the DC side, but not at the converter side, therefore producing DC bus oscillations. The average active power exchange with the grid in this example is zero, as shown in Fig. 4-54(e), while the reactive power average value is nearly 1pu. It is possible to see how at the grid side the active power oscillations are minimized, but the reactive power oscillations are not exactly zero. On the other hand, Fig. 4-54(c) and (d) show the current exchange with the grid by the converter in *abc* coordinates and in *dq* coordinates. This unbalanced current exchange, results from the minimization of grid power oscillations strategy. Finally, Fig. 4-54(f) shows the grid *dq* voltage decomposition in positive and negative sequences.

4.5 Single-phase grid-connected converter model and modulation

4.5.1 Model

This section describes the model of the single-phase grid-connected converter. The system configuration is presented in Fig. 4-55. The grid side connected converter is connected to a single phase through a filter, as depicted. The converter configuration which is studied in this section consists of two, two-level converter branches. Note that it is possible to find some other converter configurations, but these are beyond the scope of this section. The converter is composed of bi-directional switches that in this case are modeled as idealized IGBTs with diodes in antiparallel. Connected to the DC side of the converter, it is possible to find different types of converter configurations. In this section, a three-phase converter and a single-phase converter are mostly considered, as provided by the simulation-based examples presented at the end of the section.

The nature of the filter, as shown in the three-phase systems studied in previous subsections, can be a pure inductive filter (L), or the combination of inductive and capacitive filters (LC or LCL). This section will be mainly focused on the pure inductive filter. From another viewpoint, the grid is modeled as a unique and ideal sinusoidal voltage source, v_g.

Another aspect to bear in mind, by considering the equivalent circuit shown in Fig. 4-56, of the single-phase converter system presented in Fig. 4-55, the whole converter is considered as the voltage source v_{ab}. Thus, depending on the order commands of the controlled switches S_{a_g} and S_{b_g}, the output voltage of the converter v_{ab} can take different values. Three possible output voltage values can appear at the output: 0, V_{bus} and $-V_{bus}$ as

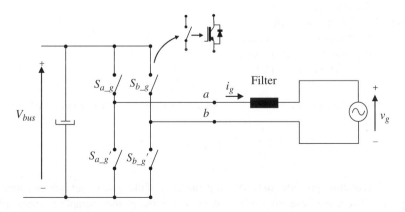

Fig. 4-55 *Grid-connected converter power circuit*

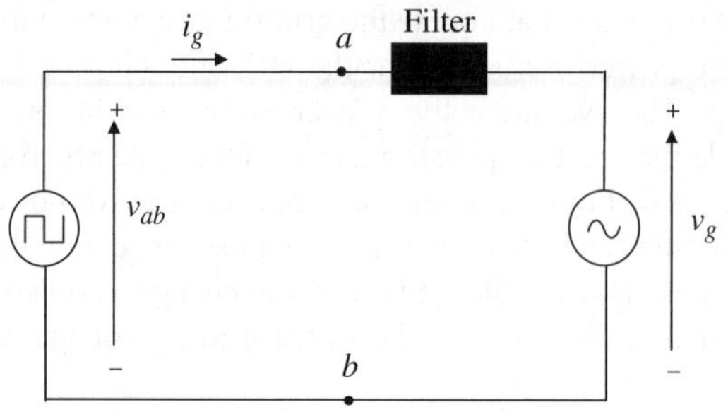

Fig. 4-56   *Simplified equivalent single-phase circuit*

Table 4-3   *Different output voltage combinations of single-phase 2L-VSC*

S_{a_g}	S_{b_g}	v_{ab}
0	0	0
0	1	$-V_{bus}$
1	0	V_{bus}
1	1	0

illustrated in Table 4-3. Ideally, the order commands of the complementary switches S_{a_g}' and S_{b_g}' are the opposite of their respective order commands of S_{a_g} and S_{b_g}.

In this equivalent electric circuit of Fig. 4-56, it is possible to simply derive the electric differential equation of the system, considering a pure inductive filter (an additional parasitic resistance is also modeled as R_f):

$$v_{ab} = R_f \cdot i_g + L_f \cdot \frac{di_g}{dt} + v_g \qquad (4.220)$$

Being:

- L_f: Inductance of the grid side filter (H)
- R_f: Resistive part of the grid side filter (Ω)
- v_g: Grid voltage (V), with ω electric angular speed in (rad/sec)
- i_g: Current flowing thorough the grid (A)
- v_{ab}: Output voltage of the converter (V)

Consequently, for modeling purposes, it is necessary to isolate the first derivative of the current as:

$$\frac{di_g}{dt} = \frac{1}{L_f} \cdot \left(v_{ab} - R_f \cdot i_g - v_g \right) \qquad (4.221)$$

This last differential equation provides the modeling equation of the system, for instance for computer-based simulation purposes. Therefore, the simplified block diagram of the complete single-phase converter

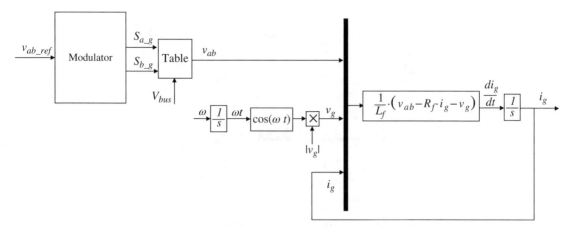

Fig. 4-57   *Simplified converter, filter, and grid model of the single-phase system*

connected to the grid is served in Fig. 4-57. Model equation (4.221) is included, together with the model of the converter itself (v_{ab}) and the grid voltage model (v_g). Note that the output voltage of the converter is created by the modulator (which is described in following subsections) that creates the order commands of the controlled switches shown in Table 4-3.

Finally, it must be noted that the model of the DC-link voltage converter is the same as the model of the DC-link for three-phase converters described in Section 4.2.3; consequently, it is not repeated here.

4.5.2   Sinusoidal PWM

This section presents two possible sinusoidal PWM strategies for the single-phase converter. With the modulation strategies, the order commands of the controlled switches of the converter are generated. For simplicity in the exposition of the modulations, the grid-connected converter system is now represented in a more compact way, as illustrated in Fig. 4-58. The grid and the filter have been grouped as a general load system. The first modulation described is the bipolar sinusoidal PWM.

This bipolar modulation strategy utilizes a sinusoidal reference which is compared with a triangular carrier signal, as depicted in Fig. 4-59(a). The controlled switches are commutated according to the following law:

$$S_{a\_g} \text{ and } S_{b\_g}' \text{ in ON state when } v_{sin} > v_{tri} (v_{ab} = V_{bus})$$

$$S_{b\_g} \text{ and } S_{a\_g}' \text{ in ON state when } v_{sin} < v_{tri} (v_{ab} = -V_{bus})$$

Consequently, the output voltage v_{ab} takes the shape as shown in Fig. 4-59(b). Note that, in this case, the output voltage only presents two voltage levels: V_{bus} and $-V_{bus}$. On the other hand, the amplitude modulation index is defined as:

$$m_a = \frac{v_{sin\ amplitude}}{v_{tri\ amplitude}} \tag{4.222}$$

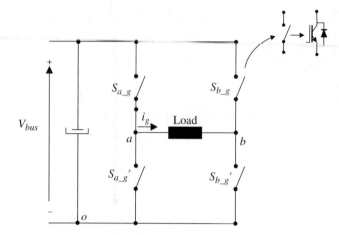

Fig. 4-58   *Simplified representation of the single-phase grid-connected converter system*

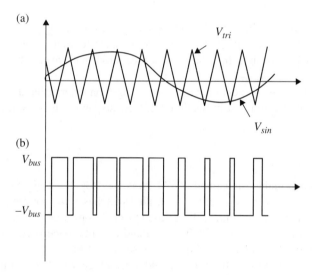

Fig. 4-59   *Bipolar sinusoidal PWM strategy operating at* $m_a = 0.8$ *and* $m_f = 9$: *(a) triangular and reference and (b) output voltage signal*

With this definition, the fundamental component of the output voltage can be calculated as:

$$<v_{ab}>_1 = m_a \cdot V_{bus} \tag{4.223}$$

Therefore, when not going to the over-modulation region, at maximum amplitude modulation index which is equal to 1, the maximum fundamental component of the output voltage is equal to V_{bus}. At lower m_a, the output voltage is proportional to V_{bus}. Another parameter typically defined in PWMs is the frequency modulation index, which relates the frequencies of the sinusoidal signal and the triangular signal:

$$m_f = \frac{f_{tri}}{f_{sin}} \tag{4.224}$$

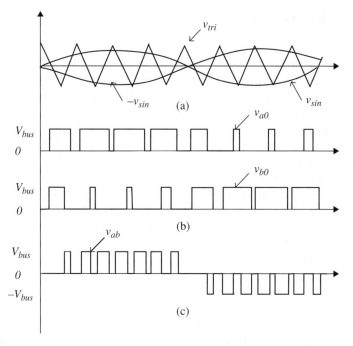

Fig. 4-60 *Unipolar sinusoidal PWM strategy operating at* $m_a = 0.8$ *and* $m_f = 8$: *(a) triangular and sinusoidal reference signals, (b) output voltage signals referred to the zero of the DC bus, and (c) output voltage signal* v_{ab}

On the other hand, the unipolar sinusoidal PWM is a modulation method that is commonly employed as an alternative to the bipolar modulation described above. It uses two complementary sinusoidal waveforms for creating the commutation of the controlled switches, as depicted in Fig. 4-60(a), by comparing them with the triangular carrier waveform. The commutation orders of the switches are created according to the following law:

$$S_{a\_g} \text{ in ON state when } v_{sin} > v_{tri}$$

$$S_{b\_g}{}' \text{ in ON state when } -v_{sin} < v_{tri}$$

$$S_{b\_g} \text{ in ON state when } -v_{sin} > v_{tri}$$

$$S_{a\_g}{}' \text{ in ON state when } v_{sin} < v_{tri}$$

Consequently, the output voltages referred to the zero of the DC bus (v_{ao} and v_{bo}) take the shape depicted in Fig. 4-60(b). Then, if the subtraction between the voltages v_{ao} and v_{bo} is performed, the output voltage v_{ab} of the converter is obtained, as illustrated in Fig. 4-60(c).

Note that for this unipolar sinusoidal PWM the voltage levels obtained at the output are V_{bus}, 0 and $-V_{bus}$. The voltages V_{bus} and 0 are obtained in the positive half-cycle of the fundamental voltage, while voltages $-V_{bus}$ and 0 in the negative half-cycle. In this case, the fundamental component of the output voltage can be calculated with (4.223). Note that the same definition for the amplitude and frequency modulation indexes is maintained for this modulation, as utilized in the bipolar modulation (expressions (4.222) and (4.224)).

Finally, it is important to highlight that the steady-state model equations are equivalent to those deduced for three-phase systems in Section 4.2.5, and therefore they are not shown here.

4.6   Single-phase grid-connected converter control

This section analyzes the control of the grid-connected converter. It is possible to find different control strategies for single-phase grid-connected converters; however, this section only describes a general control with two representative current control loops. Thus, the general control strategy is shown in Fig. 4-61. In general, this grid-connected converter is the AFE of another three-phase or single-phase converter; therefore, in general it is in charge of controlling the DC bus voltage of the DC-link. For that purpose, a V_{bus} loop control is implemented as depicted, whose regulator's output creates the amplitude of the grid current reference. The regulator can be a simple PI controller. Then this current reference amplitude is converted into a sinusoidal current reference. This is achieved by multiplying it with the sinusoidal grid voltage in per unit as depicted. This sinusoidal grid voltage can be measured directly from the grid voltage, or it can also be obtained with a single-phase PLL or similar. Therefore, thanks to this mathematical operation, the grid voltage and the grid current tend to be in phase, achieving a unitary power factor. The current reference is then controlled by a current control loop. More details about the current loops are provided later in this section.

This current control loop finally generates the order commands of the controlled switches of the converter. Note that for a proper operation of the converter the following relation must be held:

$$V_{bus} > \sqrt{2} \cdot v_g \tag{4.225}$$

Thus, the phasor diagram of the AC current and voltage magnitudes is depicted in Fig. 4-62, assuming a pure inductive filter and neglecting the voltage drop in the filter resistance. Note that the bigger the filter inductance L_f, or the bigger the transmitted current i_g, the bigger the voltage drop at the filter and, therefore, the bigger the required converter output voltage v_{ab}, implying also a bigger V_{bus} voltage requirement.

On the other hand, one possible solution for the current control loop is based on resonant controllers, as depicted in Fig. 4-63. Since the current loop operates with sinusoidal waveforms, a classic PI regulator would not provide good control performances; therefore, a resonant controller is more suitable in this case. After the

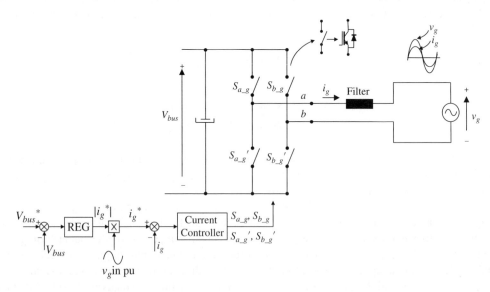

Fig. 4-61   *Single-phase grid-connected converter control*

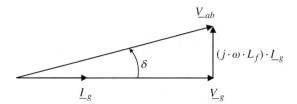

Fig. 4-62   *Phasor diagram of the single-phase converter magnitudes, assuming a pure inductive filter and neglecting the voltage drop in the filter resistance*

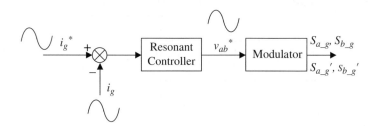

Fig. 4-63   *Instantaneous current control with resonant controller*

current resonant controller, the voltage reference is created which is later transformed into pulses for the controlled switches by means of the modulator. The modulator can employ any of the modulation strategies seen in previous subsections.

On the other hand, the typical transfer function of the resonant regulator is:

$$C(s) = k_p + \frac{2 \cdot k_i \cdot s}{s^2 + \omega_0^2} \tag{4.226}$$

It is made up of a proportional and an integrator term, which contains a resonant pole aimed at obtaining an infinite gain at the resonance frequency (ω_0), which in this case corresponds with the grid constant frequency. This controller is especially suitable when it must operate with sinusoidal currents. Thanks to this resonant controller, it is possible to eliminate the tracking error of the currents at steady state, obtaining improved performance results compared to the classic PI controller. Thus, this controller is able to work with zero error of currents at its input and then generates at the output a sinusoidal v_{ab} voltage reference. Note that if the frequency of the grid varies slightly from its rated value the performance of the resonant controller is expected to be deteriorated. Therefore, a synchronization method is used to adjust the resonance frequency of the regulator to the frequency of the grid. This synchronization method is proposed in the next subsection.

In a different direction, an alternative current control loop can be made up by employing hysteresis controllers. The block diagram of the instantaneous control loop is shown in Fig. 4-64. Compared with the previous current control philosophy, in this case the hysteresis controller does not require a modulator, since we can affirm that it directly creates the switching orders for the controlled switches with the help of a look-up table.

Thus, this controller operates in such a way that if the current error is bigger than the hysteresis band (HB) the output voltage v_{ab} is equal to V_{bus}:

$$\left(i_g^* - i_g\right) > \mathrm{HB} = > v_{ab} = V_{bus}$$

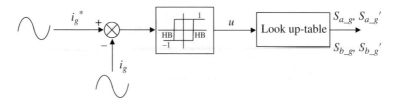

Fig. 4-64   *Instantaneous current control with HB controller*

Table 4-4   *Look up table for HB controller*

u	S_{a_g}	S_{b_g}	v_{ab}
1	1	0	V_{bus}
−1	0	1	$-V_{bus}$

On the other hand, if the current error is lower than the HB, the output voltage v_{ab} is equal to $-V_{bus}$

$$\left(i_g{}^* - i_g\right) < \mathrm{HB} => v_{ab} = -V_{bus}$$

The order commands for the controlled switches to achieve these output voltages are summarized in Table 4-4.

Hence, this controller is very simple in implementation and is able to provide a very fast transient response. Owing to its working nature, it presents practical insensitivity to DC-link voltage ripple. However, its main disadvantages are that it operates at a non-constant switching frequency (it varies within a band) and the fundamental current suffers a phase lag that increases at higher frequency. An example of how the current ripple appears at the phase current and the output voltage v_{ab} waveform is illustrated in Fig. 4-65. It will be noticed that depending on the HB the current ripple would be higher or lower. On the other hand, it is also possible to see how the output voltage v_{ab} does not present the zero voltage levels, only V_{bus} and $-V_{bus}$ as determined by Table 4-4. Note that this HB controller philosophy could be also implemented with three-phase converters.

4.6.1   Synchronization method for single-phase converters

As previously analyzed for three-phase grid-connected converters, single-phase converters need to be synchronized with the grid voltage in order to adjust the control performance in case of grid parameter variation. Therefore, there are several synchronization algorithms for single-phase systems in the specialized literature.

The synchronization methods described for three-phase systems in Sections 4.3.3 and 4.4.5 are not directly implementable in single-phase systems since they are based on space vector theory, and in single-phase systems there are no rotating vectors. However, in this section an algorithm is proposed based on the SOGI-QSG analyzed in Section 4.4.5 and whose block diagram is shown in Fig. 4-50. By means of this SOGI-QSG, a new signal will be generated using the measured grid voltage. This signal will be 90° delayed with respect to the original grid voltage and will be considered the β component of this voltage. As a consequence, using the measured grid voltage and its delayed signal as the $\alpha\beta$ components of the virtual space vector, a PLL can

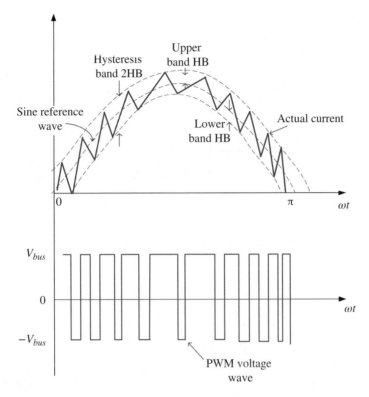

Fig. 4-65 *Instantaneous current control example and output voltage* v$_{ab}$ *waveform*

be implemented as described in Section 4.3.3 for balanced three-phase systems. This synchronization method is known in the specialized literature as SOGI-PLL [4] and its block diagram is shown in Fig. 4-66.

As can be observed, a synchronously rotating reference frame is generated as in three-phase systems which will be aligned with the virtual space vector of the grid voltage. As a consequence the rotational speed and the angle of the grid voltage are properly estimated. This rotational speed can be used, for instance, to adjust the resonance frequency of the resonant controllers described in the previous section and commonly used in single-phase converters.

4.6.2 Simulation-based examples

This section shows two examples of computer-based simulation results, presenting some of the aspects seen in grid-connected single-phase converter systems analyzed in previous sections. These simulation examples demonstrate how the single-phase connected converter behaves when it is coupled to a three-phase converter and to a single-phase converter.

4.6.2.1 Single-phase grid-connected converter with three-phase converter

This first example shows a three-phase converter coupled to the single-phase grid-connected converter, forming a single-phase/three-phase back-to-back converter system. The arrangement of converter, single-phase

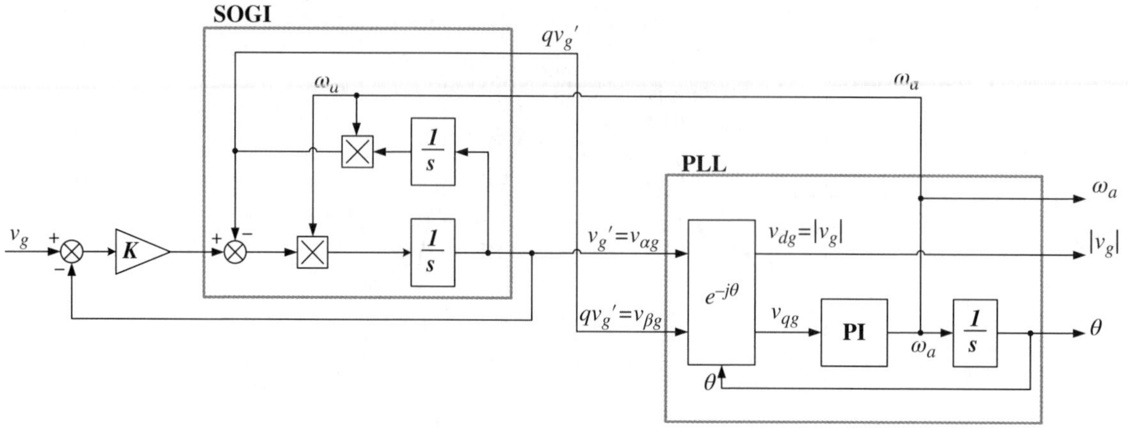

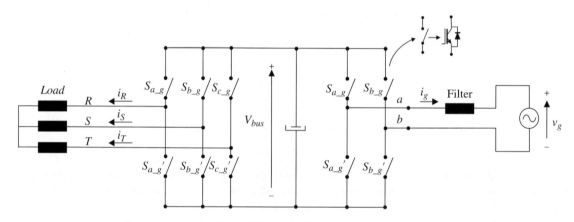

Fig. 4-66   *Block diagram of the SOGI-PLL*

Fig. 4-67   *Single-phase/three-phase back-to-back converter system*

grid, and three-phase load is illustrated in Fig. 4-67. For this particular example, the simulated conditions of the single-phase converter are as shown in Table 4-5.

The single-phase converter actuates as an AFE of a three-phase converter which is supplying a three-phase load. The grid single-phase voltage is 220V$_{RMS}$ and is connected to the converter through an inductive filter whose parameters are L_f and R_f. It could also be possible to use an LC or LCL. The single-phase converter is in charge of controlling the DC bus voltage, as described in previous section, according to the control block diagram of Fig. 4-61. The employed current regulator has been a resonant controller, with a unipolar sinusoidal PWM. Despite the fact that the single-phase converter does not need so much DC voltage for proper operation, the DC bus voltage has been set to 700V. This is due to the nature of the three-phase load that needs an extra voltage of approximately 500V$_{RMS}$ (phase to phase) for a proper operation. Thus, the single-phase converter controls the DC bus voltage, controlling the amount of active power transmitted through it, to the 700V.

Hence, the simulation results are carried out with Matlab-Simulink, resulting in the most representative system magnitudes illustrated in Fig. 4-68. There, it is possible to see how the DC bus voltage oscillates

Table 4-5   *Most representative parameters of the back-to-back converter*

V_{bus}	$700V_{DC}$
v_g	$220V_{RMS}$ (50Hz)
C_{bus}	1mF
f_{sw}	4kHz (both converters)
Single-phase converter filter	
L_f	30mH
R_f	$1m\Omega$

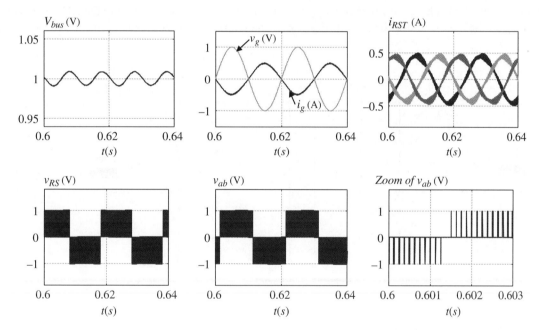

Fig. 4-68   *Most representative magnitudes of the single-phase/three-phase back-to-back converter system*

at double frequency of the grid (i.e. 100Hz). This occurs due to the fact that the single-phase converter must transmit an oscillatory (100Hz) active power. Depending on the amount of active power transmitted and the capacitor's size, the amplitude of these oscillations would vary. On the other hand, in Fig. 4-68 it is also possible to appreciate the grid voltage and the transmitted currents through the single-phase converter. Note that the opposite phases of voltage and currents, according to the convention of signs adopted, means that the active power is transmitted to the load, with a unitary power factor. In addition, it is possible to observe the shape of the v_{ab} single-phase output voltage, with a unipolar PWM switching at 4kHz. By comparison, the three-phase currents i_{RST} are shown in Fig. 4-68, presenting a higher current ripple than the single-phase converter currents, owing to the nature of the load. Finally, Fig. 4-68 depicts the phase-to-phase v_{RS} voltage of the three-phase converter, when it is being modulated by a sinusoidal PWM, as described in Section 4.2.4. The similitude between the three-phase converter voltage and the single-phase converter voltage is quite high.

4.6.2.2   *Single-phase grid-connected converter with single-phase converter load*

This second example shows a single-phase converter working as a load, coupled to the single-phase grid-connected converter, forming a single-phase/single-phase back-to-back converter system. The arrangement of converter, single-phase grid, and single-phase load is illustrated in Fig. 4-69. For this particular example, the simulated conditions are as shown in Table 4-6.

One of the single-phase converters actuates as an AFE of a single-phase converter which is supplying a single-phase load. The grid single-phase voltage is $220V_{RMS}$, as in the example of the previous subsection, and is connected to the converter through an inductive filter whose parameters are L_f and R_f. As mentioned previously, the single-phase converter connected to the grid is in charge of controlling the DC bus voltage, according to the control block diagram of Fig. 4-61. The employed current regulator has been a resonant

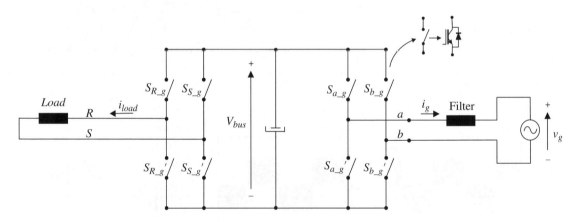

Fig. 4-69   *Single-phase/three-phase back-to-back converter system*

Table 4-6   *Most representative parameters of the back-to-back converter*

V_{bus}	$540V_{DC}$
v_g	$220V_{RMS}$ (50Hz)
C_{bus}	1mF
f_{sw}	5kHz (both converters)
Single-phase converter filter	
L_f	30mH
R_f	1mΩ
Load	
L_{load}	15mH
R_{load}	1Ω

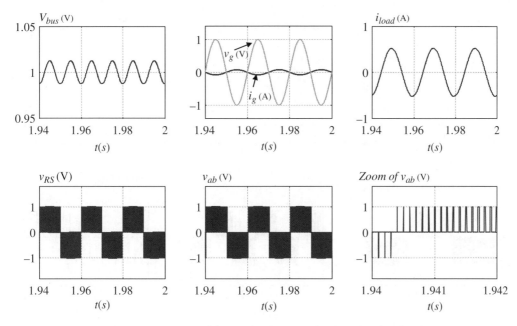

Fig. 4-70 *Most representative magnitudes of the single-phase/single-phase back-to-back converter system in per unit. Grid converter operates at 50Hz and load converter operates at 50Hz of fundamental frequencies*

controller, with a unipolar sinusoidal PWM as in the previous example. In this case, the DC bus voltage has been set to 540V, in order to be able to supply the required voltage of the load (380V$_{RMS}$).

For the load converter, it is open-loop-modulated with a unipolar PWM, with 0.4 of modulation index.

Therefore, the simulation results are carried out with Matlab-Simulink, resulting in the most representative system magnitudes illustrated in Fig. 4-70. In this case, both converters operate at 50Hz of fundamental voltage output. In this way, it is possible to see how the DC bus voltage oscillates at double frequency (i.e. 100Hz). This occurs because both single-phase converters must transmit an oscillatory active power of 100Hz. Fig. 4-70 also shows that the grid voltage and the grid current are in contra-phase, which means that they are providing active power to the load. It is also possible to observe the 50Hz current passing through the load and the V_{RS} voltage generated by the load converter to generate this load current. On the other hand, Fig. 4-70 also shows the v_{ab} voltage of the grid-connected converter and a detailed zoom of one specific time interval. It is possible to notice the similitude between both converter output voltages.

The next example shows the same working scenario and conditions, with the only difference being that the grid side converter operates at 50Hz of fundamental frequency while the load converter operates at 25Hz of fundamental frequency. In this case, the active power transmitted through the DC-link by the grid side converter presents 100Hz, while the one transmitted by the load converter presents 50Hz (double the fundamental output component in both cases). This fact is shown in computer-based simulation results shown in Fig. 4-71. It can be observed that the addition of both oscillatory powers results in a special shape of DC bus voltage behavior. On the one hand, with the simulated conditions, the grid side voltage and currents present contra-phase behavior. On the other hand, at the load single-phase current, the 25Hz fundamental frequency can be noted clearly, as illustrated in Fig. 4-71. In addition, the output voltage of

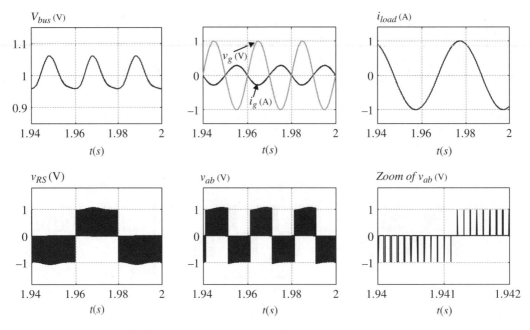

Fig. 4-71   *Most representative magnitudes of the single-phase/single-phase back-to-back converter system in per unit. Grid converter operates at 50Hz, while the load converter operates at 25Hz of fundamental frequencies*

both converters (v_{ab} and v_{RS}) can also be seen. Both voltages are very similar in shape. Finally, a zoomed version of the v_{ab} output voltage can also be appreciated.

References

[1]   Abad G, López L, Rodríguez M, et al. *Doubly Fed Induction Machine: Modeling and Control for Wind Energy Generation*. Hoboken, NJ: Wiley-IEEE Press, 2011.

[2]   Abu-Rub H, Malinowski M, Al-Haddad K. *Power Electronics for Renewable Energy Systems, Transportation and Industrial Applications*. New York: John Wiley & Sons, 2014.

[3]   Fortescue CL. Method of symmetrical co-ordinates applied to the solution of polyphase networks. *J Am Inst Elect Eng* 1918; 37: 1027–1140.

[4]   Teodorescu R, Liserre M, Rodríguez P. *Grid Converters for Photovoltaic and Wind Power Systems*. New York: Wiley-IEEE Press, 2011.

[5]   Milicua A, Abad G, Rodriguez Vidal MA. Online reference limitation method of shunt-connected converters to the grid to avoid exceeding voltage and current limits under unbalanced operation: Part I: Theory. *IEEE Trans Energy Conversion* 2015; 30: 852–863.

[6]   Milicua A, Abad G, Rodriguez Vidal MA. Online reference limitation method of shunt-connected converters to the grid to avoid exceeding voltage and current limits under unbalanced operation: Part II: Validation. *IEEE Trans Energy Conversion* 2015; 30: 864–873.

[7]   Rodriguez P, Luna A, Hermoso JR, et al. Current control method for distributed generation power generation plants under grid fault conditions. *37th Annual Conference on IEEE Industrial Electronics Society*, Melbourne, Australia, 7–11 November 2011: 1262–1269.

5

Railway traction

Xabier Agirre and Gonzalo Abad

5.1 Introduction

This chapter deals with railway traction. First of all, a general description of railway systems and railway vehicles is introduced, describing their most remarkable electric components. After that, a physical approach to the train is developed to enable the reader to obtain an understanding of simulation-based models, among other things. Then the variable speed drive of railway traction is studied, analyzing different configurations and their relation to the railway power supply system. Much of the chapter is dedicated to presenting and studying energy storage systems (ESSs) for railway applications. Among other aspects, energy storage technologies, trackside energy storage, on-board energy storage, power converters for energy storage, and some details about norms related to energy storage are studied. After this, ground level supply and auxiliary supply systems are studied. Finally, real examples are described and a short historical evolution of railway traction is provided and new tendencies are also enumerated.

5.2 General description

5.2.1 Railway systems in Europe

Due to historical reasons, railways today use widely different types of electrification systems. Often, based on what was state of the art in a particular country or area, the electrification system was developed, giving today's DC and AC electrification systems of different voltage amplitude and frequencies, as summarized in Table 5-1.

Power Electronics and Electric Drives for Traction Applications, First Edition. Edited by Gonzalo Abad.
© 2017 John Wiley & Sons, Ltd. Published 2017 by John Wiley & Sons, Ltd.

Table 5-1   *Catenary voltage characteristics for long-distance trains*

DC: 750V, 1500V, 3000V
AC: (15kV, 16.7Hz), (25kV, 50Hz, and 60Hz)

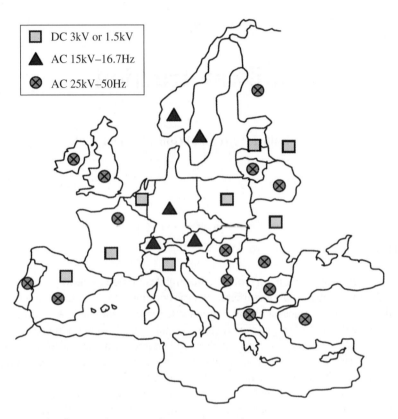

Fig. 5-1   *Railway main-line power supply systems in Europe [1]*

Fig. 5-1 shows the approximated railway main-line power supply systems in Europe, and at the same time demonstrates the wide diversity of voltages employed. The chosen electrification system type has determined over the years the evolution of the trains' technology. This chapter studies the associated electric traction technology for all AC or DC catenary characteristics.

For DC electrifications, 3kV and 1.5kV are the predominant voltages in long-distance trains. However, shorter-distance railway vehicles, such as trams or city trains, use 750V as the main power supply. With regards to the AC systems, 15kV and 16.7Hz are typical options in Europe, while the 25kV at 50Hz or 60Hz is also commonly employed in many countries. However, for security reasons, these 15kV or 25kV AC voltages are reduced inside trains by means of a power transformer, to lower 1kV, 1.5kVs or 3kV AC levels.

Note that in railway vehicles within the cities, such as trams and metros, lower speed thus power is needed so typically smaller voltage ranges are used. In addition, low-voltage catenary lines are of interest because of the difficulty of achieving high degrees of voltage isolation with respect to surrounding buildings and people. On the other hand, outside cities catenary voltage can be safely increased. This increase means less current is needed throughout for generating the high power levels necessary to achieve higher speed profiles.

Generally speaking, AC is easier to transmit over long distances because it is easier to boost than DC voltage, so it is an ideal medium for electric railways. In urban areas, where short lines are usually used, DC is the preferred option. In DC catenaries, power is carried by a third rail or by a thick wire. DC catenary lines suffer from energy loss as the distance between the electric substation and the vehicle increases; this is overcome by placing electric substations at close intervals – every two or three kilometers on a 750 V DC system, compared with every 20 kilometers for a 25 kV AC line.

5.2.1.1   *Advantages and disadvantages of AC power systems (15kV, 16.7Hz), (25kV, 50Hz and 60Hz)*

Advantages

- As mentioned before, the high voltage allows for sending high power levels with low current circulation, this helps to reduce ohmic losses in the line.
- 25 kV–50Hz systems can use the same 50Hz conventional distribution grid.

Disadvantages

- Higher voltage isolations levels are required, making AC catenary lines more expensive than DC ones.
- Railway vehicles connected to 25 kV–50Hz AC systems, as opposed to 15 kV–16 2/3Hz, generate disturbances on the three-phase main grid as they are mono phase loads. These imbalances can be minimized by alternating each time the phase from which the train is fed.

5.2.1.2   *Advantages and disadvantages of DC power systems (750V, 1500V, 3000V)*

Advantages

- Lower isolation requirements are needed and rectifiers can be smaller than in AC systems.
- Very well suited for short railway vehicles at relatively low speeds.

Disadvantages

- Even at 3kV DC voltage, current limitation through catenary line limits the use of heavy and fast railway vehicles.
- Electric substations have to be placed relatively close to each other, catenary wire has to be thick and heavy, and the pantograph needs to generate a lot of force to make a good electrical contact.
- The number of vehicles running on the same catenary line and the power that they can generate is limited by the load capability of the catenary line.
- The permissible range of voltages allowed for the standardized voltages is as stated in standard EN50163 (Railway Applications: Supply voltages of traction systems; Table 5-2). These take into account the number of trains drawing current and their distance from the substation.

Table 5-2   *Supply voltages of railway traction systems*

Electrification system	Voltage				
	Min non-permanent	Min permanent	Nominal	Max permanent	Max non-permanent (5 min)
600V DC	400V	400V	600V	720V	800V
750V DC	500V	500V	750V	900V	1000V
1500V DC	1000V	1000V	1500V	1800V	1950V
3000V DC	2kV	2kV	3kV	3.6kV	3.9kV
15kV AC, 16.7Hz	11kV	12kV	15kV	17.25kV	18kV
25kV AC, 50Hz	17.5kV	19kV	25kV	27.5kV	29kV

5.2.2   Railway vehicles classification

5.2.2.1   Tram

Main typical data

- Supply voltages: 600V DC, 750V DC, 900V DC
- Electric motors power: 75–150kW
- Power converter power: 150–300kW
- Number of electric motors per power converter: 2–4
- Number of cars: 3–6
- Total weight: 20–50 tons depending on number of cars
- Top speed: 50–70km/h
- Distributed traction: Yes. The power inverters are at different locations on the vehicle feeding the electric motors and are remotely controlled.

Fig. 5-2 shows a typical modern tram in an urban route. Nowadays, trams are being commercialized under 70–100% low-floor configurations. A 100% low-floor tram has no steps between the entrances and the passenger cabin. The low-floor design improves the accessibility of the tram for the public, and may provide larger windows and more airspace (Fig. 5-3).

5.2.2.2   Metro

A typical Metro is shown in Fig. 5-4.

Main typical data

- Supply voltages: 750V DC, 1500V DC
- Electric motors power: 150–250kW
- Power converter power: 350kW–1MW
- Number of electric motors per power converter: 2–4
- Number of cars: 4–8.
- Total weight: Around 40 tons/car
- Top speed: 80km/h
- Distributed traction: Yes.

Fig. 5-2 *Urban tramway (Source: CAF. Reproduced with permission of CAF)*

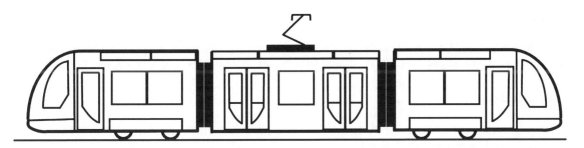

Fig. 5-3 *100% Low-floor tram configuration (Source: CAF P&A. Reproduced with permission of CAF P&A)*

Fig. 5-4 *Metro vehicle (Source: CAF. Reproduced with permission of CAF)*

5.2.2.3 Train

Trains are classified into EMUs (electrical multiple unit), DMUs (diesel multiple units), and high-speed trains, which can reach a maximum speed of between 200 and 350km/h (Fig. 5-5).

Main typical data

- Supply voltages: 750V DC, 1500V DC, 3000V DC, 15kV AC (16.7Hz) and 25kV AC (50Hz)
- Electric motors power: 200–600kW
- Power converter power: 200kW–1.4MW
- Number of electric motors per power converter: 2–8
- Number of cars: 3–12
- Total weight: 120–160 tons
- Top speed: 120–350km/h
- Distributed traction: Sometimes depending on manufacturer. The tendency is to move toward distributed traction to reduce vehicle weight per bogie and increase available adherence.

5.2.2.4 Locomotive

There are three kinds of locomotives, the first ones used to handle goods at station docks, the second ones were used for passenger transport, and the third ones to transport commodities over long distances (Fig. 5-6).

Main typical data

- Supply voltages: 750V DC, 1500V DC, 3000V DC, 15kV AC (16.7Hz) and 25kV AC (50Hz)
- Electric motors power: 340–1.4MW

Fig. 5-5 *Train (Source: CAF. Reproduced with permission of CAF)*

Fig. 5-6   *Locomotive (Source: CAF. Reproduced with permission of CAF)*

- Power converter power: 500kW–1.4MW
- Number of electric motors per power converter: 1–2
- Number of cars: 1
- Total weight: 90–120 tons
- Top speed: 100–200km/h.

Locomotive bogies are classified into various types according to their configuration in terms of number of axle and the design and structure of the suspension (Fig. 5-7).

B-B   The two axles in each one of both bogies are coupled together, either by a coupling rod (once common but now obsolete) or because they are both driven by the same motor.

Bo-Bo   Locomotive with two independent four-wheeled bogies with all axles powered by individual traction motors. Bo-Bos are mostly suited to express passenger or medium-sized locomotives.

Co-Co   Code for a locomotive wheel arrangement with two six-wheeled bogies with all axles powered by a separate motor per axle. Co-Cos are most suited to freight work, as the extra wheels give them good adhesion. They are also popular because the greater number of axles results in a lower axle load to the track.

5.2.3   Railway vehicles power architecture

Modern trains which take the energy from a catenary can be classified into two major groups: trains with DC catenary and trains with AC catenary. The energy is taken directly through the catenary and transmitted to different motors, by means of one three-phase power electronic converter. The return path is the ground through the rails. Later in this chapter it is shown that some other types of train use diesel engines as prime movers, located within the train itself.

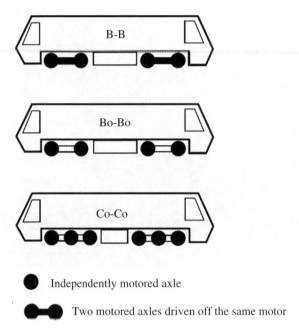

Fig. 5-7   *Locomotive wheel arrangements*

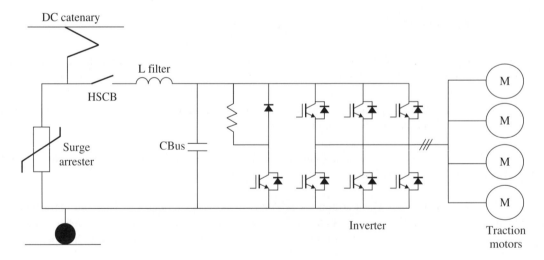

Fig. 5-8   *Electric drive configuration with DC catenary*

Fig. 5-8 and Fig. 5-9 show power circuit topologies for both DC and AC catenary-powered railway vehicles. In AC catenaries, the input alternating voltage is reduced usually by a low-frequency transformer, and sometimes (depending on the leakage magnitude of the transformer) an additional input filter is also needed to reduce the input AC current ripple. There have also been prototypes developed in recent years using

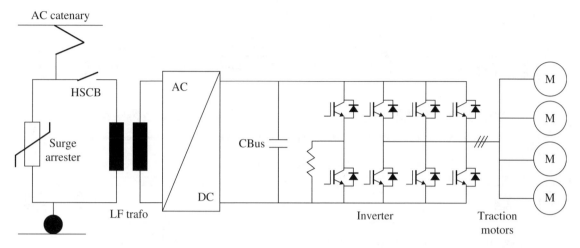

Fig. 5-9 *Electric drive configuration with AC catenary*

medium-frequency power electronic transformers for the same purpose. This subject is analyzed in detail later in this chapter.

The topology commonly used for the main power inverter is the classic two-level converter, although some examples of even more complex multilevel converters can also be found.

Fig. 5-10 depicts some examples of the distribution of the main elements for DC-based and AC-based bogie locomotives. The bogie of the train is a framework carrying the traction wheels and is attached to the locomotive. Above it is located the chassis of the locomotive. There are a wide range of solutions to transmit the torque of the motors to the traction wheels within the bogie.

5.2.4 Electric power system components classification

5.2.4.1 Pantograph

The pantograph is an electro-mechanical device capable of being raised to touch the catenary cable when electrical energy needs to be taken from the electric substation, and it has to be capable of maintaining good electric contact under all running speed conditions between the catenary cable and two existing graphite and copper-based strips. As the speed increases, it becomes increasingly difficult to maintain a good electric contact, so the pantograph's electrical contact is maintained either by spring or by air pressure. In high-speed vehicles compressed air pantographs are preferred because they can maintain the pantograph in a raised condition better than spring-based ones.

A typical tramway pantograph working at 750V DC catenary voltage can handle a nominal current of 1500A with two independently cushioned strips and is shown in Fig. 5-11.

Pantograph manufacturers recommend a tightening force of 45N per collector strip, so a total of 90N for the complete system is necessary. A combination of an incorrect tightening force and speed will cause an oscillating bouncing of the pantograph; this will cause electric sparks on catenary cables and could do serious damage. Arcing at the pantograph will interrupt the voltage and current feed to the vehicle and will cause electromagnetic interferences along the entire catenary line, possibly affecting other vehicles and electric substations.

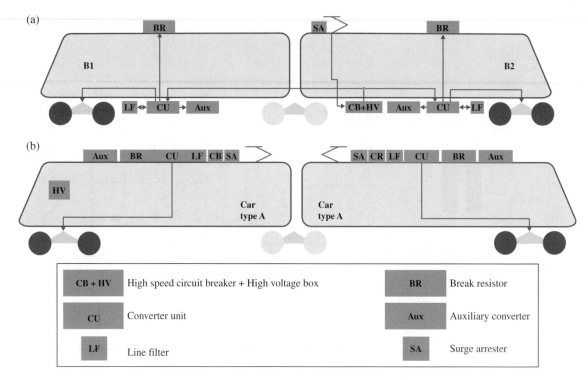

Fig. 5-10 *Main elements location for (a) DC and (b) AC locomotives (Source: CAF P&A. Reproduced with permission of CAF P&A)*

Fig. 5-11 *Pantograph raised on vehicle roof (Source: CAF P&A. Reproduced with permission of CAF P&A)*

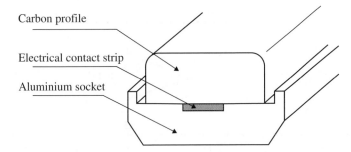

Carbon profile

Electrical contact strip

Aluminium socket

Fig. 5-12   *Cross sectional view of pantograph collector strip*

Table 5-3   *Pantograph manufacturers for the railway industry*

Pantograph manufacturer	Web
Schunk Group	www.schunk-group.com
Faiveley Transport	www.faiveleytransport.com
Stemmann-Technik	www.stemmann.de
Austbreck	www.austbreck.com.au
Melecs	http://melecs.com

The pantograph collector strips are composed of a mixture of graphite (carbon) and copper (Fig. 5-12). To increase the electrical conductivity, more copper than graphite is recommended, but the shoes wear early with the increase of copper, so a compromise between graphite and copper is necessary.

Catenary wire can be composed of one single cylindrical copper wire of 150mm^2 or two cylindrical copper wires of 107mm^2 each. It has to allow for the fact that the current passes through the collector strips while the train is in motion. In case of a static high-current static scenario (e.g. during the charge of an on-board ESS, at station stops), the current through each one of the existing four contact points (if only one single pantograph exists) can reach almost 250A. This high current can damage the copper wires if the temperature on the electrical contact between the copper wire and the collector strip reaches 120°C.

To avoid any damage to collector strips during a high-current static scenario, it is recommended to replace the catenary copper cylindrical wires with copper bus bars to improve the electrical contact.

Table 5-3 indicates the main pantograph manufacturers for the railway industry.

5.2.4.2   Surge arrester

The surge arresters are installed on the train roof between the pantograph and the main power breaker (the HSCB, or high-speed circuit breaker), as can be seen in Fig. 5-13. Their objective is to protect all electrical equipment against catenary incoming over-voltages. These over-voltages can be caused by external (lightning) or internal (switching) events. One side of the surge arrester is connected to ground, giving a path for the current but only for when an over-voltage occurs. The surge arrester uses a varistor internally, which shows different resistance values at different voltages [2].

The surge arrester mounted on the roof should be able to withstand operating conditions such as weather influences, airstreams, and existing mechanical stresses due to train movement (IEC 61373 Railway applications – Rolling stock equipment – Shock and vibration).

Table 5-4 indicates main surge arrester manufacturers for the railway industry.

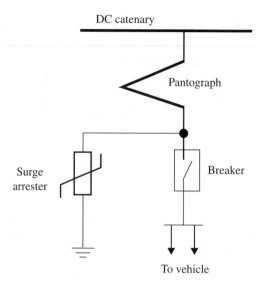

Fig. 5-13 *Arrangement of the surge arresters on the in-connector of the traction vehicle*

Table 5-4 *Surge arresters manufacturers for the railway industry*

Surge arrester manufacturer	Web
ABB	http://new.abb.com/high-voltage/surge-arresters
Tyco Electronics	http://energy.tycoelectronics.com
Siemens	www.energy.siemens.com
Tridelta	www.tridelta.de
Microelettrica Scientifica	www.microelettrica.com

5.2.4.3   HSCB (high-speed circuit breaker)

The HSCB is a device that safely interrupts extra-high fault current that is generated by a failure of an electric system during the operation of a train. The HSCB is usually located on the roof in series with the pantograph. There are three types: the vacuum-operated circuit breaker, the air-blast circuit breaker, and the electro-magnetically operated circuit breaker. The air or vacuum part is used to extinguish the arc which occurs as the two tips of the circuit breaker are opened. So long as the circuit breaker contacts are in the closed position, the electric energy flows into the train.

As an example the HSCB type IR6000 SV can operate under 1500V DC to 3000V DC catenary voltages with a short current breaking capacity of up to 60kA.

Table 5-5 indicates main HSCB manufacturers for the railway industry.

5.2.4.4   Input filter

An LC input filter's main objectives are, first, to disengage the high-frequency harmonics generated by the switching of the power inverter and, second, to present certain inductive behavior between the converter and catenary.

Table 5-5 *HSCB manufacturers for the railway industry*

HSCB manufacturer	Web
ABB	www.abb.com
Tyco Electronics	http://energy.tycoelectronics.com
Microelettrica Scientifica	www.microelettrica.com

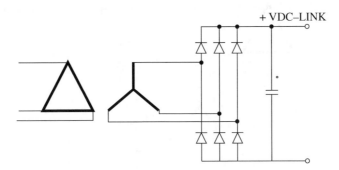

Fig. 5-14 *Six-pulse catenary rectifier*

In railway applications the admissible harmonics current levels injected in catenary lines due to the switching behavior of the power inverters need to be limited to avoid any interference with signals operating in the infrastructure.

The LC filter capacitor will absorb high-frequency harmonic currents, whereas only low harmonic current harmonics will be injected into the catenary line.

Main parameters during input LC filter sizing criteria are:

- Selection of filter cut-off frequency to comply with forbidden current harmonics injected to the catenary;
- To limit voltage ripple on filter/DC-link capacitor; and
- To limit current ripple on filter inductor.

The first step during LC filter determination is to know the topology for the main electric substation rectifier. This rectifier imposes the low-frequency voltage ripple on the catenary.

Railway electric substations are composed mainly of 6- or 12-pulse rectifiers.

Six-pulse rectifier A six-pulse rectifier is a simple configuration but the primary line current presents harmonics that includes 5th, 7th, 11th, 13th, 17th, 19th, 23rd, 25th, etc. The 5th harmonic is typically the largest harmonic component, followed by the 7th harmonic. The total input harmonic current distortion (THD) will typically vary between 100 and 35%. The THD for 12-pulse rectifiers will typically vary between 10 and 20%. See Fig. 5-14.

A delta primary requires three mains lines, without neutral, and avoids the so-called excitation imbalance.

Twelve-pulse rectifier A 12-pulse rectifier can be constructed from two six-pulse rectifiers connected in series. In this configuration each six-pulse rectifier generates one-half of the DC-link voltage. These two

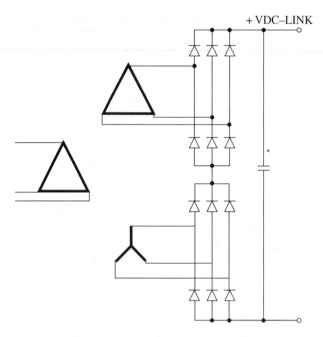

Fig. 5-15 *12-pulse catenary rectifier*

Table 5-6 *First four harmonic components for three-phase, six-pulse, and 12-pulse rectifiers (fmains = 50Hz)*

6-pulse rectifier		12-pulse rectifier	
Harmonic no.	Frequency (Hz)	Harmonic no.	Frequency (Hz)
5	250	11	550
7	350	13	650
11	550	23	1150
13	650	25	1250

individual pulse bridge rectifiers are combined in such a way that each is fed from a separate transformer winding and each of these windings is phase shifted by 30 electrical degrees from each other, then the 5th and 7th harmonics are theoretically cancelled. See Fig. 5-15.

This 12-pulse configuration will experience the harmonic stream containing the 11th, 13th, 23rd, and 25th harmonics. The total input harmonic current distortion for 12-pulse rectifiers will typically vary between 10 and 20% THD.

Series rectifier configuration in the 12-pulse rectifier allows lower transformer voltage that sums to the total rated DC-LINK input voltage. This reduces the applied voltage to individual diodes by either 50% (12-pulse) compared with the six-pulse rectifier improving their immunity to voltage transients and increasing the rectifier life. In addition, 12-pulse rectification is advantageous because output voltage is more smoothed in comparison with six-pulse rectification [3].

Looking at Table 5-6, it is quite evident that a 12-pulse rectifier introduces less disturbance on the mains AC current than a six-pulse one.

The next step is to select the cut-off frequency of the LC filter. The cut-off frequency is selected near but below the main harmonic (50Hz or 60Hz depending in the country [4], [5]).

$$f_{CUT-OFF} = \frac{1}{2\pi \cdot \sqrt{L \cdot C}} \tag{5.1}$$

The final step is to choose the desired ripple current on the inductor and ripple voltage on the filter capacitor. The next equations show current and voltage ripple values for a six-pulse rectifier (50Hz).

$$\Delta i_L = \frac{V_{\max}}{L \cdot \omega} \cdot 18,07e-3 \tag{5.2}$$

$$\Delta V_c = \frac{\Delta i_L}{C \cdot 100\pi \cdot 6} \tag{5.3}$$

being Vmax the peak value of the secondary rectifier voltage and ω:$2\pi f$.

Usually the current ripple is selected to be around 10% of the nominal current, and voltage ripple around 1% of the DC-link nominal voltage. An iterative process will be necessary to reach L and C values that comply with impedance and ripple limitations.

5.2.4.5   Filter inductor

Filter inductors for railway power converters are usually air core inductors and can be made using copper or aluminum (see Fig. 5-16). Air core inductors are fully linear and don't experience magnetic saturation like iron core inductors do. As opposite iron cored inductors are generally smaller than the air core ones. Depending also on the electrical performance required, the designer may opt for one or the other. In an air core inductor, the magnetic flux exists in its surrounded area. Due to that it is convenient to leave free space between the inductor and other elements susceptible of being affected by this stray magnetic flux.

Fig. 5-16   *Air core filter inductor (Source: CAF P&A. Reproduced with permission of CAF P&A)*

Inductive value for the filter inductor can vary from 2mH up to 10mH depending on the filter impedance requirement and allowable current ripple.

Filter inductors have to be cooled by forcing air through winding layers, especially when the inductor is located internally in a power box. Filter inductors located under the train frame are cooled by incoming air while the vehicle is running.

As railway filter inductors are cooled using forced air, they are exposed to rain, water, and dust, so special winding isolation is mandatory to avoid an electrical short circuit between inductor windings and the metallic structure (owing to the train cabling configuration, the metallic structure is at the same time negative pole of the DC-link).

A filter inductor manufacturing process involves:

1. Use NOMEX isolation paper on copper/aluminum conductors [6].
2. Conductors are wound to form a round inductor in shape and dimensions determined by the desired inductive value. Filter current shows low-frequency ripple value (300Hz–600Hz), so winding can be done with copper rectangular conductors because no negative skin effect is present.
3. Apply VPI (vacuum pressure impregnation) process to assure maximal protection against water, dust, and humidity. VPI technology is one of the most reliable methods performing good insulating conditions for electrical circuits and windings based on resins. During VPI, process resin is implied as the main insulation substance. Varnishing of circuits and inductors is common for mechanical robustness and extra environmental protection. Varnishing is implemented also widely for fixing the electrical circuits in systems exposed to shocks and mechanical vibrations. In addition, VPI eliminates dead air spaces that cause hot spots within the inductor coils and provides a low thermal resistance path that lowers the average operating temperature of the inductor.

The first step of the VPI process is to inject resin into the electrical circuit in order to fill all gaps, and then the temperature increases in order to cure all elements impregnated with the resin. Resins are produced and presented to market by several producers; they show diverse electrical, thermal, and mechanical qualities.

For the VPI process, different resin chemistries can be found in epoxy resin, polyester, polyesterimide, or silicone, but for railway inductive components it is recommended to use only epoxy resins without solvents.

During inductor mechanical construction, it is recommended to use H-class materials. H-class materials are based on mica or glass fiber with silicone compounds with a high thermal stability characteristic. H-class materials are prone to withstand 180°C of permanent temperature without degradation for the specified lifecycle of the component [7], [8].

Table 5-7 indicates main resin and varnish manufacturers for the railway industry, and Table 5-8 indicates the main filter inductor manufacturers.

Table 5-7   *Resin and varnish manufacturers for railway industry*

Resin manufacturer	Web
Dolph's	www.dolphs.com
Von Roll	www.vonroll.com
Elantas	www.elantas.com
Royal Diamond	http://electricalinsulationrd.com/
Saini Electrical	www.sainielectricals.com
Trafomec Industries	www.trafomec.com/transport-systems/

Table 5-8   *Filter inductor manufacturers for the railway industry*

Inductor manufacturer	Web
Trasfor	www.trasfor.com
ABB	www.abb.com/transformers
AQ	http://aqg.se/en/business-areas/inductive-components/about
Elettromil	www.elettromil.com
REO	www.reo.de
Schaffner	http://www.schaffner.com/products/power-magnetics/

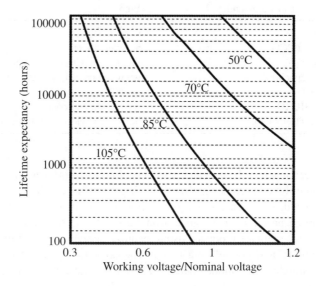

Fig. 5-17   *Lifetime expectancy vs voltage and hot spot temperature [11]*

5.2.4.6   DC-link capacitor

The main requirements for the DC-link capacitors used in railway power converters are: long life (generally 30 years), reliability, low internal stray inductance, and compact design. To achieve these requirements, railway DC-link capacitors are designed with self-healing properties and with a dry polypropylene film separator.

The self-healing ability assures that the capacitor remains operational even if an internal failure happens. Excessive DC voltage above maximum rating or transient voltage spikes higher than the dielectric withstanding voltage (DWV) will trigger internal failures in film capacitors. Other causes of failures are an excessive ambient temperature of above 85°C and excessive shock or vibration levels [9], [10].

Self-healing technology shows a linear and predictable capacitance decrease, but as in other kinds of capacitors, temperature and voltage are the key degradation acceleration factors, as can be seen in Fig. 5-17.

Film separator polypropylene high-voltage capacitors (PP-film) do not require serialization, so balancing resistors are not necessary. In addition, compared to aluminum capacitors, PP-film capacitors have lower internal resistance and stray inductance, so a higher ripple current can be managed.

Table 5-9   *DC-link capacitor standards for the railway industry*

IEC 61071:   Power electronic capacitors
IEC 61881:   Railway applications, rolling stock equipment, capacitors for power electronics
IEC 60068-2: Environmental testing
NF F 16-101: Rolling stock – Fire behavior – Materials choosing
NF F 16-102: Rolling stock – Fire behavior – Materials choosing, application for electric equipment
EN 45545-2: Railways applications – Fire protection on railway vehicles. Part 2

Table 5-10   *Capacitor manufacturers for the railway industry*

Capacitor manufacturer	Web
AVX	www.avx.com
TDK - EPCOS	http://en.tdk.eu/
Vishay	http://www.vishay.com/capacitors
API Capacitors	www.api-capacitors.com
ICAR	www.icar.com
ZEZ SILKO	www.zez-silko.com
ELECTRONICON	http://www.electronicon.com/en/

Power film capacitors can reach high capacitance values, of up to 48mF, whereas lower voltage and lower current capabilities of electrolytic-based designs mean having to use several in series or parallel for the same function. Dry film capacitors do not use acids; that's why they are considered environmentally friendly and can be stored without concern because they don't suffer from "dry out" mechanisms, as electrolytic versions do.

Self-healing, dry-type capacitors are produced using metalized polypropylene film. This ensures high rupture resistance, high reliability, and low self-inductance. The capacitor has a rectangular case which is filled with oil-based gel that is nontoxic and biodegradable. There is no need to install an overpressure vent. The capacitor terminals are specially designed to show low self-inductance [11], [12].

Working voltage, current, and temperature are the key factors when selecting a DC-link capacitor for a power inverter application. Maximum expected DC voltage has to be below the maximum permissible voltage and current RMS value through the capacitor has to be calculated in order to evaluate electrical losses and estimate the component thermal behavior. More details on capacitor selection can be found in [13], [14].

The capacitors used for railway application have to comply with the special standards listed in Table 5-9. And Table 5-10 indicates main film capacitor manufacturers for the railway industry.

5.2.4.7   *Power semiconductors*

In state-of-the-art traction applications, such as locomotives, long-distance trains, metros, and trams, which run on a worldwide variety of railroad networks, power semiconductors are imperative. Fig. 5-18 presents current and voltage limits for controllable power semiconductors, common switching frequency ranges for various power semiconductors, and current key application fields and limits.

IGBTs (insulated-gate bipolar transistors) are the most used power semiconductors in the range of some kW and several MW, owing to their medium and high switching performance. As we mention briefly in Chapter 1, IGBTs' main advantages are: 15V driven, self-turn-OFF even in the event of a short circuit, no need for snubbers, and low switching losses due to short switching times. Its package is divided into two parts: electrical

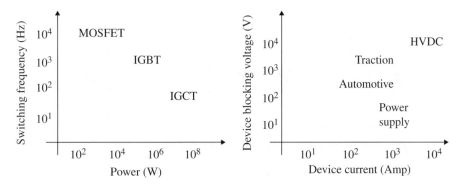

Fig. 5-18   *IGBT high-voltage module power vs railway application*

Fig. 5-19   *Traction IGBT high-voltage module packages (Source: Infineon. Reproduced with permission of Infineon Technologies AG)*

connections at the top (power and drive) and cooling plate at the bottom. This package helps in the optimization and simplification of the power converter construction.

The medium/high voltage levels stand from 1.7kVces to 6.5kVces blocking voltages. Fig. 5-19 shows different medium voltage IGBT module packages for traction applications. The internal electrical configuration can lead to double, single, or chopper configuration.

Fig. 5-20 shows the typical internal electrical configurations for the medium/high voltage packages for power module semiconductors.

Table 5-11 gives a guide to selecting the blocking voltage capability of an IGBT power module depending on the vehicle DC-link nominal voltage value. Trams have a nominal voltage of 750V DC so 1700V DC blocking voltage IGBT has to be selected. For the Metro, with a nominal voltage of 1500V DC, a 3000Vces blocking voltage IGBT is necessary, whereas for trains and locos with 3000V DC nominal DC-link 6500V IGBT are mandatory.

In railway applications the typical duty cycle for a traction inverter that feeds the traction electric motors starts with a high power delivery to accelerate the vehicle followed by low power demand during coasting and finishes again with a high power situation during regenerative braking. Owing to this working condition, the

Single   Double   Chopper

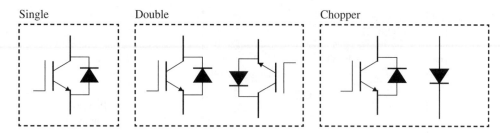

Fig. 5-20   *IGBT high-voltage module internal electrical configurations*

Table 5-11   *IGBT blocking capability as a selection criterion of the supply voltage*

AC output voltage (nominal DC-link voltage)	Preferred IGBT blocking voltage
$400V_{RMS}$ (600V DC-620V DC-900V DC)	1200V
$690V_{RMS}$ (750V DC)	1700V
$1000V_{RMS}$ (1500V DC)	3300V DC
$1600V_{RMS}$ (2500V DC)	4500V DC
$1900V_{RMS}$ (3000V DC)	6500V DC

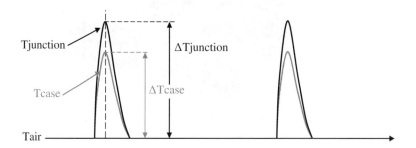

Fig. 5-21   *IGBT junction and case thermal stress on each vehicle start/stop cycle*

power semiconductor's base plate and junction suffer a high number of thermal load cycles during their lifetime. Fig. 5-21 shows the thermal behavior of the case and junction of an IGBT.

The most common failure mechanisms in an IGBT's internal structure due to thermal stress are:

- heel cracks
- bond lift-offs (increase of module saturation voltage)
- corrosion of wires
- increase of thermal resistance (increase of junction temperature)
- solder fatigue and shrink of thermal interface due to mismatch in CTE (coefficient of thermal expansion) between different layers on module internal construction.

Fig. 5-22 shows the internal layered construction of a semiconductor chip that shows a cracked wire bond due to thermal stress over time.

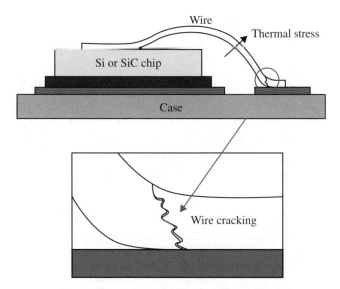

Fig. 5-22   *Wire crack in a thermally overstressed power semiconductor*

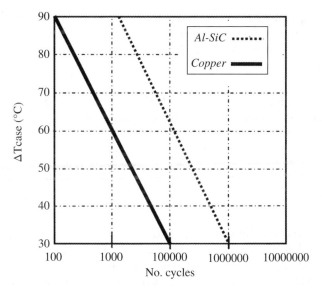

Fig. 5-23   *Thermal cycling capability of standard (Cu) and traction (Al-SiC) base plates semiconductor modules*

To withstand this thermal stress and guarantee the required lifecycle, the power semiconductors oriented to railway traction applications are made using an Al-SiC base plate combination (70% SiC + 30% Al) for the highest reliability, no thermal fatigue, and high power cycling capability. The main disadvantage of using an Al-SiC base plate is a 50% reduction of thermal conductivity with respect to copper and higher pricing.

Fig. 5-23 shows the difference in number of estimated cycles until an end of life (EOL) condition of power module regarding base plate material. The lifespan of Al-SiC base plate modules is higher than standard copper ones.

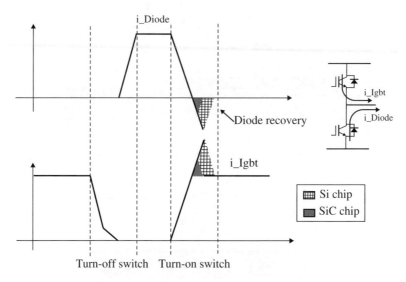

Fig. 5-24   *Switching behavior for Si and Sic chips*

To increase the power capability it is necessary to reduce the conduction losses and switching losses. These goals can be achieved by using power semiconductors based on SiC (silicon carbide) or GaN (gallium nitride) instead of Si (silicon) ones. The main advantages that power semiconductors made of SiC or GaN have over conventional Si components are:

- low conduction and switching losses (so higher-efficiency converters are possible);
- higher blocking voltages (so simpler inverter topologies like two-level VSIs are possible instead of multilevel ones);
- higher possible power densities (so smaller converters are possible);
- higher permissible operating temperatures (so less cooling effort is necessary); and
- shorter switching times, higher switching frequencies (so smaller inductive components can be made for DC/DC converters).

Fig. 5-24 shows the current behavior under switching circumstances for Si and SiC chips. Si chips experience larger current excursions than SiC chips, so higher switching losses are expected with Si devices.

In 2012, Hitachi launched a hybrid power inverter made of Si IGBT and a SiC diode. Hitachi employed SiC to reduce equipment size and weight, and to cut power loss. Using these SiC devices, Hitachi claimed to have invented a converter that was 40% smaller and suffered 35% less power loss than the standards in the market [15], [16].

Table 5-12 indicates the main power semiconductor manufacturers for the railway industry.

5.2.4.8   *Braking resistor*

There are two kinds of braking processes involved during the stopping phase of a railway vehicle: friction braking and electrical braking.

Table 5-12   *Power semiconductor manufacturers for the railway industry*

IGBT manufacturer	Web
ABB	http://new.abb.com/semiconductors
Infineon	http://www.infineon.com/cms/en
Mitsubishi Electric	http://www.mitsubishielectric.com/semiconductors
Hitachi	http://www.hitachi-power-semiconductor-device.co.jp/en
Powerex	www.pwrx.com
Fuji Electric	http://www.fujielectric.com/products/semiconductor
Toshiba	http://www.toshiba.com/taec/Catalog/Family.do?familyid=1912643
IXYS	www.ixysuk.com
Dynex	http://www.dynexsemi.com/dynex-power-semiconductors

Friction braking is simple to understand because it is based on the speed reduction of the vehicle applying mechanical friction between the wheels and brake shoes, as in a conventional car. That way the kinetic energy is converted into heat energy that contributes to the speed reduction of the vehicle [17].

Electrical braking is based on the ability of the three-phase traction inductor motor to behave as an electrical generator. The power converter has to be bi-directional, like the three-phase, two-level converter, to be able to transfer to the DC-link the electrical energy when, during a braking process, the induction motor behaves like a generator. This incoming electrical energy charges the DC-link capacitor and causes a voltage rise. If during electrical braking the DC-link voltage reaches a very high value, the power semiconductor will break or be seriously damaged [18], [19], [20].

Usually the vehicle control system controls the electrical braking effort via the power converter until the speed decreases to 2km/h; at lower speed only the friction braking works. To avoid dangerous DC-link values during electrical braking the electrical energy can be:

- sent to the catenary: only possible if another train is consuming energy or if the railway electric substation is reversible and can send this energy back to the main AC grid
- burnt in on-board resistors: electrical energy is converted to heat to avoid overcharge of the DC-link capacitor
- stored in on-board ESS: this is the most efficient way to manage electrical energy. In addition this energy can be used to boost the vehicle during acceleration. The on-board ESS adds complexity to the vehicle managing strategy, and special storage technologies like batteries, capacitors, or flywheels have to be used. This subject is examined in detail in Section 5.6.

Fig. 5-25 shows a typical railway-braking resistor which is usually located on the vehicle roof and cooled naturally (no air force is used). The mechanical construction has to guarantee a high degree of voltage isolation even for the worst climate scenarios (rain, snow). Ni-Cr and Ni-Cu alloys are the preferred materials for the braking resistor because they can withstand temperature levels of up to 600°C.

It can be supposed that during the braking process the kinetic energy of the train is converted to heat energy on the braking resistor. The kinetic energy of the train just before braking starts can be determined as:

$$E_{KINETIC}(J) = \frac{1}{2} \cdot Train\_mass(kg) \cdot V_o^2 \qquad (5.4)$$

Fig. 5-25   *Railway braking resistor (Source: GINO ESE GmbH, 2015. Reproduced with permission of GINO ESE GmbH)*

The average power on the braking resistor during braking time will be:

$$P_{BRAKING}(W) = \frac{E_{KINETIC}(J)}{t_{BREAKING}(s)} \tag{5.5}$$

Fig. 5-26 shows the heat experienced by the braking resistor during a typical train working profile. Mission profile and kinetic energy levels are necessary to correctly size the braking resistor.

The braking resistor can be commanded by two different configurations. There will be one or two IGBTs whose duty cycle determines the amount of average power managed by the braking resistor during the braking time.

It is very common for sound problems to arise during braking processes. This annoying sound is generated in the braking resistor cables owing to the elevated switched current and the fixed switching frequency. A common strategy to reduce the switching sound is to use a variable switching frequency pattern on the braking converter. This switching pattern is centered near the desired switching frequency.

Fig. 5-27 shows two possible configurations for the braking converter. In the first case there is only one single switch driving the braking resistor, whereas in the second circuit the braking power is increased by using two switches whose switching orders are shifted 180°. Both configurations are possible and its use depends on the amount of average braking power needed.

where:

$$T = \frac{1}{fsw_{igbt}(Hz)} \tag{5.6}$$

Finally, Table 5-13 shows a representative list of braking resistor manufacturers for the railway industry.

5.2.4.9   *Power converter box*

The power inverter and its cooling media are located inside the power converter box. In some cases the battery charger and the auxiliary power converter are placed together with the power inverter inside the power

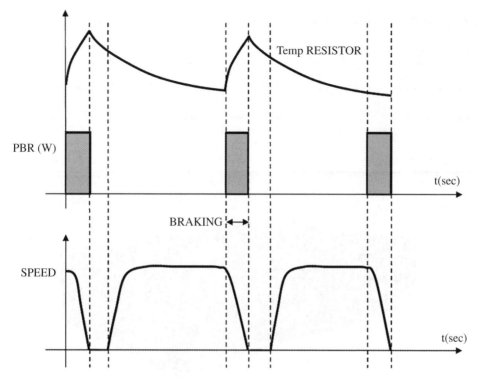

Fig. 5-26   *Railway braking resistor working profile in steady state*

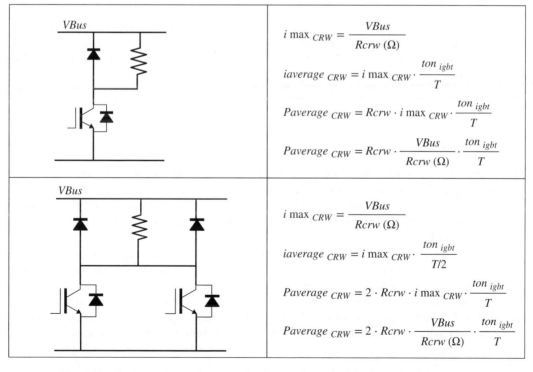

Fig. 5-27   *Braking power determination for single or double driven braking resistor*

Table 5-13   *Braking resistors manufacturers for railway industry*

Braking Resistor Manufacturer	Web
GINO AG	www.gino.de
Microelettrica Scientifica	http://www.microelettrica.com/en/
Cressal	www.cressall.com
REO	www.reo.de
Fulintech Science & Technology	http://www.fe-resistor.com/resistors-for-railway-application-p36.html

Fig. 5-28   *Train application 700kW/16.5 kV AC Power Converter Box (Source: CAF P&A. Reproduced with permission of CAF P&A)*

converter box. For trains and metros the power converter box is located under the vehicle frame but for tramways, especially as they are 100% low floor, it is located on the roof attached to the vehicle's body.

Fig. 5-28 shows a 700kW/16.5kV AC – 31kV AC power converter box and Fig. 5-29 shows a 240kW/300kW power converter box for a tramway application with nominal voltage of 750V DC. As is explained later in this chapter, if on-board ESSs exist, they will be placed in these kinds of boxes attached to the vehicle's body.

5.2.4.10   *Electric traction motor*

Traction motors enable the mechanical propulsion of the train. Nowadays, the most commonly employed machine is the squirrel cage induction motor, although permanent synchronous machines are starting to be employed as well. These traction motors are claimed by some manufacturers as modular motors because of their versatility. They can be self-cooled types or also externally forced air or water-cooled. An illustrative example is shown in Fig. 5-30.

Table 5-14 indicates main electric traction motor manufacturers for the railway industry.

The traction electric motors are located in the bogie itself in order to minimize mechanical transmissions. The axle is driven with a gear unit driven by the electric motor, as can be seen in Fig. 5-31.

In some applications no gear unit is used and the axle of the electric motor is at the same time the axle of the running wheels. This is the case of Siemens' SYNTEGRA integrated bogie, where the asynchronous electric motor and the gear unit are replaced by a synchronous traction motor becoming an integral component of the bogie [21].

Fig. 5-29 *Tramway application 240kW/300kW power converter box (Source: CAF P&A. Reproduced with permission of CAF P&A)*

Fig. 5-30 *Electric traction motor (Source: Traktionssysteme Austria (TSA). Reproduced with permission of TSA)*

Table 5-14 *Electric traction motor manufacturers for the railway industry*

Traction Motor Manufacturer	Web
Traktionssysteme Austria	www.traktionssysteme.at
ABB	www.abb.com/motors&generators
Mitsubishi	www.mitsubishielectric.com
Sherwood Electromotion Inc.	www.sherwoodelectromotion.com/traction-motors
Siemens	www.mobility.siemens.com
Skoda	http://www.skoda.cz/en/products/traction-motors/

Fig. 5-31   *Traction axle with gear unit and braking disks (Source: CAF. Reproduced with permission of CAF)*

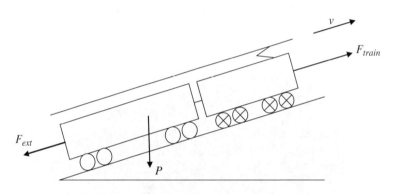

Fig. 5-32   *Forces affecting the train*

5.3   Physical approach

5.3.1   Forces

This section models train dynamics. Train dynamics is in a simplified way considered one-dimensional. The train one-dimensional dynamic here is very similar to the dynamic that is presented in more detail in Chapter 7 dedicated to road vehicles. There are some details and terms which are different, but in essence both models can be considered the same. Nevertheless, in this chapter we briefly describe the mathematical model of the train. We take into account forces which are in the same direction as the movement. Such forces can be grouped into two types [22]:

1. F_{train}: Forces created within the train, which can be positive (when the train is accelerating) or negative (when the train is braking). This force is typically created by the traction motor.
2. F_{ext}: External forces created outside the train, which in general oppose the train's movement

The graphical representation of these forces is depicted in Fig. 5-32. The effect of these forces on the train is represented by the following equation of movement:

$$F_{train} - F_{ext} = m_e \cdot a \qquad (5.7)$$

with m_e being the effective mass of the train and m the real mass of the train. Both real and effective masses of the train are related according to the correction coefficient [22]:

$$\xi = \frac{m_e}{m} \tag{5.8}$$

This correction coefficient takes into consideration all the inertias of the rotation masses (wheels, motors, etc.) and takes values greater than 1.

The external forces can be classified as [22]:

- F_f: Friction forces due to mechanical transmissions, friction between rail and wheel, aerodynamic resistance, etc. This force becomes dominant at high speeds.
- F_i: Initial static force that must be created when the train first starts to counter static friction.
- F_s: Resistance force due to slope.
- F_c: Force resistance due to curves.

Providing the following relation:

$$F_{ext} = F_f + F_i + F_s + F_c \tag{5.9}$$

Thus, friction forces are difficult to calculate analytically. There is one possible empirical approximation that can represent the friction forces as [22]:

$$F_f = a + bv + cv^2 \tag{5.10}$$

being v the train linear speed and a the static friction between the rail and wheel. The other constant, b, represents the friction due to lateral forces (rail quality and stability of the train). While finally c represents the aerodynamic opposition. This c constant is influenced by the train's form factors and whether it is inside or outside a tunnel. Fig. 5-33 shows some illustrative examples of friction forces of different trains.

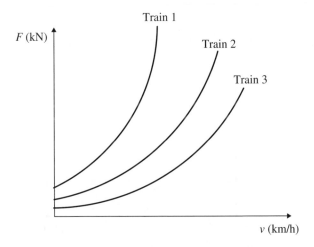

Fig. 5-33 *Friction forces of different train*

On the other hand, F_i, the initial static force, can be mathematically approximated to [22]:

$$F_i = 7.5 \cdot 10^{-3} mg \tag{5.11}$$

This force only appears at very low speeds, while at higher speeds it is zero.

On the other hand, the resistance force due to the slope is represented mathematically as:

$$F_s = m \cdot g \cdot \sin \alpha \tag{5.12}$$

with α angle representing the slope, as depicted in Fig. 5-34. This force will be positive when the train moves up and negative when the train moves down.

The force resistance due to curves depends on the curve radius, the distance between the tracks, and the mechanical construction of the vehicle. It is possible to consider the following approximated empirical expression [22]:

$$F_c = \frac{k_e}{r_v} \cdot 10^{-3} \cdot m \cdot g \tag{5.13}$$

being k_e the coefficient of distance between tracks and r_v the radius of the curve expressed in meters. An illustrative graphical example of this force is depicted in Fig. 5-35.

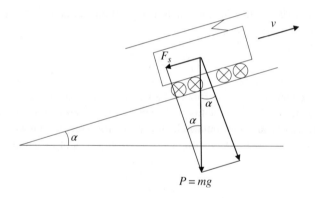

Fig. 5-34   *Force provoked by the slope*

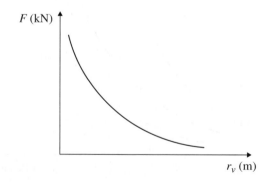

Fig. 5-35   *Example of force resistance due to curves*

5.3.2 Adhesion

The force in the train is transmitted by the contact of the driving wheel and the rail. The transmitted force can be mathematically represented as [22]:

$$F_x = N \cdot f_x \qquad (5.14)$$

being N the normal force, F_x the traction or braking force, and f_x the adhesion coefficient. The forces and speed at the driving wheel are represented in Fig. 5-36.

The adhesion coefficient f_x is not constant. It depends on factors such as the state of the rail, the linear speed, the slip speed of the wheel through the rail, the humidity, etc. In 1944, scientists Curtius and Kniffler took many experimental measurements establishing the adhesion coefficient dependency over the linear speed of the train [22]:

$$f_x = 0.161 + \frac{7.5}{3.6 \cdot v + 44} \qquad (5.15)$$

With v the train linear speed in km/h. Fig. 5-37 illustrates an example of adhesion coefficients according to Curtius and Kniffler measurements [22].

It should be noted that transferring forces via friction requires a difference of speeds (slip) between wheel and rail. Thus, the adhesion coefficient mainly depends on the difference of speeds, Δv:

$$\Delta v = v_u - v \qquad (5.16)$$

being v_u, the wheel circumferential speed that is calculated as:

$$v_u = \Omega \cdot R \qquad (5.17)$$

Thus, this adhesion coefficient is shaped as depicted in Fig. 5-38 for three different conditions: dry, wet, and greasy.

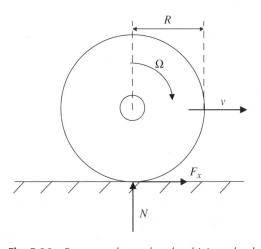

Fig. 5-36 *Forces and speed at the driving wheel*

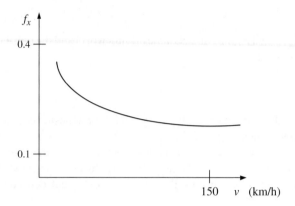

Fig. 5-37   *Adhesion coefficients according to Curtius and Kniffler measurements*

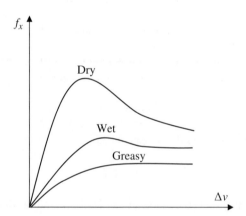

Fig. 5-38   *Adhesion coefficients $f_x = f(\Delta v)$ at different rail conditions*

Therefore, the slip is defined for driving conditions as:

$$\text{Driving: } s = \frac{\Delta v}{v}(v_u > v) \tag{5.18}$$

and for braking conditions as:

$$\text{Braking: } s = \frac{\Delta v}{v_u}(v_u < v) \tag{5.19}$$

Consequently, for a typical adhesion coefficient as depicted in Fig. 5-39, there is a stable region, A, where an increase of slip enables it to transmit a bigger force. On the other hand, there is also an unstable region, B, where a bigger slip produces less force. In region C, the slips are very small and the adhesion characteristic is nearly lineal. In region D, in order to increase a given force then a greater slip, compared to region C, is necessary.

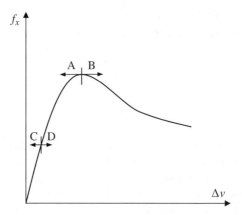

Fig. 5-39   *Typical adhesion coefficient*

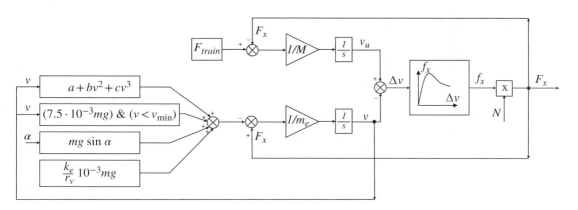

Fig. 5-40   *Model block diagram of the train*

Conversely, note that the traction motor driving the wheel produces a torque that is transmitted to the wheel and produces a force that can be calculated as:

$$F_{train} = \frac{T_{em}}{R}$$
(5.20)

being again R the radius of the wheel and T_{em} the developed torque by the traction motor, considering a gearbox transmission ratio equal to 1.

5.3.3   Model of the train

Hence, the model block diagram of the train can be represented as illustrated in Fig. 5-40. This model, in a simplified way, reproduces the behavior of the train, calculating on-line magnitudes such as forces, linear speeds, and speed difference (slips) at the wheel. The adhesion coefficient is given by a look-up table, while externally only the train force is needed (i.e. the force made by the traction motors).

Being M, the equivalent linear mass of the inertias of the rotating elements (wheels, shafts, gearboxes, etc.). Thus, it is possible to use an approximated mathematical curve to represent the adhesion coefficient. This curve could be:

$$f_x = \frac{2 \cdot \alpha}{\dfrac{\beta}{\Delta v} + \dfrac{\Delta v}{\beta}} \tag{5.21}$$

Substituting values of $\alpha = 0.3$ and $\beta = 4.8$, it is possible to obtain the adhesion characteristic represented in Fig. 5-41. By using this characteristic at the block diagram of Fig. 5-40 and assuming a total mass of the train of 2000kg, some resulting graphs can be obtained after several simulations, at different operating conditions.

Hence, in Fig. 5-42, the linear speed, forces, and f_x coefficient with a small resultant slip are illustrated. Thus, a relatively small force is applied to the wheel by means of the traction motors ($F_{train} = 750$N), obtaining a

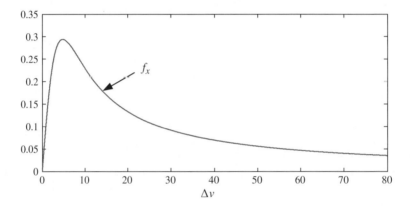

Fig. 5-41   *Detailed* $f_x = f(\Delta v)$ *curve*

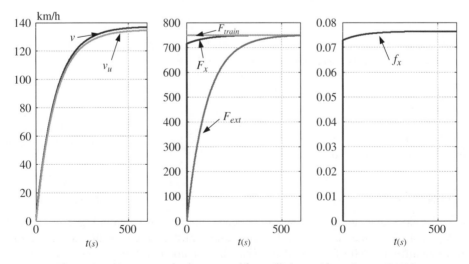

Fig. 5-42   *Linear speeds, forces, and* f_x *coefficient with an* $F_{train} = 700N$

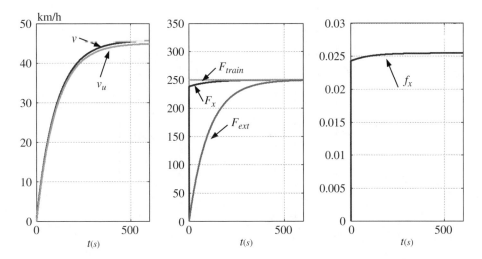

Fig. 5-43 *Linear speeds, forces, and* f_x *coefficient with a* $F_{train} = 250N$

linear speed difference of $\Delta v = 2.3$km/h. With this conditions at steady state, the operating forces affecting the train are equalized to the same value ($F_{ext} \, F_{train} = F_x = 750$N). Under these circumstances, the adhesion coefficient takes a value at steady state of $f_x = 0.76$.

On the other hand, if the traction force decreases as depicted in Fig. 5-43, the operating linear speed of the train reached is smaller, achieving also a smaller train and wheel speed difference of $\Delta v = 0.8$km/h. Therefore, the obtained adhesion coefficient also takes a smaller value at steady state of $f_x = 0.255$.

However, if we significantly increase the traction power to $F_{train} = 3000$N, the slip between the train and the wheel is uncontrollably increased, thereby losing the adhesion coefficient and operating at its unstable limits. In this way, the wheel presents a very high linear speed but the train linear speed cannot follow it. It is also possible to see how F_x force as the adhesion coefficient takes almost insignificant values, losing traction capability. It is also possible to see, in Fig. 5-44, how F_x force as the adhesion coefficient takes almost insignificant values, losing the traction capability.

As mentioned before, in essence this model of the train is equivalent to the model of the electric vehicle described in Chapter 7.

5.4 Electric drive in railway traction

This section shows the main characteristics of the variable speed electric drive of a train. It is divided into three different parts: the power electronic converter, the machine, and the control strategy employed.

5.4.1 Converters and catenaries

5.4.1.1 DC catenary

After around the year 2000, when the gate turn-OFF thyristor (GTO) was superseded by the IGBT, nowadays the typical converter configurations for DC catenary converters are the classic, simple, and robust two-level power electronic converter. For catenaries of 3kV, IGBTs of 6.5kV of blocking voltage are typically used, as they are

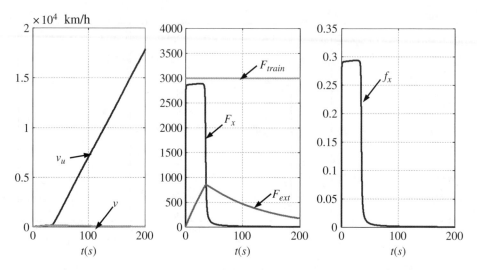

Fig. 5-44 *Linear speeds, forces, and $f_x = f(\Delta v)$ coefficient with an uncontrolled slip*

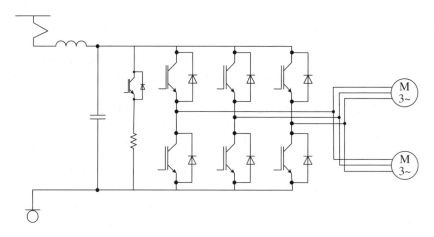

Fig. 5-45 *Electric drive configuration with DC catenary and two-level converter*

able to operate directly from 3kV catenaries with transients of up to 5kV. For catenaries of 1.5kV, IGBTs of 3.3kV or 4.5kV of blocking voltage can be used always nowadays, with two-level converter configurations.

Fig. 5-45 illustrates the typical configuration of converter for these DC catenaries. For catenaries of 750V, IGBTs of 1700V can be used.

In the past, when IGBTs of such high-voltage blocking capacity were not available, the converter configurations were more complicated. It was typical, for instance, to find input chopper converter stages, with the objective of being able to distribute the DC catenary voltage among different semiconductors. Another possible topology that was more extensively used in the past was the three-level neutral point clamped (NPC) multilevel converter topology, enabling them to be handled with high-voltage catenaries by having the semiconductors in a special configuration. For instance, for catenaries of 3kV, the motor voltages are around 2400Vrms. Then, incorporated into the converter as mentioned before, the crowbar circuit is compulsory for the dynamic braking process, since it is not always possible to ensure the regeneration of the braking energy into the grid.

5.4.1.2 AC catenary

With AC catenaries, a 16.7Hz (or 50Hz or 60Hz) single-phase transformer is required to step-down the catenary voltage (15kV or 25kV) to the drive's operating voltage. The drive, as depicted in Fig. 5-46, is first composed of a four-quadrant, single-phase converter, which converts the input AC voltage to DC. Then, the three-phase converter supplies the traction motors of the bogie. The single-phase converter at the AC side is made to operate with a unitary power factor. The typical topology for both converters is the classic two-level converter, as illustrated in Fig. 5-46. Depending on the nature of the transformer employed, sometimes it is necessary to add a filter (typically an L filter) on the AC side between the transformer and the single-phase VSC converter. If the transformer presents a sufficiently high leakage inductance, it can reduce the current ripple exchanged with the catenary and coping with the norms. If not, an additional filter must be placed accordingly following the dimensions described in Chapter 4 to reduce the current harmonics.

It should be noted that often the input single-phase converter is composed of two parallelized converters supplied by a double secondary transformer, as depicted in Fig. 5-47.

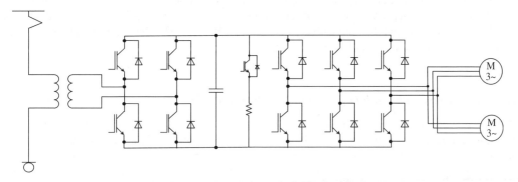

Fig. 5-46 *Electric drive configuration with AC catenary and line frequency transformer. A low-pass filter may be necessary in the DC-link in some cases, to reduce DC bus oscillations which naturally appear from the power transmission*

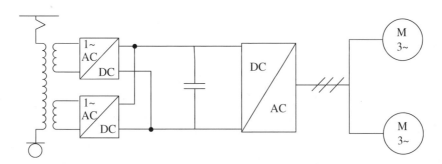

Fig. 5-47 *Electric drive configuration with AC catenary and line frequency transformer with double parallelized input converters*

This fact permits the balancing of the power rating of the input of two single-phase converters with the second three-phase converter supplying the traction motors. The number of four-quadrant converters also depends on the desired harmonic distortion at the line current and also of the redundancy that is to be achieved.

Although it is not represented in Fig. 5-47, the crowbar is still necessary. Again, as with the previous case, the typical converter configuration is the classic two-level converter. The ratio of the transformer is normally chosen in order to obtain a desired DC-link voltage of the converter. A commonly employed DC-link voltage is 1800V, allowing the use of IGBTs of 3.3kV available from many semiconductor manufacturers. In some cases, a lower DC-link voltage of 700V allows an IGBTs of 1.2kV of blocking voltage to be used producing lower switching power losses and low insulation requirements at the terminals of the motor. However, for output powers higher than 1MW, this would lead to too high currents.

It can be said that in these types of configurations the manufacturer of the electric equipment selects the DC bus voltage. For instance, if a DC bus of around 3 kV is chosen, a rated voltage of the motor is chosen of around 2.2kV, with IGBTs of 4.5kV. Or a choice of 3.6kV of DC voltage is also typical, with a motor of 2.6kV of rated voltage and IGBTs of 6.5kV.

On the other hand, with regards to the crowbar necessity, often the following design criterion is adopted: the input converter is not dimensioned with braking capacity. Then, the crowbar is installed to mitigate the DC bus fluctuations caused by the acceleration and braking processes of the train.

In addition, often a low-frequency filter (lower than 50Hz) is necessary at the DC bus to mitigate the natural voltage oscillations that are caused by the single-phase and three-phase converter connections. In general, the filter can be designed in such a way that the DC bus voltage oscillations are always equal or smaller than 1% in the worst case. It must be said that alternatively, or even complementarily, there are special control strategies which are able to mitigate this oscillation phenomenon.

5.4.1.3   *AC catenary with medium frequency transformer (MFT)*

One trend to reduce the large weight of the low frequency transformer in AC-catenary-based systems is to use MFT of less volume and weight, as depicted in Fig. 5-48. It can be observed that for this purpose a frequency converter must be placed before the transformer. Railway traction is an application which strongly benefits from avoiding large transformers. On the secondary side of the transformer, a rectifier converts the AC voltage to the DC-link voltage supplying the three-phase converter feeding the traction motors.

This topology brings a big challenge which consists of locating a frequency converter, at the high voltage side (i.e. at the AC catenary). Nowadays, semiconductors are not able to work directly with such medium

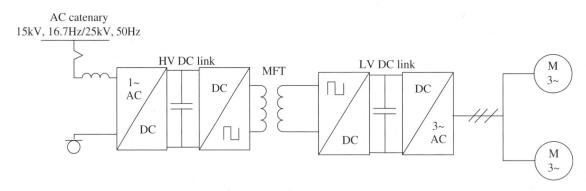

Fig. 5-48   *Electric drive configuration with AC catenary and MFT*

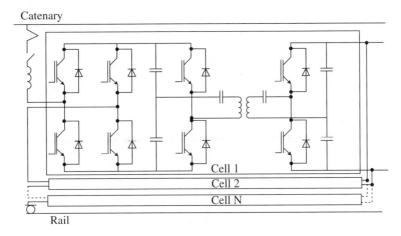

Catenary

Cell 1
Cell 2
Cell N

Rail

Fig. 5-49 *Electric drive configuration with AC catenary and cascade converter modules on the primary side and MFTs*

voltages of the AC catenary. Therefore, some kind of series connection of cells or modules is employed to reach the medium AC voltages of the AC catenary, resulting in a kind of multilevel converter structure. Thus, the input converter of Fig. 5-48 is not typically constructed with a single converter but rather as a number of converters operating at several kHz of switching frequency.

One possible configuration is based on using series cascade converter modules on the high voltage side, with outputs connected in parallel on the DC side, as illustrated in Fig. 5-49. This solution enables the system to be scalable and provides redundancy in case of failure.

Recent reports have shown that this topology is composed of several cascaded H-bridges at the input of resonant DC/DC converters. The cascaded H-bridges can be switched independently with interleaved strategy, obtaining improved quality waveforms of the input line current and therefore reducing the necessity of input filtering. The DC/DC converter is realized using a half-bridge configuration, with an LLC resonant tank where the leakage inductance and the magnetizing inductance of the transformer are involved in the resonance. In this way, zero voltage switching (ZVS) is achieved in turn-ON of the IGBTs, while near zero current switching is achieved during turn-OFF of the IGBTs. This configuration has already been prototyped by the company ABB, which calls it a power electronic transformer (PET). This MV prototype is designed for a 15kV, 16.7Hz railway network. Its rated output voltage is 1.5kV and its power rating is 1.2MVA. It is configured with eight cells (+1 for redundancy) and the transformer's switching frequency is 1.8kHz. The prototype was developed for a field trial with Swiss Federal Railways. The MF transformer fulfills three basic actions: first, it provides galvanic isolation, then it scales the voltage levels between the input and the output and, finally, the LLC resonance produces the soft switching. Some more alternative configuration solutions are described in Section 5.11.

5.4.1.4   *Diesel electric traction vehicles*

Another commonly used configuration is the diesel-electric system. In this case, as depicted in Fig. 5-50, the DC-link is fed by a three-phase synchronous generator via a diode bridge rectifier. This type of system does not require a catenary, as it uses the diesel engine as prime mover. The DC-link voltage of the three-phase converter supplying the traction motors can be adjusted by the generator's design. A voltage for the DC-link of

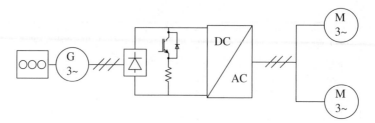

Fig. 5-50   *Diesel electric locomotive with alternator*

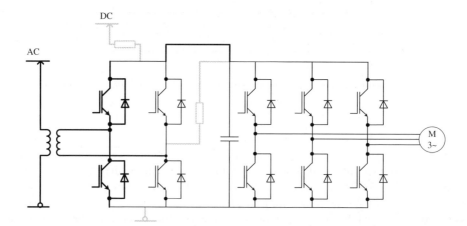

Fig. 5-51   *Multisystem converter with input chopper in DC mode. DC configuration in gray and AC configuration in black*

1.8kV with IGBTs of 3.3 kV of blocking voltage is a commonly employed choice. One disadvantage of these types of systems is the increased weight of the system due to the presence of the engine.

In principle, although the diesel electric traction is widely employed around the world, mainly because it does not require a heavy investment in catenaries and power supply networks, it presents many disadvantages compared to pure electric traction philosophy, such as: it is noisier, environmentally less friendly, heavier since the generator and diesel engine must be carried, has no possibility of energy recovery while braking, and has more expensive traction units.

5.4.1.5   *Multisystem traction vehicles*

After World War II, there arose a new demand for vehicles prepared to operate at catenaries of different voltages and characteristics. One example of this demand was in Western Union in Europe, where traffic crossing the borders started to rise enormously, demanding trains prepared to be supplied from different power line systems [23]. This, amidst many other causes, prompted the creation of multisystem traction systems. Many different configurations and designs have come and gone throughout the history of the train. However, this section only shows some illustrative examples of multisystem configurations that have been used.

For instance, in Fig. 5-51 a multisystem topology with input chopper is depicted [24]. When the system operates in AC mode, the input four-quadrant converter and the motor side converter are directly connected

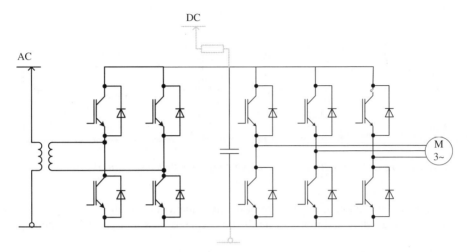

Fig. 5-52 *Multisystem converter without input chopper in DC mode. DC configuration in gray and AC configuration in black*

by the DC-link, forming a simple and robust converter system. Conversely, at the DC mode the DC-links are separated and the system is slightly more complex. The DC-link of the four-quadrant converter is directly connected to the DC catenary. An additional choke is placed and the input chopper operates as a step-down converter. This step-down converter feeds the DC-link of the motor side inverter. Since the input chopper is directly connected to the 3kV DC catenary, it needs 6.5kV IGBTs. However, for the motor side inverter, since the DC-link voltage has been stepped down, 6.5kV or 4.5kV of blocking voltage IGBTs can be used.

In a simpler fashion, Fig. 5-52 illustrates a multisystem configuration without an input chopper. In AC mode, the DC-link is simply supplied directly by the four-quadrant converter. Conversely, in DC mode, the DC catenary is directly connected to the DC-link. This configuration is typical, for instance, for trams (750V DC), which also operate in main AC lines. In this case, IGBTs of 1700V are used. It is also quite common for locomotives, which must operate at 1500V DC lines and AC lines. In this case, IGBTs of 3.3kV are employed. Finally, with IGBTs of 6.5kV, this configuration is also possible for DC catenaries of 3kV.

5.4.2 Electric machines

Associated with the concept of electric machines comes the bogie concept of the train. The electric machine can be placed at the bogie with a gearbox or in a direct drive concept (without a gearbox) as depicted in Fig. 5-53.

The main advantages of eliminating the gearbox can be summarized as:

- Reduction of the vibrations and noise produced by the gearbox, owing to a simplified mechanical transmission system.
- Improved efficiency, since the gearbox introduces some mechanical loss (typically resulting in 95–97% efficiency).
- Increased life of the system. Since there are fewer rotating elements, there are fewer elements that tend to deteriorate.
- Reduced maintenance requirements: The gearbox needs oils and to be repositioned regularly.

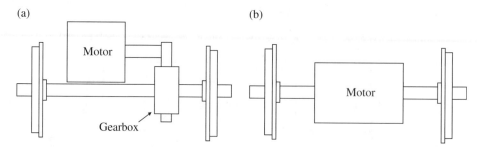

Fig. 5-53 *Simplified representations of bogies: (a) bogie concept with gearbox and (b) direct drive bogie concept*

- Improved use of space. Although the volume of the machine in a direct drive concept can sometimes be greater than a given gearbox and motor design's, this depends on a train's configuration.

On the other hand, the main disadvantages of eliminating the gearbox can be summarized as:

- The direct coupling implies that it is necessary to employ motors of higher torques, for the same traction force. This normally obliges motors with a higher active volume to be used. For railway traction applications, the necessary motor to provide high torque at low rotational speed is the permanent magnet synchronous machine (PMSM).

  The employment of PMSMs obliges one converter to be used for each motor, increasing therefore the cost of the system compared to parallelized motors. Note that, in general, for instance, one converter of 200kVA is cheaper than two converters of 100kVA. On the other hand, the reason why the parallelization of synchronous machines is not obvious in railway systems is because non-equal wheel diameters oblige the machines to be supplied at different frequencies; therefore, one converter per machine is needed. However, with induction machines, this fact is naturally solved by reaching a different slip ration of parallelized motors without any problem.

For these reasons induction machines, rather than synchronous machines, are more commonly used in railway applications. The main advantages and disadvantages of both machines are summarized in Table 5-15.

On the other hand, as an example, ABB has developed a commercial modular traction motor concept that presents several characteristics and modularized features, such as [25]:

- Scalable design: the flexible house design allows the motor's performance to be scaled to the customer's requirements.
- Predefined cross-sections: rating according to design requirements ensures high operating efficiency while providing maximum flexibility for the bogie layout.
- Cooling arrangements: the motor can present different cooling methods: closed self-ventilated, open self-ventilated, open forced-ventilated, or water-cooled.
- Bracket attachment: the brackets are adaptable to support a large number of bogies and attachment methods. The brackets are made as separate parts in order to reduce lead time and increase quality, reliability, and flexibility.
- Terminal box positions: customers can choose to place the terminal box in different predefined positions on the house. The design also supports flying leads to an externally mounted terminal box.
- Sensor arrangements: thermal sensors can be placed optionally, e.g. in the winding, stator core, or bearings. Speed sensors are integrated to keep the motor compact, while allowing sensors to be replaced without de-assembling the motor from the bogie.

Table 5-15 *Advantages and disadvantages of IM and PMSM in railway applications*

Motor	Induction motor	PM synchronous motor
Advantages	Robustness	Adequate for direct drive
	Overload capacity	Better efficiency (lower losses)
	Various motors can be supplied from an unique converter	Lower heating, simplifies the cooling system
	Low cost	
Disadvantages	Higher losses at the rotor, more complex cooling system for closed motors	High cost
	Maintenance needed of the associated gearbox	Demagnetizing of the magnets risk owing to overcurrents or over-temperatures
		Limited flux-weakening region
		Have to add a breaker between the motor and the converter*
		Needs an encoder that considers the zero position of the magnets
		Needs one converter per machine

* The traction system may be distributed (various traction converters and various bogies), so the train can operate with a converter deactivated. At high speeds, the contra-electro-magnetic-force of the machine due to the magnets can provoke the diodes of the converter to operate in an uncontrolled way, operating also as a brake.

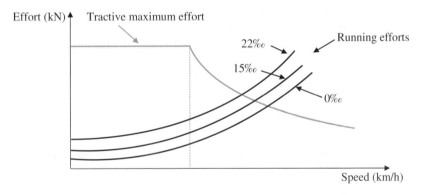

Fig. 5-54 *Traction diagram*

This example of traction motors is available in the low to high power ranges – suitable for use in suburban traction applications such as tramways and metros, intercity applications such as electric multiple units, and high-speed trains.

It must be mentioned that what have been shown here are just some examples of traction motors for railway applications. Many other examples and motor concepts can be found, but it is hoped this gives an idea of how practical and versatile designs capable of adapting to different necessities or scenarios are very useful and attractive to manufacturers.

On the other hand, given a specific machine, the maximum capability theoretical curves (see Chapter 2) and the train's load torque curves typically match, as represented in Fig. 5-54. Note that the figure shows linear forces instead of torques. As seen in Section 5.3 related to the mathematical model of the train, the different loads represent exponential curves which present a higher opposition force with a higher slope of the train track. Note also

Table 5-16 *Electrical parameters of a train's induction motor*

Supplying phase voltage (RMS)	230V
Rated torque	300Nm
Rated frequency	50Hz
Maximum power	47.1kW
Rated speed	1475rpm
Pole pairs	2
Stator resistance (20°C)	$R_s = 154.94\text{m}\Omega$.
Rotor resistance (20°C)	$R_r = 59.49\text{m}\Omega$
Magnetizing inductance	$L_m = 23.64\text{mH}$.
Stator leakage inductance	$L_{\sigma s} = 1.114\text{mH}$
Rotor leakage inductance	$L_{\sigma r} = 0.526\text{mH}$
Mechanical parameters	
Wheel mean radius	$D_{mean} = 555\text{mm}$
Gearbox ratio	$i = 7.44$

that the maximum force (torque) at constant torque region is much higher typically than the load force (torque), allowing a big available gap of force (torque) for acceleration. In addition, it can be observed how it is typical to reach a high speed in the flux-weakening region, far from the constant torque region.

Table 5-16 illustrates an example of an induction machine for railway traction applications. It is prepared to operate with catenaries of 750V, so the rated stator phase voltage is 230V. This is a relatively low power motor example, being able to provide a maximum power of only 47.1kW. It presents two pole pairs and the electrical parameters provided in the table.

Hence, from these electrical parameters, by following the procedure described in Chapter 2, it is possible to derive the maximum capability curves of the motor, as represented in Fig. 5-55.

5.4.3 Control strategy

5.4.3.1 General control strategy

The overall control strategy of a train can be divided into three different control levels, as depicted in Fig. 5-56. At each control level, the following is performed:

- Level 1: At this level, the pulses for the IGBTs of the inverter are generated. For this purpose, several modulation strategies can be performed, such as: PWM, space vector modulation (SVM), and selective harmonic elimination. At the same time, the orders for the crowbar activation and deactivation are also created.
- Level 2: At this level, the AC voltage references for control level 1 are created. These voltage references are a function of the torque and flux references, so basically a torque control is performed. The typical torque control strategy employed is a rotor flux oriented vector control, as described in Chapter 2. However, some special improvements are also typically added to the classic vector control in order to improve the performance, owing to the presence of high current ripples, for instance.
- Level 3: The highest-level control strategy consists of a sequence of different control actions. As it is later described, several functionalities are performed, which actuate on the torque reference that will be subsequently passed to the control level 2. The control level 3 described in this section is mostly oriented to drive trains with DC catenaries. In systems with AC catenaries some of the described control actions may be slightly different.

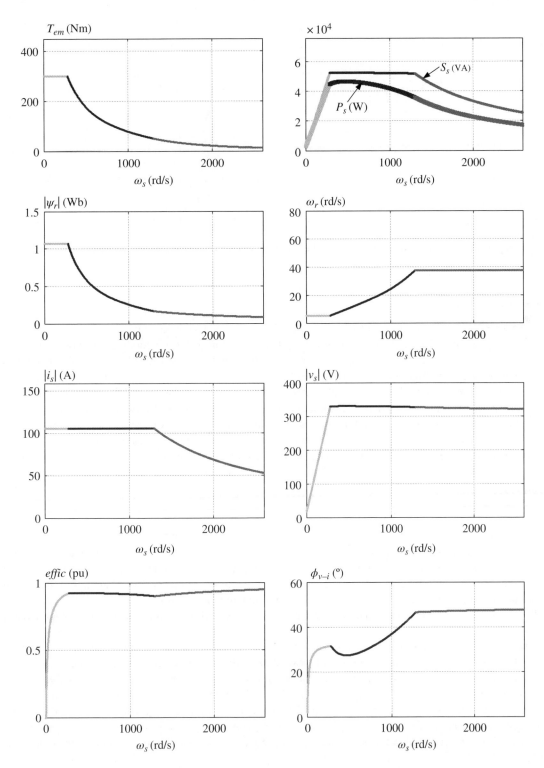

Fig. 5-55 *Maximum capability curves of the induction motor from Table 5-16*

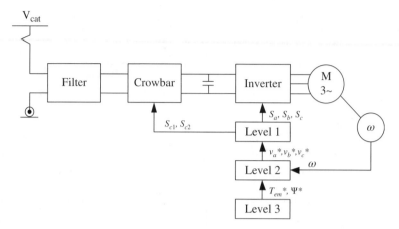

Fig. 5-56   *Simplified representation of the different levels of the overall control strategy of a train*

5.4.3.2   *Control level 1*

As mentioned before, at control level 1 basically the orders for the IGBTs are generated by means of what is normally called modulation. In this section it will be seen that different modulation strategies can be implemented into the train traction control algorithm. But it must be said first that the main objective of the modulation can be highlighted as the algorithm which calculates the ON-OFF times of the switches of the IGBTs. These ON-OFF times are calculated in order to generate the desired fundamental component waveform at the AC output, together with some secondary objectives, such as minimization of the generated harmonics, minimization of the switching losses, generation of a constant or variable switching frequency, and implementation simplicity. In the particular case of railway traction applications, it must be remarked that two main objectives of the modulation are: the minimization of the harmonic content of the output wave-form (for also later reducing the torque and current ripples) and the limitation of the switching frequency (typically between 500 and 1500Hz), in order to reduce the thermal stress and losses of the switching semiconductors.

For railway applications too, since the semiconductors typically employed are of high power, there are two special characteristics which normally impact on the modulation algorithm; first, the considerably high time of switching ON and switching OFF of the semiconductors and second, the considerably high dead time required for the switching. Thus, since the semiconductors require some time to commutate, because they do not do so instantaneously, a minimum conducting time and a dead time for the semiconductors is required, which must be compensated for by, or at least coordinated with, the modulation algorithm.

The following subsections illustrate the typical modulation algorithms employed in railway applications.

Sinusoidal PWM modulation with third harmonic injection   A classic modulation technique in this type of applications is the sinusoidal pulse-width modulation (PWM), with a third harmonic injection, for increasing the available fundamental component amplitude at the AC output side. One possible structure for this modulation is depicted in Fig. 5-57. Thanks to this injected harmonic, the theoretical amplitude increase is 15% in respect to a sinusoidal PWM without harmonic injection. This modulation solution is quite commonly employed because of its simplicity of implementation and the reasonably good performance obtained. Minimum conduction times and dead times produce a reduction of the maximum achievable output voltage.

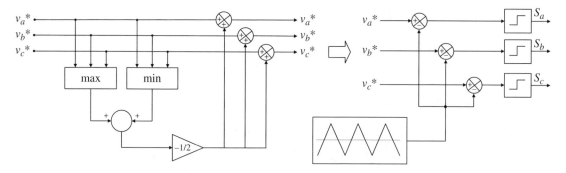

Fig. 5-57   *Sinusoidal PWM with third harmonic injection*

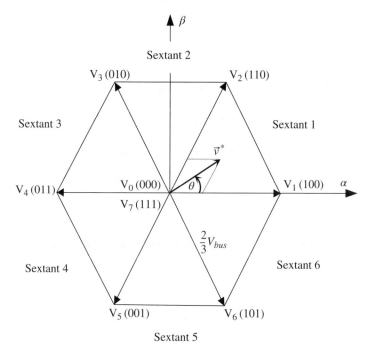

Fig. 5-58   *Space vector representation for SVM*

Space vector modulation (SVM)   The SVM generates the AC output converter voltage based on the space vector representation principle. In an alternative way to the PWM method seen in the previous section, the conduction times are generated according to the representation depicted in Fig. 5-58.

Thus, for instance, when the reference voltage vector is located in the first sextant, the active voltage vectors used are V_1 and V_2, as illustrated in Fig. 5-59. Each voltage vector is injected t_1 and t_2 times, in a constant switching period.

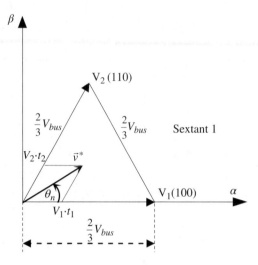

Fig. 5-59 *First sextant of the space vector diagram*

The synthesis of the required output voltage vector is carried out as:

$$v^* = V_1 \cdot d_1 + V_2 \cdot d_2 + V_0 \cdot (1 - d_1 - d_2) \tag{5.22}$$

The duty cycles (normalized injecting time) for each voltage vector can be calculated as:

$$d_1 = m_n \cdot \left(\cos(\theta_n) - \frac{\sin(\theta_n)}{\sqrt{3}} \right) \tag{5.23}$$

$$d_2 = m_n \cdot 2 \cdot \frac{\sin(\theta_n)}{\sqrt{3}} \tag{5.24}$$

Therefore, the injecting time for the zero vectors is (the remaining time):

$$d_{0,7} = 1 - d_1 - d_2 \tag{5.25}$$

being m_n: normalized voltage amplitude of the reference space vector:

$$m_n = \frac{|\overrightarrow{v}|^*}{\frac{2}{3} V_{bus}} \tag{5.26}$$

There is an additional degree of freedom for choosing the zero vector. It is possible to split the injection of the zero vector into different parts, generating different sequences of applied vectors. As with the previous case, the minimum conduction times and dead times affect the output voltage quality and fundamental amplitude. It can be said that this modulation is more often used than the previous one, since it brings the possibility of providing more improvements.

Six-pulse generation   Another modulation possibility that achieves the maximum output voltage waveform is the six-pulse generation modulation. The idea is that each switch operates half of the period. Therefore, the switching frequency obtained is half of the fundamental output period. The switching orders for the semiconductors as well as the output voltage waveforms are depicted in Fig. 5-60. With this modulation technique, it is not possible to modify the output voltage amplitude waveform, only the frequency. The amplitudes of the odd n harmonic components of this output waveform are given according to the following expression (line-to-line voltage):

$$Amplitude_{n,LL} = \frac{4}{n\pi}\frac{V_{bus}}{2}\cos\left(n\frac{\pi}{6}\right) \tag{5.27}$$

As can be deduced, this type of modulation is employed when the output voltage waveform must be maximum and constant, since there is no possibility to control it, but the frequency can be variable.

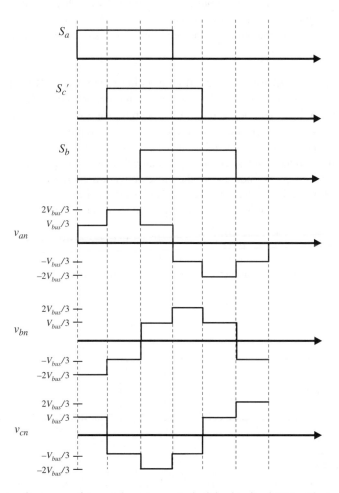

Fig. 5-60   *Output voltage waveforms (phase to neutral of the load) of VSC with six-pulse generation*

Selective harmonic elimination   Selective harmonic elimination (SHE) employs an off-line switching angle calculation with the aim of eliminating undesired harmonic content in the output AC voltage spectrum. In order to perform this, the switching angles are classically calculated for the first 90° of the output waveform, α_1, α_2, and α_3, as illustrated in Fig. 5-61. Accordingly, symmetry is applied for the rest of the period, in order to eliminate even harmonics in the resultant waveform. These angle sets are normally calculated off-line and then stored in look-up tables, for later use, then in the on-line generation of the voltage waveform. Assuming that we want to obtain k calculated angles (α_1, ... α_k), a set of solutions can be achieved equaling *k-1* harmonics to zero and providing a specific value to the amplitude of the fundamental component. Thus, at the output waveform *k-1* harmonics are eliminated while the fundamental component results in the amplitude that we want to achieve. The switching frequency generated by this type of modulation depends on two factors: the number of calculated angles and the frequency of the fundamental component of the output voltage.

With these basics of SHE modulation, it is possible to obtain output voltage waveforms in which some undesired key harmonics are eliminated. This solution is useful in high-power applications, such as railway traction, where a reduced switching frequency is desirable and the harmonic content of the obtained waveform must be minimized. Thus, in general, this modulation philosophy is typically employed at high amplitude output voltage waveforms, when the consumed current is high and therefore the thermal stress of the semiconductors is increased and the number of commutations of the semiconductors should be decreased.

Modulation combinations   Now that the typical modulation strategies for railway traction applications have been briefly described, it must be said that, in general, more than one modulation is used in the whole operation regimen of the train. Often, depending on the operation speed of the train, two, three, or even four different operation regions are distinguished and, at each region, a different modulation is used to improve the performance of the train. Thus, for instance, at low speeds, and therefore at low output voltage amplitudes, the typical employed modulation techniques are the sinusoidal PWM with third-harmonic injection or the SVM indistinctively. Then, above a predefined speed, therefore above a predefined output voltage amplitude, the modulation strategy can typically be changed to an SHE philosophy. Finally, at higher speeds (i.e. at highest output voltage waveforms), a six-pulse scheme is typically used, providing the maximum output voltage amplitude.

In this way, improvements such as: power losses, thermal behavior of the semiconductors, quality of the output voltages and currents, and so on. In this way, it is possible to obtain improvements such as: power losses, thermal behavior of the semiconductors, and quality of the output voltages and currents.

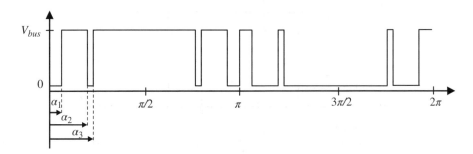

Fig. 5-61   *Output voltage waveform according to three calculated angles (α_1, α_2, α_3) of SHE*

5.4.3.3   *Control level 2*

At control level 2, the electro-magnetic torque and the rotor flux are typically controlled. Thus, the typical control strategy adopted is the indirect vector control strategy described in detail in Chapter 2. As is explained in a subsequent subsection, the speed is controlled in control level 3, so here in control level 2 only the torque control and rotor flux control are performed. Note that it has been supposed that the employed electric machine is the induction machine, which is probably the most commonly employed one in traction applications nowadays, as noted in the previous section.

It must be highlighted that variants of this control strategy are to be found. These variants try to employ basically the same vector control philosophy, but they incorporate some special improvements in order to minimize the effect of the torque and current ripples, which are normally considerably high in railway traction applications, owing to the low switching frequency employed.

If one single inverter drives two traction motors, both have to be of the same manufacturer's reference so their electric performances are the same. Both induction motors will be controlled as if they were a single traction motor, but with double electric and torque characteristics.

In terms of controllability, the inverter output current is measured and it is considered that half of this currents feeds each traction motor. The speed is measured in both electric motors and the average is considered in the vector control philosophy.

5.4.3.4   *Control level 3*

The different actions taken at control level 3 are represented in a simplified manner in block diagram of Fig. 5-62. It can be appreciated that, from the speed or acceleration references, the torque and flux references are finally

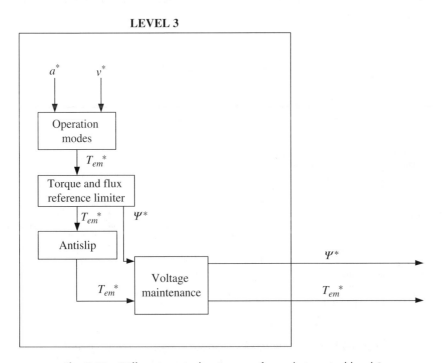

Fig. 5-62   *Different control actions performed at control level 3*

created, after performing several control functionalities. There is one task dedicated to the operation mode performed, another for the torque and flux references limitation, another for the anti-slip, another for the filter stabilization, and finally one for the voltage maintenance.

Operation modes   Fig. 5-63 illustrates the operational mode of speed and acceleration control. It is seen that two possible operational modes are possible: the speed control mode and the manual mode, or acceleration control mode. With the speed control mode, the objective is to maintain the speed at a given speed reference. A cascaded acceleration control loop is intermediately introduced, in order to obtain smooth control variations. It generates the force reference, for the subsequent control blocks.

In manual mode, the system operates at open-loop acceleration control. The objective is to impose a force which implies an acceleration equal to that demanded.

Jerk limitation   Then, once the force reference is created by one of the operation modes, as depicted in Fig. 5-63, the jerk limitation action is taken. The jerk, which is also defined in Chapter 8, is the first derivative of the acceleration:

$$jerk(t) = \frac{da(t)}{dt} \tag{5.28}$$

Thus, for guaranteeing the comfort of the passengers in all conduction modes, there is a jerk limitation which reduces the force applied, in case of overshooting a pre-established maximum jerk value. Hence, once the jerk is calculated, if this is bigger than the pre-established value, the force reference will be recalculated so the jerk does not take higher values than the maximum. Consequently, from the force reference, it is possible to calculate the train acceleration as:

$$a(k) = \frac{F^*}{M_{train}} \tag{5.29}$$

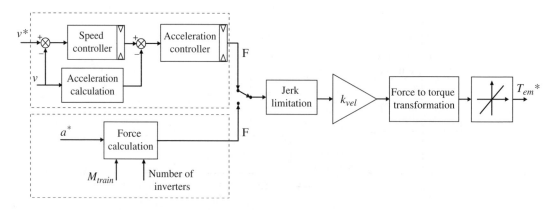

Fig. 5-63   *Operation modes of speed and acceleration control*

From the acceleration calculated at the previous sample time, it is possible to calculate the jerk according to:

$$jerk(k) = \frac{a(k) - a(k-1)}{T_{sample}}$$

(5.30)

If this calculated jerk is bigger than the maximum permitted jerk, a new force reference is calculated as:

$$a_{new}(k) = jerk_{max} \cdot T_{sample} + a(k-1)$$

(5.31)

$$F^*_{new}(k) = a_{new}(k) \cdot M_{train}$$

(5.32)

Force limitation   On the other hand, for security reasons after the jerk limitation, there exists a force limitation constant whose behavior is as depicted in Fig. 5-64. Thus, if the linear speed is higher than the rated speed, the force is reduced according to the curve shown in Fig. 5-64.

Transformation from force to torque   If it becomes necessary to reduce the force, the force must be transformed into torque. As depicted in Fig. 5-63, there is one block which is in charge of this task. Thus, the equation that produces this relation in traction mode is:

$$T_{em} = F \frac{1}{N} \frac{D_r}{2} \frac{1}{R \cdot R_t}$$

(5.33)

being:

- T_{em}: electromagnetic torque of each motor [Nm]
- F: force [N]
- D_r: wheel diameter [m]
- R: reduction gearbox ratio
- R_t: gearbox efficiency
- N: number of motors.

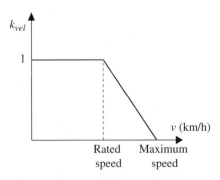

Fig. 5-64   *Force limitation in function of the linear speed*

However, for braking conditions, the efficiency of the gearbox offers an inverse effect, so the relation yields:

$$T_{em} = F\frac{1}{N}\frac{D_r}{2}\frac{R_t}{R} \tag{5.34}$$

Torque and flux references limitations   The torque reference must be limited, according to its maximum capability curves. The torque provided by the motor cannot go further than specified by the maximum capability curve. However, the maximum capability curves depend on the available DC bus voltage. Therefore, the maximum available torque in function of the rotating speed is in function of the available catenary voltage. In Fig. 5-65, a schematic block diagram of this limiter is depicted. This block is in charge of limiting the torque reference and of generating the appropriate flux reference, including the flux-weakening region. The torque reference is limited in function by the available DC bus voltage, according to five different zones, since the torque curves for traction and braking are different. The limiter actuates as a reduction gain in function of the DC bus voltage provided due to the available catenary voltage.

Finally, there is also an additional reduction gain (K_{temper}) at the end, which reduces the torque reference due to temperature restrictions of different elements of the drive of the train.

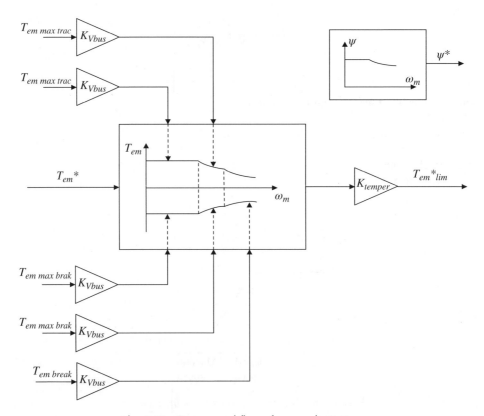

Fig. 5-65 *Torque and flux references limitations*

Anti-slip control   The slip occurs when the adherence level between the wheel and the rail are not sufficient. In order to avoid the slip, it is necessary to actuate from a control point of view. First of all, it is necessary to detect that the wheels are slipping. This detection is carried out by supervising the accelerations and speeds of the different shafts of the train: the traction shafts and the non-tractive shafts (the non-tractor shafts do not slip). The anti-slip control will start actuating whenever slip is detected. The actuation will consist of the progressive reduction of the torque reference, if the train is accelerating. Conversely, if the train is decelerating, the torque reference will be progressively increased. Once the anti-slip control is actuating, it will operate until the slip completely disappears. Fig. 5-66 illustrates progressive reductions of the torque reference due to consecutive anti-slip actions.

DC bus voltage maintenance   This control action is related to the fact that sometimes the pantograph is transitorily disconnected from the catenary. As mentioned above, the pantograph disconnection sometimes happens because of vibrations caused by the vehicle traveling at high speed. This happens because the pantograph is not producing sufficient pressure to the cables. The time interval in which the pantograph can be disconnected is around 100ms. Apart from that, pantograph disconnections produce overvoltage that can damage all the electric equipment.

   Thus, the objective is to maintain a certain DC bus voltage level, avoiding its discharge during this transitory pantograph disconnection. The control algorithm is activated when it is detected that the voltage is lower than the desired minimum DC bus voltage. This value is normally lower than the minimum catenary voltage. If the disconnection of the pantograph is produced when the train is braking, the DC bus voltage will tend to increase and, therefore, the algorithm will not be activated. In this case, the braking chopper control will maintain the DC bus voltage to the desired value. Conversely, if at the moment of disconnection of the pantograph the train is in traction mode, the DC bus voltage will tend to decrease and, therefore, the DC bus voltage maintenance algorithm will be activated, reducing the torque reference and even making it slightly negative (braking torque). In this way, it is possible to extract energy from the kinetic energy of the train, being possible to maintain the DC bus voltage at a fixed value. The classic type of controller employed is a PI regulator, with the DC bus voltage as input and the torque reference decrease as output, as depicted in Fig. 5-67. Note that the torque output created by the PI controller is negative, producing a decrease of the total torque reference. The PI control is deactivated, when the DC bus voltage is higher than the reference value, which means that the pantograph has been reconnected. The reconnection of the pantograph is delicate, since it can produce overcurrent at the catenary.

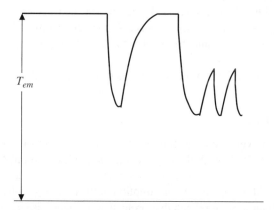

Fig. 5-66   *Progressive reductions of the torque reference due to consecutive anti-slip actions*

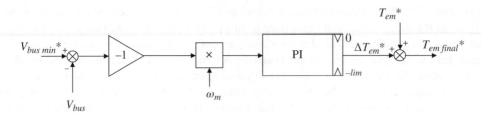

Fig. 5-67   *Algorithm for torque decreasing when disconnection of the pantograph occurs*

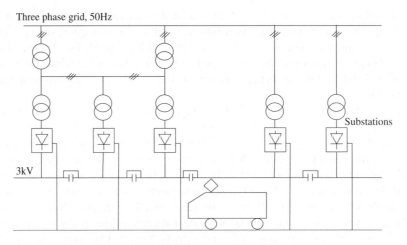

Fig. 5-68   *Railway DC power supply system*

5.5   Railway power supply system

5.5.1   DC supply system

As mentioned before, DC supply systems in general are fed by six- or 12-pulse diode rectifiers directly connected to the 50Hz grid. Fig. 5-68 illustrates a basic layout of how the DC catenary is obtained from the three-phase 50Hz grid. Substations generating the DC supply are basically composed of transformers and rectifiers as depicted. The DC supply voltage can range from 600V to 3kV. The negative pole is usually connected to the earthed rails.

The reader is referred to Chapter 6 for more details about possible rectifier configurations.

5.5.2   AC 50Hz supply system

In 50Hz AC supply systems, power electronic equipment is not used. The AC catenary is obtained by taking two phases from the public MV 50Hz grid, thereby reducing the voltage amplitude with the help of transformers. Thus, separated feeding sections are connected to subsequent phases of the three-phase MV grid, in order to mitigate the imbalance caused by the power consumption of the train. Fig. 5-69 illustrates an example of an AC 50Hz power supply configuration. It is seen that protective insulation is necessary to isolate catenary sections fed from different phases. This means that there exist certain sections of the catenaries where the train is

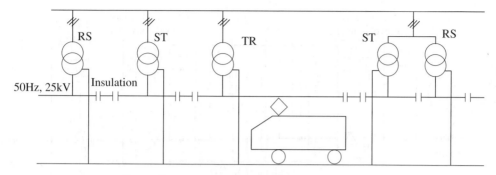

Fig. 5-69   *Railway 50Hz AC power supply system*

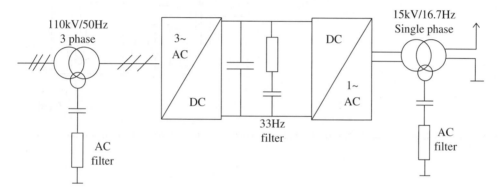

Fig. 5-70   *Schematic of a 16.7Hz AC power supply system obtained from the 50Hz grid*

not connected to any supply voltage and it continues running owing to the inertia of the train itself, losing a short amount of traveling speed. This power supply configuration presents a limited recuperative capability of energy, owing to separation segments of the catenary.

Some countries work at 60Hz, there not being any great difference with what has been described in this subsection, since both cases can be treated the same. There are some exceptions where power electronic converter-based supply systems are used, such as in the Channel Tunnel and in Tokyo–Osaka.

5.5.3   AC 16.7Hz supply systems

With regards to 16.7Hz AC power supply systems, the situation is slightly different from 50Hz supply systems. In this case, the catenary is not segregated in different parts, providing a very good-quality supply system with a high capacity of regeneration of the brake energy. This regenerative power normally is provided by the four quadrant converters that are in trains.

In the past, rotatory converters were commonly used to obtain the 16.7Hz AC catenaries. However, their usage is being increasingly displaced by static converters. A typical solution is based on single-phase bridge inverters with IGCTs, in a three-level NPC configuration, as depicted in Fig. 5-70. The AC filter in the DC-link compensates the second harmonic power fluctuation. Additional filters are needed on both AC sides in order to

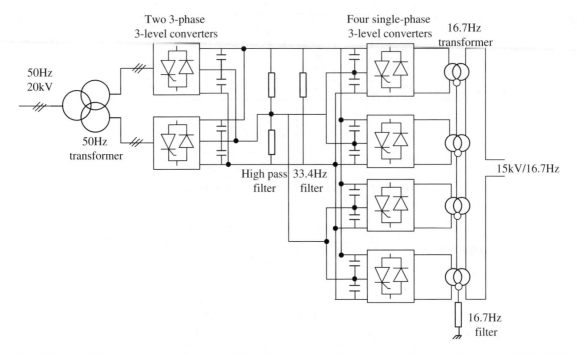

Fig. 5-71   *15MW static converter based 16.7Hz AC power supply system configuration, based on IGCTs (ABB Switzerland) [23]*

meet the standards. The efficiency of the converter is very high, but further losses in the filters and transformers are generated.

A more specific illustrative example is given in Fig. 5-71. The single-phase inverters operate with SHE angles, in order to reduce the power losses, among other reasons. Thus, by means of a power electronic converter, basically from the 50Hz grid side, a 16.7Hz is obtained for the catenary side. It will be noted that, compared to the 50Hz catenary configurations, a much more complex conversion system is needed.

An alternative design is based on the modular multilevel converter configuration. As illustrated in Fig. 5-72, this topology is conceived as a transformerless configuration, being one of its main advantages. The voltage waveforms at AC sides of the converter, owing to the considerably high number of submodule utilizations, present a high-quality multilevel waveform ($2n + 1$ stepped, being n the number of submodules). A voltage-balancing algorithm of such capacitors is needed for a proper operation of the converter. Within this configuration, there exist several alternative modular multilevel converter configurations suitable for this application.

5.6   ESSs for railway applications

5.6.1   Introduction

The energy consumption of rail vehicles like trams, light rail, metros, and commuter trains, depends on speed and mass. The speed is dependent on driving schedule and service requirements and mass depends on the

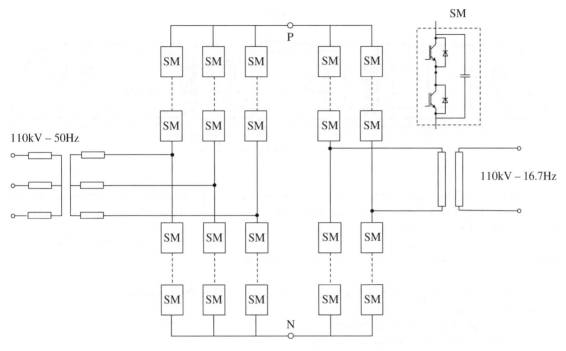

Fig. 5-72 *Modular multilevel converter based transformerless 16.7Hz AC power supply system*

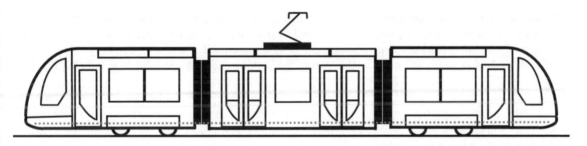

Fig. 5-73 *A 100% low-floor tram*

vehicle body, the number of people being transported, and on-board installed equipments. Is it considered that for a tramway vehicle life of about 30 years, the energy consumption represents about 30% of the lifetime cost.

A mass reduction is mandatory to reduce the amount of energy required to propel the vehicle. The vehicle's metallic structure is designed to be reliable enough to comply with the desired operational lifetime, which is typically 25–30 years. Vehicle mass depends on the material used for the structure and can be made of aluminum or steel. Aluminium-made bodies will weigh less that steel-made ones, but aluminum body repair costs will be higher than steel-based ones in the event of a collision with a road vehicle.

Tramways have become very popular in the urban mobility scenarios; 100% low-floor (Fig. 5-73) tramways construction in general makes the vehicle bodies slightly heavier than 70% low-floor ones (Fig. 5-74), because the vehicle under frame has to be designed with a number of steps which still need to be able to withstand the required longitudinal forces.

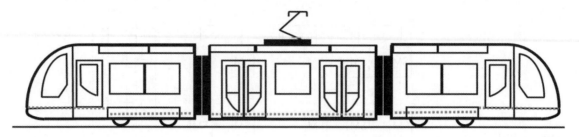

Fig. 5-74   *A 70% low-floor tram*

For a 100% low-floor tram the whole passenger saloon is accessible to passengers with reduced mobility, and steps and ramps inside the vehicle are avoided. With a 70% low-floor tram, one step is necessary to access some sections of the tram floor.

Generally speaking, the more population is required, the higher the vehicle mass will be in the end. The factors that lead to an increase of vehicle weight and mass are:

- comfort requirements for driver and passenger saloon (e.g. 100 or 70% low-floor trams);
- multiple unit operation;
- air conditioning power level requirements;
- higher acceleration of the propulsion system;
- passenger seat numbers;
- on-board energy storage units.

Knowing that the traction electric motors behave like electric generators when the vehicle starts braking, a great potential exists if the braking energy is recovered and used again in the next accelerating phase. Thus one important conclusion is that the energy consumption from the overhead catenary for a railway vehicle can be decreased, thereby reducing the mass and/or storing the braking energy to be used later to help propel the vehicle.

Fig. 5-75 shows typical traction and braking processes in a railway vehicle that doesn't have an on-board ESS. In this case during the braking process, only a small portion of the regenerative energy is used for feeding other vehicles in the same line, most of the regenerative energy is lost as heat in the on-board braking resistor to prevent an overvoltage in the catenary.

When an on-board energy system is used in the vehicle, its energy can help during the acceleration process to reduce the consumption from the catenary line. In addition, almost all the regenerative energy can be recharged in the storage system, minimizing the resistive losses. Fig. 5-76 shows energy recovery in a tramway.

As a general rule, it can be considered that any railway vehicle shows the following working conditions:

- acceleration process
- constant speed driving, coasting
- braking process
- stop.

During the acceleration process, the power consumption from the electric substation has to be large enough to accelerate the vehicle's total mass to reach a constant speed. During the constant speed driving mode the power needs are less than during acceleration. When the vehicle starts to brake, the kinetic energy is converted

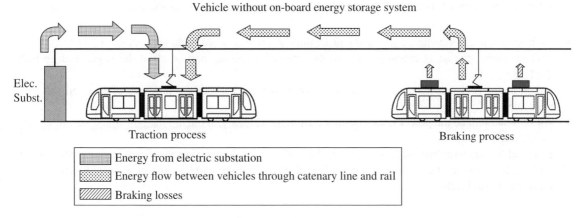

Fig. 5-75   *Braking energy losses in a tramway with no on-board ESS*

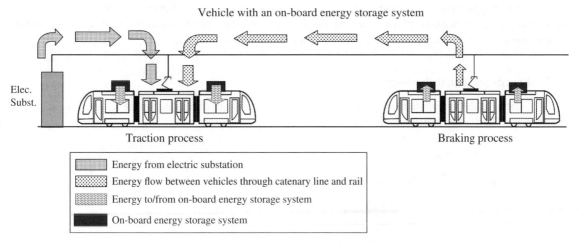

Fig. 5-76   *Energy recovery in a tramway with an on-board ESS*

into electric energy, which is sent back to the common DC voltage bus where the rest of the powertrain power converters coexist.

As described before, there are two kinds of electric substations: the ones that feed the vehicles with AC voltage (up to 25kV) and the ones that feed the vehicles with DC voltage (up to 3kV). In the case of AC electric substations, these are connected to the main grid so their receptivity is greater than in DC electric substations. The regenerative energy in AC electric substations can be sent to the main grid to be used for other vehicles or loads connected to it. The DC electric substation characteristic is that they are usually made of non-bi-directional AC/DC converters/rectifiers, so the energy can flow only from the utility grid to the railway network. As a consequence during periods when regenerated power is higher than consumed power this energy has to be sent to vehicles' on-board electrical resistances to dissipate as heat.

The need for wasting electrical energy in electric substations can be avoided using an ESS located on-board in a vehicle or in a fixed location close to the electric substation (trackside energy storage).

The main objectives of using an ESS (on-board or fixed) in a railway scenario are:

- To recover the kinetic energy that takes place during a regenerative braking process.
- To help during the acceleration process of the vehicle supplying part or all of the stored energy from the ESS.
- An on-board ESS can manage to drive the railway vehicle without the presence of an overhead catenary line.
- An ESS can reduce the power peak consumed from the catenary, reducing the power and size of electric substations.
- If hybrid locomotives are used, an on-board ESS allows for downsizing the diesel engine, reducing gas emissions, and gives the chance to operate in all-electric mode with the diesel engine switched off to minimize noise levels.

Later in this chapter, the main energy storage chemistries that had been used for conforming ESSs in the railway industry are analyzed. Second, two kinds of energy storage applications are identified: ones that make use of on-board ESSs and ones that place the storage system close to an electric substation. Third, and regarding the on-board ESSs applications, their use in tramways and in locomotives is studied and existing energy management strategies are reviewed for both scenarios. Finally a classification of commercially existing railway applications that make use of ESSs is made, both for tramways and for trackside ESSs.

5.6.2   Energy storage technologies

Fig. 5-77 shows a classification of existing energy storage technologies depending on the storage media, power vs time capability, and application reference.

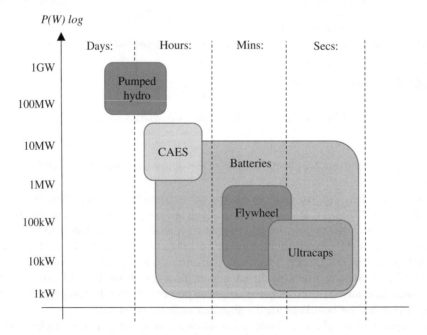

Fig. 5-77   *Energy storage technology classification (CAES stands for compressed air energy storage system)*

Even if there are some developments with SMES and flywheel storage technologies, today commercially available ESSs used in railway applications are usually made using supercapacitors and batteries. This chapter focuses on the study of these two storage technologies.

Supercapacitors store energy in an electrical way like common capacitors, and batteries store energy in an electro-chemical way, changing the metallic properties of elements attached to the anode and to the cathode. Supercapacitors' main characteristic is their high power capability (same in charge and discharge) compared with batteries, which can store much more energy than supercapacitors. This is why in Fig. 5-77 the super-capacitors' power capability (supercapacitors are also known as ultracapacitors, or ultracaps) is in the range of seconds and batteries' in the range of minutes to hours.

A detailed analysis of supercapacitors, NiMH, and lithium-ion batteries is made in the following three sub-sections respectively, taking into account historical developments with ESSs on commercial railway vehicles.

5.6.2.1 *Supercapacitors*

A supercapacitor, also called an ultracapacitor or electric double-layer capacitor, is a kind of electrical energy stor-age device in which energy is stored in an electric field between two plates, as with a normal capacitor. In com-parison with electrolytic or metalized polypropylene capacitors where the capacity is measured in μF and in mF respectively, the supercapacitor's value is given in farads up to some kilo-farads in one single device.

The supercapacitor is the series combination of two double-layer capacitors coexisting back to back in the same package. A packaged cell consists of two porous carbon deposited aluminum electrode films with a paper separator foil between them. This assembly is immersed in a liquid organic acetonitrile electrolyte. Cell inter-nal resistance is composed of the aluminum films resistance (Re) plus the electrolyte ionic resistance (Ri). Each one of both capacitors C(U) is very high due to the big equivalent surface, atomic charge separation distance and is dependent on element voltage (U) [26], [27], [28]. Fig. 5-78 shows the internal structure of a supercapacitor.

Equation (5.35) expresses the energy stored in a supercapacitor.

$$E(J) = \frac{1}{2} \cdot C(F) \cdot V^2 \tag{5.35}$$

For example, a 3000F cell fully charger at 2.7V contains 3Wh of energy, 8100 coulombs of charge, which equals $1.3*10^{23}$ electrons.

Nevertheless, during normal working conditions a supercapacitor can't be discharged down to 0V because at very low voltage the associated DC/DC power converter can't manage the necessary power flow between the supercapacitor-based energy storage system and the DC-link. The useful energy that a supercapacitor can supply is determined by (5.36):

$$E(J) = \frac{1}{2} \cdot C(F) \cdot \left[V_o^2 - V_f^2 \right] \tag{5.36}$$

with V_o being the initial high voltage and V_f the final voltage after discharge. If V_f is half of V_o the energy extracted from the supercapacitor will be 75% of the maximum energy. Nevertheless, the maximum and min-imum working voltage values are dependent on the power and energy needed in the application and on the desired lifecycle of the component.

Commercial supercapacitors have an energy density range of around 0.5–10Wh/kg. For comparison, a con-ventional lead-acid battery is typically 30–40Wh/kg and modern lithium-ion batteries are about 100Wh/kg. This compares with a net energy density of automotive fuels of over 10,000Wh/kg.

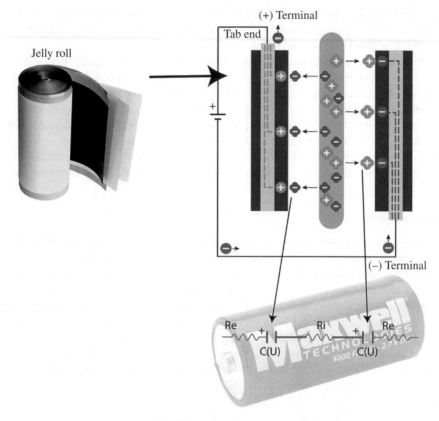

Fig. 5-78   *Internal structure of a supercapacitor (Source: MAXWELL Tech. Reproduced with permission of MAXWELL Tech)*

Fig. 5-79   *A BCAP3400 supercapacitor from MAXWELL Tech (Source: MAXWELL Tech. Reproduced with permission of MAXWELL Tech)*

Table 5-17   *Cylindrical 3000F supercapacitor manufacturers*

Manufacturer	Web link
MAXWELL Tech	www.maxwell.com
BATSCAP	www.batscap.com
LS MTRON	www.lsmtron.com
IOXUS	www.ioxus.com
NESSCAP	www.nesscap.com
KEMET	www.kemet.com
WIMA	www.wima.com
EPCOS-SIEMENS	www.epcos.com
CORNELL DUBILIER	www.cde.com
SAMWHA	www.samwha.com
EATON	www.eaton.com

Compared to batteries, supercapacitors have a higher lifecycle and are able to sustain rapid deep discharge and recharging cycles. The main advantages of supercapacitors are high power density, high efficiency because of their low internal resistance, fast charging and discharging speed, long lifecycle of about 1 megacycle of charges/discharges at 25° cell temperature, and a wide working operating temperature range from –40°C to 65°C.

The main supercapacitor characteristic that makes them attractive for using in ESSs is their ability to fast charge and discharge without loss of efficiency and for thousands of cycles. Supercapacitors can recharge very quickly and can deliver high and frequent power peaks; that's why supercapacitors are extensively used in high-power applications, such as hybrid power systems, and regenerative energy systems where high power levels are necessary for short time intervals.

Fig. 5-79 shows the cylindrical shape of a BCAP3400 supercapacitor device from MAXWELL Tech. Its rated capacity is 3400F and the maximum working voltage can reach 2.85V DC.

The cylindrical shape shown in Fig. 5-79 (in threaded or welded version) has become a standard for the industry in developing around 3000F supercapacitors, and it has been used by the main developers of supercapacitors. Table 5-17 shows the main manufacturers of medium and high capacity supercapacitors.

Cylindrical supercapacitors are often matched in capacity and in resistance, and connected in series or parallel arrangements to build up industrial usable modules with high energy and power capabilities. These modules are electro-mechanically designed to work under hard environmental, electrical, and shock conditions in applications like electric or hybrid vehicles, trucks, forklifts, UPS systems, windmills, cranes, and tramways. Fig. 5-80 shows some examples of commercially available supercapacitor modules.

Working voltage, working temperature, cycling current, vibrations, and storage conditions are the main degradation factors affecting the life of supercapacitors. Supercapacitor degradation is apparent from a gradual loss of capacitance and a gradual increase of internal resistance (Ri). The EOL of a supercapacitor cell is considered when the capacitance and resistance is out of the application range and will differ depending on the application.

Working voltage   A supercapacitor can be discharged down to 0V but going beyond 2.7V leads to the liquid organic electrolyte decomposition, gas emanation, and internal pressure increasing. This, in combination with elevated temperature, can lead to a possible breakdown of the component following harmful and flammable electrolyte venting.

Higher average working voltage leads to faster capacity decrease. As a general rule, supercapacitor manufacturers estimate that 0.1V of cell voltage increment reduces its life by a half.

Fig. 5-80   *Examples of commercially available supercapacitor modules (Source: MAXWELL Tech. Reproduced with permission of MAXWELL Tech)*

Working temperature   Theoretically the working temperature range for a supercapacitor is between –45°C and 65°C. At temperatures as low as –45°C the capacity doesn't experience any great change but the internal resistance is high, owing to the viscosity of the electrolyte. Even at very low ambient temperatures, the current through the supercapacitor warms the cell enough to decrease the internal resistance value to the nominal one. As a general rule, supercapacitor manufacturers estimate that 10°C of cell temperature increment reduces its life by a half.

Cycling current   Cycling current has a direct effect on the heating of the cell. For the same cooling strategy, a higher cycling current leads to a higher cell temperature. This leads to faster capacity fade or supercapacitor degradation. Unless it is said that the current per se doesn't provoke any degradation phenomena in the electrochemistry of the cell and the degradation is related only to the combination of voltage and temperature, very high cycling currents lead also to a faster degradation phenomenon in the device even if the body is at room temperature.

Fig. 5-81 shows the capacity degradation of two different constant cycling currents. The higher the current level, the higher the capacity fall or degradation of the supercapacitor.

Storage conditions   It is very usual that supercapacitor modules have to be left charged until their next duty time as with UPS systems where there is no possibility to completely discharge the ESS. Under storage conditions, the higher the temperature and voltage applied to the supercapacitor, the faster the capacity and internal resistance increases, as can be seen in Fig. 5-82. In terms of self-discharge, the stored energy of a supercapacitor decreases by 50% within 30–40 days.

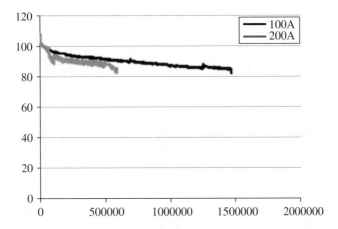

Fig. 5-81 *Cycling behavior under two different current values (Source: MAXWELL Tech. Reproduced with permission of MAXWELL Tech)*

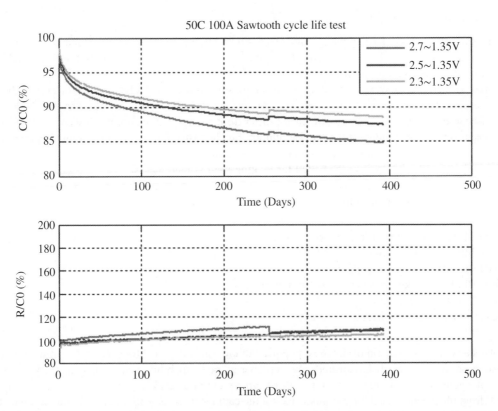

Fig. 5-82 *Capacity and internal resistance increase on stored supercapacitor (Source: MAXWELL Tech. Reproduced with permission of MAXWELL Tech)*

5.6.2.2   NiMH batteries

The use of NiMH batteries for hybrid applications in hybrid electric vehicles started in the 1990s with the Toyota Prius hybrid car.

More recently, in 2005–2006, this chemistry was used for railway applications for back-up systems or as traction batteries, like in the case of the cities of Nice (France) and L'Aquila (Italy). In Nice, the French railway manufacturer Alstom installed an on-board ESS composed of Saft NiMH batteries in Citadis trams to be able to drive with no overhead catenary for 500m at 20mph. Lohr Industries developed in 2007 a rubber-wheeled tram systems with Saft NiMH batteries for its Translohr vehicle for the city of L'Aquila. In this case a 576V-80kWh Saft NiMH storage system was installed [29], [30]. In 2012, Kawasaki tested its Swimo catenary-free LRV in the Japanese city of Sapporo. It used Kawasaki Gigacell NiMH batteries, which could be fully charged in five minutes through the 600V DC overhead catenary. This configuration allowed the vehicle to operate for up to 10km on non-electrified lines.

The NiMH cell has a low internal impedance and flat discharge characteristic. Owing to its relatively low nominal voltage of 1.2V, a lot of cells would have to be connected in series to make up high-voltage ESSs. NiMH cells are cheaper and safer than lithium-ion ones. Nevertheless, its poor energy and power density relation in comparison with new promising lithium-ion batteries has led to the latter's being used widely in on-board ESSs applications. In addition, NiMH batteries typically suffer from a capacity loss of 20–50% after six months of storage [31]. More information about Saft NHP NiMH batteries can be found in [32].

5.6.2.3   Lithium-ion batteries

A lithium-ion cell consists of a positive electrode (made with lithium metal oxides), a negative electrode (made with carbon or titanate materials), an electrolyte (liquid, gel or solid, composed of organic solvents and lithium salts), and separators (see Fig. 5-83).

The most attractive characteristic of lithium-ion batteries for its use in portable applications is that they have higher energy and power density than the rest of battery chemistries. Lithium-ion chemistry has the lower rate of self-discharge of approx. 5% per month, do not suffer from memory effect, accepts fast charge (but less than supercapacitors) and depending on the precise lithium-ion chemistry family, the operating temperatures stay between –20°C to 60°C. In addition high cell nominal voltage of 3.6V means that a reduced number of electrical connections and electronics are needed for making high voltage ESSs.

Lithium-ion batteries are classified depending on the mixture of metal oxides used in the cathode, anode material, and electrolyte composition:

- LFP or LiFePO4 (lithiated iron phosphate)
- LCO (lithiated cobalt oxide)
- LMO (lithiated manganese oxide)
- NMC (lithiated nickel manganese cobalt oxide)
- LTO (lithiated titanate oxide)
- NCA (lithiated nickel cobalt aluminum oxide).

Fig. 5-84 shows a classification of the most popular lithium-ion cell chemistries. Each one of the lithium-ion chemistries shows different behavior under power, energy, rate capability, state of charge (SOC) vs voltage profile, lifecycle, manufacturing cost, and working temperature range [33], [34].

Apart from the material used for the anode and for the cathode, the electrolyte can be liquid or not (ceramic, polymer, or gel). If a gelled electrolyte is used, the cells are named as lithium-ion polymer cells. Cells that use liquid electrolyte need a rigid case to press the electrodes together to ensure ionic movement throughout the

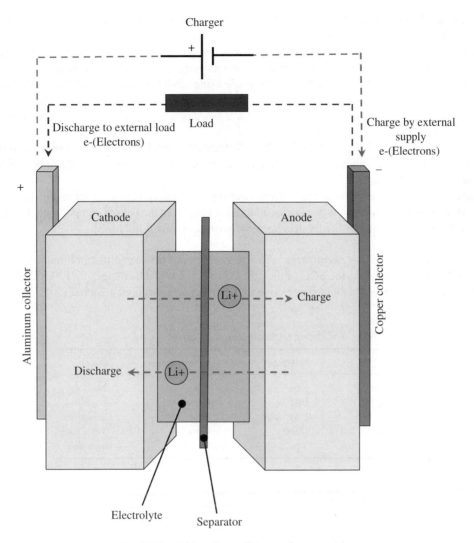

Fig. 5-83 *Lithium-ion cell internal composition*

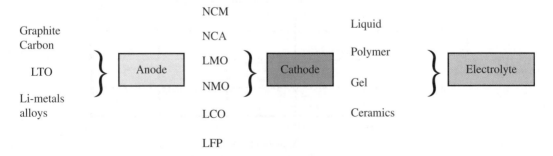

Fig. 5-84 *Lithium-ion cell compositions [34]*

intermediate liquid electrolyte. Both the liquid and the gelled electrolyte are ionic conductive materials wherein a salt and a solvent are dissolved [35].

The main advantage of using a non-liquid electrolyte is that the cell never leaks electrolyte if the cell case is damaged. Another benefit of using lithium-ion polymer cells is that the polymer is laminated to the electrodes, permitting stacked cell structures, and they can be made using laminated foils that do not need compression. This manufacturing process allows for the production of pouch formats with foil type enclosure. A reduction of approximately 20% of the weight is reached when using a pouch format compared with the classic hard (prismatic or cylindrical) shells [36], [37].

Table 5-18 compares the three existing packaging technologies. Prismatic cells can be packaged more efficiently than cylindrical ones, giving a high packaging density but less in comparison with pouch cells because they do not need a rigid case.

Nevertheless, the pouch format can suffer from swelling if internal pressure increases due to over-voltage and over-temperature, so pouch cells need some alternative support in the battery compartment to be protected against mechanical stress, which finally decreases the energy density a little [38].

Brief comparison of lithium-ion chemistries   The main difference between different lithium-ion chemistries is based on the voltage profile vs the SOC of the cell. Fig. 5-85 shows cell voltages for LTO, LFP, NMC, and LMO chemistries; discharge behavior is sensitive to the selection of cathode material [39].

Table 5-18   *Packaging technologies comparison*

	Prismatic	Cylindrical	Pouch
Stiffness	Yes	Yes	No
Packaging	Bad	Good	Excellent
Cost	Economic	High	Economic
Thermal	Good	Good	Very good
Safety	Yes	Good	Swelling

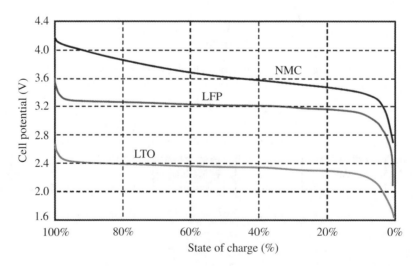

Fig. 5-85   *Voltage profiles of lithium-ion chemistries [39]*

As can be seen in Fig. 5-85, LTO chemistry has comparatively less cell voltage than the others. This implies less energy content even at the same cell capacity.

The stored energy in a cell is determined as:

$$Energy(Wh) = Vcell(V)_{f(SOC)} \cdot Capacity(Ah)_{f(SOC)} \tag{5.37}$$

Both, the cell voltage (V) and the capacity (Ah) or stored coulombs (Q) are dependent on the cell's SOC. As the SOC varies, the voltage and capacity of the cell vary, thus varying the stored energy.

When determining the SOC of lithium-ion cells, some techniques use a combination of coulomb counting and open circuit voltage (OCV) measurement [40], [41], [42]. During the daytime working hours, the coulomb counting technique integrates the current going through the cell to know the electrical charge input and output balance. At night, at rest when there is no current through the cell, the OCV technique reads the cell voltage to know what will be starting SOC when the cycle starts again. The average SOC calculation error for this technique is not less than ±5%. In addition, for some special lithium-ion chemistries, like in the case of LFP, the SOC calculation using the combination of coulomb counting and OCV measurement can lead to a higher SOC estimation error because the voltage profile of this chemistry is almost flat. New advanced techniques based on Kalman filtering algorithms, artificial neural networks and fuzzy logic have been developed to reach SOC estimation errors between ±1.5% and ±3% SOC [43], [44], [45], [46], [47], [48], [49].

Fig. 5-86 shows a comparison of lithium-ion chemistries regarding safety, cost, specific energy, specific power, and life span, where higher values are located farthest from the center of the graph [50].

Lithium-ion chemistries using cobalt oxides in the cathode like NMC and LCO are less tolerant than other chemistries in terms of temperature and voltage, even if their specific power is greater. Catastrophic failures like explosion or fire may result if over-voltage (>4.2V) or over-temperature (>60°C) are maintained in cobalt-based cathode chemistries [51]. Cobalt is also toxic, so recycling can be a problem.

LTO chemistry accepts higher charging currents than the rest of lithium-ion chemistries because of the titanate anode. Lithium-ion chemistries having a graphite anode can be damaged if an excessive charging current is applied or if the charging process is done at low temperatures (<0°C). Graphite-based anode lithium-ion chemistries show a strong aging dependency on the charge current rate so lifecycle degrades very quickly with an

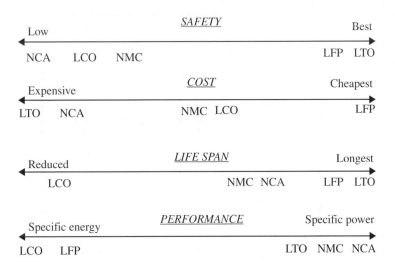

Fig. 5-86 *Lithium-ion chemistries comparison [50]*

increasing charge rate. LTO anodes show almost no influence from the charge rate if the cell is maintained below 40°C, so lifecycle is maintained even at high charge/discharge rates. Conversely and because of its lower nominal voltage (approximately 2.3V), LTO chemistry suffers through a lower energy density than the rest.

In terms of lifecycle, the lower the depth of discharge (DOD), the higher the lifecycle of the cell. It can be said that LFP and LTO are the two chemistries which show the highest lifecycle for the same depth of discharge, LTO being the longer-lasting one.

Lithium-ion batteries degradation factors   Voltage, temperature, and DOD are the main factors affecting the lifecycle of lithium-ion batteries. The higher the voltage, temperature, or DOD, the faster the degradation.

Voltage   Both, the maximum and minimum working voltages lead to faster degradation. Lithium-ion cell manufacturers usually recommend not going above 90% SOC or below 20% SOC. That's why for hybrid applications with frequent charge and discharge processes the average working SOC is kept close to 50%. If lithium-ion batteries are to be stored for long periods, it is recommended to store them with an SOC of approximately 50% and in a dry place free from large temperature swings.

High-voltage and low-voltage values trigger anode and cathode degradation processes inside the cell. High voltage causes lithium plating, electrolyte decomposition, followed by gas generation, an increase of internal pressure, and finally the emanation of flammable gases. Low-voltage operation causes the copper anode current collector to dissolve and the cathode to break down, creating the possibility of an internal short circuit. Overvoltage can be avoided with safety electronics implementation that takes care of maximum voltage opening any charging converter connected to the cell, but low voltage can't be avoided, because of cell self-discharge phenomena. To avoid permanent damage caused by low voltage, it is mandatory to perform a recharge task periodically (at least twice a year). Fig. 5-87 shows the range outside of which lithium-ion cells are liable to be damaged.

Temperature   The optimal working temperature range for lithium-ion cells chemistries ranges between 20°C and 35°C (Fig. 5-88). Low temperatures lead to an internal resistance increase with a subsequent decrease in

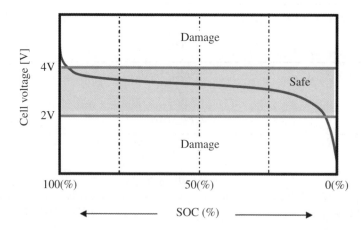

Fig. 5-87   *Lithium-ion cells are very susceptible to damage outside the allowed voltage range*

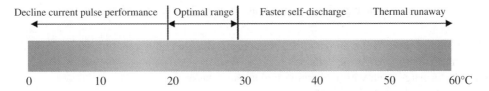

Fig. 5-88 *Optimal temperature range for lithium-ion battery chemistries*

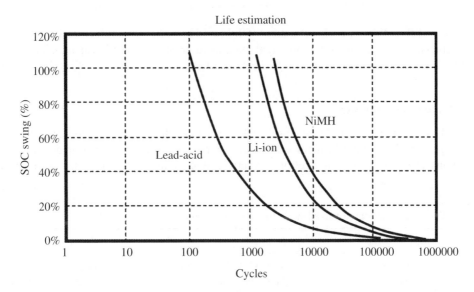

Fig. 5-89 *Lifecycle estimations vs SOC swing for different chemistries [52]*

power pulse performance and lower charging current capability. Charging at low temperatures can trigger lithium-plating formation on the anode with the risk of an internal short circuit generation. High temperatures provoke irreversible reactions in the electrolyte, which, in turn, will result in the venting of internal gases and catastrophic thermal runaway [51]. LTO chemistry is intrinsically more stable than cobalt-based NMC or NCA ones because LTO reaches thermal runaway when the cell temperature increases above 105°C, while NMC goes into thermal runaway at cell temperatures higher than 80°C.

Storage temperature is also very important because the self-discharge rate of lithium-ion cells doubles for every 10°C increase.

When working with lithium-ion batteries in very demanding applications (RMS current above 1C), an active cooling system capable of controlling cell temperature to below 40°C is mandatory.

Depth of discharge Fig. 5-89 shows the estimated lifecycle of different chemistries [52]. It will be noted that for the same SOC swing the lifecycle for lithium-ion batteries is almost the same as for NiMH batteries but their performance in terms of energy and power is better.

Fig. 5-89 relates SOC swing (DOD) to lifecycle estimation, but the average SOC level can be the major life-dependent factor. The same SOC swing with a higher average SOC value, and so a higher average voltage, leads to faster degradation.

The main recommendations to prolong as much as possible the life of lithium-ion batteries are:

- Avoid cell temperature rising above 40°C, both during storage and operation.
- Do not unnecessarily charge the cells.
- Avoid full discharge even with recharging maintenance tasks.
- Store the cells at 30–50% charge level.
- Store the cell at low temperatures in a dry and well-ventilated depot.

Market analysis　Table 5-19 is a classification of lithium-ion battery manufacturers, chemistries, and packaging formats. As will be seen, the principal manufacturers are located in Japan, China, and Korea. LFP and NCM are the prominent chemistries because of their high power-handling characteristic.

5.6.2.4　*Cells management system*

As is explained earlier in this chapter, care must be taken when handling supercapacitors and lithium-ion batteries. With supercapacitors, the maximum voltage and temperature have to be measured and checked with predefined limits in order to prevent electrolyte venting. With lithium-ion batteries, maximum/minimum

Table 5-19　*Lithium-ion cells market worldwide by main competitors* [53]

	Manufacturer	Chemistry	Packaging
JAPAN	GS YUASA	LMO, NCA, NMC, LFP, LCO	Prismatic
	HITACHI	LMO-NMC, NMC	Cylindrical, Prismatic
	SAYO-PANASONIC	NMC, LMO-NMC, LMO-NCA, LFP	Cylindrical, Prismatic
	TOSHIBA	LCO (LTO anode)	Prismatic
CHINA	BAK	LFP	Cylindrical, Prismatic, Pouch
	CALB	LFP	Pouch
	WINSTON BATTERY	LFP	Prismatic
	BYD	LFP	Prismatic
South KOREA	KOKAM	NMC, NMC-LTO, LCO	Pouch
	LG-CHEM	LMO, LFP, NMC	Pouch
	EIG LTD.	LFP, LMO-NMC	Pouch
	SAMSUNG	LCO, NMC, LMO-NMC, LFP	Cylindrical, Prismatic
	SK ENERGY	LMO	Pouch
USA	A123	LFP	Cylindrical, Pouch
	AESC	LMO, NMC, LMO-NMC, NCA	Pouch
	ENERDEL	NMC, LMO	Pouch
	JOHNSON CONTROLS	NMC, NCA	Cylindrical, Prismatic
	VALENCE TECH	LFP	Cylindrical
	XALT ENERGY	NMC, LTO	Pouch
EUROPE	LECLANCHE	NMC, LTO	Pouch
	HOPPECKE	LFP	Pouch
	VARTA AG.	LMO	Cylindrical, Prismatic, Pouch
	SAFT	LFP, NCA	Cylindrical,
	GAIA-LTC	LCO, NCA, LFP	Cylindrical

voltages, cell maximum/minimum temperatures, and charging/discharging current levels have to be monitored and controlled.

In addition, even for supercapacitors and batteries, balancing strategies have to be implemented for assuring same working SOC between cells. During the cell manufacturing process, slight differences in cell active material quantity or electrolyte filling lead to different cell characteristics. This is usually reflected in capacity (farads or Ah) variation among ideally equal cells. This inequality between cells' capacities provokes different cell voltages for the same current in series-connected cells.

A balancing system is implemented in the battery management system (BMS) even for supercapacitors and lithium-ion battery systems. The balancing system objective is to balance and minimize any differences in the charged state between cells in order to keep the cells' voltages controlled inside a narrow window [54], [55], [56]. In addition, the BMS has to control the heat generated in the cells due to continuous operation to keep the cells at their optimum temperature level, homogenize cell temperature in the pack, protect the cells from overvoltage, sub-voltage, or over-current, isolate the battery pack from the rest of the system in case of an error, and communicate with the vehicle control system, giving information about the SOC of the complete ESS.

Fig. 5-90 shows the typical configuration of a BMS associated with a battery pack in an ESS.

5.6.2.5   *Lithium-ion batteries vs supercapacitors comparison*

Lithium-ion batteries and supercapacitors are complementary in the frequency space because the first ones have the best energy density values and supercapacitors have a lower energy density but a high power density. Furthermore, lithium-ion batteries are lifetime-limited devices providing few cycles in comparison with supercapacitors. In order to optimize the lifetime of lithium-ion batteries it is recommended to use them with slow dynamic current profiles, thus a hybrid combination with supercapacitors is interesting.

Table 5-20 shows the main differences between supercapacitors and lithium-ion batteries. As a general rule it can be said that supercapacitors are the preferred devices for managing high-power transient demands, while lithium-ion batteries are the preferred solution for portable applications to store large amounts of energy with medium power capabilities.

5.6.3   Trackside energy storage applications

5.6.3.1   *Introduction*

In trackside or wayside applications, an ESS can be installed close to the electric substation that feeds the catenary line, near station stops, or at another point on the route but always connected to an existing catenary infrastructure.

Unlike on-board ESSs, stationary ESSs are installed at ground level, where weight and space is not a major problem. They cannot allow for catenary-free driving and don't reduce catenary line losses.

Fig. 5-91 shows a railway application where an ESS is connected at the end of the line. In this scenario both the electric substation and the ESS can deliver energy to the railway vehicles existing in the line, but only the ESS can recover the braking energy that is not used by the rest of the vehicles because the electric substation is not reversible (three-phase rectifier bridge).

The main objective for a trackside ESS is to be able to give the extra power to the vehicles that can't be given it by the electric substation. The second requirement for a trackside ESS is to be able to avoid catenary overvoltages, assuming the braking energy generated by one vehicle is not used by the rest of the vehicles connected to the same line. In addition, the energy generated by the braking vehicle would simply be converted into waste heat by its braking resistors if no other vehicle was accelerating exactly at the same time. The trackside ESS would be able to store that energy and discharge it again when another vehicle needed accelerating

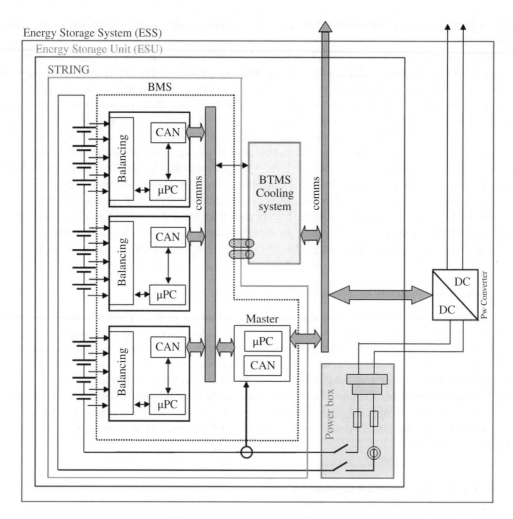

Fig. 5-90   *BMS in a complete ESS*

Table 5-20   *Comparison between supercapacitors and lithium-ion batteries*

Supercapacitor	Lithium-ion batteries
Power application	Mainly energy application
Electrostatic behavior	Electrochemical behavior
Abuse and rapid charge/discharge tolerant	Sensitive to high currents
High lifetime expectation	Limited lifetime vs SOC window and temperature
High efficiency: 92–98%	Moderate efficiency: 85–95%
Operating temperature: –40°C to 45°C	Operating temperature: 15°C to 35°C
Wide SOC window operation	DOD swing has a large influence on cycle life
Ease of SOC and SOH monitoring	Difficult algorithm for SOC and SOH determination

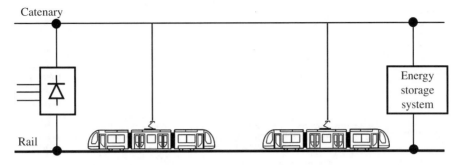

Fig. 5-91 *ESS connected at the end of the line [57]*

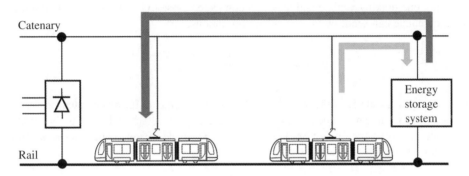

Fig. 5-92 *When accelerating and braking vehicles coexists at the same time the trackside ESS will stabilize catenary voltage charging or discharging its stored energy*

power, consequently enabling a complete exchange of energy between vehicles, even if they were not braking and accelerating at the same time. Fig. 5-92 illustrates the coexistence of braking and accelerating vehicles.

In applications where station stops are located far from the electric substation, the catenary line shows high equivalent ohmic resistance (especially in DC catenary lines). When a vehicle accelerates when leaving the station, the draining power demanded by the vehicle and the electrical resistance of the catenary from the vehicle to the station provoke voltage drops in the station itself. An ESS located in between stations helps to reduce these voltage drops, feeding the catenary line with the amount of power needed to overcome this problem.

Passenger evacuation availability can be an interesting feature for a railway metro application in case the underground electric substation fails. In this scenario, an ESS connected to a metro line network can assure the displacement of the vehicles to the closest station at very low speeds and without auxiliary loads like air conditioning systems in order to minimize energy consumption. This self-rescue feature will avoid panic situations inside wagons stopped in tunnels [58], [59].

Generally, when braking, in regenerative mode, the vehicles can send energy back to the network so long as their voltage ($V_{vehicle}$) does not go over a predefined limit, V_{REGEN_MAX}. This voltage limit is imposed for safety reasons to assure the integrity of the power converters on the vehicle. When this value is exceeded, the current being sent to the network is reduced [60]. Fig. 5-93 illustrates this.

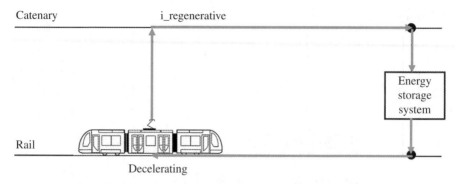

Fig. 5-93 *Regenerative energy being sent back to the wayside ESS*

If there is only one braking vehicle and one ESS connected close to the electric substation, the maximum current that the vehicle can send, owing to voltage limitations, is:

$$I_{REGEN\_MAX} < \frac{V_{REGEN\_MAX} - V_{ESS}}{R(d) \cdot d} \tag{5.38}$$

The further away from the ESS the vehicle is, the lower the maximum regenerated power capability is, owing to added catenary line resistance, $R(d)$. Owing to braking processes that are usually composed of high power levels with a short duration (<20sec), a trackside ESS made of supercapacitors is worth considering because of its fast charge/discharge ability.

5.6.3.2 *Control strategy on trackside ESSs*

The trackside ESS control strategy consists of sending energy (discharge) from the ESS to the vehicles connected to the line when the line or catenary voltage drops below a predefined voltage (V0), and on feeding the ESS (charging) again when the catenary voltage goes beyond V0 again. The catenary voltage goes below V0 because there is an imbalance between the vehicles that are accelerating and braking, because acceleration demands greater power from the electric substation. The catenary voltage goes above V0, however, when there is more braking power than accelerating power, which means power is being sent from braking vehicles to the electric substation.

The main idea is to select the voltage trigger values for charging/discharging the ESS based on the catenary voltage value. The selection of these voltages, VCHARGE and VDISCHARGE, will govern the behavior of the ESS. There will be always a power converter between the catenary and the ESS to control its charge/discharge based on catenary voltage conditions.

The electric substation OCV (V0) will be higher than VCHARGE. When a vehicle in the catenary line starts braking, the catenary voltage starts increasing. If the value VCHARGE is selected close to V0, when the catenary voltage goes above VCHARGE the ESS will start charging. In this situation, most of the braking energy of a nearby braking vehicle will be stored in the ESS.

During discharge, the voltage falls below V0, and the trackside control strategy will discharge the ESS in an attempt to maintain the catenary voltage at the level of the VDISCHARGE.

The basic actions performed by the trackside ESS controller in the different states are explained in [60]. Reference [61] presents a lithium-ion-based trackside ESS. The current delivered to and absorbed from the

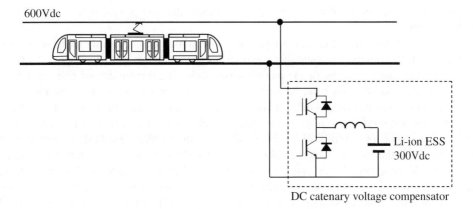

Fig. 5-94   *Lithium-ion trackside ESS [61]*

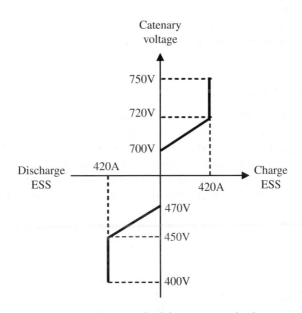

Fig. 5-95   *Control strategy for lithium-ion trackside ESS [61]*

storage system depends on the catenary voltage. The systems will be switched off when the catenary line is close to its nominal voltage of 600V DC. Fig. 5-94 shows a lithium-ion trackside ESS, as adapted from [61].

Fig. 5-95 shows the trackside ESS control strategy. The energy storage system will feed the catenary through the DC/DC converter when the line voltage will equal or be less than 470V DC. This can happen if some vehicles accelerate at the same time. For catenary voltages lower than 450V DC, the discharge current on the ESS is limited to 420A based on manufacturer recommendation. If the catenary voltage goes down to 400V DC the voltage compensator is switched OFF. If the catenary voltage increases farther than 700V DC the ESS starts charging with 420A as a maximum current.

After each charge or discharge, the control strategy balances the energy system to reach 50% SOC to be ready for the next event.

Instead of one ESS placed at the electric substation, there is also the possibility to connect several ESSs along the line at different points. By doing this, the total energy savings achieved will grow with the number of ESSs and with the energy content of the modules [60]. Selecting the number of ESS modules and their placement in the line is not an easy task, because the increase in the number of ESSs on the line decreases the amount of energy saved by each module and because the vehicles located near the middle of the line have a greater potential to send their braking energy to other vehicles compared with ones near the end of the line.

The vehicles moving near the middle of the line are close to other vehicles and braking energy can be sent to accelerating vehicles in the vicinity. Catenary ohmic resistance shows its maximum between vehicles moving near the end of the line and vehicles moving around the middle. When a vehicle that is close to the end of the line brakes, the energy is sent through the catenary line to vehicles placed in the middle. Owing to high ohmic resistance on the catenary line, its voltage will grow with regenerated current and over-voltage problems can occur on the powertrain of the braking vehicle. To avoid these over-voltage problems, the regenerated power will have to be reduced. In these cases, it is better to place a trackside ESS at the end of the line, because the line resistance between the braking vehicle and the ESS will be lower, allowing for an increase in the power regenerated to the line and thus in the ESS.

Wherever the ESS is located in the line, its strategy will be to charge or discharge according to the measured line voltage at that point, comparing it with predefined switching voltage levels V_{CHARGE}, $V_{DISCHARGE}$.

5.6.3.3 Trackside ESS developments

One of the first trackside ESS developments was done in 2007 by the Korean Railroad Research Institute, where the storage system was installed close to an existing electric substation [62]. The ESS was installed inside a substation of LRV test track located in Gyeongsan City. The length of line was 2.37km and there were four stations. The feeder voltage was 750V. The ESS connection is shown in Fig. 5-96. The supercapacitor-based ESS stores braking energy and delivers it later when needed to help power the vehicle.

The trackside ESS configuration reported in [62] was made of supercapacitors (EDLC) with a total energy of 2.34MJ, 13.7F, 250A of maximum current, and a 583V to 230V voltage swing.

The feeder voltage will fall a little bit, because of traction, during a vehicle's acceleration; at that point the ESS will discharge energy in order to keep the feeder voltage stable at its nominal voltage of 750V DC. During

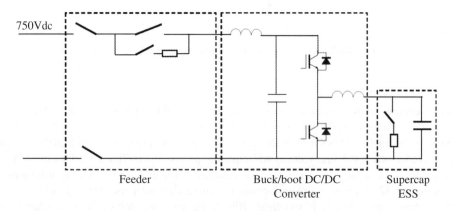

Fig. 5-96 *ESS power architecture [62]*

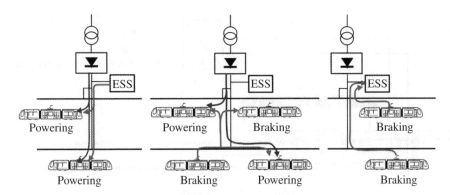

Fig. 5-97 *Energy flow from the electric substation and trackside ESS during theoretical casuistic powering and braking situations among several trains*

the regeneration processes if the catenary line increases, the ESS will charge again trying to maintain the feeder voltage at 750V DC. Fig. 5-97 shows different scenarios where the trackside ESS gives or receives energy depending on whether the trains near the substation are in accelerating or braking working mode.

The Japanese manufacturer of lithium-ion batteries GS YUASA developed some examples of trackside ESSs denoted as E3-Solutions (energy, ecology, economy) for the West Japan Railway Company that has been working from 2005 to 2007 for 1500V DC/1080kW lines.

GS YUASA's E3 solution provided stationary energy storage for railway applications by using high-power and low-impedance lithium-ion batteries with associated bi-directional DC/DC power converters, allowing the line/rail voltage to be stabilized, capturing and storing braking energy, and reducing the number of substations required [61].

In 2012, ABB in collaboration with MAXWELL, the supercapacitor manufacturer, launched onto the market the wayside ENVISTORE™ regenerative ESS which was connected to the electric substation in order to recover >90% of the kinetic energy, stabilize the demand on the grid avoiding potential voltage disturbances, and provide power support during acceleration [63].

ABB's ENVISTORE product is divided into two cabinets named ENVISTORE 750 and ENVISTORE 1500 ESSs. ENVISTORE 750 is for use in catenaries up to 1000V DC, whereas ENVISTORE 1500 is intended for use up to 1850V DC. Both ENVISTORE 750 and 1500 can handle up to 16.2kWh, 750Kw of nominal power, and 4.5MW of maximum power.

Siemens Transportation Systems developed in 2002 the SITRAS SES stationary ESS made up of supercapacitors [64]. SITRAS SES can work in voltage stabilization mode or energy-saving mode. In the voltage stabilization mode the ESS is completely charged and is discharged when the catenary voltage falls below a specified limit, while in energy-saving mode it absorbs the energy regenerated by braking vehicles and stores it until it is released again to feed future accelerating vehicles.

The system can switch automatically between these two operating modes so that it can adapt perfectly to the prevailing operational requirements.

Fig. 5-98 shows the B-Chop trackside lithium-ion battery ESS developed by Hitachi. It was able to connect to 750V DC, 600V DC, or 1500V DC catenary lines, giving up to 2MW of power during time intervals of between 18 and 20 seconds [65].

In 2012, ABB was part of a consortium that installed the world's first pilot lithium-ion trackside ESS to capture the excess energy produced during train braking in the city of Philadelphia. The solution allows the train operating company SEPTA (South East Pennsylvania Transit Authority), to store the braking energy

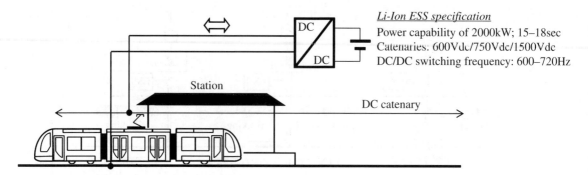

Li-Ion ESS specification
Power capability of 2000kW; 15–18sec
Catenaries: 600Vdc/750Vdc/1500Vdc
DC/DC switching frequency: 600–720Hz

Fig. 5-98   *B-Chop trackside ESS [65]*

Table 5-21   *Trackside ESSs developed by international manufacturers [67]*

Technology	Voltage	Energy 1–4kWh	Energy 0.8–16.5kWh	Energy > 400kWh	Locations
Supercapacitor	600–750V DC	SIEMENS Sitras® ADETEL Adeneo®			Madrid, Cologne, Beijing, Toronto Lyon
	500–1850V DC		ABB Envitech Energy®		Warsaw, Philadelphia
		MEIDENSHA Capapost®			Hong Kong
Lithium-ion	600V DC, 750V DC, 1500V DC			HITACHI B-Chop®	Kobe, Osaka
				SAFT Intensium Max®	Philadelphia

and sell it back to the regional transmission grid when it was economically more convenient. The ESS was performed using ABB's ENVILINE™ Regenerative ESS composed of high-performance SAFT lithium-ion batteries. Electric energy prices were monitored to sell stored energy back to the grid during peak hours and in hot weather conditions, because that is when energy is more expensive [66].

The SEPTA system was located close to an electric substation. The ESS, made of a large-scale lithium-ion battery (the INTENSIUM MAX20 from SAFT), was charged when the track voltage was above 800V. The MAX20 could store and deliver power at a rate of 1.5MW, having a storage capacity of about 500kWh.

Table 5-21 shows trackside EDLC and lithium-ion-based ESS developments by international manufacturers.

5.6.4   On-board energy storage applications

This section focuses on the use of on-board ESSs in railway vehicles. Some studies [68], [69], [70] have demonstrated that traction energy consumption could be reduced by approximately 15–35% in existing railway applications. On-board ESSs also help to minimize the power peaks during the acceleration of vehicles, which reduces losses in catenary lines. They can also help to stabilize the network voltage and run the vehicle without overhead catenary in routes going through historic city centers. However, vehicles with an on-board ESS typically require large spaces on the vehicle to locate it, usually on the roof in the case of a low-floor tramway. This increases the weight by about 2–4 tons depending on energy storage size.

The main objectives when installing an on-board ESSs in a railway vehicle are:

- storing braking energy when decelerating;
- reducing the power drain from the electric substation when accelerating;
- being able to run without overhead catenaries in the case of tramways and trains or without switching on the diesel engine in the case of shunting locomotives.

On-board ESS classification and analysis can be split into two different applications: shunting locomotives and tramways, each one has its own peculiarities and energy management strategies.

5.6.4.1 ESSs in diesel-electric shunting locomotives

Fig. 5-99 shows the classic powertrain configuration of a diesel-electric shunting locomotive, which includes a diesel engine, electric generator, main power inverter, and crowbar resistive system. The diesel electric configuration doesn't allow any braking energy to be stored, and so it has to be wasted as heat on the crowbar resistors. These kinds of vehicles are used to transport very heavy loads, such as ship containers, on non-electrified lines.

In general, shunting locomotives can be classified into two groups: heavy and light shunting locomotives. Heavy shunting locomotives operate on yards and usually move high load train wagons. Light shunting locomotives operate near industrial railway areas picking-up or dropping wagons at sidings for loading and unloading.

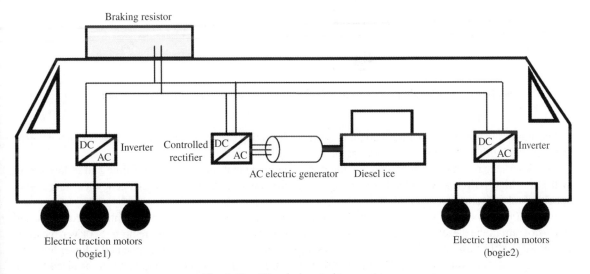

Fig. 5-99 *Diesel-electric locomotive*

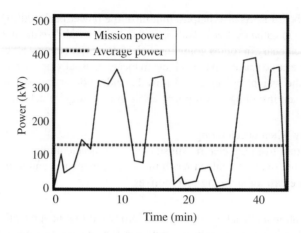

Fig. 5-100 *Diesel generator operation histogram in a locomotive [71]*

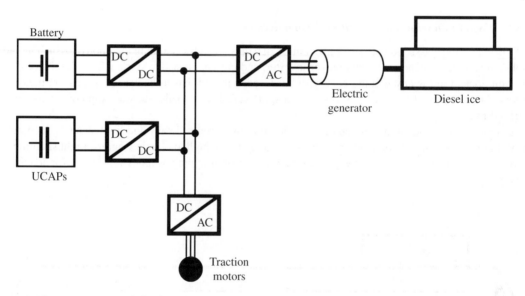

Fig. 5-101 *Series-hybrid powertrain configuration in Green Goat shunting locomotive [72]*

One important aspect to take into account in these two shunting diesel-electric locomotives is that they operate very randomly with very low diesel engine power utilization.

On diesel-electric shunting locomotives, the diesel engine operates at idle most of time and at low power with short high-power peaks. This leads to higher specific consumption and low efficiency, causing higher specific fuel consumption and emissions.

Fig. 5-100 shows a real power profile on the diesel engine in a heavy diesel-electric shunting locomotive and the histogram of the diesel generator operation [71]. Even if the peak power levels are high (570kW), the average power level is low (200kW). Overall, the diesel generator operates most of the time at the lower load and only briefly at its nominal power. It has to be taken into account that the diesel engine has to be sized for the maximum peak power level demanded even if most of the time the power level demanded is less. This leads to

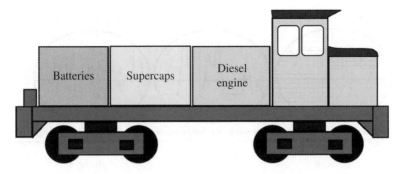

Fig. 5-102   *Hybrid shunting locomotive*

the non-efficient operation of the diesel engine and a high degree of pollution. Thus, it is clear that often the diesel engine is ideally not used in these applications.

An on-board ESS can be used to avoid the non-efficient nature of the shunting locomotive by adding another energy source to the powertrain composed of batteries, supercapacitors, or both.

Fig. 5-101 shows a series-hybrid diesel-electric shunting locomotive powertrain where two ESSs are connected to the common DC bus by power converters. The supercapacitor ESS gives power for boosting the vehicle and the battery ESS in combination with the diesel engine gives energy for average power level missions. The main objective of this powertrain configuration is to undersize the diesel generator to the mission average power level. Although additional cost and space is necessary, because of accumulators, other benefits like lowering fuel consumption, recuperative braking, and silent or zero emission operation like the "electric last mile run operation" are possible. This configuration can be found in the RailPower project Green Goat [72].

5.6.4.2   *Energy management strategy for diesel-electric shunting locomotives*

The main energy management strategies in the series-hybrid powertrain of the diesel-electric locomotive consist of replacing the diesel generator with a smaller one capable of providing the average power, giving to the on-board ESS the responsibility to deliver the rest of the fluctuating power [72], [73], [74]. When the loading power is lower than the generator capacity, the storage devices will charge from the diesel generator to be ready again the next time the loading power is higher than the diesel generator capacity. The idea is to operate the diesel generator under these conditions close to its nominal power, thus reducing diesel oil consumption and pollution. Fig. 5-102 illustrates this configuration.

With this series-hybrid philosophy, the on-board battery ESS will deliver slow dynamic power cycles to avoid fast dynamic currents, which will be delivered by the supercapacitors (as shown in Fig. 5-103).

Fig. 5-104 shows the principle of the energy management strategy based on a frequency approach. $P_{LOAD}(t)$ is the load power profile needed to run the vehicle under the track requirements. $P_{ESS\_SC}(t)$ is the discharging power demanded by the supercapacitor energy storage system, which is determined by subtracting to $P_{LOAD}(t)$ the slow dynamic behavior of $P_{LOAD}(t)$ itself. The slow dynamic behavior is obtained using a low-pass filter. The diesel generator is supposed to provide its nominal power $P_{DIESEL\text{-}NOM}$ continuously. Therefore the mission of the battery's $P_{ESS\_BATT}(t)$ is obtained by subtracting the diesel generator power from the lower frequency part of the locomotive power mission (filtered $P_{LOAD}(t)$).

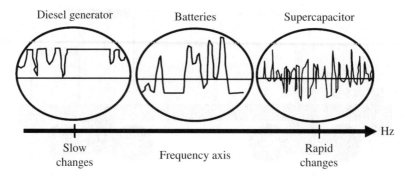

Fig. 5-103   *Energy sources in the frequency space*

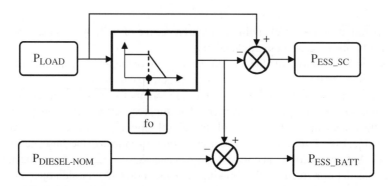

Fig. 5-104   *Energy management strategy in a series-hybrid diesel-electric locomotive*

For the same traction requirements, $P_{ESS\_SC}(t)$ and $P_{ESS\_BATT}(t)$ could be changed depending on the low-pass filter's cut-off frequency (fo) and the $P_{DIESEL-NOM}$.

The diesel engine does not usually work during stops. It works and charges the ESSs while the vehicle drives at constant speed.

Another useful application of the on-board ESS is to use its energy for starting the diesel engine. Starting these big diesel engines is not an easy task, so powerful accumulators are used. The task is complicated by cold temperatures, owing to the increase in the internal resistance of the accumulators, which reduces their available power. In this scenario the on-board energy could feed the power inverter to drive the electric generator as an electric motor and start the diesel generator.

Some considerations have to be taken into account in order to guarantee the required lifetime for the batteries and supercapacitor cells used for this kind of series-hybrid diesel-electric locomotives:

- maximum charging/discharging current;
- maximum nominal charging/discharging RMS current;
- maximum/minimum working SOC;
- maximum storage and working temperatures;
- cooling strategy for ESS (natural, forced air, active air cooling, passive liquid cooling, active liquid cooling);
- SOC and temperature at depot commissioning.

All of these aspects greatly influence the cell's behavior and degradation during the lifecycle of the component on the ESS. The selection of the energy management strategy and the design of the ESS itself shall combine the required load profiles and the desired lifecycle of the battery and supercapacitor ESS.

The same strategy has been found in the literature but sometime with, instead of batteries, an ESS composed of a flywheel and a fuel cell. In both cases their objective was to deliver smooth and low-frequency power cycles to the powertrain [74], [75], [76].

5.6.4.3   *Hybrid diesel-electric locomotive developments*

A classification of existing hybrid locomotives that incorporate some kind of on-board ESS can be found in [77]. It can be observed that the hybridization has been done mainly for locomotives weighing between 60 and 130 tons, Bo-Bo axle configuration, and under the series-hybrid topology. In hybridized locomotives, the diesel engine average power is around 200kW. If the on-board ESS is composed of lithium-ion batteries, its maximum power is around 240kW, except for the special case of the RailPower GG20B unit, where the maximum power of the lithium battery is around 1.2MW, whereas the maximum power delivered by nickel cadmium batteries can be as high as 330kW.

The main objectives in the development of these projects were to reduce toxic exhaust gases, lower the exterior noise level, and reduce CO_2 emissions.

In the case of, for example, the Toshiba HD300 hybrid shunting locomotive, NOx emission was about 62% lower, whereas the fuel consumption was reduced to around 36% in comparison to the conventional diesel locomotive. For the TEM9H SINARA hybrid, diesel fuel consumption went down by 40% and exhaust emissions were reduced to 55%. RailPower's GG20B locomotive achieved large savings in fuel consumption (40–70%) and significantly reduced emissions (80–90%). General Electric's hybrid loco design reduced fuel consumption by 15% and emissions by as much as 50%; a 40% fuel saving and an approximately 50% reduction of emission were measured on Alstom's BR203H, and a 20–30% lower fuel consumption in comparison to the standard diesel-electric shunters was experienced by Raba's 718.5 hybrid locomotive.

5.6.4.4   *ESSs in AC/AC shunting locomotives*

First-generation electrified lines for trolleybuses, trams, underground, and trains were exclusively of DC current with voltages of 600, 750, 1500, and 3000V DC respectively. These catenary voltages are relatively low and limit the power capability of the railway infrastructure. For powerful shunting locomotives connected to overhead lines, it is preferable to use high AC voltages up to 15/25kV AC to increase the power, the speed, and/ or the mass being transported.

Fig. 5-105 shows the electrical power scheme of an AC/AC shunting locomotive which includes an on-board ESS in a series-hybrid configuration. This power configuration allows for sending back to the main AC network the regenerative energy in order to be useful to other vehicles connected to the AC catenary line. The power inverter is a classic two-level converter configuration which enables the regeneration of energy during braking. If the electric substation is made by a simple rectifier, the regenerative energy during braking cannot be sent back again to the AC main grid, but this topology allows this regenerative energy to be sent back to the on-board ESS. The ESS can also be charged through the AC main line and assists later during departure to boost the vehicle.

5.6.4.5   *ESSs in Tramways*

Tramways are very popular railway vehicles that transport people at different locations through city avenues. The main characteristics of trams are that they suffer from frequently stop/start conditions, the average speed can vary from 30 to 50km/h, and they are fed from non-reversible DC catenary lines.

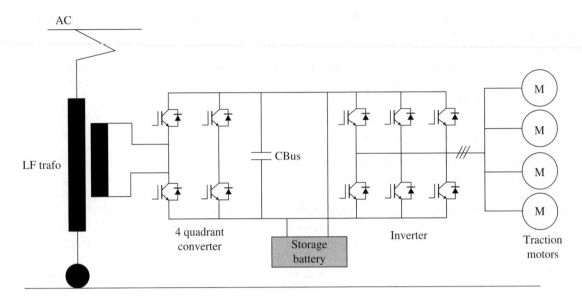

Fig. 5-105   *AC/AC series-hybrid shunting locomotive*

Under a standstill condition, the tramway regenerates the kinetic energy into electric energy back to the catenary line. This electric energy can be absorbed by other vehicles connected to the same line if they are powering. However, when the density of vehicles is low, the probability that the regenerative energy will be absorbed by other vehicles is low too. This way, to prevent dangerous over-voltage values in overhead catenary line during braking, the energy has to be lost as heat in on-board resistive elements and in the mechanical friction brake.

Here again, an on-board ESS is an interesting solution for a tramway in order to recover the braking energy to prevent over-voltages and losses on resistive elements. Braking processes are characterized by high momentary power levels, so an ESS made of supercapacitors is the preferred choice for using in these applications thanks to its unique characteristic of being capable of handling high-power pulses. Later on, the energy stored can be provided for acceleration processes. This way both peak power in the electric substation and catenary losses will be reduced.

More energy-dense on-board ESSs like the ones made of batteries allow for both, recovering the braking energy and running on routes where no overhead catenary is present. In fact, in this last application, the size of the on-board ESS depends on the catenary-free route length taking into account that a typical tramway demands between 3 and 5 kWh/km.

5.6.4.6   *ESS management strategies for LRVs*

References [78], [79] present a practical sizing of a supercapacitor-based on-board ESS for a tramway application, where a uni-directional catenary is always present. The objective of the sizing is to evaluate the energy savings that can be achieved by using the supercapacitor's storage system. The main objective here is to recover all braking energy, and the second is to split the power needs of the tram during vehicle acceleration between the storage system and the overhead catenary line. Study has shown that there is no other vehicle running on the same catenary that could take the braking energy, so all of this has to be stored in the ESS.

The supercapacitor energy storage sizing depends on several factors:

- speed cycle of the tram route;
- deceleration factor or stopping time;
- tram weight taking also into account people/m^2 factor;
- altitude;
- auxiliary load consumption;
- supercapacitor electric characteristics and beginning (BOL) and end of life (EOL) condition.

As a general rule it can be said that the higher speed, deceleration factor, tram weight, altitude, and less auxiliary load just before the tram stopping process lead to higher braking energy being stored in the ESS. In addition, supercapacitor cells at the end of their journey have less capacity, which means less energy storage capability. This issue, depending on energy storage size, could mean it becomes impossible to manage all of the braking energy present during the braking process.

Fig. 5-106 shows two speed profiles considered during energy storage sizing reported in [78]. The upper speed profile has higher maximum speed values than the second but lower acceleration factors 0.7m/sec^2 vs 0.84–1.17m/sec^2.

The simplified powertrain configuration is shown in Fig. 5-107. The supercapacitor ESS is composed of a number of strings connected in parallel in combination with a bi-directional DC/DC buck-boost converter. In the example of [78] the auxiliary converter power is set up to 23kW, corresponding to measured tram average values.

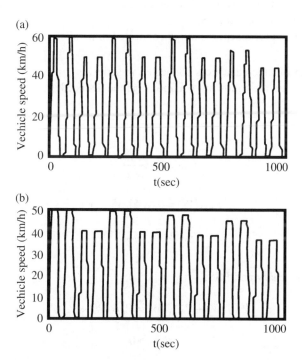

Fig. 5-106   *Tram driving cycles [78] (a) higher maximum speed values, (b) lower maximum speed values*

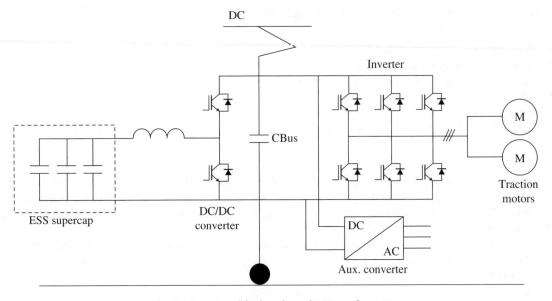

Fig. 5-107   *Simplified on-board ESS configuration*

being:

- n1: number of supercapacitors cells in series per string;
- n2: number of strings in parallel on ESS;
- nSTRINGS: number of supercapacitors strings connected in parallel.

A string is a fixed number of supercapacitors connected in series.
   Based on the power scheme the following equations can be deduced:

- On ESS:

$$i_{Cell}(A) = \frac{i_{ESS}(A)}{n_{STRINGS}} \tag{5.39}$$

$$P_{ESS}(W) = i_{ESS}(A) \cdot V_{ESS}(V) \tag{5.40}$$

$$V_{ESS}(V) = V_{string}(V) \tag{5.41}$$

- When assisting from catenary and ESS:

$$P_{INV}(W) + P_{AUX}(W) = P_{ESS}(W) + P_{CAT} \tag{5.42}$$

$$P_{CAT} > 0 \tag{5.43}$$

$$P_{ESS} > 0 \tag{5.44}$$

- When braking:

$$P_{CAT}(W) = 0$$

$$P_{INV} < 0 \tag{5.45}$$

$$P_{ESS} < 0$$

$$PI_{NV}(W) = P_{AUX}(W) + P_{ESS}(W) \tag{5.46}$$

The useful energy of the supercapacitor ESS depends on the amount of supercapacitor strings connected in parallel and the maximum and minimum cell working voltage. It is very usual to work between 100 and 50% of SOC at EOL condition on cells; that way the available energy of the supercapacitor will be 75% of the total energy stored. The following equations describe the useful energy on supercapacitor ESS:

$$E_{CELL\_MAX}(J) = \frac{1}{2} \cdot CAP_{CELL}(F) \cdot V_{CELL\_MAX}^2 \tag{5.47}$$

$$E_{CELL\_USEFUL}(J) = \frac{1}{2} \cdot CAP_{CELL}(F) \cdot \left(V_{CELL\_MAX}^2 - V_{CELL\_MIN}^2\right) \tag{5.48}$$

Voltage and capacity on ESS:

$$V_{STRINGMAX}(V) = n1 \cdot V_{CELL\_MAX} = V_{ESSMAX}(V) \tag{5.49}$$

$$V_{STRINGMIN}(V) = n1 \cdot V_{CELL\_MIN} = V_{ESSMIN}(V) \tag{5.50}$$

$$CAP_{ESS}(F) = n2 \cdot \frac{CAP_{CELL(F)}}{n1} \tag{5.51}$$

Useful energy on complete ESS being d, the discharging factor, $d = 0$ *full charge*, $d = 100\%$ *full discharge* (considered 50% at EOL):

$$E_{ESS\_USEFUL}(J) = \frac{1}{2} \cdot CAP_{ESS}(F) \cdot V_{ESSMAX}^2 \cdot \left(1 - \left[\frac{d}{100}\right]^2\right) \tag{5.52}$$

$$E_{ESS\_MAX}(J) = \frac{1}{2} \cdot CAP_{ESS}(F) \cdot V_{ESSMAX}^2 \tag{5.53}$$

SOC on cell and on ESS:

$$CELL_{SOC}(\%) = \frac{V_{cell}^2}{V_{cellMAX}^2} \cdot 100 \tag{5.54}$$

$$ESS_{SOC}(\%) = \frac{V_{ESS}^2}{V_{ESSMAX}^2} \cdot 100 \tag{5.55}$$

The following points are recommended to take into account during sizing strategy:

- *Point 1*: In [78] a maximum cell working voltage of 2.5V is selected but it depends on the desired lifecycle of the component [26], [27], [28]. For longer lifecycles, a lower value has to be selected.

- *Point 2*: During the sizing criteria it is very important to limit the current through the supercapacitor cell. In the case of [78] $0.12 \cdot I_{SHORTCIRCUIT}$ is selected (576A) but other authors limit the maximum current to 400A [80].
- *Point 3*: The maximum ESS voltage has to be selected for safe operation. In [78] a maximum of 700V is fixed but in [81] not more than 500V is allowed on ESS, owing to DC/DC converter architecture and possible uncontrolled catenary voltage variations based on EN5163 standard.

ESS sizing criteria [78]   The recoverable energy in the braking phase of the train is the sum of the kinetic and potential energy; if the altitude difference is negligible, only the kinetic energy is taken into account. The recoverable energy is lower than the kinetic energy of the vehicle, owing to internal losses of the vehicle, rolling resistances, aerodynamic drag, and so on.

$$E_{KINETIC\_RECOVERABLE}(J) = K1 \cdot [E_{KINETIC}(J) + E_{POTENTIAL}(J)] \, 0.5 < K1 < 0.6 \qquad (5.56)$$

The vehicle controller measures the vehicle speed and altitude and estimates the amount of energy that would be generated in a sudden deceleration from the current speed to standstill and determines the $^*ESS_{SOC}$ set point.

$$^*ESS_{SOC}(\%) = 1 - \frac{E_{KINETIC\_RECOVERABLE}(J)}{E_{ESS\_MAX}(J)} \qquad (5.57)$$

The $^*ESS_{SOC}$ set point decreases as the relation $\dfrac{E_{KINETIC\_RECOVERABLE}(J)}{E_{ESS\_MAX}(J)}$ increases, to have plenty of room to recovering the energy. $^*ESS_{SOC}$ set point decreases lead to a decrease in V_{ESS} too. This is done by discharging the ESS via the DC/DC convert.

$$V_{ESS} = \sqrt{\frac{^*ESS_{SOC}(\%) \cdot V_{ESSMAX}^2}{100}} \qquad (5.58)$$

Several ESS supercapacitor configurations have been developed using this strategy in [78]. The conclusion is that none of all the storage systems developed is optimal in all conditions of load, speed profiles, weight, cost, BOL, and EOL condition. Nevertheless, an ESS of an average of 1.2kWh has been considered a good option that behaves well in all circumstances for a typical tramway application.

Fig. 5-108 shows that ESS SOC set point is different for different speed profiles, being lower at higher vehicle speed values. Also as it can be seen the use of an on-board ESS during vehicle acceleration reduces catenary voltage drop. And as can be seen in Fig. 5-109, the use of an on-board ESS during vehicle acceleration reduces catenary voltage drop.

Main conclusions obtained from [78] are:

- The energy savings range from 23 to 26% and increase with supercapacitor size.
- Savings are higher when the vehicle is loaded with passengers.
- Energy savings decrease at the EOL of the supercapacitor.
- Catenary voltage drops are significantly reduced during acceleration.

In [80] the authors present a procedure to size an on-board supercapacitor ESS with the objective to reach a compromise between the minimum number of supercapacitors, ESS efficiency, and cost for a specified braking

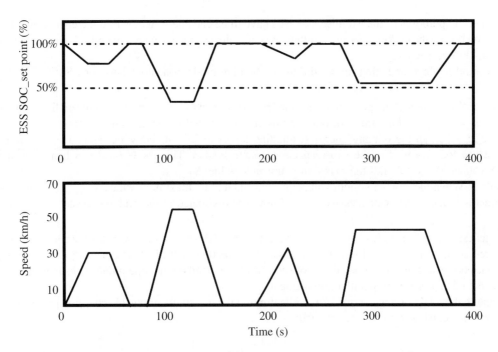

Fig. 5-108 *Speed and supercapacitor energy storage SOC set point evolution [78]*

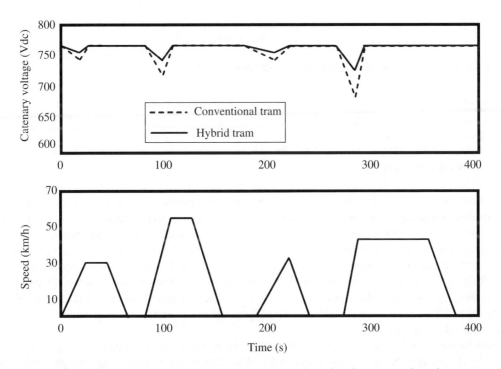

Fig. 5-109 *Catenary voltage comparisons for a tram with and without an on-board ESS [78]*

process. For the purposes of the exercise, the cells have to be capable of recovering the energy and the maximum power during the braking process, limiting the cell voltage and current to the desired maximum values of 2.5V and 400A respectively.

There could be three operating modes during the braking mode of a supercapacitor ESS:

- Mode 1: When the braking power exceeds the ESS power capability restricted by current limitation. In this mode $i_C = I_{CMAX}$, and a portion of braking power is dissipated in the braking resistors.
- Mode 2: When all the braking energy (excluding losses in DC/DC converter and supercapacitor series resistance RC) is saved in the supercapacitor storage system. This mode finishes when supercapacitor ESS voltage, V_{ESS}, reaches the maximum allowable value, V_{ESS_MAX}.
- Mode 3: When the storage system voltage is controlled at its maximum value, V_{ESS_MAX}. The charging current is reduced and the excessive braking power is dissipated in the braking resistor.

The highest energy losses are observed during time intervals $0 - t1$ and $t2–t3$ when some energy is lost in the braking resistors. To avoid these losses it is necessary to correctly size the ESS and select its discharge depth, d. The usable energy and the power capability of such an ESS depend on the total number of supercapacitors, N, independent of their series or parallel arrangement.

To capture the total braking energy, Ebr, while limiting the maximum allowable charging power, $PbrMAX$, the following inequalities should be satisfied:

$$N \geq \frac{2 \cdot E_{br}}{V_{CELL\_MAX}^2 \cdot CAP_{CELL} \cdot (1 - d^2)} \tag{5.59}$$

$$N \geq \frac{P_{brMAX}}{V_{CELL\_MAX} \cdot i_{CELL\_MAX} \cdot d} \tag{5.60}$$

The main conclusions from the sizing exercise developed in [80] are:

- The optimum discharge depth, d, is greater than 0.5, in most cases being in the range of 0.75 to 0.85.
- A number of supercapacitors greater than the optimum given value leads to slightly more saved energy, owing to a reduced supercapacitor current but ESS cost will be higher.
- Fewer cells than the optimum number might be an option for when the ESS payback time is more important than the amount of energy saved.

Three more different control strategies for an on-board ESS composed of supercapacitors are presented in [82]. These control strategies are named Mean Power, Peak Shaving, and Proportional. In the Mean Power control strategy, the control tries to take from an overhead catenary line the mean power managed by the vehicle, supplying this power as a constant value and letting the rest of the varying power (acceleration and all braking) be managed by the supercapacitor storage system. With the Peak Shaving control strategy, the ESS discharge starts after the power that has been taken from the catenary reaches a predefined maximum absolute value. The Proportional control strategy is based on discharging the storage system with a power level proportional to the traction power. In addition, all of the strategies presented recharge supercapacitor storage system with braking energy.

The conclusion is that for the control strategy selection to achieve low substation losses depends on the ESS size, efficiency, and catenary wire resistance, and the Mean Power and Peak Shaving strategies being the preferred ones. The Mean Power control strategy has advantages over other strategies at high capacities and efficient ESSs and high catenary wire resistances, while the Peak Shaving control strategy is best for low-efficiency catenary lines independently of ESS capacity.

It has to be noted that the size of the ESS on the applications where the overhead catenary is always present is much less than on tramways running on non-catenary lines, because in the second case no overhead catenary can help to supply energy to the vehicle so all the energy has to come from the on-board ESS.

Reference [81] is a practical example where a low-floor tramway has to deal with a combination of sections with overhead line and catenary-free sections on the same route. In this case, the on-board energy storage sizing is directly dependent on the largest catenary-free section on the route based on the worst-case scenario, which is present when the auxiliary loads are at full power and at EOL condition for the cells. In the case of [81] a supercapacitor-based on-board ESS has been developed which has been sized to be able to comply with vehicle requirements (power, speed) even in the worst-case conditions that are present when the auxiliary loads are at full power and the cells have reached their EOL condition.

At EOL condition, the DOD of the supercapacitor-based on-board ESS will be 50% aiming to use 75% of stored energy. This sizing criterion leads to a DOD lower than 50% at BOL on the on-board ESS. If greater DOD than 50% is wanted at the EOL condition, vehicle power performances will have to be reduced to prevent over-currents through supercapacitor cells, owing to lower working voltage in the ESS.

In [81] the on-board supercapacitor ESS is recharged completely at each station stop to be prepared for helping the acceleration of the vehicle on the catenary sections or to be able to source alone the vehicle if no overhead catenary is present. See Fig. 5-110.

The energy management strategy for the application presented in the preceding paragraph is:

- The energy storage SOC doesn't have to be maximum when the tram is running on an existing catenary section; in fact, on existing catenary lines, the supercapacitor ESS maximum SOC has to be selected high enough to help with accelerating the vehicle with the desired power. On existing catenary lines and during acceleration, the power needed for vehicle acceleration is taken from the electric substation and from the supercapacitor ESS, so the higher SOC on the energy storage, the higher the power that can be delivered for acceleration.
- When the vehicle approaches a station stop, the SOC of the on-board supercapacitor ESSs has to be low enough to be prepared to receive the maximum braking energy as possible that will be generated during the stopping process. The braking energy mainly depends on vehicle mass, deceleration factor and speed just before the braking process starts. If the SOC of the supercapacitor storage system is high, part of the braking energy will be used to charge it completely and the rest of the braking energy will have to be dissipated as heat on the crowbar resistive system leading to the non-efficient management of the braking energy. Usually, the higher the speed, the minimum SOC on the supercapacitor energy storage because more kinetic energy is intended to be stored during the braking process.

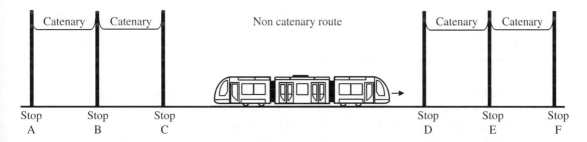

Fig. 5-110 *Low-floor tramway in combined catenary and non-catenary sections on the same route [81]*

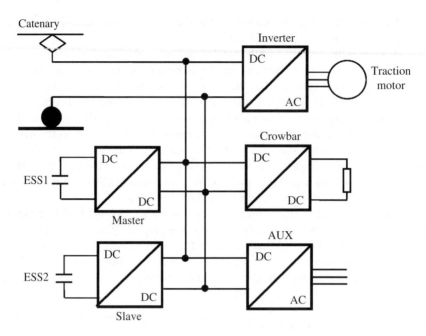

Fig. 5-111   *LRV powertrain architecture developed in [81]*

Fig. 5-111 shows the powertrain configuration for the low-floor tramway developed in [81]. As can be seen, there are two on-board supercapacitor ESSs. Each ESS has an associated DC/DC converter, one being the master and the other the slave.

At station stops both storage systems are charged simultaneously, avoiding any deviations between supercapacitor ESS voltages.

- Catenary support mode: When the tramway is accelerating in a catenary existing line, both ESSs are discharged at the same power level. The DC-link voltage is established by the catenary regarding EN50163 standard (500V DCmin–750V DCnom–900V DCmax). When the tramway starts braking, the DC/DC converters change the control mode from power control to voltage control mode thus charging both storage systems to control the DC-link voltage higher than the catenary one but below a dangerous value.
- Autonomous mode: In this mode the tramway operates in a catenary-less section and consequently the ESS must supply all of the tramway power needs during acceleration and braking periods (voltage control mode). The master DC/DC is working in voltage control mode discharging ESS1 to keep the bus voltage at 850V. The slave DC/DC controls ESS1 discharging power in order to keep both ESSs voltages equal [81]. In this way, as the tramway accelerates, both supercapacitor storage system voltages decrease. Again, when the tramway starts braking, the DC/DC converters change the control mode from power control to voltage control to control the DC-link voltage value.

The on-board ESS is always calculated to fulfill the nominal operating conditions of the tramway. However, the system also has the capacity to deal occasionally with exceptional situations, such as unexpected stops, unsuitable driving, failures, traffic jams, or accidents. These unexpected stops are an issue if they occur when the tram has been stopped on a non-catenary section because the energy is lost only on feeding auxiliary loads and could not be high enough to move the vehicle once the issue has been resolved. Moreover, a

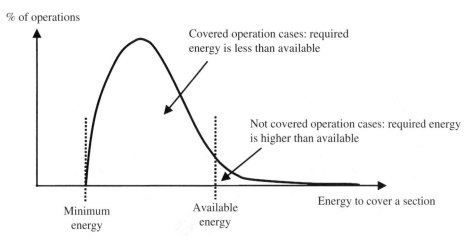

% of operations

Covered operation cases: required
energy is less than available

Not covered operation cases: required energy
is higher than available

Minimum
energy

Available
energy

Energy to cover a section

Fig. 5-112 *Probability curve of the tramway operation*

supercapacitor-only-based tramway stores much less energy than a battery-based one so long unexpected stops are critical in this case.

Fig. 5-112 represents a typical probability curve for the real operation of a tramway. It is always some minimum energy consumption to cover a section and the dimensioning of the ESS has to cover almost all the cases but, as explained before, some unexpected cases cannot be covered with the available energy.

Preventive measures will have to be taken if during an unexpected stop the ESS SOC decreases below a certain value. These preventive measures can lead to the reduction or complete switching off of the air conditioning unit, which is the most demanding auxiliary load.

Reference [83] presents a lithium-ion-battery-driven low-floor tram. The battery-driven tramway is 31.2m length with five cars and three-bogie low-floor articulated, with a total storage energy of 50kWh/600V DC. The vehicle can cover up to 30km running catenary-free after a full charge of the on-board ESS. During stop conditions on stations, the vehicle can charge the on-board ESS using an existing pantograph, but for the rest of the route the battery ESS has to manage the power demanded from the vehicle.

For assuring sufficient boost power capability during vehicle acceleration, the maximum SOC of the battery system is set to 80%, and to accept the regenerative energy during braking the SOC can go down to 50%. This is due to charging power capability increases as the DOD increases, and the discharging power increases as the DOD reduces.

On this kind of vehicle running on catenary-free sections, the auxiliary loads (heating, ventilation, air conditioning system) are a big issue because they are mandatory but can demand over 20–30% of vehicle driving energy consumption during regular operation.

The energy management strategy implemented in the battery-powered tramway aims to increase overall energy efficiency and to maximize driving distance; to reach both objectives during departure the non-critical auxiliary loads like the air conditioning unit is minimized or directly switched off, during coasting the battery will supply all auxiliary loads, and during braking and subsequent stopping time the air conditioning system will work at its maximum power. This strategy implementation in controlling in an intelligent way the air conditioning system developed in [83] has saved vehicles' overall energy consumption by at least 10%.

Fig. 5-113 shows the hybrid powertrain configuration of a lithium-ion-battery-driven tram [84]. In a catenary existing line, the energy storage SOC is maintained between 45 and 55% to be capable of managing high power values both during acceleration and braking, but before a non-electrified section starts, the battery

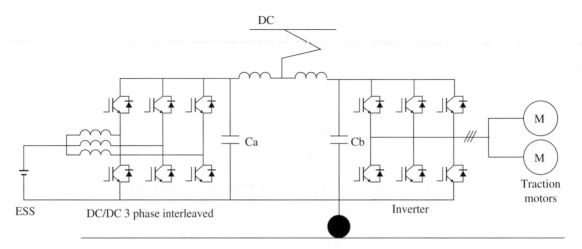

Fig. 5-113   *Traction circuit configuration of hybrid electric vehicle [84]*

storage system is fully charged from overhead catenary. During catenary-free operation, battery SOC is allowed to decrease to 30% to maximize the available energy. Energy storage charging up to medium SOC will be done again on the catenary section.

As explained before, railway vehicles using on-board storage systems based only on supercapacitors are limited in energy so they can work well if an overhead catenary is always present but have a limited range when a non-catenary route has to be traveled. On the other hand, battery-based on-board storage systems have enough energy to travel several kilometers without a catenary, but the DOD and excessive heat due to charge and discharge currents are the main stress factors that lead to an accelerated degradation. Some other applications use a combination of two on-board ESSs, one composed of supercapacitors and the second of lithium-ion batteries to make a hybrid powertrain configuration.

All the energy management strategies implemented for this hybrid topology split the power and energy between both ESSs. The supercapacitor storage system takes the responsibility for managing high-power transitory values, while the battery storage system takes care of lower average power value, reducing the charging and discharging current level and reducing the SOC swing [85], [86]. Fig. 5-114 shows a hybrid powertrain configuration with two on-board ESSs, one composed of supercapacitors and the other of batteries.

For a non-catenary tram application, the energy needed to comply with the longest driving distance will determine the size of the battery storage system, and the maximum expected power profiles during acceleration and braking will define supercapacitor storage. It has to be taken into account that the size of the battery storage system has a direct influence on the expected DOD, so it is usual to oversize the battery system to comply with the desired lifecycle, which makes it necessary to work at reduced DODs.

5.6.4.7   *ESS developments in tramways*

This section presents three real on-board applications for tramways being the ones currently commercially available and in service since 2007. Siemens with its SITRAS MES, Bombardier with its MITRAC^{TM} and Caf with its Greentech energy storage family products [87].

Siemens Transportation Systems developed in 2008 its first on-board ESSs concepts SITRAS HES and SITRAS MES. SITRAS MES (mobile energy storage unit) was the name given by Siemens for when the on-board ESS is made up of supercapacitors and SITRAS HES (hybrid ESS) where there are two separate

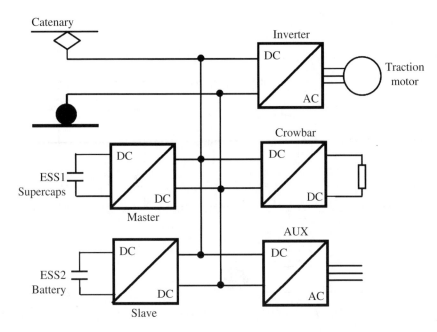

Fig. 5-114 *On-board supercapacitor and battery hybrid ESS architecture*

storage boxes, one made of supercapacitors and the other of NiMH batteries. SITRAS MES uses air force to cool the supercapacitors, and the NiMH traction battery is cooled using an active liquid cooling system [88], [89]. SITRAS MES and the NiMH traction battery are usually installed on the roof of a tramway vehicle.

SITRAS MES usable energy is around 0.8kWh, with a maximum power of 280kW and voltage swing between 190V DC to 480V DC. The NiMH traction battery has the capability to handle 18kWh of energy at 85kW of nominal power with 528 nominal voltage [88].

One of the advantages of the SITRAS HES configuration is that the supercapacitor storage system can be used to assist during accelerations/braking conditions and the traction battery system can allow catenary-free working scenarios and in case of emergency (catenary failure) use the energy stored in the traction battery to reach the next station.

The MITRAC Energy Saver is the name given by the Canadian manufacturer Bombardier to its supercapacitor made ESS for tramway applications (Fig. 5-115). Bombardier started in 2003 with the studies for adopting on-board ESSs using only supercapacitors [90].

The MITRAC Energy Saver is composed of one power inverter, a supercapacitor ESS, a braking resistor, and associated power converters in one unique compact electro-mechanical solution.

The MITRAC supercapacitor ESS could be sized accordingly to vehicle operation, depending on whether it is used for vehicles working with or without overhead catenary (catenary-free operation). For catenary-free operation the aim is to be able to run for 500m as an average length being the numbers of MITRAC units installed the unique sources of energy in the vehicle.

The usable energy of the MITRAC energy saver is around 1kWh with a maximum power capability of 300kW and 432V DC nominal voltage. Depending on vehicle size and requested performances more than one can be installed on the roof [91]. For example, in a 30m tramway two independent on-board MITRAC Energy Savers coexist. Recharging of the energy saver could be done during station stops taking up to 600kw from the overhead catenary over a 20-second stop.

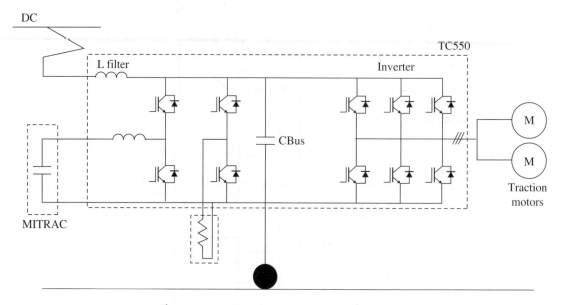

Fig. 5-115   *MITRAC Energy Saver configuration*

When working with an overhead catenary, at the beginning of the day, the supercapacitor energy modules installed on the vehicle are charged completely but the rest of the day they are recharged using only the regenerated energy during braking processes. In this scenario the MITRAC Energy Saver helps with accelerating to reduce the power demand from the line and charges again from the regenerated energy. Bombardier claims that during braking around 13% of energy is returned to the overhead line and another 28% goes into the supercapacitor storage boxes, meaning that around 40% of the braking energy can be recovered in total.

For a non-catenary working scenario, the supercapacitor modules discharge their 75% of energy down to 200V if the vehicle goes above 60km/h. This is done to locate all the braking energy, which increases with vehicle speed, to recharge the ESS. The rest of the energy to complete the charge of supercapacitor energy modules will be done from the overhead catenary in the station. By using this strategy and depending on the sizing of the ES510 module the tramway can run from 500m up to 800m in catenary-free lines.

The Spanish CAF railway equipment manufacturer developed in 2007 its first on-board energy storage solution made of supercapacitors. CAF has developed a family of on-board energy storage named Greentech. This family comprises ACR-Evodrive solutions (made only of supercapacitors that allow for an energy saving of up to 20% and 100m driving on non-electrified lines) and ACR-Freedrive (made up of a combination of supercapacitors and batteries that allows for an energy saving of up to 30% and up to 1400m driving on lines with no overhead catenary) [92].

The ACR-Freedrive system comprises two DC/DC converters, one associated to n1 number of supercapacitor branches connected in parallel and the second one associated to n2 number of battery branches. Sometimes more power than energy is required (e.g. for existing catenary lines), but there can be other scenarios where a large section with non-catenary (some km) has to be managed where energy is the most demanding factor. The ACR-Freedrive system is modular in terms that it allows for scaling the number of supercapacitor branches (n1) and battery branches (n2) connected in parallel depending on the vehicle operating needs for each different application.

Fig. 5-116 shows the ACR-Freedrive system architecture philosophy for LRV applications. Depending on the size, more than two ESSs can be placed on the floor of the tramway.

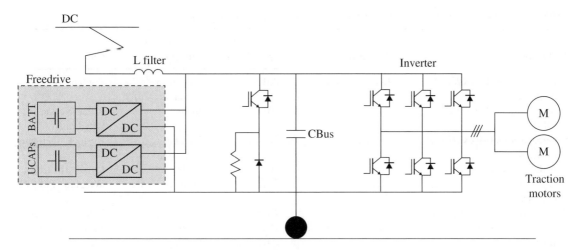

Fig. 5-116 *CAF ACR-Freedrive system architecture*

Table 5-22 *CAF ACR-Freedrive (2014) solution data (Source: CAF P&A)*

	Supercapacitor branch	NiMH battery branch
Max Voltage	550V DC	160V DC
Max Energy	580Wh	15kWh
Useful Energy	435Wh	12kWh
I_{RMS}	120Amp	60Amp
Max Power (20seg)	100kW	48kW
Average Power	50kW	10kW
Cooling	Forced air	Forced air

Table 5-22 gives the data for the ACR-Freedrive system solution (2014). There is a combination of several supercapacitor branches and a NiMH branch. Associated DC/DC bi-directional power converters can be configured in terms of average and maximum power level depending on the number of branches connected in parallel to each one. This configuration depends on the requirements of each application. More than one ACR-Freedrive can coexist in the same vehicle depending on the vehicle's power/speed/non-catenary route requirements.

In applications where higher values of energy and power are required, NiMH chemistry can be replaced with lithium-ion batteries because of their unique higher energy and power density characteristics.

In 2014, Alstom developed Citadis Ecopack on-board supercapacitor- and battery- based ESS that allows for catenary-free driving on urban routes [93].

Table 5-23 shows real on-board ESS applications both for energy saving (routes where catenary is always present) and for partially or totally catenary-free routes. As can be seen, for energy saving applications a supercapacitor on-board ESS does the job, but more energy is needed to travel on sections with no catenary. That's why in these last applications lithium-ion batteries are preferred above other battery chemistries.

Table 5-23 *On-board ESSs developments for tramways by railway manufacturers*

Technology	Application	Year operational	Vehicle Manufacturer	City
Supercapacitors (MAXWELL modules)	*Energy saving with catenary*	2012	American Maglev Co.	Portland (USA)
Supercapacitors (Bombardier MITRAC)		2012	Stadler	Rhine-Neckar (Germany)
Supercapacitors (MAXWELL modules)		2014	Vossloh	Rostock (Germany)
Supercapacitors (CAF Evodrive)		2015	CAF	Cuiaba (Brazil)
Supercapacitors (MAXWELL modules)		2015	Pesa	Wrocław (Poland)
Supercapacitors (CAF Evodrive)		2015	CAF	Tallin (Estonia)
Supercapacitors (CAF Freedrive)	*Totally or partially*	2011	CAF	Seville (Spain)
Lithium-Ion batteries	*catenary-free*	2012	Stadler	Munich (Germany)
Supercapacitors (Voith SuperCap)	*operation*	2013	CNR	Shenyang (China)
Supercapacitors (CAF Freedrive)	*(≈less than 1 mile)*	2013	CAF	Zaragoza (Spain)
Supercapacitors (Siemen SITRAS)		2014	Siemens	Guangzhou (China)
Lithium-Ion batteries (Bombardier Primove)		2014	Csr Puzhen	Nanjing (China)
Supercapacitors (CAF Freedrive)		2015	CAF	Granada (Spain)
Supercapacitors (CAF Freedrive)		2015	CAF	Kaohsiung (Taiwan)
Supercapacitors (ALSTOM Ecopack)		2015	Alstom	Rio de Janeiro (Brazil)
Lithium-ion batteries		2015	Brookville	Dallas (USA)
Lithium-ion batteries		2015	Skoda	Konya (Turkey)
Lithium-ion batteries		2015	Vossloh	Santos (Brazil)
Lithium-ion batteries		2015	Inekon	Seattle (USA)
Lithium-ion batteries		2016	Brookville	Detroit (USA)
Lithium-ion batteries (Siemen HES)		2016	Siemens	Doha (Qatar)
Supercapacitors (Alstom Ecopack)		2016	Alstom	Nice (France)

5.6.5   Power converters for ESSs

As discussed before, on-board ESSs can be composed of batteries and/or supercapacitors. The ESS has to be linked to the rest of the power converters in the vehicle powertrain by means of a bi-directional DC/DC power converter.

The DC-link is the voltage to which all on-board power converters are connected. In case the vehicle is connected to an overhead catenary, the DC-link is imposed by the catenary voltage characteristics that follow the standard EN50163. This standard specifies that, for a tramway vehicle, the catenary nominal voltage has to be 750V DC, with a minimal permanent of 500V DC and maximum permanent of 900V and 5 minutes at 1000V DC.

Fig. 5-117 shows the voltage profile of batteries and supercapacitors. As can be seen, the battery voltage seems to decay a little as the SOC and supercapacitor voltage decrease theoretically linearly when discharged.

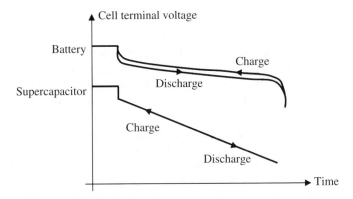

Fig. 5-117 *Battery and supercapacitor voltage profiles*

Knowing the voltage profiles of both supercapacitors and batteries, it is clear that a DC/DC converter is necessary to be able to charge/discharge the ESS independently from the DC-link voltage. If catenary-free operation is present, the DC/DC converter associated with the ESS is the one to impose the DC-link voltage of 750V DC (in the case of a tramway vehicle). If the ESS is used in other kinds of vehicles the DC/DC converter will have to impose a different value of voltage on the DC-link, 1500V DC in the case of a metro, and 3000V DC for a regional train.

Bi-directional DC/DC power converters can be classified into non-isolated and isolated types:

- Non-isolated converters are generally used where the voltage needs to be stepped up or down by a relatively small ratio (less than 4:1), and when there is no problem with the output and input having no dielectric isolation. There are three main types of converter in this non-isolated group, usually called buck, boost, and buck-boost converters. The buck converter is used for voltage step-down, the boost converter is used for voltage step-up, while the buck-boost converter can be used for either step-down or step-up.
- Galvanic isolation is required in some applications and mandated by different standards. Isolated DC/DC converters need an AC link in their structure to enable power transfer via a magnetically isolating medium (i.e. a transformer). The volume and weight of this AC transformer used in isolating DC/DC converters depends on the power and switching frequency; as a general rule, the higher the power capability of the converter, the lower the switching frequency, owing to switching losses on switching semiconductors. Low switching frequencies lead to bulky AC transformers, while high frequencies lead to smaller transformers. Resonant isolated DC/DC converters combine high power capability with high switching frequency leading to a compact and small ac transformer. Nevertheless, resonant topologies have a big impact on the cost of the power converter.

In this section the most commonly used bi-directional DC/DC converters are presented:

- buck-boost converter
- dual active bridge (DAB) converter
- dual half-bridge (DHB) converter
- resonant DAB converter
- UP/DOWN power converter.

5.6.5.1 *Buck-boost converter*

Fig. 5-118 shows the non-isolated buck-boost power converter. This converter is the simpler one to control the energy flow between two DC buses when isolation between them is not required. It consists of two IGBTs and an inductor. To be able to control the bi-directional power flow between both DC buses *Vbus*1 must be always higher than *Vbus*2. The ESS is usually connected to the low voltage side so *Vbus*2 = VESS, and *Vbus*1 is connected to the same DC-link where the remainder of the existing power converters on the application are connected.

The minimum DC value for *Vbus*1 thus determines the maximum value of VESS and in this way the maximum cells of batteries or supercapacitors that can be connected in series for each string. If more energy is needed, the ESS will have to increase the number of strings connected in parallel.

Although some studies have concluded that the high switching frequency ripple coming from the current output of the buck-boost does not have an accelerating impact on the cell degradation [94], [95], [96], it is especially recommended to reduce the ripple voltage when working with batteries, owing to the high internal resistance compared to supercapacitors. A high current ripple in batteries can lead to false over-voltage detections, owing to this high internal resistance. The reduction of the output current ripple can be obtained by using the same converter topology with an LCL filter instead of a unique output inductor L. Fig. 5-119 shows the current waveform through the output inductor of the buck-boost converter during the transition from buck mode to boost mode of operation.

Buck mode   The duty cycle of IGBT S1 controls the average power transferred from *Vbus*1 to *Vbus*2. IGBT S2 will stay switched OFF.

$$V_{bus2} = D_{S1} \cdot V_{bus1} \tag{5.61}$$

$$D_{S1} = \frac{ton_{S1}}{T} \tag{5.62}$$

Boost mode   The duty cycle of IGBT S2 controls the average power transferred from *Vbus*1 to *Vbus*2. IGBT S1 will stay switched OFF.

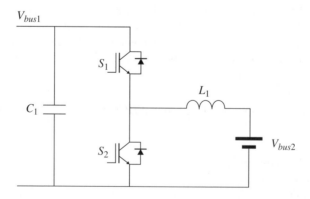

Fig. 5-118   *Non-isolated buck-boost converter*

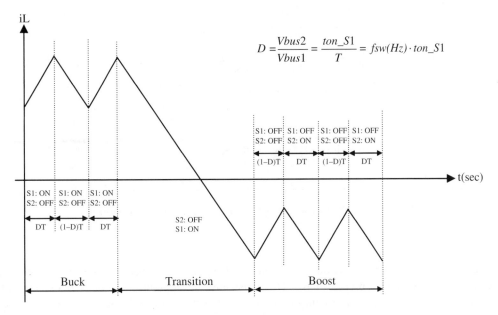

Fig. 5-119 *Buck-boost converter waveforms*

$$V_{bus1} = \frac{V_{bus2}}{1 - D_{S2}} \tag{5.63}$$

$$D_{S2} = \frac{T - ton_{S1}}{T} \tag{5.64}$$

where:

$$T = \frac{1}{fsw(Hz)} \tag{5.65}$$

The power capability of the buck-boost converter can be higher, increasing the current capability of the switching semiconductors, and the current ripple in the inductor can be lower, increasing the inductive value of the output inductor or by increasing the switching frequency. Both aspects can be reached using a multiphase buck-boost converter [97]. Fig. 5-120 shows a multiphase buck-boost converter composed of three phases and three inductors.

Multiphase buck-boost converter main features:

Power capability sharing: Each one of the phases manages 1/*N* the instantaneous and average power of the whole converter, being *N* the number of buck-boost phases connected in parallel in the multiphase converter. Thus the current capability of the switching semiconductors can be lower.

Output current ripple: In order to have low current ripple in the output inductor in the case of one single-phase buck-boost converter, one way is to increase the switching frequency, but it leads to the increase of the switching losses, and the second way is to increase the inductor self-inductive value. With the

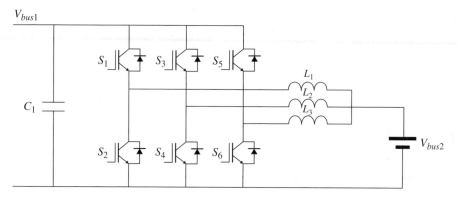

Fig. 5-120   *Multiphase three-phase buck-boost DC/DC converter*

multiphase buck-boost converter, the output current ripple can be low, even at low switching frequencies and low self-inductive values in each phase. This is because in the multiphase buck-boost converter the firing orders between switches of adjacent phases are phase shifted 360°/N degrees, leading to a reduced output ripple current with a frequency N times higher than the switching frequency of each phase. For example, a three-phase buck-boost multiphase converter is driven by gate signal at 0, 120, and 240 degrees. If the switching frequency of each phase is 1KHz, the output current ripple will be at 3kHz. In addition, transient response is higher and input and output capacitor RMS currents are lower with this multiphase topology [97].

The duty cycles for S1, S2, and S3 are:

$$D_{S1} = D_{S2} = D_{S3} = \frac{V_{bus2}}{V_{bus1}} \tag{5.66}$$

The average current in each inductor is:

$$i_{L1\,AVERG} = i_{L2\,AVERG} = i_{L3\,AVERG} = \frac{i_{OUT\,AVERG}}{N} \tag{5.67}$$

Fig. 5-121 shows the phase currents and the total output current for a three-phase buck-boost multiphase converter. It will be noticed that the output current average value is three times higher than on each phase and its frequency is three times the switching frequency.

Redundancy: Another characteristic of the multiphase buck-boost converter is that if one phase fails the rest of the phases can still work, managing their nominal power. For example, in the above-mentioned three-phase buck-boost converter, if one phase fails the whole converter power capability will be reduced by 33%.

Inductor manufacturing considerations: The phase inductor core can be made of air or ferromagnetic material. Ferromagnetic cores suffer from saturation at high current and high temperature so for high-power applications the cross-section of the ferromagnetic core has to be designed to avoid current saturation. In addition, the choice of the ferromagnetic material is not a trivial task because current ripple frequency generates hysteresis losses and heat on ferromagnetic materials. To avoid these current or temperature saturation effects, a

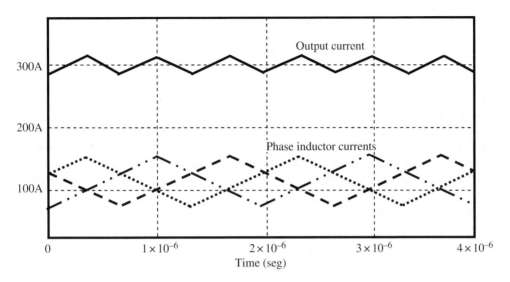

Fig. 5-121 *Current profiles on a three-phase buck-boost multiphase converter*

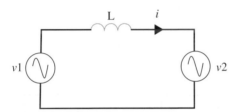

Fig. 5-122 *DAB power-transfer philosophy*

ferromagnetic core inductor would be used, but this is usually heavy, bulky, and expensive. This can be solved partially using a three-phase buck-boost converter with magnetically coupled phase inductors [98], [99].

Air core inductors don't suffer from magnetic saturation like ferromagnetic cores. Air core inductors have to be designed to thermally withstand the steady-state equivalent RMS current and for managing high current transient peaks, for example during the charge of a supercapacitor ESS.

Because of the high frequency triangular shape of the inductor current, it is recommended to use CTC cable, Litz wire, or thin copper band to minimize winding losses due to proximity effects. To avoid very high temperatures it is recommended that both ferromagnetic and air core inductors will be cooled by forced air.

5.6.5.2 *Dual active bridge (DAB) converter*

Fig. 5-122 shows the basic philosophy that informs the DAB converter. The idea consists of transferring power and energy between V1 and V2 AC voltage sources. The amount of power transferred between both AC voltages can be controlled modifying the phase shift between voltages, the RMS value, and the frequency of V1 and V2, and the path inductance value.

The active power transferred between both AC voltages is expressed as:

$$P_{SINE}(W) = \frac{V1_{RMS} \cdot V2_{RMS}}{2\pi \cdot fo \cdot L} \cdot \sin\phi \tag{5.68}$$

where *fo* is the frequency of AC voltages and Φ is the phase shift angle between them.

The same effect can be obtained when the DAB converter applies high-frequency square voltage waves to both sides of the linking inductor. Fig. 5-123 shows the DAB's working philosophy.

Fig. 5-124 shows the topology of the DAB converter, which comprises two full-bridge converters, two DC capacitors, an auxiliary inductor, and an HF transformer that provides the required galvanic isolation. The inductor can be the leakage inductor of the HF transformer or the external one connected to the primary winding of the HF transformer.

In this configuration the power transferred between both DC buses is expressed as:

$$P_{SQUARE}(W) = \frac{n \cdot V_{bus1} \cdot V_{bus2}}{2\pi \cdot f_s \cdot L} \cdot \Phi \cdot \left(1 - \frac{\Phi}{\pi}\right) \tag{5.69}$$

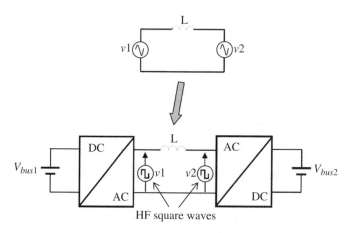

Fig. 5-123   *DAB converter working philosophy*

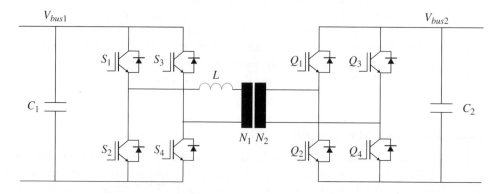

Fig. 5-124   *DAB converter topology*

where f_s is the switching frequency of both full bridges (both same), n is the transformer voltage ratio (N_1/N_2), and Φ is the phase shift angle between both generated square voltages [100], [101]. Reactive power as a consequence of the switching pattern flows between both bus capacitances, C_1 and C_2.

The maximum power transfer capability is when the phase shift between the square voltages generated by both full bridges reaches $\Phi = \pi/2$.

To transfer power from Vbus1 to Vbus2, V1 should lead V2 an angle corresponding to Φ. Conversely, power is transferred from Vbus2 to Vbus1 when V1 lags V2. This leading or lagging phase shift is simply implemented by proper timing control of converter switches [102]. The inductor is an important element which determines the maximum amount of transferable power with a given switching frequency. Therefore, apart from other practical limitations, it is possible to reach a high power density converter with a low leakage transformer.

The DAB converter was proposed in the early 1990s but because of the performance limitations of power devices the power loss of DABs was high and the efficiency unacceptable. In recent years, the advance in very low-loss power devices like silicon carbide (SiC) and gallium nitride (GaN) ones, and magnetic materials like iron-based nanocrystalline soft magnetic alloys, have made DAB topology feasible for eliminating bulky and heavy low-frequency transformers in power conversion systems. Another benefit of using the DAB topology is the ZVS behavior on the switching elements of both full bridges in certain working conditions.

Fig. 5-125 shows the normalized power of the DAB converter depending on phase shift and d parameter.

$$d = n \times \frac{V_{bus2}}{V_{bus1}} = \frac{n1}{n2} \times \frac{V_{bus2}}{V_{bus1}} \tag{5.70}$$

The soft switching condition on bridge switches is reached when the DAB converter works in the area situated between the input and output bridge boundaries.

The ZVS working condition helps to increase the switching frequency in order to reduce transformer size. Nevertheless, the transformer's magnetic core material has to be selected properly, depending on the switching frequency, to prevent extremely high hysteresis losses and heat on the magnetic core. Fig. 5-126 shows core materials used in medium- and high-frequency transformers in isolated DC/DC converters [103].

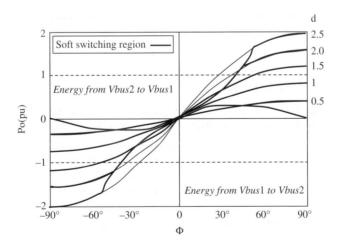

Fig. 5-125 *Normalized output power in DAB converter as a function of Φ for different values of d [100], [101], [102])*

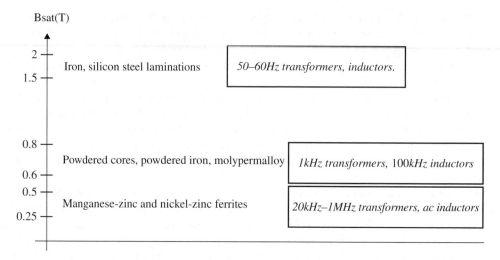

Fig. 5-126 *Core materials for medium- and high-frequency transformers [103]*

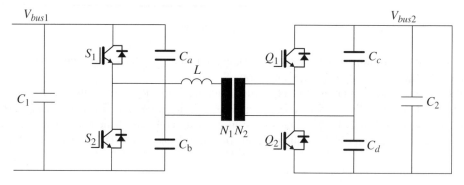

Fig. 5-127 *DHB converter topology*

5.6.5.3 *Dual half-bridge (DHB) converter*

The difference between the dual active and the DHB converter is that the latter replaces two of the inverter transistors by two capacitors in each bridge, which leads to a more economical topology. Nevertheless, with this modification the transformer's primary winding current circulates through the balancing capacitors Ca and Cb, and the secondary winding current circulates through Cd and Cc. The balancing capacitors' selection has to be done according to the RMS value of the winding currents, and can lead to a bulky converter for a high-power application, owing to large winding currents through the HF transformer [104]. Fig. 5-127 shows a DHB converter's topology.

In DHB configuration the power transferred between both DC buses is expressed as:

$$P_{SQUARE}(W) = \frac{n \cdot V_{bus1} \cdot V_{bus2}}{8\pi \cdot f_s \cdot L} \cdot \Phi \cdot \left(1 - \frac{\Phi}{\pi}\right) \qquad (5.71)$$

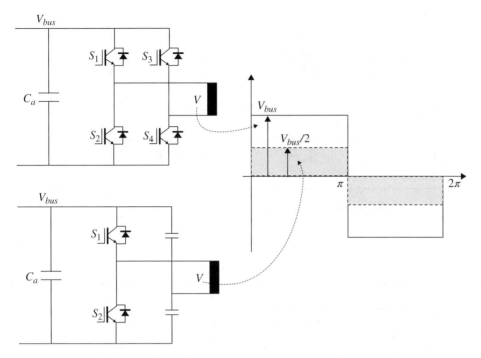

Fig. 5-128 *Transformer flux density swing in DAB and DHB converters*

where, as in the DAB converter, f_s is the switching frequency of both full bridges (both same), n is the transformer voltage ratio (N_1/N_2), and Φ is the phase shift angle between both generated square voltages.

The energy transfer philosophy is the same as in the DAB converter, when Φ is between 0 and $\pi/2$ the active power will be transferred from Vbus1 to Vbus2 and opposite when Φ is between 0 and $-\pi/2$.

The main reason for less power transfer capability with the DHB compared with the DAB is that, for the same cross-sectional area of the transformer, the transformer's flux swing is only half that of the DAB's. See Fig. 5-128.

Another variant of DC/DC converters that also use HF transformer link are the isolated DC/DC resonant topologies. Resonant topologies are very similar to DAB and DHB ones with the only difference being that a capacitor is connected in series in the transformer winding. Resonant topologies' main characteristic is the ability to increase the switching frequency with very low switching losses on semiconductors. The maximum power transference will be switching the bridge semiconductors at LC tank resonant frequency [105], [106].

5.6.5.4 *Cascaded buck-boost converter*

S1, S2, S3, and S4 IGBT semiconductor switches and the inductor L constitute the cascaded buck-boost bi-directional DC/DC converter shown in Fig. 5-129. The ESS composed of supercapacitors or batteries is connected to the DC bus managed by S_3 and S_4, while the DC bus V_{BUS1} is connected usually to a common DC-link on the vehicle powertrain [107].

Table 5-24 summarizes the working modes for the cascaded buck-boost DC/DC converter.

One important characteristic of this DC/DC converter is that, even with energy storage voltage higher or lower than V_{BUS1} a bi-directional energy flow is possible between both DC buses.

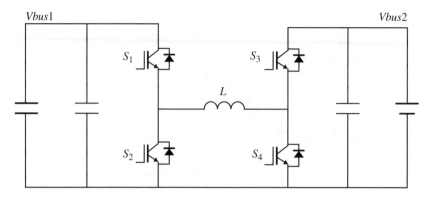

Fig. 5-129 *Cascaded buck-boost DC/DC converter [107]*

Table 5-24 *Working modes for the cascaded buck-boost DC/DC converter [107]*

		S1	S2	S3	S4
Energy from Vbus1 to Vbus2	Vbus1 > Vbus2	PWM	OFF	OFF	OFF
Energy from Vbus1 to Vbus2	Vbus1 < Vbus2	ON	OFF	OFF	PWM
Energy from Vbus2 to Vbus1	Vbus1 > Vbus2	OFF	PWM	ON	OFF
Energy from Vbus2 to Vbus1	Vbus1 < Vbus2	OFF	OFF	PWM	OFF

This DC/DC power converter architecture is used in [107] and [108] to perform an on-board emergency self-traction system metro vehicle combining batteries and supercapacitors.

Same as in the buck-boost converter, and because of the high frequency triangular shape of the current, the inductance shall be manufactured using CTC or Litz wire to minimize inductance losses.

5.6.6 EN 62864-1 standard for railway applications

This standard applies to railway vehicles which incorporate on-board ESSs in a series-hybrid configuration when the storage systems are composed of supercapacitors and lithium-ion batteries. It intends to establish the basic system configuration, tests methods to verify effective use of energy, as well as to provide railway operators and manufacturers with guidelines for manufacturing and evaluating hybrid systems, and specifies basic requirements and characteristics [109]. See Fig. 5-130.

5.7 Ground level power supply systems

GLPS (ground level power supply) systems are characterized by supplying energy to the vehicle from the track itself instead of from the overhead catenary. The energy can be transferred while the vehicle is in motion or only at stations (in this latter case an on-board ESS is mandatory in order to capture the energy and to travel later to reach the next station).

Level 1/2: Vehicle/System interface

IEC 62864-1: Railway applications – Rolling stock. Power supply with onboard energy storage system. Part1: Series hybrid system

IEC 61133: Railway applications – Rolling stock. Testing of rolling stock before entry on service

IEC 61377: Railway applications – Rolling stock. Combined test method for traction systems

Level 3: Components

IEC 61287-1: Railway applications. Power converters installed onboard rolling stock. Test methods

IEC 60349: Electrical Traction. Rotating electrical machines for rail and road vehicles

Level 4: Subcomponents

IEC 62928: Railway applications – Rolling stock equipment. Onboard lithium-ion traction batteries

IEC 61881-3: Railway applications – Rolling stock equipment. Capacitors for power electronics. Part3: Electric double-layer capacitors.

Fig. 5-130 *EN 62864-1 Standard [109]*

GLPS systems can be classified into two types:

- contact type, using an embedded third rail;
- contact-less type, using an underground induction coil.

In both types of GLPS systems, segmented power supplies between rails are used. These segments are energized only when the vehicle is over them. This is to prevent any electrical injuries to pedestrians who would otherwise come into contact with a live part of the system.

These GLPS systems need specialized infrastructure and vehicle equipment, leading to significantly higher technical problems than on-board ESSs, their reliability is an issue, owing to water, snow, and ice build-up conditions, and they are highly proprietary being manufactured by only three companies, Alstom, Ansaldo, and Bombardier.

5.7.1 Contact type GLPS systems

Fig. 5-131 shows the classic third rail solution for feeding the on-board power converters from the energized rail that stays beside the running rails along the track. The third rail is shielded, and on one side, for safety reasons, a flat shoe collects current from the top of the rail, which is supported by insulators every 2.5 to 5m. Third rail system speed is limited to 70km/h and it is used mainly in metro applications.

Alstom developed in 2003 its APS (Alimentation Par le Sol) GLPS solution for the French city of Bordeaux. The APS system was installed through the center of the historic city of Bordeaux, where the overhead catenary was seen as a threat to the look of the city. The APS system consists of short segments of conductor rail separated by insulated segments that are installed between the running rails along the length of the track. The sections are 11m long and are only powered when the tram passes over them [110].

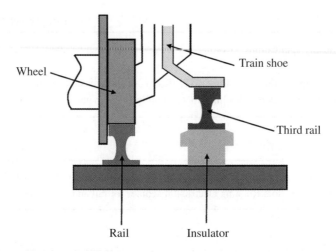

Fig. 5-131   *Third rail conductor*

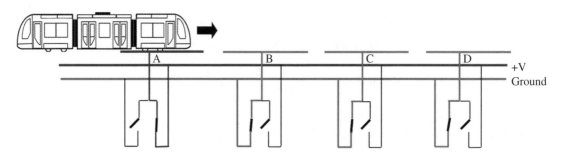

Fig. 5-132   *Alstom's APS working philosophy*

Each conductor rail is surrounded by a loop embedded in the track. When the vehicle is over each segment it is energized from a "power box" embedded under the track. See Fig. 5-132.

The APS system operates over 13.4km of the 43km system of Bordeaux with 74 trams fitted with the technology.

The APS system requires significant investment in infrastructure, although this is offset by the lack of renewal costs that might accompany an on-board supercapacitor or battery-based system. More APS systems are installed in Reims (2km of a 12km system), Angers (1.5km of a 12km system), Orleans (2.1km of a 12km system), Tours 2km of a 15km system), Brasilia (Brasil), Cuenca (Ecuador) and Dubai (19km fully catenary-free). Nevertheless, APS is an expensive solution; [111] claims that its cost is at least more than a catenary-based powering system.

The Italian railway manufacturer Ansaldo presented in 2009 its GLPS TramWave solution. The TramWave system consists of short segments of conductor rail installed between the running rails along the length of the track. Each segment is made by 50cm long conductive steel plates that are insulated between each other and release energy only when the vehicle passes over them [112]. When the vehicle is not just over each segment, they are not energized and are connected to ground to prevent electrical hazard to pedestrians that could cross over the track.

The vehicle bogie has a power collector that generates a magnetic force when activated. When the power collector shoe is retracted in the rest position whilst in the upper position it is unable to activate the power line segments because its magnetic pulling force is not enough to lift up the buried energized belt on the track.

When the shoe is released down, the power collector shoe gets in contact with the segments and its magnetic force attracts the energized buried belt, which pushes up the internal movable conductive contacts, allowing the closing of the electric circuit and power transmission from the segments, in contact with the collector, to the vehicle.

The TramWave solution has been implemented in the Italian city of Naples (600m with a non-catenary line) and in the Chinese city of Zhuzai (2014).

5.7.2   Contact-less type GLPS systems

The Primove system developed by Bombardier in 2009 transfers the energy from between the running rails to the vehicle by an inductive magnetic field. A conductive loop is buried between the runnings rails that is energized just when a vehicle passes overhead, creating a magnetic field beneath that vehicle. As the vehicle traverses the magnetic field, an onboard receptor converts that magnetic field into electrical energy to power on-board propulsion and auxiliary systems. There are no live parts exposed that could cause any electrical hazard to pedestrians and the track aspect is the same as for conventional trams [113].

Instead of providing the full route with an induction system, the infrastructure costs can be reduced by using an on-board ESS capable of storing the energy sent by the Primove system at the station stops. In this way, the tram can be driven using only the stored energy until it arrives at the next station, where the inductive system will recharge it again.

Although it is not properly a ground power supply system, the linear motor (LIM) development has gained popularity for mass transit applications as in the case of Automated Light Metros. Bombardier has delivered over 600 vehicles with this technology since 1985 to cities like Toronto (Canada), Vancouver (Canada), Detroit (USA), Kuala Lumpur (Malaysia), New York (USA), Beijing (China), and Yongin (South Korea), which form a total of 150km [114].

In the LIM philosophy, the propulsion inductor motor stator is fixed on the bogie of the vehicle and the rotor is placed in between running rails. The stator is fed by a power inverter that takes the energy from a third rail system along the route [114].

The main important characteristic of LIM propulsions systems is that there is no traction effort applied on the wheel. The vehicle is dragged by the magnetic force generated between the on-board feed primary circuit and the grounded linear rotor. Because of this feature, no gearboxes, driveshaft, or bearings are used, like in classical bogies. The bogie in an LIM railway vehicle will be reduced in height so smaller tunnels can be feasible. In addition, because of the lighter bogie in combination with the light structure, a less elevated structure is needed because the system can deal with higher steep grades.

Because there is no traction effort applied on the bogie wheels, less wear is present both on the wheels and on the rails, leading to less maintenance cost. Another interesting advantage of using LIM propulsion is that it allows the use of true radial steering bogies, which considerably reduces noise and allows very small radius curves. The small curve capability can minimize the need for demolition of existing buildings as the vehicle can move through narrow city center streets. See Fig. 5-133.

Magnetic levitation railway vehicles are another kind of ground-powered vehicle. The main characteristic of these vehicles is that they can move floating along the dedicated line thanks to magnetic forces between the vehicle and the track [115].

The vehicle contains on-board levitation electro-magnets and the stator is split along the track. Driving power is supplied by the long winding stator attached wayside. Because the stator winding and power conditioning equipment is located wayside, the vehicle is lighter than a conventional train. This permits

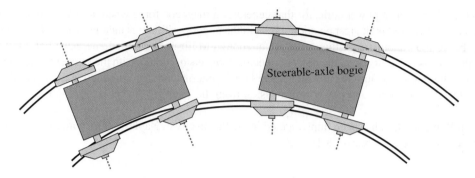

Fig. 5-133   *Steerable bogie for LIM*

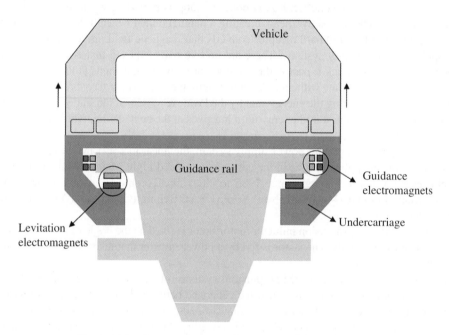

Fig. 5-134   *German Transrapid MAGLEV philosophy [117]*

operation at high speed (up to 500km/h has been demonstrated [116]). Fig. 5-134 shows a cross-sectional draft of the magnetic levitation vehicle and the guidance rail. Some levitation magnets are placed on the concrete guidance and in the vehicle itself to assure magnetic levitation on both horizontal and vertical axles.

5.7.3   Advantages and disadvantages of GLPS systems

The main disadvantage of contact and contact-less GLPS systems is that they rely on high infrastructure and maintenance costs because of the need of modifying the existing railway routes and maintaining them in wet, icy, and snowy conditions.

5.7.3.1 Advantages of GLPS systems

- The main obvious advantage of GLPS systems is that there are no overhead catenaries in the selected sections of the route.
- Applications where an on-board energy storage technology exists are limited in that the vehicle must recharge the batteries/capacitors at regular intervals. GLPS technologies are not limited in this fashion.
- In the case of the Primove system, the ground power supply elements are not visible on the surface, because there is no direct contact between the components. The cables are only energized as the vehicle passes overhead, and can be laid under any surface, including tarmac, concrete, and grass.
- Even an on-board ESS needs a charging station that can be a GLPS or the classical overhead catenary.
- Recharging time in every station stop for a supercapacitor on-board ESS is limited to around 20 seconds of dwell time. Depending of the capacity of the ESS, it cannot be recharged completely in 20 seconds so more station stops are required or a limitation in the non-catenary distance is mandatory. GLPS systems don't worry about this situation because they can be implemented everywhere on the route.
- The GLPS system can be very attractive to use when a catenary-free section starts to accelerate the vehicle without the use of the on-board ESS. In the acceleration process much more energy than during coasting is necessary because during coasting the speed is almost constant. Using the GLPS for this purpose the on-board ESS will be in charge only to supply lower demands of energy so its size—and so its cost— can be dramatically reduced.

5.7.3.2 Disadvantages of GLPS systems

- Systems which include exposed conductor surfaces like APS and TramWave may be susceptible to weather conditions of rain, ice, and/or snow. Particularly where wet conditions are accompanied by salt for ice clearing, the concern exists for stray currents from beneath the vehicle being a hazard to nearby pedestrians.
- The associated costs are very high (both initial investments and maintenance).
- There are durability issues if the system is implemented on sections also used by road traffic.
- There are only three suppliers worldwide, Alstom, Ansaldo, and Bombardier.
- LIM propulsion systems need a dedicated line to prevent damage to the grounded rotor system.
- The efficiency of the power supply, that is losses due to the internal resistance of the system and so on, is about the same as for a conventional overhead system.
- Regenerating braking is not possible unless the GLPS system is combined with an on-board ESS.
- Both on-board ESSs and GLPS systems present a limitation in the ability for the system to achieve high speeds. The battery/supercapacitor on-board ESSs limits speed, as higher speeds and acceleration require greater energy and power. Contact GLPS systems are speed limiting in that their segmented power blocks have a fixed switching speed. Generally, the speed limitations for these systems are within the normal operating parameters for a streetcar running in mixed traffic. Overhead catenary systems are mature in terms of power demand so they are not as limited as on-board energy storage and GLPS systems.

In addition, on-board ESSs and GLPS systems will have some impact upon the physical infrastructure of the system. GLPS technologies require additional excavation for the placement of underground switching components and the respective cross-section of the technology and on-board ESSs require recharging power stations at locations along the system's route. These impacts affect installation cost and, later, maintenance.

Classic overhead catenary systems need less infrastructure works, but need installation and maintenance as well.

5.8   Auxiliary power systems for railway applications

The auxiliary power system is the converter in charge of generating the on-board 400V AC/50Hz grid and charging the vehicle service battery. The on-board 400V AC/50Hz grid is necessary in order to feed compressors, fans, air conditioning and heating units, lighting, and the control for doors and breaking systems.

Auxiliary power systems and the power traction inverter share a DC-link. Depending on power configuration, one or more auxiliary power converters can coexist in the same vehicle, as shown in Fig. 5-135.

Auxiliary converter AC voltage output frequency can be 50 or 60Hz, with a maximum variation of between 5 and 10%. DC output depends on the existing service battery, the most common ones being 24V DC, 36V DC, 72V DC, and 110V DC. The maximum allowable voltage ripple is limited between 1 and 2% of a battery's nominal voltage.

The auxiliary power supply of a railway vehicle is designed with galvanic insulation to ensure safety between the high-voltage catenary line and low-voltage AC and DC grids. Electrical isolation can be guaranteed by using a power converter with transformer topology working at low-frequency or with advanced converter topology switching at higher frequency levels.

5.8.1   Low-frequency auxiliary power converter

The low AC output voltage L1, L2, L3, and N is generated from the DC-link bus using a classic three-phase inverter on the high voltage side feeding a low-frequency transformer. The low-frequency transformer first 3-phase secondary assures the galvanic insulation for L1, L2, and N. The DC battery charger is connected

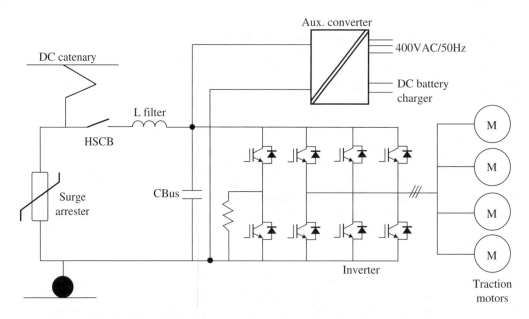

Fig. 5-135   *Auxiliary power converter location on system architecture*

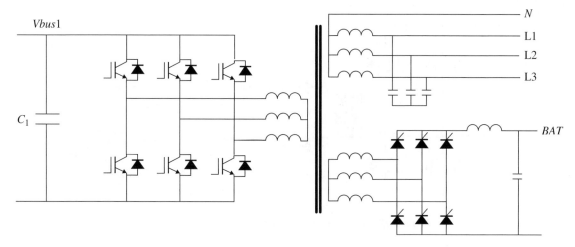

Fig. 5-136 *Low-frequency auxiliary power converter architecture and thyristor controller battery charger*

to the second three-phase secondary and controls the battery voltage by a three-phase thyristor bridge converter. Classically, this solution leads to a heavy and bulky low-frequency transformer configuration. See Fig. 5-136.

Advantages:

- simple (few components)
- reliable
- cheap.

Disadvantages:

- heavy and bulky three-phase transformer due to working frequency (50Hz, 60Hz)
- electrical noise of LF transformer.

5.8.2 MF auxiliary power converter

Fig. 5-137 shows an auxiliary converter topology that uses a full-bridge topology connected to the main Vbus1 bus. The full-bridge converter generates a second Vbus2 bus by switching at medium frequency and rectifying the AC voltages that appear on the secondary transformer. Finally a classical three-phase inverter connected to Vbus1 generates the AC voltages L1, L2, L3, and N at 400V AC/50Hz.

Later on, the DC battery charger can be connected to the generated AC grid, as shown before with a thyristor-controlled rectifier, or connected directly to the second DC bus to perform a more sophisticated and isolated topology, as shown in Fig. 5-138.

More-sophisticated auxiliary converter designs in the low- and high-frequency ranges for the railway industry can be found in specialized literature.

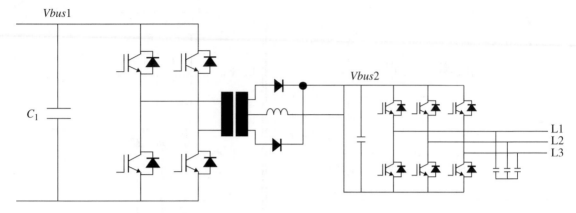

Fig. 5-137   *MF auxiliary power converter architecture*

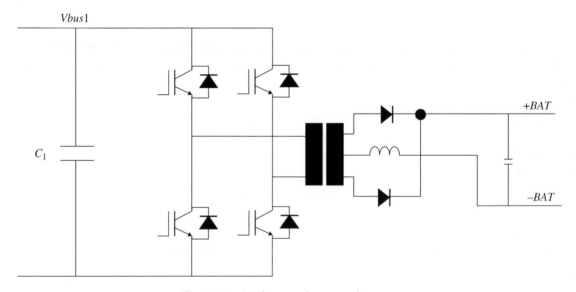

Fig. 5-138   *MF battery charger architecture*

5.9   Real examples

5.9.1   Tram-train example

This subsection briefly provides the main parameters and characteristics of a typical tram. This tram is composed of five cars. There is one cabin for the driver at each end. It has a 100% low-floor configuration, with a height of approximately 35mm above the rail. The first and last cabins present a unique traction bogie, the central car has a non-motorized bogie, and the intermediate cabins are suspended. Each driver cabin has a converter box located on the roof, as depicted in the Fig. 5-139. Each converter box has two power converters which each supply two induction motors for the traction bogies.

DC Catenary

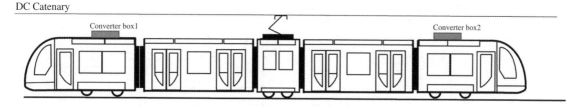

Fig. 5-139 *Simplified representation of the tram (Source: CAF P&A. Reproduced with permission of CAF P&A)*

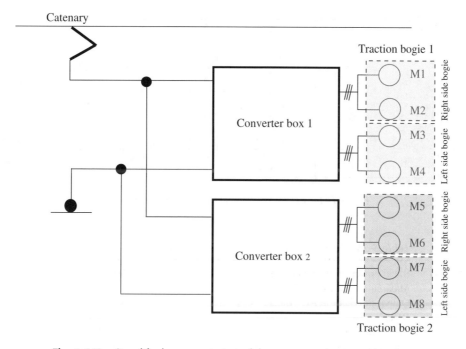

Fig. 5-140 *Simplified representation of the converter boxes and motors*

The simplified representation of the converter boxes and motors is depicted in Fig. 5-140. It is seen that eight traction electric motors are employed. On the other hand, the catenary employed is of 750V DC, with a maximum permanent DC of 900V and a minimum permanent DC of 500V. The vehicle presents a nominal mass of 57Tn (tara 42Tn), with a length of 30m.

Each converter box configuration is illustrated in Fig. 5-141. As mentioned before, each converter box is composed of two classic two-level converters with IGBTs. And each converter uses an independent LC filter for attenuating the current harmonics exchange between the converter and the catenary.

The main characteristics of the converter box are listed in Table 5-25. In addition, each converter presents its own crowbar protection, composed of a resistance (with a rated value of between 1 and 2Ω), a commutated IGBT, and a constant switching frequency when it is activated. The crowbar is dimensioned in such a way that it operates with natural cooling.

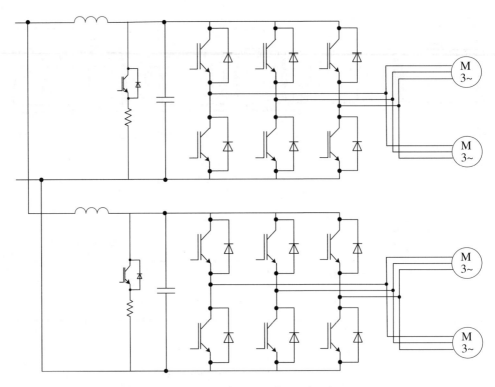

Fig. 5-141   *Converter boxes with two-level converters*

Table 5-25   *Main characteristics of the converter box*

Rated power of each converter	160kW
Maximum power of each converter	250kW
Output three-phase AC voltage	0–565$V_{LL\ RMS}$
Blocking voltage of the IGBTs	1700V DC
Switching frequency	1.5kHz
Maximum frequency of the AC output	200Hz
Filter inductance	3mH
Filter/DC-link capacitance	5mF

In addition, the acceleration and deceleration profiles of the tram at maximum load are shown in Fig. 5-142. These profiles are useful to specify the working cycles of the tram. First, the tram accelerates at a maximum acceleration of 1.33m/sec^2 until it reaches a speed of 31.5km/h. Then, it accelerates at lower regimen until reaching 70km/h. During braking, the tram decelerates from 70km/h to 55km/h. After that moment, it decelerates at 1.34m/sec^2 until it stops at the station.

The described speed profile of the tram is depicted in Fig. 5-143. The braking process is totally electric from 55km/h until approximately 5km/h. From 5km/h until the total stop of the tram, the electric brake is substituted by the hydraulic brake. In addition, from 70km/h until 55km/h, the combination of the electric brake and the hydraulic brake is employed to achieve the desired deceleration force.

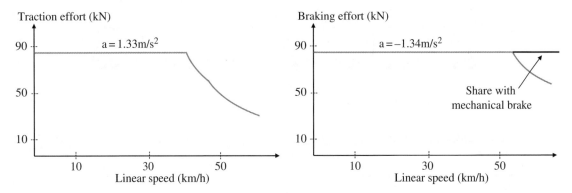

Fig. 5-142   *Acceleration and deceleration profiles of a tram at maximum load*

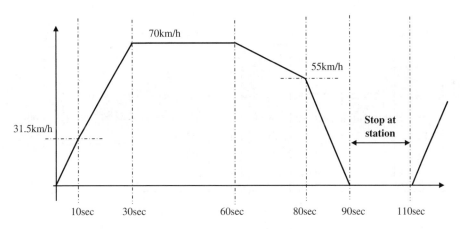

Fig. 5-143   *Speed profile of the tram*

With regards to the traction motors, they are induction motors of 60kW of rated power, $450V_{RMS}$, 1540rpm of rated speed, 52Hz of rated frequency, and two-pole pairs. The average diameter of the wheels is 550mm. The gearbox ratio of the transmission system is 5.5.

5.9.2   Recife metro

This subsection shows the example of a metro built by the manufacturer CAF [118], which operates in Recife in Brasil. The traction converter of this metro is prepared to work with DC catenaries of 3000V. The maximum available speed of this train is 90km/h, with an acceleration of 1m/sec^2 (between 0 and 60km/h). The rated power is 3120kW, while the braking performance is 1.2m/sec^2 with a service brake able to operate between 90 and 0km/h.

The train's configuration is depicted in Fig. 5-144. It is composed of two motor car/cabins (M car), a motor-train car (N car), and a train car (R car). Thus it is composed of three traction converters, distributed amongst six motor bogies (two motors per bogie).

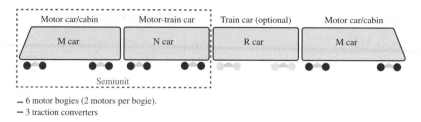

— 6 motor bogies (2 motors per bogie).
— 3 traction converters

Fig. 5-144   *Train configuration of Recife metro [118] (Source: CAF P&A. Reproduced with permission of CAF P&A)*

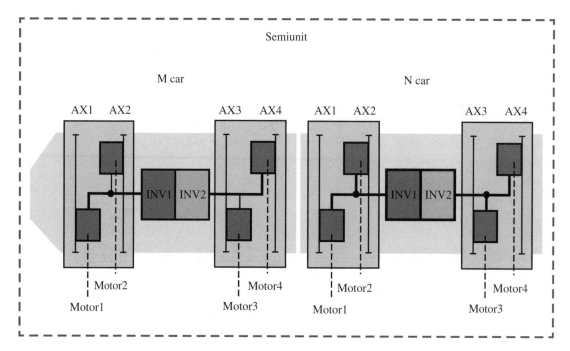

Fig. 5-145   *Train connection to a semiunit (M-N) [118] (Source: CAF P&A. Reproduced with permission of CAF P&A)*

The train connection to a semi-unit is depicted in Fig. 5-145, where it is possible to observe the disposition of the motors and inverters at each bogie for each car.

The traction converter schematic is depicted in Fig. 5-146. It is seen that each traction converter is composed of a pair of two-level converters, each mounted within a converter box. Then, each inverter supplies the pair of induction motors of each bogie. The two braking resistor branches are outside the converter, as depicted in the figure. They are naturally cooled so no forced air is used to cool them down.

In a similar way, the filter inductances also are outside the converter boxes. The converter is composed of 6500V IGBTs. It is able to provide a 2028kW (2×1014kW) of maximum power, with a rated power of 1208kW (2×604kW). Although the catenary rated voltage is 3000V DC, it is prepared to operate within the range of 2100–3600V DC. The cooling system is forced air. It presents electric regenerative braking capacity if the catenary is prepared for energy absorption. If it is not prepared, as mentioned before, it incorporates braking resistors. Its total weight is 900kg.

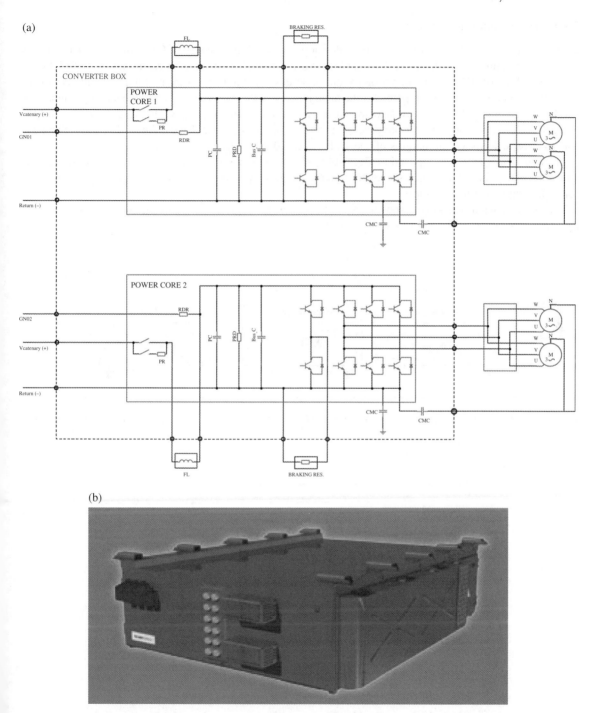

Fig. 5-146 *(a) Traction converter schematic and (b) converter box [118] (Source: CAF P&A. Reproduced with permission of CAF P&A)*

5.9.3   Tram-train Cádiz, Spain

This subsection gives details of the tram-train developed by CAF for the city of Cádiz in Spain. The traction converter configuration of this train is versatile and it is prepared to work with catenaries of 750V DC and catenaries of 3000V DC. The main reason for this is because the tram passes over a train line with a catenary of 3000V, therefore it must be prepared for the main journey of the 750V catenary and the journey catenary of 3000V for a train line. The tram is able to operate at a maximum speed of 70km/h. The rated power is 1140kW, while the acceleration capacity is 1.06m/sec^2 at margins of speeds between 0 and 40km/h. The service brake is able to achieve 1.2m/sec^2 at margins of speeds between 70 and 0km/h.

The train configuration is depicted in Fig. 5-147, which is composed of two motor car/cabins and a motor-trailer car. Therefore, it is composed of three motor bogies, with two motors per bogie and two traction converters supplying those motors.

The connections of the motor, converters, and cars are depicted in Fig. 5-148. It can be seen how each converter supplies three different motors of two bogies.

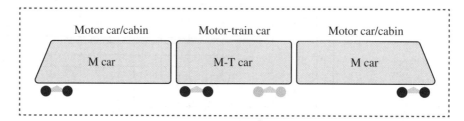

Fig. 5-147   *Train configuration of Cádiz tram-train [119] (Source: CAF P&A. Reproduced with permission of CAF P&A)*

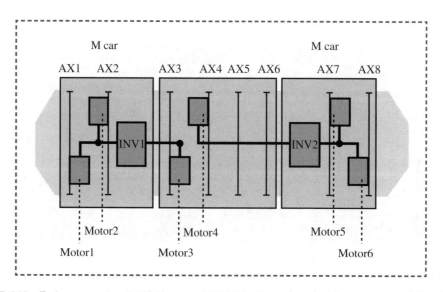

Fig. 5-148   *Train connection [119] (Source: CAF P&A. Reproduced with permission of CAF P&A)*

(a)

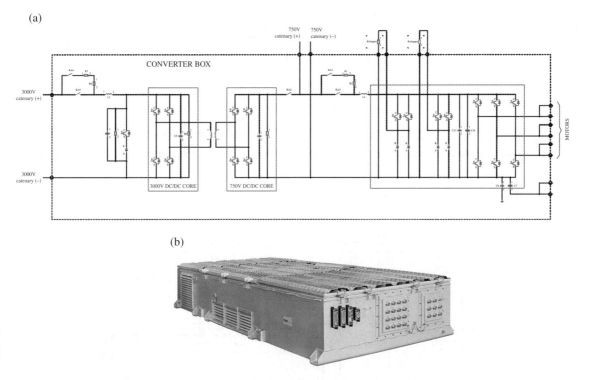

(b)

Fig. 5-149 *(a) Traction converter schematic and (b) converter box [119] (Source: CAF P&A. Reproduced with permission of CAF P&A)*

The traction converter schematic is depicted in Fig. 5-149. It can be seen that, as mentioned before, it is prepared for two different catenary voltages: 750V and 3000V. The three motors are supplied by a classic two-level converter composed of 1.7kV IGBTs. The two resistor branches for braking are in separate boxes and naturally cooled. The filter inductances (L1 and L2) are within the converter box, as illustrated in the figure. The converter side which is connected to the 3000V catenary is composed of 6.5kV IGBTs and a DC/DC converter with two half-bridges, with an MF transformer. It diminishes the voltage from 3000V to 750V. The maximum available power is 1MW, while the rated power is 400kW. The converter is able to operate with catenaries within the range 500V-900V and 2000V-3600V. The cooling system is forced air, while the electric braking can be regenerative if the catenary is prepared for absorption of the energy. The converter is fitted on the roof and its weight is 1960kg.

When the train passes from one catenary voltage to the other, there is a short way in which the tram crosses a catenary path without voltage, so that the converter configuration can be changed from one voltage to the other. During this path without catenary voltage, the tram losses a little bit of speed and continues running thanks to the its inertia.

5.9.4 Chittaranjan locomotive, India

This subsection gives example details of the Chittaranjan locomotive, India, manufactured by CAF. The rated power developed by the locomotive is 4.5MW and the stating force is 510kN. The locomotive is prepared for ambient temperatures up to 50°C.

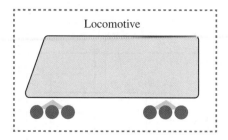

Fig. 5-150   *Locomotive configuration [120] (Source: CAF P&A. Reproduced with permission of CAF P&A)*

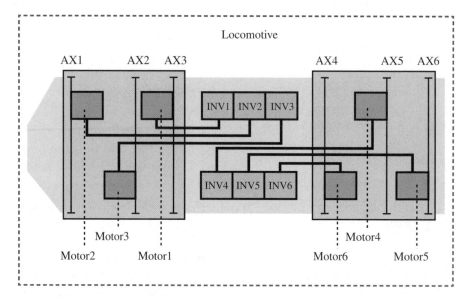

Fig. 5-151   *Locomotive connection [120] (Source: CAF P&A. Reproduced with permission of CAF P&A)*

This is a Co-Co locomotive, with two bogies as depicted in Fig. 5-150 with three traction motors per bogie. These motors are fed by two traction converters, each one composed of three independent inverters.

A detailed configuration of the locomotive converter and motors connections is illustrated in Fig. 5-151. These six traction motors located in two bogies move the entire locomotive.

A detailed schematic of one traction converter is shown in Fig. 5-152(a). Fig. 5-152(b) is the external view of the converter whose general power scheme is showed in Fig 5-152(a). It is seen that three independent parallel two-level converters supply each one motor. The IGBT technology employed is 4.5kV and 1200A. The rated power of each traction converter is 2250kW. Thus, each inverter is supplied from a common DC bus of 2800V DC. There is a small crowbar resistance inside the converter, while the filter inductance is located outside. Then, the DC bus is supplied by two parallel four-quadrant half-bridges, as depicted in the figure, which are directly connected to the transformer secondaries. The catenary rated voltage is 25kV and 50Hz. However, the converter is prepared to properly operate within the ranges of 16.5–31kV and 46–54Hz. The electric braking is regenerative. The converter incorporates a water-cooling system and the converter is mounted inside the locomotive. Its total weight is 2500kg.

(a)

Traction converter schematic

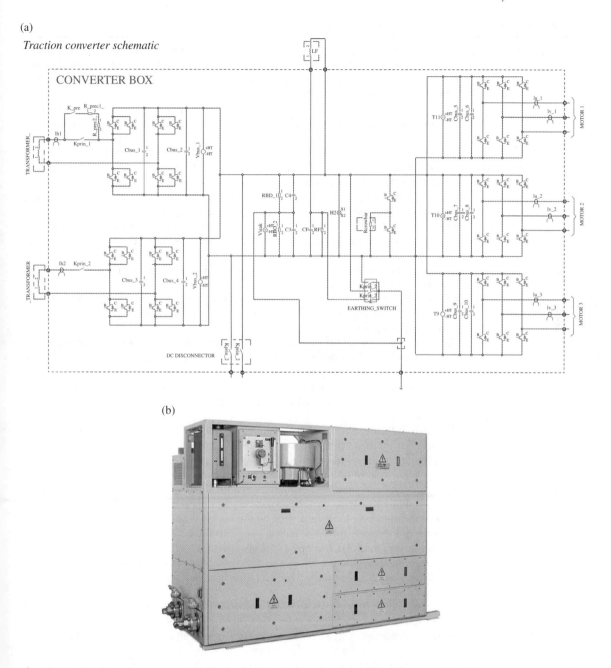

Fig. 5-152 *(a) Traction converter schematic, (b) Converter box [120] (Source: CAF P&A. Reproduced with permission of CAF P&A)*

Table 5-26   *Synthetized evolution of railway traction, [121], [122], [123]*

Around 600 BC	The first-known evidence of a wagonway (a predecessor of the railway), was the 6–8.5km long Diolkos wagonway. It was used to transport boats across the Isthmus of Corinth in Greece.
Around 1350	The earliest-known record of a railway in medieval Europe is a stained-glass window in the Minster of Freiburg in Breisgau.
1515	There is a description of the Reisszug, a funicular railway at the Hohensalzburg Castle in Austria, made by Cardinal Matthäus Lang von Wellenburg. The line originally used wooden rails and a hemp haulage rope. It was tractioned by human or animal power, through a treadwheel.
Middle of 18th century	Coal and mining companies in England replace wooden tracks with iron tracks, increasing the durability of the rails and improving the rolling friction.
1804	Richard Trevithick of Cornwall develops the first steam locomotive for the Welsh Penydarran Railroad in England, employing a high-pressure steam engine.
1812	First commercially successful steam locomotives, using the Blenkinsop rack-and-pinion drive. They started to operate on the Middleton Railway. This was the world's first regular revenue-earning use of steam traction.
1825	The first public transport railway is developed by Stephenson, built for the Stockton & Darlington Railroad in England.
1836	An electric railcar was tested in Scotland, tractioned by electric motor similar to today's switched reluctance machines. These trials enjoyed little success, basically owing to the non-mature technology employed.
1879	Siemens develops the first electrically powered locomotive in Berlin, for the transportation of passengers. It was operated by a 2.2kW DC motor. An engine and a generator supplied the rails from where the energy was taken.
1895	The first mainline electrification was in Baltimore, USA. A rigid overhead conductor supplied 675V DC via a one-sided tilted pantograph to the 96-ton four-axle, four-motor locomotives.
1903	H. Behn-Eschenburg develops ohmic commutation pole shunts to the series-wound commutator motor, thus obtaining compatibility with the AC supply.
1908	The advent of the diesel locomotive.
1912	German Länderbahnen of Prussia, Hesse, Bavaria, and Baden sign an agreement setting the standard for single-phase alternating voltage at 15kV and 16.7Hz. This special low frequency was useful for supplying the series-wound commutator motor.
1917	The first diesel-electric locomotive prototype in the US is built by General Electric.
After the second World War	In Lorraine, northern France, coal railways and 50Hz and 25kV electrification is introduced. From thereon, the 50Hz is adopted in many countries.
1971	First successful diesel-electric locomotive with three-phase drive technology is produced by BBC and Henschel. Frequency converters allow the robust squirrel cage induction motor to be supplied.
1979	Three-phase drive technology is applied to overhead system locomotives, the class 10 of the DB (Deutsche Bundesbahn). Since then, this technology is common for most trains.
1980–2000	Massive introduction of IGBT-based power converters in all types of trains. Starting with the opening of the first Shinkansen line between Tokyo and Osaka in Japan in 1964, high-speed rail transport, functioning at speeds of up to 300 km/h has been built in Spain, France, Germany, Italy, China, Taiwan, the UK, South Korea, Scandinavia, Belgium, and the Netherlands. Linear-motor-based traction put into trains. Magnetic-levitation trains introduced.
2000–2016	Permanent-magnet-based synchronous motors used in some trams. Introduction of on-board ESSs. Catenary-less solutions. Advanced DC electrification with regenerative substations.

5.10   Historical evolution

This section tries to synthetize the evolution of railway traction. The most remarkable events related to the evolution of the trains are summarized in Table 5-26. Thus, from the first early attempts reportedly made in ancient Greece, many improvements have been carried out to give us our modern-day trains. Table 5-26 tries to provide details of the most important events that have occurred in railway traction.

The table shows that at the beginning of the 19th century the steam turbine was originally introduced as the propulsion system for trains. Later on, electric and diesel propulsion systems were adopted. This evolution of the train's propulsion system is similar to that of the ship, as will be seen in next chapter. In recent years, the main developments have been focused on electric train traction systems, with different complexity levels being at the heart of the availability of the semiconductors employed for the conversion of power circuits. Thus, the lack of semiconductors of sufficient power and blocking voltage capacities obliged the use of complex converter structures for many years. Thanks to the advance in semiconductor technology, it has been possible to achieve semiconductors with higher blocking voltages, enabling the use of simpler and more efficient converter topologies (i.e. the classic two-level converter topology). There now follows an illustrative synthesized summary of various employed converter topologies.

Fig. 5-153 shows the input chopper circuit which was commonly used in DC catenaries of 750V. Since the catenary line voltage can present large voltage fluctuations of up to 1000V, two inverters were connected in series by these input choppers. A possible difference at the loads of the two choppers was compensated by different modulation ratios of the input choppers. Once IGBTs of 1500V appeared, this complex converter structure was no longer necessary and the simple two-level converter directly connected to the line DC catenary was used.

Something similar occurred with catenary voltages of 1500V. At the beginning, three-level NPC converters were used, as depicted in Fig. 5-154, with semiconductors of 2.5kV. However, with the apparition of IGBTs of 3.3kV, these converters were substituted by classic and simple two-level converters.

With regards to catenaries of 3kV, the system must be prepared to operate in the range of line voltages of 2.7–4kV. The converter structure depicted in Fig. 5-155 is an example of the complex converter topologies that were used. It consists of a two-level converter with a two-step input chopper. With this topology, GTOs were a commonly used semiconductor, since at that time they were suitable semiconductors prepared to operate at 4.5kV.

A different possibility for catenaries of 3kV could be the three-level converter, with semiconductors of 3.3kV. In addition, another possibility would be to use two inverters in series with a special motor of double supply in the configuration illustrated in Fig. 5-156. Many other converter configuration possibilities have been used over the years of railway traction, considering also the necessity to supply different machines and the usage of different AC catenary systems; however, they are beyond the scope of this section.

5.11   New trends and future challenges

This section provides an insight into the new trends and future challenges in railway traction, from the electric drive point of view. It focuses on the development of three main technologies: converter topologies, semiconductors, and ESSs. Although they are not described in this book, there are also advances taking place in the development of traction motors with permanent magnets for the railway industry.

5.11.1   Converter topologies

With regards to the power electronic converter, it seems that many manufacturers are focused on medium frequency transformers (MFTs). There are several options of MFT already researched [123]. This subsection tries

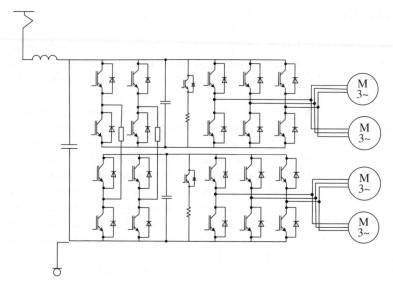

Fig. 5-153   *Series connected two-level converters with input chopper of 750V line voltage and 1200V IGBTs*

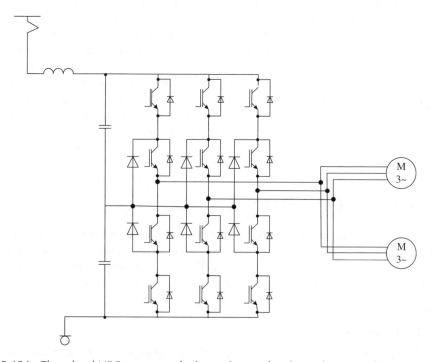

Fig. 5-154   *Three-level NPC converter, for line voltages of 1.5kV and semiconductors of 2.5kV*

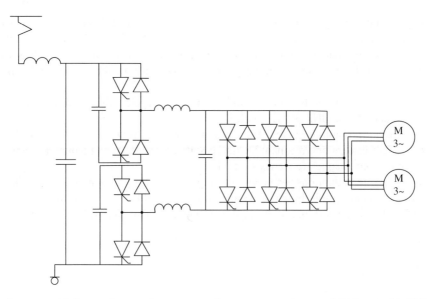

Fig. 5-155 *Inverter with input chopper in two steps, for catenary voltages of 3kV and with GTOs of 4.5kV*

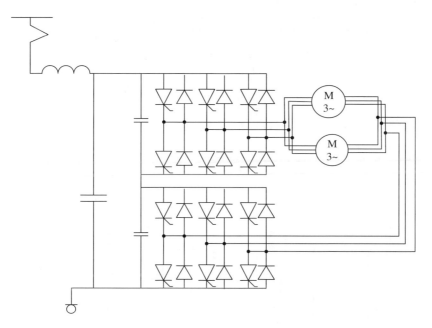

Fig. 5-156 *Inverter with double stator motor in way configuration*

to give a short overview of the most remarkable MFT-based configurations that have been researched in the last few years.

Low-frequency transformers advantages include its low-cost, robustness, reliability, and efficiency (98–99% depending on power rating). Nevertheless, it suffers from several weaknesses, the most important being: its weight, volume, voltage drop under load, sensitivity to DC offset load imbalances, and sensitivity to harmonics. All the drawbacks of LF transformers can be avoided by using MFTs, the main advantages being seen in their weight, volume, and harmonic influence reduction. MFT solutions, nevertheless, use power semiconductors switching at frequencies in the range of 2–10kHz, so reliability, EMI problems, and cost are affected.

Thus, Fig. 5-157 shows some early examples of MFT based on cycloconverter configurations, with switching frequencies around 200–400Hz of MFT [123].

In a different direction, Alstom built, in 2003, a prototype MFT as illustrated in Fig. 5-158. The topology is based on cascaded half-bridges and resonant DC/DC converters [123]. A single multiwinding MFT was used of ferrite core with a conmutating frequency of 5kHz.

Some other earlier different options can be grouped together, as illustrated in Fig. 5-159. The configurations (a) and (b) are cascaded half-bridge-based converters, with resonant DC/DC converters (LC or LLC) and they use multiple MFTs. The range of frequencies of MFTs is between 4 and 10kHz. On the other hand, topology (c)

(a)

(b)

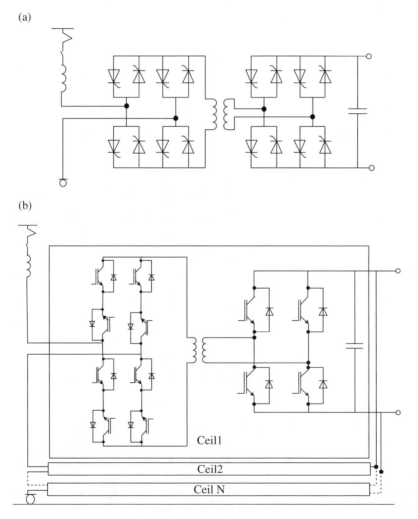

Fig. 5-157 *(a) Cycloconverter (SCR-based), (b) Cascaded cycloconverters (IGBT-based) [123]*

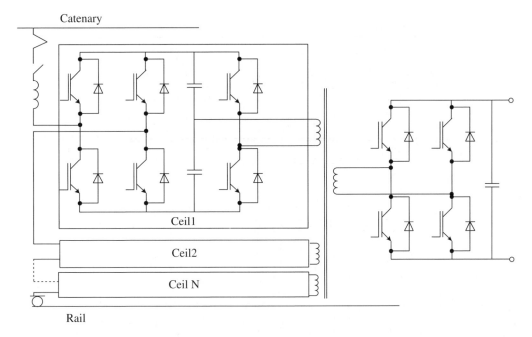

Fig. 5-158 *Alstom prototype based on cascaded half-bridges and resonant DC/DC converters [123]*

presents a different configuration based on the modular multilevel converter proposed by Siemens, only using a unique MFT.

With regards to the MFTs, the challenges of building them are identified in [123]: insulation requirements, cooling efforts, skin and proximity effects, weight and volume restrictions, integration aspects, cost, and so on. On the other hand, the degree of freedom during the design stage is exemplified by: magnetic materials (nanocrystalline, amorphous, ferrites, etc.), windings (copper, aluminum), insulation (air, solid, oil), and cooling (air, oil, water).

5.11.2 Power semiconductors

Power semiconductors used in all industry markets, including railway, can experience a change from silicon chips to silicon carbide ones (SiC). SiC chips show less loss than Si-based ones. SiC MOSFTEs and SiC diodes enable systems switching frequencies to be increased up to six times compared to silicon-based power semiconductors. With this higher switching operation, reductions in component ratings for magnetic and capacitor elements often enable SiC-based power systems to be shrunk dramatically.

The first step of power semiconductor manufacturers for the railway industry was to replace silicon diodes with SiC diodes, with the aim of reducing switching losses, but efforts are also being made to develop all-SiC power modules too. In fact, in February 2015 Odakyu Electric Railway of Japan ordered 3.3kV, 1.5kA, all-SiC power modules to be used in its traction inverters. Owing to lower losses, the company anticipates an energy saving of between 20 and 36% and weight savings of up to 80% [124].

During 2014–2016, Infineon, ABB, Hitachi and Semikron all launched new power module references denoted as XHP™ [125], LinPak™ [126], nHPD2™ [127], and Semitrans20™ [129] respectively. These power modules' main characteristics are their low internal stray inductance, half-bridge configuration, blocking voltages from 1200 to 6.5kV, a simplified structure allowing the connection of similar parts in parallel and strategic terminal placement. These three modules have the same housing dimensions with plus and minus DC

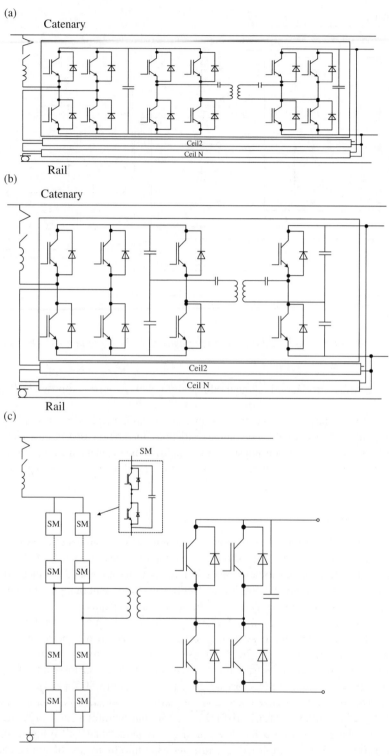

Fig. 5-159 *Several options of MFT [123]*

terminals located at one side and AC terminals on the opposite side. This module configuration simplifies the power converter design and increases its power density.

5.11.3 ESSs

With respect to supercapacitors new developments aim to increase the working voltage and the capacity for the same can shape. A larger-voltage reduces the number of series cell connections in an ESS for the same operating voltage, so fewer balancing circuits and electronics are needed. Changes in electrolyte composition are needed to avoid its degradation for cell operating voltages above 3V. As an example, MAXWELL Technologies launched the DURABLUE® 2.85V 3400F supercapacitor cell in July 2015. This cell increases the range of available specific power and stored energy in the industry-standard 60mm cylindrical K2 form factor.

Regarding lithium-ion batteries, huge research work is focused on developing Li-air and Li-sulfur chemistries. Lithium-air battery technology was first studied in the 1970s. This lithium cell chemistry takes and releases the oxygen from the atmosphere for the charging and discharging processes respectively. Lithium-air battery chemistry could increase the energy density by up to 11.2kWh/kg compared with the current 1.4kWh/kg for existing lithium-ion chemistries. Theoretically, replacing the cathode with air will reduce the weight, but hazards related to lithium flammability and electrolyte degradation have made its development difficult. Another point of interest is in the development of Li-sulfur chemistry, which, from the point of view of chemistry researchers, could replace lithium-ion particularly in pure EVs. This is because the manufacturing process of Li-sulfur cells is cheaper than that for lithium-ion cells [128].

New hybrid chemistries combining the advantages of supercapacitors (power density, cycling, and low discharge rate) and lithium-ion batteries (energy) are starting to be developed and launched onto the market for real applications. These chemistries are denoted as LICs, or lithium-ion capacitors. The peak power can be of several kW and cell voltage range is the standard for an lithium-ion cell. The use of this promising cell chemistry will allow only one kind of on-board storage system to be used, instead of two (one for power capability made of supercapacitors and the second made of lithium-ion batteries). The use of one ESS will greatly simplify energy management and will only require one power converter [128].

The cost of supercapacitors and lithium-ion batteries is one of the most important parameters that limit the widespread utilization of these devices in ESSs. Nowadays, as a rule of thumb, it can be estimated that lithium-ion battery packs (cells + cooling arrangement + control electronics BMS) can vary from 1000€/kWh for LFP chemistries up to 2000€/kWh for LTO-based ones. For supercapacitor energy systems, the cost could be around 15k€/kWh. It is supposed that these costs will be reduced in the future with the incoming increase of electric and hybrid vehicles sales in the automotive market.

On-board ESSs started to be present in tramways and metros since 2007 but nowadays they can be seen also in battery-assisted locomotives, EMUs and DEMUs. As an example, a battery/electric EMU was trialed for a passenger service in the UK in January 2015. The target was to operate a 185-ton four-car Bombardier Electrostar Class 379 train solely on battery up to 120km/h over a distance of 50km. A total of 500kWh of energy was stored in its on-board lithium-ion batteries [93].

References

[1] ABB Group. *Power Electronic Transformer for railway on-board applications: An overview*. Zurich, Switzerland: ABB Group. 2013.
[2] http://www.railway-technology.com/contractors/electrification/abb-surge-arrestors/abb-surge-arrestors2.html.
[3] Visintini R. *Rectifiers*. Elettra Synchrotron Light Laboratory, Trieste, Italy.
[4] http://www.school-for-champions.com/science/ac_world_volt_freq_list.htm.
[5] http://en.wikipedia.org/wiki/Mains_electricity_by_country.
[6] http://www.dupont.com/products-and-services/electronic-electrical-materials/electrical-insulation/brands/nomex-electrical-insulation.html.

[7]   http://www.engineeringtoolbox.com/nema-insulation-classes-d_734.html.

[8]   http://www.aemt.co.uk/technical-info/general-engineering/classification-of-insulation-systems.

[9]   Vetter H. *Dry MKK Capacitors for Modern Rail Traction*. Munich, Germany: Siemens Matsushita Components.

[10]   http://www.windpowerengineering.com/maintenance/safety/films-in-capacitors-let-them-self-heal-and-more/.

[11]   Ritter A. Capacitor reliability issues and needs. *Sandia National Laboratories Utility-Scale Grid-Tied PV Inverter Reliability Technical Workshop, Kirtland AFB*, Albuquerque, NM, 27–28 January 2011.

[12]   Makdessi M. Modeling, ageing and health monitoring of metallized films capacitors used in power electronics applications. *Journée scientifique du dept. MIS*, 11 July 2012.

[13]   Medium power film capacitors. FFLC AVX.

[14]   Salcone M, Bond J. *Selecting Film Bus Link Capacitors for High Performance Inverter Applications*. Eatontown, NJ: Electronic Concepts Inc, 2009.

[15]   http://www.hitachi.com/New/cnews/120419b.html.

[16]   http://www.hitachi-rail.com/products/on-board/sic/.

[17]   Cruceanu C. *Train Braking*. University Politehnica of Bucharest, 2012.

[18]   Sharma RC, Dhingra M, Pathak RK. Braking systems in railway vehicles. *IJERT* 2015; **4**(1): 206–211.

[19]   Okada Y, Koseki T, Hisatomi K. Power management control in DC-electrified railways for the regenerative braking systems of electric trains. *WIT Transactions on State-of-the art in Science and Engineering* 2010.

[20]   Liudvinavičiusa L, Lingaitis LP. New technical solutions of using rolling stock electrodynamical braking. *Transport Problems* 2010; **4**(2): 23–35.

[21]   Löwenstein L, Jöckel A, Hoffmann T, et al. Syntegra: The intelligent integration of traction, bogie and braking technology, http://www.uic.org/cdrom/2008/11_wcrr2008/pdf/R.1.1.3.1.pdf, 2009.

[22]   Allenbach JM, Chapas P, Compte M, Caller T. *Traction Electrique*. Lausanne, Switzerland: Presses polytechnique et universitaires romandes, 2008.

[23]   Steimel A. *Electric Traction Motive Power and Energy Supply*. Munich, Germany: Oldenbourg Industrieverlag Gmbh, 2008.

[24]   Eckel H-G, Bakran MM, Krafft EU, Nagel A. A new family of modular IGBT converters for traction applications, http://ieeexplore.ieee.org/xpl/articleDetails.jsp?arnumber=1665438, 2005.

[25]   ABB. *Traction Motors for Railway Rolling Stock*, Zurich, Switzerland: ABB, 2010.

[26]   Burke A. Ultracapacitors: Why, how, and where is the technology? *J Power Sources* 2000; **91**: 37–50.

[27]   Yao YY, Zhang DL, Xu DG. A study of supercapacitor parameters and characteristics. *IEEE Conference on Power System Technology* 2006.

[28]   Wei T, Wang S, Qi Z. A supercapacitor based ride-through system for industrial drive applications. *IEEE Conference on Mechatronics and Automation* 2007.

[29]   http://www.greencarcongress.com/2005/12/saft_providing_.html.

[30]   http://www.railway-technology.com/projects/nice-trams/.

[31]   http://data.energizer.com/PDFs/nickelmetalhydride_appman.pdf.

[32]   http://www.hellopro.fr/batterie-rechargeable-ni-mh-nhp-2008187-271056-produit.html.

[33]   AXEON. *Our Guide to Batteries*, www.axeon.com, 2001.

[34]   FOCUS. *Batteries for Electric Cars: Challenges, opportunities and the Outlook to 2020*. Boston, MA: The Boston Consulting Group, Inc. 2010.

[35]   http://www.buchmann.ca/article6-page1.asp.

[36]   Kalhammer FR, Kopf BM, Siwan DH, et al. *Status and Prospects for Zero Emissions Vehicle Technology*. Sacramento, CA: Report of the ARB Independent Expert Panel 2007. State of California Air Resources Board, 2007.

[37]   Wohlfarhrt-Mehrens M. *Energiespeicher für die Elektromobilität: Stand der Technik und Perspektiven*. ZSW Zentrum, 2011.

[38]   http://batteryuniversity.com/learn/article/types_of_lithium_ion.

[39]   Gross O. Introduction to advanced automotive batteries. *IEEE Vehicle Power and Propulsion Conference*, Chicago, IL, 6–9 September 2011: 1–79.

[40]   Soon K, Moo C, Chen Y, Hsieh Y. Enhanced coulomb counting method for estimating state-of-charge and state-of-health of lithium-ion batteries. *Applied Energy* 2009; **86**:1506–1511.

[41]   Coleman M, Lee C K, Zhu C, Hurley W G. State-of-charge determination from EMF voltage estimation: Using impedance, terminal voltage, and current for lead-acid and lithium-ion batteries. *IEEE Transactions on Industrial Electronics* 2007; **54**(5): 2550–2557.

[42] http://liionbms.com/php/wp_soc_estimate.php.

[43] Yan J, Xu G, Qian H, Xu Y. Robust state of charge estimation for hybrid electric vehicles: Framework and algorithms. *Energies* 2010; **3**(10): 1654–1672.

[44] Codecà F, Savaresi SM, Manzoni V. The mix estimation algorithm for battery State-Of-Charge estimator: Analysis of the sensitivity to measurement errors. *Dipartimento di Elettronica e Informazione, Politecnico di Milano*. Milan, Italy, 2009.

[45] Knauff M, Dafis C, Niebur D. A new battery model for use with an extended Kalman filter state of charge estimator. *Conference Paper in Proceedings of the American Control Conference (ACC)* 2010.

[46] Eichi, HR, Chow, M-Y. Adaptive Parameter identification and state-of-charge estimation of lithium-ion batteries. *Annual Conference of the IEEE Industrial Electronics Society* 2012.

[47] Hongwen H. State-of-Charge estimation of the lithium-ion battery using an adaptive extended Kalman filter based on an improved Thevenin Model. *IEEE Trans Veh Technol* 2011; **60**(4): 1461–1469.

[48] Fleischer C, Waag W, Bai Z, Sauer D. Adaptive on-line state-of-available-power prediction of lithium-ion batteries. *JPE* 2013; **13**(4): 516–552.

[49] Sauer DU, Waag W, Gerschler J. Monitoring and state-of-charge diagnostics of lithium-ion batteries and supercapacitors. *Energy Management & Wire Harness Systems*. Essen, 2011.

[50] http://batteryuniversity.com/learn/article/types_of_lithium_ion.

[51] Wang Q, Sun J, Chu G. *Lithium Ion Battery Fire and Explosion*. Beijing, China: State Key Laboratory of Fire Science University of Science and Technology of China, 2005.

[52] Markel T. Plug-in hybrid-electric vehicles current status, long-term prospects and key challenges. *Clean Cites Congress and Expo*, Phoenix, Arizona, 7–10 May 2006.

[53] Sarasketa Zabala E. *A Novel Approach for Lithium-Ion Battery Selection and Lifetime Prediction*. PhD thesis, Mondragon University, Spain, 2014.

[54] http://www.mpoweruk.com/bms.htm.

[55] Kyung-Hwa P, Chol-Ho K, Hee-Keun C, Joung-Ki S. Design considerations of a lithium ion battery management system (BMS) for the STSAT-3 Satellite. *JPE* 2010; **10**(2): 210–217.

[56] http://www.autoblog.com/2009/12/11/ricardo-and-qinetiq-announce-new-reduced-cost-lithium-ion-batter/.

[57] Flavio C, Iannuzzi D, Lauria D. Supercapacitors-based energy storage for urban mass transit systems. *Power Electronics and Applications (EPE)* 2011.

[58] Dengke Y, Shuai G, Junjun L, et al. Design and analysis of emergency self-traction system for urban rail transit vehicles. *International Conference on Future Energy, Environment, and Materials* 2012; **16**(Part A): 585–591.

[59] Liu J, Yuan D, Zhou Q, et al. Simulation and analysis of emergency self-traction system for urban rail vehicles. *Mechatronics* 2011; **17**(7): 20–23.

[60] Barrero R, Tackoen X, Van Mierlo J. Improving energy efficiency in public transport: Stationary supercapacitor based energy storage systems for a metro network. *IEEE Vehicle Power and Propulsion Conference* 2008.

[61] https://www.gs-yuasa.com/us/technic/vol4_2/pdf/004_2_030.pdf.

[62] Lee H, Kim G, Lee C. *Development of ESS for Regenerative Energy of Electric Vehicle*. Korea Railroad Research Institute, 2007.

[63] http://www.abb.es/cawp/seitp202/2092703c1a48b4c2c1257aef004b2626.aspx.

[64] http:///cn.siemens.com/cms/cn/Chinese/TS/Mobility/media/Media%20pool%20content/Documents/Sitras%20SES_en.pdf.

[65] Takahash H, Kato T, Ito T, Gunji F. Energy storage for traction power supply systems. *Hitachi Review* 2008; **57**: 28–32.

[66] http://www.septa.org/sustain/pdf/Wayside%20Storage%20-%20Press%20Coverage.pdf.

[67] González-Gil A, Palacin R, Batty P, Powell JP. Energy-efficient urban rail systems: Strategies for an optimal management of regenerative braking energy. *Transport Research Arena (TRA)*, Paris, France, 14–17 April 2014.

[68] Barrero R, Tackoen X, Van Mierlo J. Stationary or onboard energy storage systems for energy consumption reduction in a metro network. *J. Rail Rapid Transit* 2010; **224**(3): 207–225.

[69] Domínguez M, Cucala AP, Fernández A. Energy efficiency on train control: Design of metro ATO driving and impact of energy accumulation devices. *9th World Congress on Railway Research (WCRR)*, Lille, France, 22–26 May 2011.

[70] Chymera M, Renfrew A, Barnes M. *Analyzing the potential of energy storage on electrified transit systems*. University of Manchester, http://www.railway-research.org, 2008.

[71]   Akli CR, Roboam X, Sarenil B, Jeunesse A. Energy management and sizing of a hybrid locomotive. *European Conference on Power Electronics and Application (EPE)*, 2007: 1–10.

[72]   RailPower project Green Goat, http://railindustry.com/coverage/2002/2002g02a.html.

[73]   Liudnavicius L, Povilas L. *Locomotive Energy Saving Possibilities*. Department of Railway Transport, Vilnius Gediminas Technical University Lithuania, 2012.

[74]   Jaafer A, Rockys Akli C, Roboam X. Sizing and energy management of a hybrid locomotive based on flywheel and accumulators. *IEEE Trans Veh Technol* 2009; **58**(8): 3947–3958.

[75]   Thiounn M, Jeunesse A. *A Platform for Energy Efficiency and Environmentally Friendly Hybrid Trains*. Paris, France: SNCF, 2008.

[76]   Destraz B. *Power Assistance for Diesel-Electric Locomotives with Supercapacitive Energy Storage*. Lausanne, Switzerland: Institute of Technology, 2004.

[77]   Konarzewski M, Niezgoda T. Hybrid locomotives overview of construction solutions. *Journal of KONES Powertrain and Transport* 2013; **20**(1): 127–134.

[78]   Barrero R, Tackoen X. *New technologies (supercapacitors) for energy storage and energy recuperation for a higher energy efficiency of the Brussels public transportation company vehicles*. Vrije Universitet Brussel (VUB), 2008.

[79]   Barrero R, van Mierlo J. Energy savings in public transport. *IEEE Veh Technol Magazine* 2008; **3**(3): 26–36.

[80]   Latkovskis L, Sirmelis U, Grigans L. *On-Board Supercapacitor Energy Storage Sizing Considerations*. Riga, Latvia: Institute of Physical Energetics, 2012.

[81]   Mir L, Etxeberria-Otadui I, Perez de Arenaza I, et al. A supercapacitor based light rail vehicle: System design and operations modes. *IEEE Energy Conversion Congress and Exposition*, San Jose, CA, 20–24 September 2009: 1632–1639.

[82]   Grigāns L, Latkovskis L. Study of control strategies for energy storage system on board of urban electric vehicles. *Proceedings of EPE-PEMC 2010*, Macedonia, Ohrida, 6–8 September 2010.

[83]   Wootae J, Soon-Bark K. *Efficient Energy Management for Onboard Battery-driven Light Railway Vehicle*. Korea Railroad Research Institute, 2011.

[84]   Ogasa M, Taguchi Y. *Power Electronics Technologies for a Lithium Ion Battery Tram*. Tokyo, Japan: Railway Technical Research Institute, 2007.

[85]   Vulturescu B. Ageing study of a supercapacitor-battery storage system. Versailles, France, 2010, DOI: 10.1109/ICELMACH.2010.5608197.

[86]   Allègre AL, Bouscayrol A, Trigui R. *Influence of Control Strategies on Battery/Supercapacitor Hybrid: Energy storage systems for traction applications*. University of Lille Nord de France, 2009.

[87]   Meinert M. *New Mobile Energy Storage System for Rolling Stock*. Munich, Germany: Siemens AG, Industry Sector, Mobility Division, 2009.

[88]   http://www.siemens.com/press/en/pressrelease/2009/mobility/imo200903024.htm.

[89]   http://www.railway-technology.com/news/news52360.html.

[90]   Steiner M, Scholten J. *Improving Overall Energy Efficiency of Traction Vehicles*. Berlin, Germany: Bombardier Transportation, 2005.

[91]   http://www.bombardier.com/content/dam/Websites/bombardiercom/supporting-documents/BT/Bombardier-Transport -ECO4-MITRAC_Energy_Saver-EN.pdf.

[92]   www.cafpower.com.

[93]   http://insideevs.com/battery-power-passenger-train-trial-now-underway-in-uk-video/.

[94]   De Breucker S. *Impact of DC-DC Converters on Lithium-ion Batteries*. Thesis, Katholieke Universiteit Leuven, Germany, 2012.

[95]   Sritharan T. *Impact of Current Waveforms on Battery Behaviour*. Thesis, University of Toronto, 2012.

[96]   De Breucker S. *Impact of Current Ripple on Lithium-ion Battery Ageing*. Belgium, Kristof Engelen, 2012.

[97]   Hegarty T. Benefits of multiphasing buck converters. National semiconductor: Part 1 and Part 2. *EE Times*, http://www.eetimes.com, 2007.

[98]   Cha H, Rogers C, Lu X, Peng FZ. Design of high efficient and high density integrated magnetics for interleaved DC/DC boost converter for series hybrid electric bus. *5th IEEE Vehicle Power and Propulsion Conference, VPPC '09*. Dearborn, MI: 1428–1433.

[99]   ADETEL Equipment. ADEP 101 758 000. Datasheet. http://www.adetelgroup.com/.

[100] De Doncker RWAA, Divan DM, Kheraluwala MH. A three-phase soft-switched high-power-density dc/dc converter for high power applications. *IEEE Trans. Ind. Appl.* 1991; **27**(1): 63–73.

[101] Kheraluwala MH, Gascoigne RW, Divan DM. Performance characterization of a high-power dual active bridge dc-to-dc converter. *IEEE Trans. Ind. Appl.* 1992; **28**(6): 1294–1301.

[102] Inoue S, Akagi H. A bi-directional DC/DC converter for an energy storage system. *Applied Power Electronics Conference and Exposition (APEC)*, Anaheim, CA, 25 February to 1 March 2007: 761–767.

[103] http://ecee.colorado.edu/copec/book/slides/Ch12slide.pdf.

[104] Ngo T. A single-phase bidirectional dual active half-bridge converter. *Twenty-Seventh Annual IEEE Applied Power Electronics Conference and Exposition (APEC)*, Orlando, FL, 5–9 February 2012: 1127–1133.

[105] Gerekial W. *Bi-directional Power Converters for Smart Grids: Isolated bidirectional DC/DC converter.* Norwegian University of Science and Technology (NTNU), 2014.

[106] Zhao B, Song Q, Liu W, Sun Y. Overview of dual-active-bridge isolated bidirectional DC–DC converter for high-frequency-link power-conversion system. *IEEE Trans Power Electron*, 2014; **29**(8): 4091–4106.

[107] Yuana D. Design and analysis of emergency self-traction system for urban rail transit vehicles. *International Conference on Future Energy, Environment, and Materials*, 2012, Suzhou, China.

[108] Gu S. Research on control strategy of the emergency self-traction system in urban rail transit. *International Conference on Electrical and Control Engineering*, 2011 Yichang, China.

[109] https://www.vde-verlag.de/standards/1100277/e-din-en-62864-1-vde-0115-864-1-2014-12.html.

[110] http://www.alstom.com/products-services/product-catalogue/rail-systems/Infrastructures/products/aps-ground-level-power-supply/.

[111] Diguet M, Periot R, Moskowitz J. ALSTOM transport experience with onboard energy storage system. *VDE Congress*, 2010 Leipzig, Deutschland.

[112] http://www.ansaldo-sts.com/en/arc/press/ansaldo-sts-unveils-tramwave-new-catenary-free-power-supply-system-trams.

[113] http://primove.bombardier.com/application/light-rail/.

[114] Woronowicz K, Safaee A. *Linear Motors for Mass Transit Systems, their merits, Controls and Drive Aspects*, http://www.apec-conf.org/wp-content/uploads/2014/03/IS3-1-6.pdf.

[115] http://www.transrapid.de/cgi/de/basics.prg?session=42f947334d4bcc56_420615&a_no=121&r_index=2.

[116] https://www.pinterest.com/pin/166492517446915337/.

[117] Heinrich K, Kretzchmar R, eds. *Transrapid International, Transrapid Maglev System*. Darmstadt, Germany: Hestra Verlag, 1989.

[118] Caf Power & Automation. *Recife Metro, 3000VDC Converter.* Catalog available on the Internet, http://www.caf-power.com/en/electric-traction-systems/electric-dc-traction-converter/3000v-electric-traction-converter.

[119] Caf Power & Automation. *Train-Tram Cádiz, 750VDC/3000VDC.* Catalog available on the Internet, http://www.cafpower.com/en/electric-traction-systems/electric-dc-traction-converter/750v-3000v-electric-traction-converter-dual-voltage.

[120] Caf Power & Automation. *Chittaranjan Locomotive: 25kVAC traction converter.* Catalog available on the Internet, http://www.cafpower.com/en/electric-traction-systems/locomotive-traction-converter/electric-traction-converter-for-locomotives.

[121] Bose BK. *Modern Power Electronics And AC Drives*. Upper Saddle River, NJ: Prentice Hall, 2002.

[122] http://en.wikipedia.org/wiki/History_of_rail_transport#Diesel_and_electric_engines.

[123] Dujic D. *Power Electronic Transformer for Railway On-board Applications: An overview*. Zurich, Switzerland: ABB.

[124] http://www.railwaygazette.com/news/traction-rolling-stock/single-view/view/odakyu-electric-railway-to-test-all-sic- power-module.htm.

[125] https://www.infineon.com/dgdl/Infineon-XHP-PB-v01_00-N.pdf?fileId=5546d4624cb7f111014d6a9e59946d58.

[126] Schnell R, Hartmann S, Trüssel D, et al. *LinPak: A new low inductive phase-leg IGBT module with easy paralleling for high power density converters designs*. Nuremberg, Germany: PCIM Europe, 2015.

[127] http://www.hitachi-power-semiconductor-device.co.jp/product/igbt/pdf/nHPD2_.pdf.

[128] Mueller M. Life after lithium. *Electric & Hybrid Vehicle Technology International*, 2015: 67–72.

[129] https://www.semikron.com/products/new-products/semitrans-20.html.

6

Ships

Iñigo Atutxa and Gonzalo Abad

6.1 Introduction

This chapter is dedicated to the study of ship propulsion, from an electric and electronic point of view. First of all, a general description about ships is presented. General definitions, types of ships, components of ships, and different types of propulsions systems are briefly described. After that, the subject of ship propulsion is approached using mathematical equations describing the movement of the ship and the forces and torques involved in that movement. It is here that the dynamic model for computer-based simulation is presented. Then, once the physical basics are established, a detailed description of the electric drives employed for ship propulsion is given. Related to this, the complex but necessary power generation and distribution systems of ships are also studied. This includes a description of the generation itself, together with power cables, fault protections, harmonic distortion, and voltage droop analysis. Finally, as with the rest of the chapters, further information is also included: computer-based simulation examples, design and dimensioning of the electric propulsion, description of some real examples, and historical evolution and future challenges. In addition, the dynamic positioning (DP) concept of ships is also briefly introduced.

6.2 General description

6.2.1 Definition and basic concepts

In this chapter, when we refer to a ship, we understand this to be a vehicle employed to transport goods and persons from one point to another over water [1]. The movement of ships normally occurs thanks to the propulsion system, of which an important part is the propeller. Nowadays, the most extensively used system is

Power Electronics and Electric Drives for Traction Applications, First Edition. Edited by Gonzalo Abad.
© 2017 John Wiley & Sons, Ltd. Published 2017 by John Wiley & Sons, Ltd.

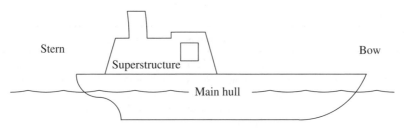

Fig. 6-1 *Basic picture of a ship*

based on electric propulsion. The concept of electric propulsion originated more than 100 years ago, in the late 19th century. Nowadays, electric propulsion using a gas turbine or diesel engine as the primary source is used in hundreds of ship types. It must be emphasized that much of the analysis and description in this chapter is inspired by the well-recognized and established reference [2] about ships.

The floating of a ship is described by the Archimedes' principle. This states that when a solid is immersed in a liquid it experiences an up-thrust equal to the weight of the fluid displaced. When the ship is in a specific load condition, it floats at an arbitrary water line. Normally, each ship has a load limit that is in relation to the depth to which the ship is submerged. The part of the hull that is under the water line (depending on the loading conditions) defines the propulsion requirements for the ship. The dimensions and shape of the hull also determine floating, stabilization, and other characteristics of the ship and so they must be carefully designed, but this is beyond the scope of this chapter. Fig. 6-1 shows a basic picture of a ship.

6.2.2 Types of ships

There are a wide range of ships, the type depending on what the ship is designed for. Some of the ships are dedicated to passenger transportation, while others for merchant transportation. Some ships are used for both. Also, a wide range of ships are used in the energy sector from basic tank ships to extremely complex industrial vessel such as dynamically positioned deep-water drilling vessels and include other vessel such as offshore crane-ships, pipe-laying vessels, icebreaking supply vessels, semisubmersible drilling units, and so on.

Ships can be classified according to many different criteria: number of hulls (mono-hull, catamaran, trimaran), hull material, shape, size, and function or use, the type of propulsion system, or the manufacturer.

Table 6-1 shows a classification and rough outline of different ships types according to their use. It is not definitive, since many other types of ships can be found. Ship type can be based on many different requirements in terms of propulsion power, noise-producing level, redundancy in power generation, reliability, or other requirements may be needed. Most of these vessels are typically powered propellers driven by a diesel engine or, less usually, by gas turbine engine, combined in both cases with electric propulsion. Fig. 6-2 illustrates a pair of examples of the mentioned types of ships of Table 6-1.

On the other hand, in a general perspective, attending on the type of propulsion system that can be found on a modern ship, Fig. 6-3 shows a general classification. It will be seen that there are mainly two types of propellers: with fixed or variable pitch. More details about this fact are provided in subsequent sections. With regards to the main energy source, the most popular ones are: steam turbines, gas turbines, and diesel engines. It is possible to find ships with propulsion systems not covered in Fig. 6-3, as well as different main sources such as the nuclear power. However, this book focuses its analysis only on the dashed type, which is the diesel engine with electric transmission (variable speed drive, or VSD) and fixed-pitch propeller, which is probably the most extensively used configuration in modern ships.

Table 6-1 *Summarized ship types that use electric propulsion [1], [2]*

Category	Short description of typical characteristics
Passenger ship • Cruise ships and vessels • Ferries • Yachts and leisure boats	High requirement of comfort in terms of absence of noise and vibrations The propulsion power can range from less than 1MW (yachts and leisure boats), up to more than 40MW for large cruises In large cruises, the total service and hotel load can be a significant part of the installation
Tanker • Oil tanker • Gas tanker • Chemical tanker • Shuttle tanker	This category is one of the largest, typically being able to reach powers of 20–55MW Shuttle tankers are used to transport oil from offshore facilities to onshore. Most of them need to maintain an accurate position at sea, incorporating often DP with a high degree of redundancy
Bulk carrier	Merchant ship specially designed to transport unpackaged bulk cargo, such as grains, coal, ore, cement, etc. Larger bulkers, usually have a single low-speed diesel engine directly coupled to a fixed-pitch propeller On smaller bulkers, one or two diesels are used to move either a fixed or controllable pitch propeller via a reduction gearbox, which may also incorporate an output for an alternator
Container ship	Are cargo ships that carry the load in containers The propulsion is typically diesel
Oil and gas exploitation and exploration • Drilling units • Production vessels	Thanks to sophisticated DP, deep-water drilling, and floating production is possible They present large installed thruster power of 20–50MW, that then is increased for utilities and service loads High redundancy to increase availability is typically required
Field support vessels and offshore construction vessels	For instance, diving support vessels, crane-ships, pipe-layers, etc. Typically DP is required The total installed power can range from 8 to 30MW
Dredges and construction vessels	Typically require to do accurate maneuvers with economized fuel consumption A big part of the installed power can be designated not only to the propulsion of the ship but also to the works that the ship is designed for: dredge, drilling, etc.
Icebreakers and ice going vessels	Prepared for high load variations, which means that the propulsion system must be prepared to provide fast dynamic responses A redundant power generation and distribution system is often required No dynamic positioning requirement Installed propulsion power: 5–55MW
Warships	Requirements of reliability and redundancy normally stricter than in merchant vessels They can be surface warships, submarines, and support and auxiliary vessels
Research vessels	Very strict underwater noise requirements They can perform hydrographic surveys, oceanographic research, fisheries research, naval research, polar research, or even oil explorations
Fishing vessels	Fishing vessels are generally small and often not subject to strict regulations and classifications Fishing boats are in general small, often no longer than 30m, but up to 100m for large tuna or whaling ships

(a)

(b)

Fig. 6-2 *Examples of different ships: (a) platform supply vessel* Blue Storm *and (b) seismic research vessel* Polarcus Adira *(Source: Ulstein. Reproduced with permission of Ulstein)*

6.2.3 Components of a ship

A ship is a complex construction composed of many different elements. The mechanical structure, the floating physics, hull design, and so on are some few examples of the many aspects that form a ship's design. In this chapter, only the electrical point of view of the ship is studied, focusing on ships with electric propulsion.

Thus, in a similar way as in a land-based electric power system, in a ship the electric energy must also be generated to be later distributed to different loads and to the propulsion system. Therefore, from an electrical point of view, it is possible to distinguish four main parts of a ship:

- electric power generation
- electric power distribution
- VSD for propulsion units or thrusters
- propulsion units or thrusters.

Fig. 6-4 shows a simplified single line diagram of an electric power system of a ship. The electric power generation unit is composed of prime movers and electric generators. The most commonly used prime mover is typically the medium- or high-speed diesel engine, owing to its good balance between cost and energy efficiency. In some ship applications, gas turbines or steam turbines are also used, as mentioned above.

The mechanical power generated by the prime mover is converted into electricity by the generator. Thus, the electric generator produces electric energy in AC (constant frequency and amplitude voltage) that is transmitted to the main feeder or electrical bus by means of the main switchboard. The electrical bus is composed of parallel conductor bars where the loads and generators are connected. Through the distribution switchboards, this electric energy is distributed to the loads of the ship (hotel, ship service, etc.) and to the VSD, which is in

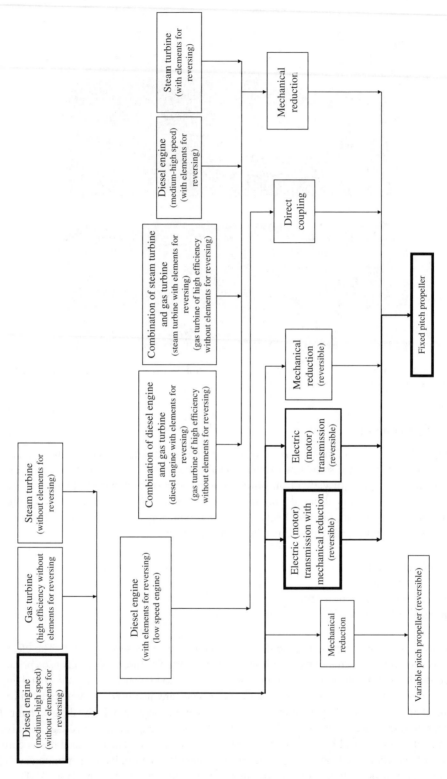

Fig. 6-3 *General classification of ships attending to the propulsion type employed [3]*

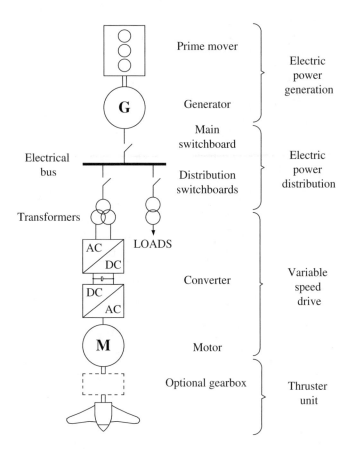

Fig. 6-4 *Simplified electric power system of a ship (in real ships redundancy is applied to this basic schema)*

charge of moving the propeller. The propeller generates the movement of the ship. The VSD is composed of the transformer, the power electronic converter, and the electric motor, which converts the electric energy into mechanical rotation driving the propeller. Between the thruster or propulsor and the electric motor, occasionally there can be a gearbox that adapts different rotational speeds of the motor and the propeller.

The power system schema shown in Fig. 6-4 is a simplified version of what is really used in real ships. In general, redundancy with physical separation is applied to ships in order to avoid the possibility of a "blackout" situation. A blackout is when all the electrical systems of the vessel are lost. In electric propulsion vessels, this situation leads to a loss of all thrusting capability, leaving the vessel in a very dangerous situation. The problem with non-splitted power systems is that a short circuit in any part of the system could lead to a total blackout. This could be avoided by installing two power systems (usually both port and starboard) which are connected by means of a so-called bus tie, a circuit breaker designed to detect and open when a fault occurs on one side, thus allowing the operator to use the undamaged side. Two physically separated power systems are especially important for vessels that work close to other vessels or to offshore platforms (DP operation), where propulsion loss could lead to a catastrophic accident.

Therefore, going even further, power generation and distribution is split into two, three, or even four parts by using different electrical busses and bus ties. In this way, it is possible to split the ship generation, distribution, and propulsion into different sections that, with a well-thought-out operational strategy, give the flexibility to

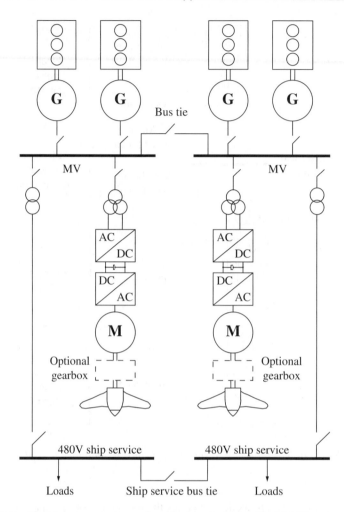

Fig. 6-5 *Electric power system of a ship (single line) with ABS-R2 redundancy*

enable the necessary sections and generators to receive the correct amount of energy at specific moments, thus increasing the efficiency of the system in terms of fuel consumption. In addition, this ability to self-split by applying redundancy makes it possible to isolate faulty parts, and thereby increases system reliability and availability, which is an important issue for a ship.

Hence, Fig. 6-5 shows a power system schema of a ship with ABS-R2 redundancy [4]. It is composed of four generators, divided into two groups and connected by a bus tie. The ship service distribution is also distributed into two sections, obtaining redundancy. The propulsion system is also composed of two thrusters.

Other elements that can be found in a ship are:

- Transformers which are used to adapt different voltage levels of generation, distribution, VSD, and service.
- Cable routing between different elements in a ship which is of shorter distance than in land-based electric power systems.

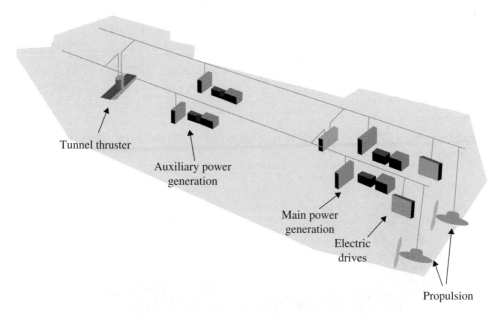

Fig. 6-6 *Example of system layout for a cruise vessel [2]*

- Fuses and protective elements that must be located at different places of the system so they protect the system against faults. The location of the fuses is often determined by detailed fault analysis.
- In many ships, there are specific loads or auxiliary applications, depending on the ship's purpose: hotel loads for passenger vessels, drilling systems for drilling rigs, dredging equipment and pumps for dredgers, etc.
- Finally, the marine automation system is in charge of the control and coordination of all the elements of the ship's navigation. Thus it performs tasks such as: power management system (PMS) for handling generators, human–machine interface between ship and operator, communication between different control levels, emergency shutdown control, and so on.

Later sections describe many of these components of the ship in more detail. Fig. 6-6 shows an example of a system layout for vessels.

Electric propulsion, as mentioned before, is nowadays practically employed in all modern sophisticated ships. The most important reasons for this selection can be summarized as:

- very economical approach in relation with the provided performance;
- enables good comfort, low noise and vibration levels, apart from meeting requirements of safety and availability;
- enables flexibility design;
- provides very good energy efficiency.

Fig. 6-7 illustrates an approximated breakdown of the efficiency achieved by an electrically propelled ship.

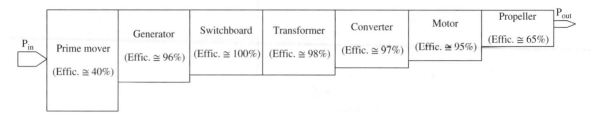

Fig. 6-7   *Approximated breakdown of the efficiency achieved by an electrically propelled ship (other elements may also be included, such as gearboxes)*

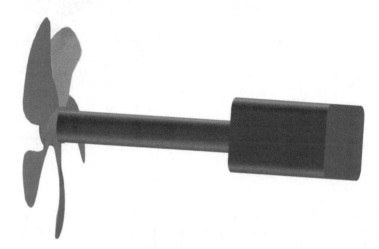

Fig. 6-8   *Example of a fixed-pitch propeller*

6.2.4   Propulsion system

The traditionally employed element used to move a ship is the propeller. Most ships generally have either one or two propellers, depending on their characteristics. However, there are also cases where more than two propellers are employed. A wide range of propeller types have been employed over the years. In this section, only the most commonly employed propeller types in modern ships with electric propulsion are briefly described [5]. Propellers are commonly located at the stern of the ship, but ships that require accurate maneuverability also use propellers or thrusters located at the bow, as is shown in subsequent sections.

6.2.4.1   Fixed-pitch propeller

In later sections it will be seen that the pitch, P, is a constructive characteristic of a propeller that, combined with other elements, such as the diameter or blade number, determine the propulsive capacity of a propeller. The fixed-pitch propeller, as its name denotes, presents a constant pitch characteristic, and therefore it is not possible to modify its propulsive characteristics during operation. The only way to modify its propulsion thrust is by modifying its speed of rotation by the electric drive. The pitch of the propeller is associated with the position of its blades, which cannot be changed in fixed-pitch propellers. In general, ships that do not require especially accurate maneuverability are equipped with fixed-pitch propellers.

Mono-block propellers made of a copper alloy are the most commonly used nowadays. They cover a wide spectrum of types, weights, and sizes. Fig. 6-8 shows an example of a fixed-pitch propeller.

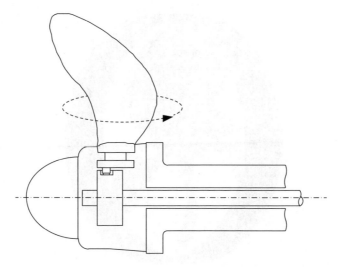

Fig. 6-9 *Schematic of a variable pitch propeller [5]*

For these types of propellers a gearbox is typically used that adapts the rotational speed of the propeller itself and the driving motor.

6.2.4.2 *Variable pitch propeller*

As the name indicates, these types of propeller can modify the angle of the blades and, therefore, their pitch during the normal operation of the ship. For this purpose, variable pitch propellers typically incorporate a hydraulically activated mechanism. This improves the maneuverability of the ship. Thus, variable pitch propellers, in principle, present two modifiable variables during normal operation: pitch and rotational speed. However, it has to be remarked that, in many cases, variable pitch propellers have been used at fixed speed, only their pitch being modifiable. Over the years, many ships' variable pitch and fixed-pitch propellers have been used indistinctively. Fig. 6-9 depicts a schematic example of a variable pitch propeller.

6.2.4.3 *Ducted propellers*

As their name denotes, in general they are composed of two main components:

- an annular duct having an airfoil cross-section;
- specially designed propellers whose blades take into account the presence of the duct.

The propeller itself can be either variable or fixed pitch. Their application area has been found principally at applications requiring high thrusts at low speeds. Fig. 6-10 shows an example of a ducted propeller.

6.2.4.4 *Podded and azimuthing propulsors*

Azimuthing (360° rotatable) thrusters are those that can be rotated in order to produce thrust in any direction. Azimuthing thrusters can be found either in ducted or non-ducted propellers and in variable or constant pitch propellers. In addition, they can be further classified as pusher or puller units. Puller units are

Fig. 6-10   *Example of a ducted propeller*

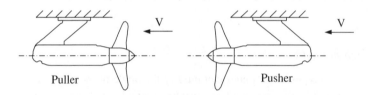

Fig. 6-11   *Pusher and puller azimuth propellers*

able to reduce the wake and hull resistances. A graphical representation of puller and pusher units is given in Fig. 6-11.

In azimuthing propellers, the motor is located inside the hull of the ship and the mechanical power is transmitted to the propulsor using shaft and gearbox arrangements. The mechanical shaft from the motor to the propeller can have a Z or L shape. The main difference of azimuthing and podded propulsors is the location of the motor. In podded propulsors, the motor is directly mechanically coupled to the propeller shaft, avoiding the necessity of gearboxes. Note that, in this case, the motor must be specially designed to operate at same-speed range as the propulsor itself and is submerged. In both podded and azimuthing propulsion systems, the propeller can rotate 360° around the vertical axis, being able to provide propulsion thrust in any direction. Therefore it is also able to perform the propulsion and steering functions of the ship, eliminating the need for rudders. Azimuth pod propulsion, not having a traditional propeller shaft, can be located further below the stern of the ship in a clearer flow of water, increasing the hydrodynamic efficiencies. In addition, it saves space on the ship and makes repairs easier to perform, since it is outside the ship. Azimuthing pod propulsion is commonly a pulling type with fixed pitch. Fig. 6-12 and Fig. 6-13 show some examples of podded and azimuthing propulsors. In general, the underwater shape of azimuthing is optimized for low hydrodynamic resistance, for higher propulsion efficiency. Azimuthing propulsion systems have become popular for other reasons, too, since they provide improved efficiency and maneuverability, low noise and vibration levels, flexibility in the engine room design, and are easy to maintain.

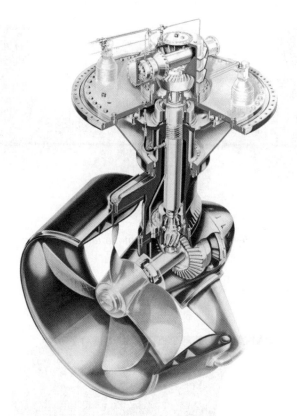

Fig. 6-12 *Azimuth propulsion pulling systems with 360° turning capability and main components (Source: Schottel. Reproduced with permission of Schottel)*

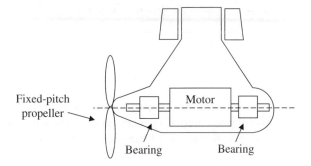

Fig. 6-13 *Podded propulsion system with 360° turning gear [2]*

6.2.4.5   *Contra-rotating propellers*

The contra-rotating principle is based on two coaxial propellers, placed one in front of the other, rotating in opposite directions (Fig. 6-14). It takes advantage of recovering part of the slipstream rotational energy that would otherwise be lost using a conventional propeller.

Fig. 6-14   *Example of a contra-rotating propeller*

Fig. 6-15   *Tunnel thruster examples*

6.2.4.6   *Tunnel thrusters*

These produce side forces and turning movement to keep a ship on station and for low-speed maneuvers. They can be found as fixed- and controlled pitch propellers. They present low efficiency at higher speeds of the vessel. They can also often produce cavitation noise onboard the vessel. Fig. 6-15 graphically illustrates some tunnel thruster examples.

6.2.4.7   *Vertical axis propellers*

It can be seen that the above propellers present the axis parallel to the direction of movement. However, there are also some types of propellers whose axis of rotation is perpendicular to the direction of movement. These propellers are not as prevalent as those we discuss in the previous sections, but they do still exist. They are sometimes called cycloidal propellers [5]. These types of propellers contain between five and eight blades perpendicularly located to the direction of movement. The Kirsten-Boeing is one type, but the Voith Schneider

Fig. 6-16   *Voith Schneider propeller*

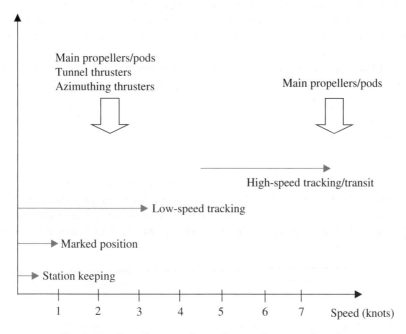

Fig. 6-17   *Speed ranges depending on the control modes*

is probably the most extended vertical axis propeller used nowadays and is shown in Fig. 6-16. It is typically located at the stern, but it is also possible to find them located at the bow to assist with and improve the maneuverability of the ship. In general, the efficiency of this type of propeller is lower than the horizontal axis propellers seen in the previous sections.

Finally, it must be mentioned that there are many other types of propellers which have not been described here, for example tandem propellers and overlapping propellers. They can be studied in the specialized literature, such as [5].

Depending on the speed range and actuation of the vessel required, the type of propeller(s) may be different. Fig. 6-17 illustrates some indicative values of the type of propulsion employed depending on the control mode of the ship.

6.3   Physical approach of the ship propulsion system

In this section a physical analysis of the ship propulsion that is useful for modeling purposes is presented. The basic principles describing resistance forces interacting with the ship, together with the basics of propeller thrust forces, are described first, followed by the dynamic model of the ship that is specifically designed to perform computer-based simulations. With the help of this model, it is possible to obtain and analyze operating conditions of the ship, such as resistance and thrust forces, advance speed, consumed power, efficiencies, and even electric variables of the machine and converter generating the thrust. Many parts of this analysis are based on the well-recognized reference [1].

6.3.1   Ship resistance

In order to study the movement of a ship, it is necessary first to study the nature of the resistance forces that actuate against the movement. In this context these forces in general are called resistances, R. The resistance forces affecting the ship depend on how it has been designed and are influenced by many different factors, both internal and external to the ship, such as hull form, speed, mass (displacement of the water of the ship), wind conditions, water conditions, etc.

In order to avoid working with overly complicated models while still being able to achieve reasonable accuracy, three main resistance forces affecting a ship are generally considered [1]:

- frictional resistance
- residual resistance
- air resistance.

The first two resistances are mainly affected by the length of the hull below the water. Conversely, air resistance is affected by the length of the hull above the water. Note that air resistance can eventually vary in ships that carry elements such as containers, because these modify the surface of the hull above the water.

The origin of these forces can be explained by Bernoulli's principle of fluid dynamics. It states that a fluid with a speed V and density ρ generates a force to a static body provided by:

$$F = \frac{1}{2}\rho V^2 A \tag{6.1}$$

being A, the area of the body perpendicular to the fluid speed. In general, Bernoulli's principle expressed by this equation, combined with dimensionless resistance coefficients C, is used to mathematically describe the above-mentioned frictional, residual, and air resistances.

6.3.1.1   Frictional resistance

The frictional resistance, R_f (expressed in newtons), of the hull can be described by the following equation:

$$R_f = \frac{1}{2}\rho V^2 A_s C_f \tag{6.2}$$

being ρ the density of water in kg/m^3, V the speed of the ship in m/s, A_s the hull's submerged area in m^2, and C_f the dimensionless frictional resistance coefficient. The seawater density is normally considered as $\rho = 1025$ kg/m^3. The dimensionless parameters such as C_f are typically obtained by experimental tests of ships in towing tanks, whenever that is possible. The physical phenomenon associated with this resistance is that, when the ship is moving, hydrodynamic friction is caused in the hull.

In low speed ships (bulk carriers, tankers, etc.), frictional resistance can be in the range of 70–90% of the total resistance. However, in high-speed ships (e.g. passenger ships) it can be in the region of 40%. Living organisms fouling the hull can increase the frictional resistance. This fact can be avoided with anti-fouling hull paints. Thus, from a mathematical point of view, the frictional resistance is proportional to the square of the ship speed.

6.3.1.2 Residual resistance

The residual resistance, R_r (expressed in newtons), is modeled by an expression equivalent to frictional resistance:

$$R_r = \frac{1}{2}\rho V^2 A_s C_r \tag{6.3}$$

being, in this case, C_r the residual resistance coefficient. The physical phenomenon describing this resistance is associated with the wave resistance and the eddy resistance. The wave resistance is generated by the waves created by the ship when it moves through the water. The eddy resistance is generated by flow separation which creates eddies, especially at the aft of the ship.

The residual resistance for low-speed ships typically represents 8–25% of the total ship resistance. For high-speed ships it increases to 40–60%. Thus, from a mathematical point of view again, the residual resistance is proportional to the square of the ship speed. Detailed experimental measurements have shown that at higher speeds the residual resistance increases much higher than in squared speed relation. However, for simplicity, this fact is not considered and is left for an advanced model.

6.3.1.3 Air resistance

Air resistance, R_a (expressed in newtons), is generated by the movement of the ship through air. Normally it is not a big fraction of the total resistance. The air resistance can be described by the following expression:

$$R_a = \frac{1}{2}\rho_{air} V^2 A_{air} C_{air} \tag{6.4}$$

being ρ_{air} the density of air in kg/m^3, V again the speed of the ship in m/s, A_{air} the hull's cross-sectional area in m^2 above the water, and C_{air} the dimensionless air resistance coefficient. The density of air is typically considered as $\rho_{air} = 1.225$ kg/m^3. This resistance normally represents around 2% of the total resistance and, again, it is proportional to the square of the ship speed. However, as stated before, with ships with suprastructures, which are especially big or variable, the air resistance can take higher and different values.

6.3.1.4 Total towing resistance

The total ship resistance, R_t (expressed in newtons), is the addition of the three described resistances:

$$R_t = R_f + R_r + R_a \tag{6.5}$$

It corresponds to the total resistance at constant speed, to tow the ship in calm water. One picture representing these forces is depicted in Fig. 6-18.

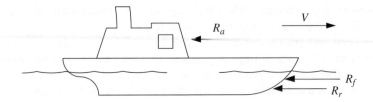

Fig. 6-18   *Schema of the resistances affecting the ship's linear movement*

Table 6-2   *Indicative resistances [1]*

Resistance	% of R_t	
	High-speed ship	Low-speed ship
Friction	45	90
Wave	40	5
Eddy	5	3
Air	10	2

Common indicative values of resistances are summarized in Table 6-2. It has to be noted that according to the resistances described in previous sections the total towing resistance is proportional to the square of the speed. Therefore, the necessary effective towing power to move the ship at constant speed is calculated as:

$$Power = V \cdot R_t \qquad (6.6)$$

Which expressions (6.2), (6.3), and (6.4) reveal that is proportional to the cubic ship's speed:

$$Power \propto V^3 \qquad (6.7)$$

However, as stated before, at higher speeds the wave resistance increases very quickly, leading to a phenomenon called "wave wall" ship speed barrier. Once speed increases above the wave wall, the squared speed relation is no longer valid because this increased wave resistance means that more power is needed at higher speeds to increase the traveling speed of a ship. This fact is graphically illustrated in Fig. 6-19, where it can be seen that, obviously, the ship is designed in such a way that the speed barrier is still far from the service ship speed.

6.3.2   Propeller propulsion

6.3.2.1   *Basic principles*

Ship propulsion in most of today's ships is achieved by propeller propulsion. This section describes some basic principles of propeller propulsion. The design of modern propellers is not an easy task. It requires knowledge

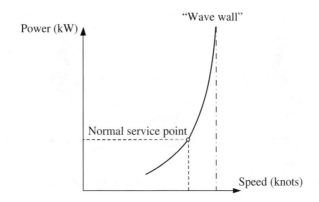

Fig. 6-19   *"Wave wall" ship speed barrier*

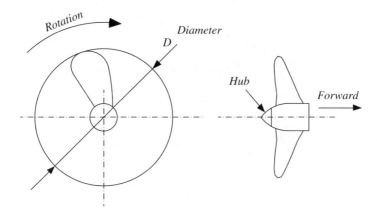

Fig. 6-20   *Propeller geometry*

of many disciplines. In general, complex propeller geometry is characterized by a few key parameters, as illustrated in Fig. 6-20 [6]:

- Diameter: D
- Blade number: Z
- Propeller pitch: P

By designing propellers with different D, Z, and P parameters, different propulsive forces and powers can be obtained. The shape and form of a propeller can also influence its performance. Blade number Z, for instance, is an important parameter for a propeller. In general, a high number of blades reduces vibration problems but increases manufacturing costs. In large ships, the typical blade number can be between four and seven. Whereas for small ships, blade numbers of between two and four are commonly employed.

A propeller can be understood as a helicoid surface which in rotation screws its way through the water. Considering a line AB perpendicular to AA′, as illustrated in Fig. 6-21, AB rotates around the axis of AA′

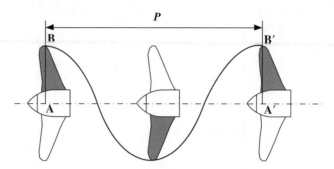

Fig. 6-21   *Helicoid surface illustrating propeller. Later it will be seen that the advance does not normally correspond with the pitch,* P

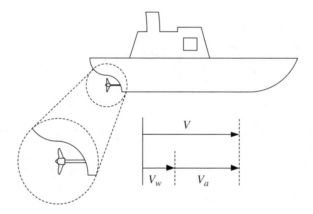

Fig. 6-22   *Different speeds of the water in a ship when moving at constant speed*

with constant angular speed, while moving along AA′ with uniform speed. Therefore, AB forms a helicoid surface and the pitch distance is AA′.

Both constant pitch and variable pitch propellers incorporate a hydraulically activated mechanism in the hub to modify the pitch of the blades as described before. Thanks to this, variable pitch propellers achieve increased maneuverability, for instance during heavy seas, by changing their performance curves during an operation. In general, variable pitch propellers are used in ships that require a high degree of maneuverability. Ordinary ships moving most of the time in open sea at constant speed would not need to install a variable pitch propeller.

Hence, when the ship moves through the water at constant speed, V, the friction of the hull with the water creates a boundary layer of water around the hull. The ship drags the surrounding water with it. At the stern of the ship, this produces the wake to follow along with the ship at a wake speed, V_w [1], which has the same direction as the ship's speed, V, as illustrated in Fig. 6-22. Thus, the propeller is experiencing a flow velocity less than the ship velocity. In this way, the flow velocity through the propeller is called the speed of advance, V_a:

$$V_a = V - V_w \qquad (6.8)$$

Therefore, the effective wake speed is:

$$V_w = V - V_a \tag{6.9}$$

and may be expressed by a dimensionless wake fraction coefficient, w:

$$w = \frac{V_w}{V} = \frac{V - V_a}{V} \tag{6.10}$$

or

$$1 - w = \frac{V_a}{V} \tag{6.11}$$

6.3.2.2 *Force and torque*

On the other hand, the thrust force, F (in newtons), described by the propeller, is commonly described by a non-dimensional equation:

$$k_f = \frac{F}{\rho n^2 D^4} \tag{6.12}$$

being: k_f: thrust coefficient, ρ: water density, and n: propeller shaft rotation speed in rev/sec; k_f is a coefficient being the function of the advance coefficient J:

$$J = \frac{V_a}{nD} \tag{6.13}$$

Or:

$$J = \frac{V(1-w)}{nD} \tag{6.14}$$

In a similar way, the torque T produced by the propeller (in Nm) is described by:

$$k_t = \frac{T}{\rho n^2 D^5} \tag{6.15}$$

being also k_t a dimensionless torque coefficient. As occurs with k_f, k_t is also in function of J. For determination of propeller force and torque, one must first determine k_f and k_t coefficients. The typical coefficient curves shapes are illustrated in Fig. 6-23. They depend on its geometry and how the propeller has been designed. The general shape of these coefficients will remain roughly the same, regardless of the propeller. It can be experimentally obtained, for instance, from propeller open-water tests.

Hence, once thrust and torque are described, it is possible to obtain the propeller open-water efficiency, η_o, as:

$$\eta_o = \frac{F V_a}{(2\pi n) T} \tag{6.16}$$

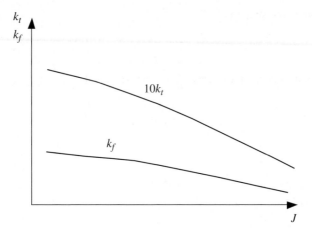

Fig. 6-23 k_f and k_t coefficient curves in function of J

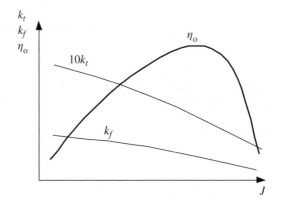

Fig. 6-24 η_o curve in function of J

This relates the applied rotational power and the obtained thrust power. It can be further developed to:

$$\eta_o = \frac{k_f \rho n^2 D^4 V_a}{k_t \rho n^2 D^5 (2\pi n)} = \frac{k_f}{k_t} \frac{J}{2\pi} \tag{6.17}$$

Typically, the propeller open-water efficiency curve takes the shape shown in Fig. 6-24.

Thus, systematic tests with similar propellers are typically compiled in a propeller series. Perhaps the most well-known series is the Wageningen B-screw series. Over the years, many model series of propellers have been added so as to provide a fixed-pitch propeller series. They assume a set of common geometry characteristics of propellers and the k_f and k_t coefficients are provided by equations:

$$k_f = \sum_{n=1}^{39} C_n (J)^{S_n} (P/D)^{t_n} (A_E/A_O)^{u_n} (Z)^{v_n} \tag{6.18}$$

$$k_t = \sum_{n=1}^{47} C_n(J)^{S_n}(P/D)^{t_n}(A_E/A_O)^{u_n}(Z)^{v_n} \tag{6.19}$$

The coefficients for these equations are typically provided in tabulated form [5], but they are not shown here, for simplicity. In these equations describing the coefficients: P/D is defined as the pitch ratio, A_E/A_O is defined as the extended blade area ratio (depends on two geometric parameters of the propeller A_E and A_O), and Z the number of blades. The rest of the coefficients—C_n, S_n, t_n, u_n, and v_n—as stated above, are covered in one table for each specific propeller type. For instance, B5-75 propeller would have a dedicated table of coefficients ($Z = 5$ and $A_E/A_O = 75$). Hence, it is seen that the thrust of a marine propeller working in open water may be expected to depend on geometry parameters (D, P, Z, and A_E/A_O), the density of the fluid, and the variables such as rotational speed, n, and speed of advance, V_a.

In this model deviation, only one quadrant operation of the ship is being considered: $V_a > 0$ and $n > 0$ (ahead in motor mode). For four quadrants of operation models, the reader is remitted to specialized references, such as [7], [8].

Thus, Fig. 6-25 illustrates η_o, k_f, and k_t curve shapes at different P/D values of a Wageningen B5-75 propeller [5].

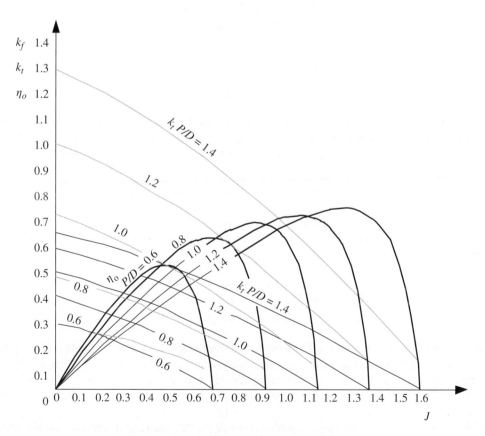

Fig. 6-25 *Open-water diagram for Wageningen B5-75 screw series*

It must be highlighted that with these characteristic curves contain all the necessary information to define the propeller performance at a particular operation condition. In fact, these curves can be also used for design purposes of the propeller and ship. Thus, if a given diameter (D) and pitch (P) is chosen, for example $P/D = 1.4$, the maximum efficiency operation point determines $\eta_o \approx 0.72$, $k_t \approx 0.042$ at $J \approx 1.18$, which means that if we select the rotational speed, n, (rev/sec) we have already selected V_a and power:

$$\begin{cases} V_a = JnD \\ Power = Tn\dfrac{2\pi}{60} = k_t \rho n^3 D^5 \dfrac{2\pi}{60} \end{cases} \tag{6.20}$$

There exist several propeller choice methodologies, described in specialized literature, but they are beyond the scope of this chapter. However, it must be emphasized that in a preliminary basic design for a desired V_a and power of the ship, n, P, and D must be carefully chosen, in order to make the propeller work at maximum efficiency.

On the other hand, the rotation of the propeller causes a phenomenon which consists of water in front of it being "sucked" back toward the propeller. This has the effect of augmenting the resistance of the ship. Thus, the thrust deduction factor, t, owing to this phenomenon, is defined as:

$$t = \frac{V_w}{V} = \frac{F - R_t}{F} \tag{6.21}$$

Or

$$1 - t = \frac{R_t}{F} \tag{6.22}$$

being R_t, the total resistance as seen in previous sections. The thrust deduction factor t can be obtained by calculation models, and the shape of the hull has a significant influence. The typical values of t are often in the range of 0.05 and 0.3. More detailed models for this coefficient can be found in [5].

6.3.2.3   Efficiencies

With regards to the efficiencies, in general there are several coefficients which are taken into account in a ship [1]. First of all, the hull efficiency, η_h, is defined and relates the effective power and the thrust power that the propeller delivers to the water [1]:

$$\eta_h = \frac{R_t V}{F V_a} = \frac{R_t V}{\dfrac{R_t}{1-t} V_a} = \frac{V}{\dfrac{1}{1-t} V(1-w)} = \frac{1-t}{1-w} \tag{6.23}$$

This hull efficiency in some cases can be higher than 1, meaning that the wake is helping the ship to move. Then there is the propeller open-water efficiency, η_o, which is described in the previous subsection by means of (6.17). Intuitively, it can take values of between 0.3 and 0.75. After that, the relative total efficiency, η_r, is also defined, which often is used basically to adjust the tank tests to the theory. In general, it takes values of between 0.95 and 1.05.

There also exists the shaft efficiency, η_s, which depends on the mechanical components of the shaft and gearboxes (if there are any). Typically, it takes values of between 0.96 and 0.995.

Therefore, the total efficiency, η_t, which relates the total effective (towing) power, P_e, and the necessary power at the drive motor, P_m, is [1]:

$$\eta_t = \frac{P_e}{P_m} = \eta_h \eta_o \eta_r \eta_s \tag{6.24}$$

6.3.3 Dynamic model for computer-based simulation

This section describes a simplified dynamic model of a ship suitable for computer-based simulation purposes. It only considers first-quadrant, straight-ahead movement of the ship in calm water. Rudder action to change the ship's direction is not considered. It is assumed that the propeller is a constant pitch propeller. In specialized literature, more complex models can be found that also try to describe the ship's behavior when it is maneuvering even in strong sea. However, this chapter explores a reasonably simple but useful model that helps to understand the basic variables and magnitudes affecting the behavior of the ship's movement that will also affect the drive of the propeller.

In this way, from the basic physics of the ship studied up to now, it is possible to describe the linear movement of the ship by the following equation [7], [9], [10]:

$$(1-t)F - R_t(V) - F_{ext} = (M + M_a)\dot{V} \tag{6.25}$$

being M the ship's total mass in kg and M_a the added mass (hydrodynamic force) in kg due to the acceleration of a body in water. For simplicity, this added mass is set as a fraction of the ship's total mass. Note that this mass only affects the acceleration and deceleration dynamics of the ship, not at the steady state when traveling at constant speed.

As described before, V is the ship's speed, F is the total thrust generated by the propeller, t is the thrust deduction factor, R_t is the total towing resistance, and F_{ext} is an external force that could occasionally affect the ship (external works, pulling, etc.). A schematic illustration is depicted in Fig. 6-26.

On the other hand, the propulsive thrust is supposed to be generated by the VSD's motor. Figure 6-27 depicts the electric drive which is in charge of controlling the rotational speed of the propeller, n, and therefore also the speed of the ship, V. Note that in most of the cases the propeller is speed controlled. However, very often, the propeller can be also controlled in torque or power (in order to not "stress" the power plant).

Simply assuming that there is no gearbox or mechanical losses at the shaft, the rotational speed behavior can be described by the following differential equation:

$$T_{em} - T = I\dot{\Omega} \tag{6.26}$$

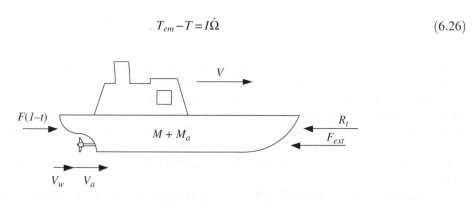

Fig. 6-26 *Schema showing the forces affecting the ship*

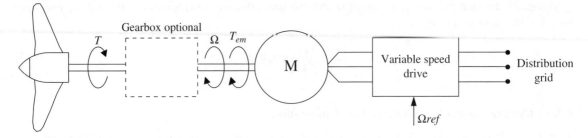

Fig. 6-27   *VSD in charge of controlling the rotational speed of the propeller*

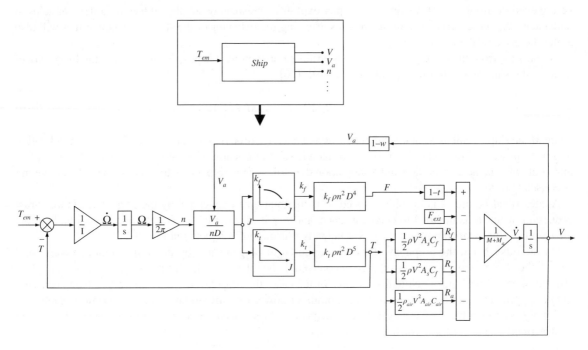

Fig. 6-28   *Computer-based simulation model of the ship*

being T_{em} the electromagnetic torque provided by the motor in Nm. T is the propulsive torque in Nm. I is the inertia of the shaft, propeller, and motor. Ω is the angular speed. The inertia should also include inertia due to the water mass caused by movement; however, modeling this effect accurately can be very difficult. Therefore, in order to simplify the analysis, this added inertia is neglected. Note that this fact would only influence the rotational speed dynamics, but not at the steady state. With this, it is seen that the drive typically modifies T_{em}, in order to keep controlled Ω (or n) all the time, to obtain the desired propulsive torque T and force F.

Hence, once the model differential equations are presented, it is possible to describe the overall ship dynamic model as illustrated in Fig. 6-28.

Note that the constant wake coefficient, w, and the constant thrust deduction factor, t, have been considered. This assumption is not far from reality, since studies reveal they do not really suffer from large variations during straight movement in calm water. Nevertheless, more accurate coefficient models can be obtained by analyzing, for instance, reference [5].

It can be seen that the model is nonlinear. The machine with its own dynamics imposes the rotational speed, n, then the advance coefficient J is calculated, from n and from V_a, which has been obtained from the ship's speed. Then the advance coefficients k_f and k_t are obtained from two look-up tables, to finally obtain the propulsive thrust and torque. The acceleration and speed of the ship are derived from the resistances and thrust, by means of (6.25).

It must be highlighted that this model assumes an error during acceleration and deceleration, since torque, thrust forces, and resistances (F, T, R_t) are strictly valid at steady state (i.e. at constant speed). Therefore, although probably during the speed variations of the ship, the modeled variables (speed, acceleration, torque, etc.) are accounting an error, once the steady state is reached with stabilization of the speed value the model is valid, allowing us to also derive correct steady-state ship's forces, speeds, etc. or even machine and converter variables, such as currents, voltages, and fluxes.

6.3.4 Brief steady-state analysis

Once the dynamic model has been presented, one can derive steady-state values from it. At steady state, without consideration of any external forces affecting the ship ($F_{ext} = 0$), the acceleration of the ship is zero, $\dot{V} = 0$, which means that:

$$F = \frac{R_t}{(1-t)} \tag{6.27}$$

This implies that the propulsive force generated by the propeller must be equal to the total towing resistance, taking into account the thrust deduction factor. In previous subsections, it has been seen that resistances in which R_t is divided can be considered proportional to the squared ship speed. Therefore, from (6.2), (6.3), and (6.4), it is possible to simplify R_t to:

$$R_t = C_t V^2 \tag{6.28}$$

being C_t a new coefficient that groups all the constants and coefficients of (6.2), (6.3), and (6.4). Thus, the shape of resistance and power curves take quadratic and cubic shapes, as depicted in Fig. 6-29.

On the other hand, the thrust has been seen that can be modeled according to (6.12):

$$F = k_f \rho n^2 D^4 \tag{6.29}$$

which means that when the ship is free sailing at constant desired speed, V, the propeller will necessitate rotating at one specific n to match both forces, as graphically represented in Fig. 6-30.

If the characteristics of the ship are known, the steps that must be followed to derive steady-state values at given operating ship speeds are:

1. With given operation ship speed V_{op}, calculate the towing resistance:

$$R_{top} = C_t V_{op}{}^2 \tag{6.30}$$

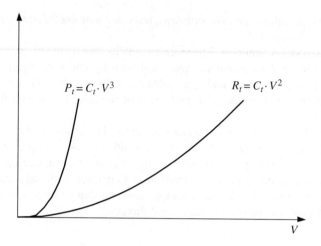

Fig. 6-29   *Resistance and power curves in function of* V

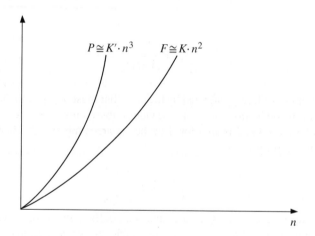

Fig. 6-30   *Force and resistance curves at constant* V

2.   Calculation of the advance speed and thrust force:

$$F_{op} = \frac{R_{t\,op}}{1-t} \tag{6.31}$$

$$V_{a\,op} = V_{op}(1-w) \tag{6.32}$$

3.   Once the thrust is known, it is possible to approximate the k_f characteristics of the propeller, to a known simplified curve, for instance a line as illustrated in Fig. 6-31 (the actual curve defined by (6.18) can, of course, also be employed).

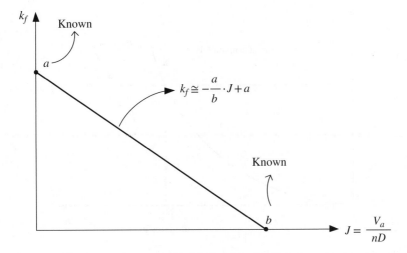

Fig. 6-31 *Simplified* k_f *characteristic*

Therefore we obtain:

$$F_{op} = \frac{R_{t\,op}}{1-t} \Rightarrow$$

$$F_{op} = k_f \rho n^2 D^4 = \left(-\frac{V_{a\,op}}{nD}\frac{a}{b} + a \right)\rho n^2 D^4 = \frac{R_{t\,op}}{1-t} \tag{6.33}$$

$$F_{op} = \left(-\frac{V_{a\,op}}{D}\frac{a}{b}n + an^2 \right)\rho D^4 = \frac{R_{t\,op}}{1-t}$$

Graphically, the rotational speed operation point, n_{op}, is given by the intersection of the corresponding line and the exponential, as illustrated in Fig. 6-32.

4. Once n_{op} is obtained, calculate the rest of the variables applying the corresponding equations:

$$J_{op} = \frac{V_{a\,op}}{n_{op}D} \tag{6.34}$$

$$k_{f\,op} = -\frac{a}{b}J_{op} + a \tag{6.35}$$

$$k_{t\,op} = -\frac{a'}{b'}J_{op} + a' \tag{6.36}$$

$$\eta_o = \frac{k_{f\,op}}{k_{t\,op}}\frac{J_{op}}{2\pi} \tag{6.37}$$

$$T_{em\,op} = T_{op} = k_{t\,op}\rho n^2 D^5 \tag{6.38}$$

$$\eta_t = \frac{1-t}{1-w}\eta_o\eta_r\eta_s \tag{6.39}$$

With a, b, a', b', η_r, and η_s supposed known.

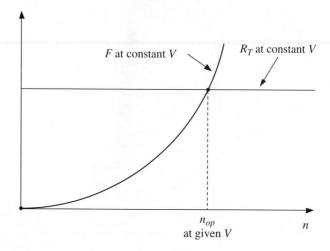

Fig. 6-32   *Operating point resulting from the intersection of a line and exponential curves*

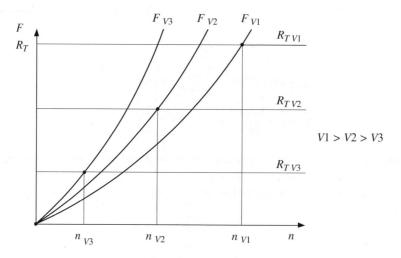

Fig. 6-33   *Operating points at different ship speeds*

On the other hand, if the speed of the ship must be modified, note that R_t changes accordingly in squared relation, resulting in the situation shown in Fig. 6-33.

This means that if the ship's speed is increased the rotational speed must also be increased. This fact allows us to deduce a simple control law that can keep control of the ship's speed, as shown in Fig. 6-34.

Connected to this, it must be remarked that, unlike a screw in a cork, for instance, the propeller does not operate in a solid material and it causes slip. If the propeller would operate in an unyielding medium, the speed of the propeller would be imposed by the pitch, P:

$$V = P \cdot n \tag{6.40}$$

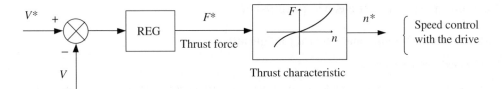

Fig. 6-34   *Ship's speed control law*

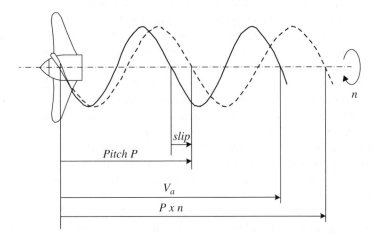

Fig. 6-35   *Propeller slip illustration [7]*

However, since the propeller operates in water, its acceleration causes the slip, which is defined as:

$$slip = \frac{P \cdot n - V_a}{P \cdot n} \tag{6.41}$$

The graphical illustration of this phenomenon is shown in Fig. 6-35.

The slip is an indicator which says how loaded the ship is, or in other words how great is its speed and, accordingly, how great the resistance is. As seen in Fig. 6-33, a small increase of the ship's speed, V (therefore also an increase at V_a), implies a higher increase at n, producing a higher slip also, as deduced from (6.41).

Finally, it must be remarked again that this developed model attempts only to be a fairly simple and reasonably accurate approximation of the real and complex behavior of the ship. Much more sophisticated models and accurate measurements have shown that, for instance, actual power consumption of the ship is often described with a higher power of 3, as modeled in this chapter (in (6.7)). It has been estimated that [1]:

- Large, high-speed ships like container vessels: *Power* $\propto V^{4.5}$
- Medium-sized, medium-speed ships like feeder containers: *Power* $\propto V^{4}$
- Low-speed ships like bulk carriers: *Power* $\propto V^{3.5}$.

This increased ship resistance is due to complex phenomena (time of service and state of the hull, conditions of sea, etc.) that are very difficult to model for the level or accuracy sought in this chapter. Consequently, the developed model is valid in a context where we assume that there are some model uncertainties that are acceptable for the accuracy required.

6.4   Variable speed drive in electric propulsion

6.4.1   General characteristics and general configurations

6.4.1.1   *General characteristics*

The typical power range of electric propulsion can vary from 1MW to 50MW. However, it is possible to also find cases that go above and below that margin. Consequently, it is seen that the electric drive configurations employed must handle powers in the high-power range. In the case of, for instance, a drive of 20MW of rated power, in general, this propulsive power is not only achieved by a single propulsion unit, with a single motor and single converter configuration. Instead, in general, several propulsors with several associated motors and converters are used, splitting the power in several drive units, making it possible to achieve the required high-power propulsive levels and, at the same time, obtaining different levels of redundancy, as is further described later in this chapter.

As described in earlier sections, this electric power is generated by prime movers moving AC electric generators, generating a constant amplitude and frequency electric grid. Part of this generated electric power is then transmitted to the propeller, with the help of the variable speed drive (VSD). The electric motor used can be asynchronous machines, wound rotor synchronous machines, or permanent magnet synchronous machines (PMSMs; more typically at the lowest power levels).

Nowadays, the most commonly used electric drive configuration feeding the motor is a voltage source converters (VSC) based on either insulated-gate controlled thyristors (IGBTs) or integrated gate-commutated thyristors (IGCTs). However, initially in the past, current source converters (CSCs) or cyclo-converters, and before them DC converters, were also employed.

In previous sections, when we describe the physical approach of the ship, we focus the analysis on only one quadrant of operation (i.e. positive rotational speed n and positive advance speed V_a). This first operation quadrant is the conventional way of working on the propeller, but in many cases, when moving astern, the propeller can work at any of the four quadrants. These four quadrants of operation are defined in Table 6-3.

Thus, from (6.15), introducing the effect of the four possible quadrants, the propeller-produced torque is defined as:

$$T = sign(n)k_t \rho n^2 D^5 \tag{6.42}$$

being k_t a strictly positive coefficient, also at the four quadrants of operation. Consequently, the consumed power is given by:

$$Power = sign(n)2\pi k_f \rho n^3 D^5 \tag{6.43}$$

Note that this power is always positive, which means that the propeller only actuates in motor mode, consuming power from the prime mover. Consequently, the motor of the propeller itself only operates at two quadrant modes on the torque–speed plane, as illustrated in Fig. 6-36.

Table 6-3   *Four quadrants operation modes of the ship*

1st quadrant	2nd quadrant	3rd quadrant	4th quadrant
Advance speed – ahead, $V_a > 0$	Advance speed – ahead, $V_a > 0$	Advance speed – astern, $V_a < 0$	Advance speed – astern, $V_a < 0$
Rotational speed – ahead, $n > 0$	Rotational speed – astern, $n < 0$	Rotational speed – astern, $n < 0$	Rotational speed – ahead, $n > 0$

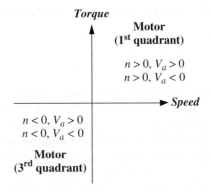

Fig. 6-36   *Torque–speed quadrants operation of the machine*

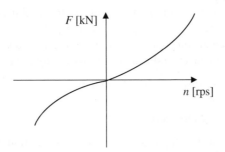

Fig. 6-37   *Torque–speed characteristic of a propeller*

This means that the converter of the drive must be prepared to deliver power from the distribution grid to the motor only. Hence, the VSC supplying the motor is typically fed by a passive front end composed of diodes, which is able to move the motor in both directions. However, in standard propulsions, the braking of the ship is accomplished by running the propeller in the opposite direction. Thus, in these cases and especially when a crash stop is performed, the deceleration of the propeller requires a small amount of regenerative power. This amount of power is typically around 15% of the rated power. This means that a braking chopper at the DC-link must be included, enabling the regenerative power in a resistance to be dumped, protecting the DC-link from damaging over-voltages. In a different way, with azimuth propellers the braking of the ship is made by rotating the entire propeller 180°, without having to change the rotating direction of the propeller. In this particular case, the braking chopper is typically not included in the installation because speed reversal will not be necessary. On the other hand, between the electric grid and the converter, a transformer is typically needed, owing to the difference of voltages between the distribution grid and the drive.

However, as commented before, the motor drive is normally speed-controlled, and the effect on the propulsion of this speed control can be interpreted as a torque control created by the propeller, as illustrated in Fig. 6-37 and described in previous sections.

Finally, redundancy for a ship is a very important issue. Any unexpected fault on the open sea or when approaching land should never avoid controlled operation, since it would have a potentially negative impact on any passengers. Hence, a certain level of redundancy is required by authorities and classification societies, in order to increase the availability of the ship. This results in special configurations of drives and generation and distribution systems, in attempts to achieve improved redundancy and availability levels compared to other (not so critical) applications.

6.4.1.2   Redundancy

The concept of redundancy applied to VSDs of ship propulsion can be understood as having a part of the VSD available if an unexpected fault occurs. Fig. 6-38 graphically illustrates this [11].

When the drive is operating without fault in normal operating conditions, the redundant part of the drive can be working, sharing part of the power with the rest of the parts or can be not working and only operate when the fault occurs. In both cases, the redundant part of the system can be dimensioned at full power level or lower. Ideally, redundancy is applied in order to obtain better availability ratios for the drive.

Depending on the ship type, the redundancy requirements can be more or less restrictive. Thus, it is possible to have, for instance, total (100%) redundancy, which means that the system can operate at full power in the event of a single fault. On a different redundancy level, for instance 50%, the system would work at half of the rated power in the event of a single fault.

In general, the level of redundancy is a compromise between availability and cost. High redundancy level is associated with high availability, but also to a higher cost and complexity of the system. Note also that the probability of a failure occurrence also increases if more elements are added to the system. Finally, it must be remarked also that without redundancy at the drive the probability to have a total loss of propulsion power is higher.

6.4.1.3   Modern electric drive configurations

In general, redundancy requirements for ship propulsion are quite common, in order to obtain higher levels of availability. As commented before, depending on the ship, this redundancy requirement can be more or less restrictive. For meeting this redundancy requirement, ships utilize more than one propeller. In addition, double

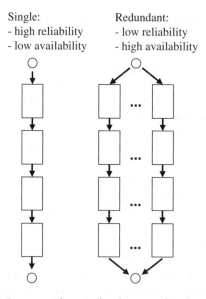

Fig. 6-38   *System without redundancy and with redundancy*

stator motors supplied by parallel converters are also commonly used in order to obtain redundancy. Note that the concept of redundancy means it is easier to attain higher power levels, since the total power is split among different parts of the drive (i.e. propellers, converters, motors, and stators). Added to this, from the drive point of view, parallelization of converters also reduces the harmonic distortion or harmonics of the system currents. Otherwise, in subsequent sections, the redundancy is not only applied to the drive itself but also is utilized at the generation and grid distribution system of the ship.

Hence, Fig. 6-39 shows nowadays typically used different VSC-based VSD configurations for ship propulsion [11]. All of these configurations obtain different levels of redundancy, depending on how the motors and converters are dimensioned with respect to the rated power.

- The simplest case is presented in Fig. 6-39(a), where no redundancy is applied and, in theory at least, is the most vulnerable configuration.
- Configurations shown in Fig. 6-39(b) and (c) use two drives in a unique propulsion unit, being in both cases able to work one drive if failure occurs in the other. Fig. 6-39(c) utilizes a gearbox prepared to a couple of two motor axes, which is very common.
- Finally, Fig. 6-39(d) consists of a motor with two parallel stator windings supplied by one converter each. The motor can operate with only one stator, in the event of failure. This configuration is also useful to reach high power levels, splitting the rated currents of the motors and converters, reaching standard acceptable levels.

Fig. 6-40 shows an example of a generation, distribution, and propulsion system of a cruise vessel of 20MW, with a distribution grid of 11kV [11]. Four diesel generator units are used to supply energy to the whole ship. The distribution system is split into two parts. The propulsion is also divided into two podded propellers, each one supplied by two independent VSDs in redundant configurations. There is no gearbox used, which means that the double stator synchronous machines are directly coupled to the shaft of the propeller. Each stator of the motors is supplied by two independent power electronic converter, with their own rectifier, transformer, and control units. This configuration provides 50% of the rated power if a fault occurs in the main switchboard, propulsion motor, or propeller. However, if a fault occurs at any of the other part of the system, 75% of power is available. Therefore, a typical propulsion system of a ship provides redundancy in all components, from the engines to the propellers. Later sections show more details of different ship configurations.

It must be highlighted that the motor with two parallel stator windings concept is often used in ship propulsion. It enables both converters to be electrically independent, simplifying the design when redundancy is necessary. With this system, it becomes very easy to connect two independent converters to the same motor, which is very attractive from a practical point of view. Nevertheless, in case of operation with only one winding, when, for instance, one converter has been damaged, it is necessary to include breakers between the converter and the windings in order to avoid affecting the induced voltage of the winding at the damaged converter and to allow maintenance of the damaged converter to be performed, while the motor is running with the safe converter.

It is important to note that to obtain total redundancy (100% of the power) with this technique of double stator winding, it would be necessary to perform a dimensioning of full current for both converters and also for both windings of the stator. From the point of view of the motor, this would imply the need to increase the copper necessity of the motor considerably, leading to a very inefficient motor design. In practice, the full current dimensioning of the windings of the motor is not usual, save for very special cases. Connected to this, in practice it is possible to find a full scale of current redundancy in the converters. This means that it must be accompanied by a special breaker arrangement, allowing short-circuiting for both windings in case one converter is damaged and while still being possible to operate at full current with the remaining safe converter.

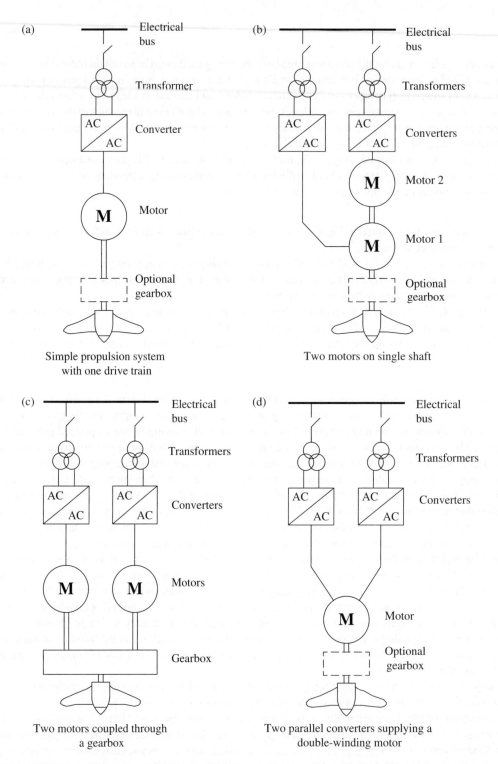

Fig. 6-39 *Different configurations of VSDs applied to propulsion units, achieving different levels of redundancy*

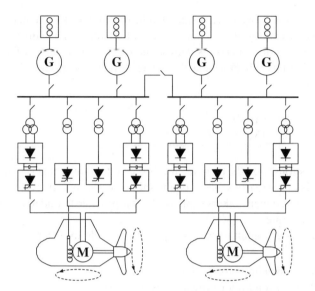

Fig. 6-40 *Example of a generation, distribution, and propulsion system of a cruise vessel of 20MW [11]*

On the other hand, sometimes the double stator winding motor employed is an asynchronous machine also. In this particular case, if one winding is damaged, the total magnetizing current must be exclusively applied at the other winding. When neither the windings nor the converters are prepared to assume the magnetizing current increase, it is not possible to reach 50% of the rated torque. In practice, with a typical motor design it is possible to obtain, approximately, 40% of the torque. We emphasize that we are speaking about torque, because the power can only reach these values if the speed can reach the rated value. Note that in fixed-pitch propeller, it is not possible to reach the rated speed with such small torques, which means that the power that can be supplied by only one winding is of lower value.

6.4.2 Electric machine

From the point of view of the electric motor, which is in charge of moving the propeller, although in the past DC motor configurations were also used, nowadays, AC machines are the most used ones. Within them, both induction machines and synchronous machines are common. Therefore, it is possible to find:

- induction motors with one three-phase stator winding or two, if redundancy and increasing availability is required;
- wound rotor synchronous machines, which can be with brushes or brushless, again with one three-phase stator winding or two, in case redundancy is required. The rotor must be externally supplied in DC, for instance by an extra converter and its own transformer, as depicted in Fig. 6-40. Additionally, it is very usual with brushes and motors of low voltage range to use a DC/DC converter to supply the rotor which is connected to the DC bus of the main converter. On the other hand, in medium voltage motors with brushes, it is also common to employ a secondary winding from the transformer, of voltage between 400V and 690 V and then to obtain the DC voltage by a thyristor based rectifier;
- PMSMs, typically with only one stator winding. In general, this machine technology is not available at highest powers or at medium voltage in ships.

In general, motors for ship propulsion are designed to be robust in order to reduce maintenance. When two separated stator windings are employed at the machine, fault tolerance is obtained because the motor can still work with only one stator winding, in case its corresponding converter fails, or the winding itself fails.

As a general approach, it is possible to say that the voltage level of the machine in order to obtain an efficient design can be selected as [2]:

- Rated power < 5MW → 690V (LV)
- Rated power > 5MW → 3kV (MV).

However, this depends strongly on the manufacturer's converter family that is going to install the drives for the ship. There are also historical reasons, based on the ship owner's country, which tend to give advantage to LV or MV. Thus, these facts and many others explain why it is possible to find a wide range of actual ships that do not follow this indicative rule.

In terms of machine topology, again a general approximated rule to obtain an efficient design could be:

- Rated power < 5MW → asynchronous machine.
- Rated power > 10MW → wound rotor synchronous machine.
- 5MW < rated power < 10MW → both asynchronous and synchronous machines are possible. At lower speeds, a multi-pole synchronous machine is more suitable.
- At lowest powers, at low voltage and at lower speeds, multi-pole PMSMs are suitable, especially where the efficiency and volume are very restrictive. However, this topology is being implanted slowly, owing to well-known drawbacks, such as: necessity of a breaker at the motor terminals, thermal care on the magnets, and availability and cost of magnets.

Here, in the machine, there are many exceptions. For instance, it is possible to affirm that there are manufacturers who install asynchronous machines at highest powers. There are already real ships, with asynchronous machines of, for instance, 14MW or 20MW.

In silent ships or submarines, DC motors were traditionally installed in order to reduce vibrations and noise. But even in these types of ships, asynchronous machines are now being introduced as well. Therefore, it can be concluded that it seems that it is a tendency to install asynchronous machines more widely in ship propulsion. As an example, Fig. 6-41 illustrates a commercially available machine design.

Thus, given a specific machine which drives a specific propeller, typically the maximum capability theoretical curves (see Chapters 2 and 3) and the propeller torque curve match as represented in Fig. 6-42. Generally speaking, the maximum operation rotational speed cannot be bigger than the crossing point of both machine maximum torque curve and propeller torque curve. In general, ω_{max} can be chosen a little bit bigger than ω_{base}, as illustrated. The torque and power curves demanded by the propeller are given by (6.42) and (6.43), which are quadratic and cubic evolution in function of the rotational speed. In this way, the maximum torque and power curves of the machine are only crossed at one point (maximum speed). Note that these represented curves are ideal. Actual machine designs can differ from this representation. It must be highlighted also that the power or torque curves of the ship can vary depending on many phenomena (sea conditions, wind, etc.). In other words, the simplified model derived in Section 6.3 gives a rough idea of the torque and power necessary, but in general ship manufacturers utilize much more accurate models, considering also many other factors and phenomena that are beyond the scope of this book.

On a different application, icebreakers present different propeller torque curves, as illustrated in Fig. 6-43, depending on the ice level found. At free running, the necessary torque at a given rotational speed is lower than with a given specific ice level.

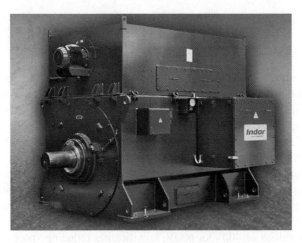

Fig. 6-41 *Ship propulsion example of an asynchronous machine (Source: Ingeteam. Reproduced with permission of Ingeteam)*

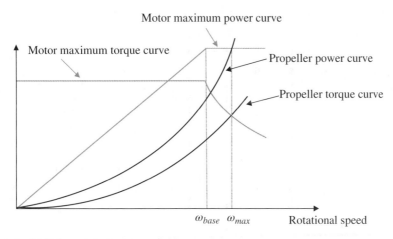

Fig. 6-42 *Maximum torque and power capability curves of the motor and the propeller (valid for both synchronous and asynchronous machines)*

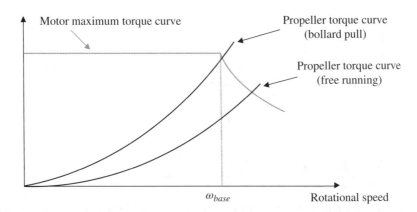

Fig. 6-43 *Maximum torque capability curves of the motor and propeller torque curves of an icebreaker*

6.4.3   Power electronic converter and transformer

As seen in previous subsections, the electric drive of a propeller typically incorporates a transformer at the input reducing the distribution voltage to the voltage of the converter itself. Then, a diode front-end converts the input AC voltages and currents to DC. Then, by means of the DC bus capacitors and the inverter, the machine is AC supplied again, at variable fundamental frequency and amplitude voltages. Fig. 6-44 shows an example of an electric drive configuration, with a six-pulse diode rectifier at the input and DC chopper protection, for sudden regenerative moments.

This simple configuration is not common in ship propulsion, mainly because the line current quality achieved by the six-pulse diode rectifier in general is poor for the "weak" distribution grid of a ship. In fact, it is more common to use this converter arrangement configuration without a transformer, in ships of low power, as depicted in Fig. 6-45. In these cases, a transformer-less configuration is suitable for employing low-voltage electric distributions of 400V or 690V, significantly reducing space and volume needs. In fact, there is also a newer tendency of some manufacturers to employ transformer-less electric drives, as shown in Fig. 6-45, but even at medium voltage levels for bigger ships. In this particular case, it is necessary to incorporate passive filters of currents harmonics, as is described in Section 6.5.5.2, in order to not deteriorate the

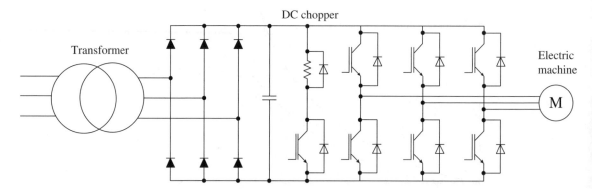

Fig. 6-44   *Electric drive of a propeller with a transformer, a six-pulse diode rectifier and a two-level inverter (although possibly not often used in ship propulsion)*

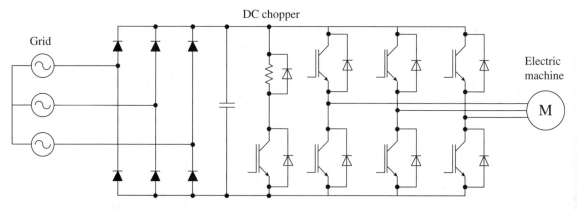

Fig. 6-45   *Electric drive of a transformer-less propeller, with a six-pulse diode rectifier and a two-level inverter suitable for low voltage distributions*

voltage quality of the distribution network. In this case also, it is necessary to install current smoothing inductances at the AC terminals to improve the voltage quality.

With regards to the DC braking chopper, its controlled switch can be activated by the following two main strategies. One would consist of activating the switch when the DC bus voltage reached a maximum pre-established value and then deactivated it, when the voltage was reduced until a minimum, also pre-established, value was reached. In this case, the switching frequency of the controlled switch does not result constant. The other strategy consists of operating the controlled switch at constant frequency when activated. These two philosophies are often used.

Continuing analyzing the quality of the voltage of the distribution network, as graphically illustrated in Fig. 6-46, the ideal line currents of a six-pulse rectifier are quite close to square shape. In general, currents near to square shape produce an important source of harmonics. These harmonics are propagated through the distribution grid and, therefore, they also produce voltage harmonics, which negatively affect the remaining ship's load. Note that the main loads (in terms of power consumption) in most of the ships are the propellers themselves.

In order to reduce the polluting effect caused by the high current harmonics consumed by the six-pulse rectifier, it is possible to use a 12- or 24-pulse rectifier, as illustrated in Fig. 6-46. In the 12-pulse case, by using a transformer with two secondaries with delta–delta and delta–wye connections, a 30° phase shift is obtained in line currents consumed by each rectifier, resulting in a better quality of total consumed current at the primary side of the transformer, therefore obtaining reduced grid distortion compared to the 6-pulse rectifier case. In the 24-pulse case, the line current quality is further improved by eliminating several current harmonics, by duplicating two 12-pulse rectifier configuration. Thus, splitting the necessary input currents into rectifier currents that have been appropriately phase shifted, the line current harmonic can be significantly reduced. However, this is obviously achieved by increasing the complexity of the converter. Note that the current harmonics consumed at the distribution grid will cause harmonic voltage distortion also at the grid seen by other loads. This undesired fact can cause malfunctioning of other electronic equipment of the ship or additional losses at the system, as is further discussed in subsequent sections.

Hence, depending on the rectifier configuration used, the line current quality consumed by the propellers' drives can be improved and, therefore, the total voltage harmonics of the distribution grid can also be improved. Thus, for instance with the example of ship electric system shown in Fig. 6-47, where four electric drives for four propellers are utilized and assuming that the propulsion is the major load of the system, by

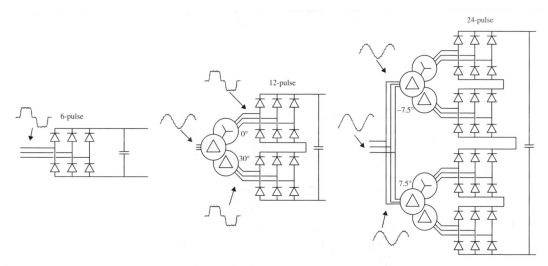

Fig. 6-46 *6-pulse, 12-pulse and 24-pulse rectifier configurations, with schematic representation of their corresponding ideal line currents [12]*

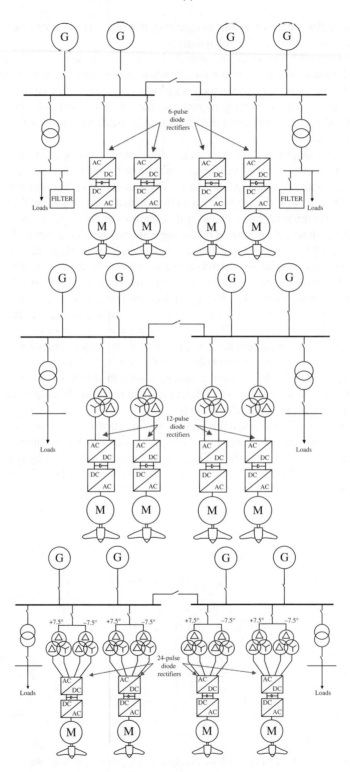

Fig. 6-47   *Simplified electric generation and distribution with electric drives with 6, 12, and 24 pulses [12]*

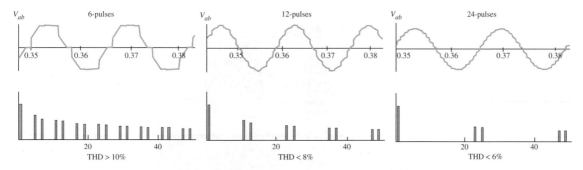

Fig. 6-48   *Simulated grid voltages (V$_{ab}$) and spectrums with 6-, 12-, and 24-pulse rectifier configurations [12]*

selecting 6-, 12- or 24-pulse rectifier configurations, the obtained voltage waveforms and approximated total harmonic distortions (THDs) of grid voltages are shown in Fig. 6-48.

It can be noticed that for 6-pulse rectifier configurations the grid voltage is distorted by the current harmonics consumed by the electric drive, it being necessary to often include additional passive filters to mitigate this undesired phenomenon. Deeper analysis about the filter is carried out in Section 6.5.5.2.

By selecting 12- or 24-pulse configurations, it is possible to further improve the grid voltage quality. The chosen configuration is normally selected in a compromise between the performance of the system, investment costs, and other factors, such as the requirements of the classification societies.

However, it can be said that the transformers basically fulfill two main functions: one is to match the converter input voltages and the distribution grid voltage, while the other is to provide the necessary number of pulses to the power converter's rectifier. Nowadays, it can be affirmed that is more common to find transformers with three windings (to obtain 12-pulses) in ship propulsion.

It must be highlighted that the diode rectifier arrangement configurations for 12 and 24 pulses in Fig. 6-46 can be either, as depicted, in series connection or in parallel. Thus, for instance in low voltage, it is much more common to use parallel connections of rectifiers, owing to the standard rated voltage of the diodes available in the market. For example, for a 690V distribution network, to build a 6-pulse diode rectifier requires diodes of approximately 2200V. Then, if a 12-rectifier is needed with parallel connection, the required diodes are of the same voltage. However, if a series connection of rectifiers is needed, the diodes required are in theory 1100V and of double current, which are not so easy to find in the market. Consequently, as mentioned before, at low voltage it is much more common to employ a parallel connection of rectifiers to obtain 12- or 24-pulse arrangements.

However, with medium voltage the situation is just the opposite. It is not easy to find 6-pulse diode rectifiers of greater output voltage than 3000V in the market. The principal reason is that the required diode is of such a large voltage that its availability in the market is reduced. Therefore, typically the MV diode rectifiers of 12 and 24 pulses are more often connected in series. Nevertheless, it is also possible to find arrangements for 24 pulses, with four 6-pulses in series or with two 12-pulses in parallel.

With regards to the inverter feeding the motor, as has been described before, it can be of low (690V) or medium voltage (typically 3.3kV) output. Fig. 6-49 shows a commercially available converter.

The commonly used inverter topology for motor supply is the classic two-level converter and the three-level neutral point clamped (NPC) converter, as depicted in Fig. 6-50. Principally, the two-level converter is used in low voltages of 400V and 690V. Then, the three-level converter is used at medium voltages of 3.3kV and 4.16kV.

It is also possible to use a different converter topology, to try to eliminate the necessity of the costly and bulky transformer, especially for small ships whose propulsive power is around 4MW or less. In these

Fig. 6-49   *Commercial converter for ship propulsion (Source: Ingeteam. Reproduced with permission of Ingeteam)*

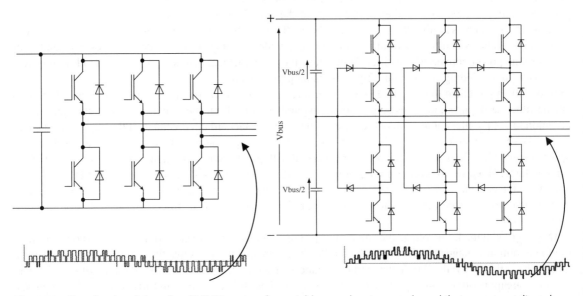

Fig. 6-50   *Two-level and three-level NPC inverters for variable speed motor supply and their corresponding phase voltages*

particular cases, there is no need of a medium-voltage generation and distribution voltage level, which means that the distribution grid can operate at low voltage (690V) or even at medium voltage, such as 3.3kV. This eliminates the necessity of the transformer, to match considerably high-generation and distribution voltage levels, such as 6.6kV, 11kV, or 13kV, with the electric drive converter's typical voltage levels of 690V or

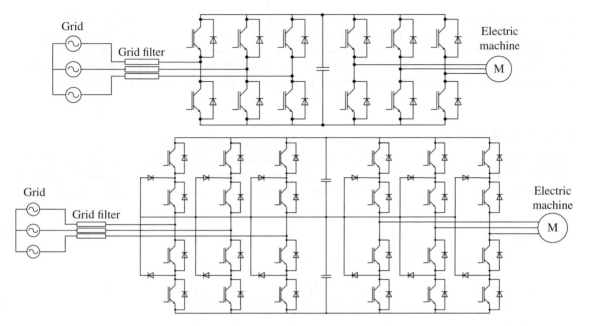

Fig. 6-51 *Two-level and three-level NPC converter-based transformer-less drive, with active front end*

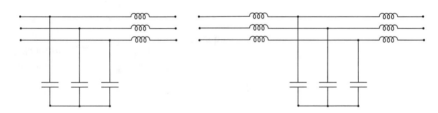

Fig. 6-52 *Ideal LC and LCL filters (damping resistances are typically needed to avoid resonance problems)*

3.3kV. However, the avoidance of the transformer would mean an unacceptable quality of line currents if a 6-pulse diode rectifier were used, as shown in Fig. 6-44 (without transformer, directly connected to the grid). Alternatively to the diode rectifier, an active front rectifier can be used, as depicted in Fig. 6-51 with its corresponding harmonic filter (LC or LCL typically). In this way, the quality of the line currents exchanged with the grid is much better, being reasonably easy to achieve line current THDs of around 3%. Thus, by eliminating the transformer and including a more sophisticated and costly IGBT-based rectifier, this solution is very competitive in the low range of powers.

On the other hand, as mentioned before, in order to achieve a good quality of line-exchanged currents with the grid, an LC or LCL grid side filter is typically used, ideally depicted in Fig. 6-52 (damping resistances need to be added to the inductive and capacitive elements).

Thus, to conclude about transformers, it can be said that the transformers are one of the bigger and heavier items of electric equipment on board a ship. There is a tendency to eliminate them. Nowadays, transformer-less concepts are becoming very common in ships. There are several options. One of them is to use a distribution of 690V, with active front ends (AFEs) as rectifiers in principal propulsions. It is very common at lower power, but it is sometimes used up to 20MW approximately. Another option also being used is a distribution of 3.3kV, without using transformers in the principal propulsion and with 6-pulse passive rectifiers. In order to improve the THD, passive filters are used, as mentioned earlier. Finally, for higher powers the voltage choice for the distribution is 6.6kV or higher. In almost all these cases, transformers are used. Some manufacturers have announced converters of 6.6kV being able to connect to the distribution grid without a transformer.

6.4.4   Control strategy

The propeller typically is either controlled in speed (rotational) reference or in power reference. Typically the references or set points are set with levers. Fig. 6-53 shows a schematic control block diagram of a propeller. When the ship incorporates more than one propeller, each one has its own control. It will be seen that the motor

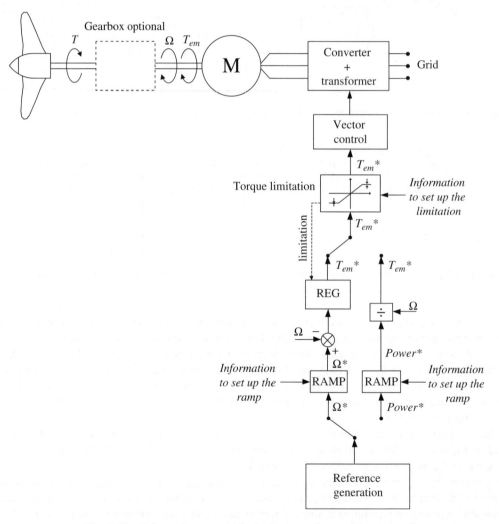

Fig. 6-53   *Schematic control block diagram of propulsion control*

moving the propeller is torque vector controlled, according to the vector control strategies presented in Chapter 2 and 3 for asynchronous machines or synchronous machines. And there is no real difference between what is discussed in those chapters and ship torque control. Instead of vector control strategy, alternatively, the manufacturer ABB, for instance, often uses direct torque control (DTC) strategies for the drive [2]. In both vector control and DTC cases, current measurements are needed as well as for the position/speed of the rotor shaft. Some manufacturers in some of their products often employ an encoder-less strategy. Note that, in this application, a not especially accurate torque control is required at low speeds, when the sensorless control is known to be more problematic. Therefore, it is possible to say that in principle the sensorless control in ship propulsion is a suitable scenario. Then, the torque reference for the motor can be created by a rotational speed control or directly from a power reference. The torque reference must be limited in order to avoid many restrictions—such as: mechanical limits or those due to cooling systems—but, clearly, the essential limitation arises from prevention of a blackout. This means it is crucial to avoid an overload demand to the generation plant. In general, there is an automation system (PMS) which supervises the load being demanded from the generation system and establishes the limits.

Thus, the torque limit is set dynamically, according to the number of generators which are connected at a specific moment, that is, according to the available power from the engines (PMS). If the ship is running in rotational speed mode, the speed regulator must know the torque limit every moment, in order to avoid control problems (in order to avoid overloads, blackouts, and power plant instability). In power control mode, the power reference is converted into torque reference by simply dividing the reference by the speed. In both control modes, the reference changes are limited by ramps. As with the torque limit, the ramps can be different depending on the available power or connected generators. Apart from this, normally three types of ramps can be distinguished for three different situations: normal ramps, emergency ramps, and crash stop ramps. Depending on the available power, these three ramps present different values. It is also possible to find ramps of different values, one for the low power demand and another for the high power demand (low speed and high speeds of the power-speed load curve). The power or speed references are generated by a higher-level control level [12].

The dynamic response of the controlled speed/power variations in a ship can last up to several minutes. This means that the ship can require several minutes to reach its steaming speed from zero speed. However, it depends on the characteristics of the ship itself (dimensions, purpose, etc.). On the other hand, we can also find restrictive DP requirements, which demand time responses of a few seconds (10 seconds or less) to reach a rated speed from zero speed.

Thus, Fig. 6-54(a) shows the acceleration, constant power running, and deceleration process of a ship in control power mode. It can be seen that, since power is the controlled variable, the ship advances at constant power, but the torque and rotational speed are modified accordingly ($Power = T_{em} \cdot \Omega$), as the ship finds different external opposite forces, owing to sea conditions or changes of direction. In these conditions, it is ensured that the prime movers work at constant power regimen. Alternatively, Fig. 6-54(b) illustrates the acceleration and deceleration process at controlled rotational speed mode. In this case, the rotational speed is maintained constant, while the torque and power change accordingly depending on the encountered external forces during the trajectory of the ship.

On the other hand, the reference generations of power and rotational speed can be made according to the schematic block diagram shown in Fig. 6-55. It can be seen that for the power reference nothing is done, but for the rotational speed reference it can be created by a ship's linear speed control. Thus, it can be seen that from the general ship's speed reference the regulator creates the thrust force reference. From the propeller force-speed characteristic (i.e. the thrust characteristic) the rotational speed of the propeller is created. Note that this characteristic has been very ideally described in previous sections, as a quadratic relation, but in reality it is influenced by many complex factors and external and sea conditions. Consequently, there is no guarantee of fulfilling the thrust commands. Conventionally, the resulting thrust characteristic is determined by stationary

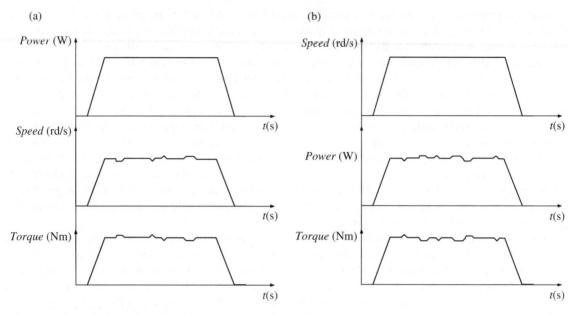

Fig. 6-54   *Acceleration and deceleration at normal conditions in (a) control power mode and (b) control speed mode*

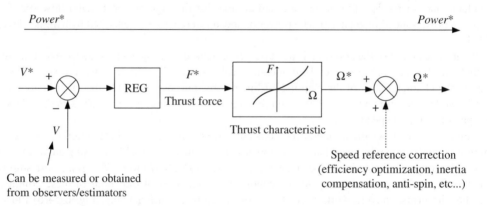

Fig. 6-55   *Reference generation strategy according to ship's speed control or direct power control (connected with block diagram of Fig. 6-53)*

propeller force-speed relations based on information about thruster characteristics found from model tests and bollard pull tests provided by the thruster manufacturer. These relations may later be modified during sea trials. However, as mentioned before, they are strongly influenced by factors such as local water flow around the propeller blades, waves and water current, etc. [10]. Thus, in order to correct these uncertainties and improve the dynamic control performance, sometimes additional speed reference correcting signals are added, which try to take into account different physical phenomena based on complex estimators or observers. It must be also highlighted that the ship speed regulator must be coordinately tuned and must share information with the ramps

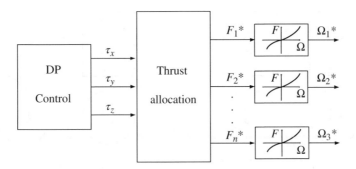

Fig. 6-56 *Basic control structure of DP control [10]*

and torque limitations, which can be modified according to the available generators' power. On the other hand, when there is more than one propeller in the ship, the thrust force reference, F^*, generated by the output of the regulator is split for each propeller proportionally. Alternatively to this control schema, several different control philosophies can be found in ship propulsion.

Finally, if the ship incorporates DP, it will typically incorporate several propellers at different locations on the hull, which allows operators to control the position of the ship with precision. Hence, Fig. 6-56 shows the general control scheme, in which the dynamic position control generates three-dimensional force references at the x, y, and z axis of the ship. Then, these force references (τ_x, τ_y, τ_z) are translated into thrust forces for each of the ship's propellers (F_1, F_2, ..., F_n), creating finally the rotational speed references for each propeller control according to block diagram of Fig. 6-56.

6.5 Power generation and distribution system

6.5.1 Power generation and distribution configurations

As noted before, the propulsion system of a ship is normally required to be fault tolerant, for instance to faults like short circuits. For that purpose, in previous sections it has been seen that redundancy is applied to the propulsion system. However, this redundancy in general is accompanied with redundancy at the power generation and distribution systems, splitting the generation and distribution into two, three, or even four units. An example of a ship's electric power system configuration is depicted in Fig. 6-57. It is seen that the power generation is split into four generation units that can fulfill redundancy requirements. In addition, it permits different numbers of generators, depending on the load requirements of the moment. On the other hand, the power distribution system is divided into two sections, enabling system operation at 50% of propulsion power, in case one fails. The availability of this system can be further increased if motors with double stator are utilized with their associated power electronic converters. The ship service power is taken from the electrical bus, also applying redundancy and creating two independent, but connected, service power grids. Note that the voltage is reduced to a lower level by using two transformers in order to achieve redundancy. Bus ties connect different electrical buses and they are opened when part of the system fails. In this specific example, all of the generated power goes to the ship service and propulsion; however, as is shown in the following sections, many ships need a significant amount of power to perform several functions (drills, dredges, etc.).

From another aspect, Fig. 6-58 shows an equivalent electric power system configuration with physical separation between the two main parts. In a different way, Fig. 6-59 show a power system configuration in ring bus

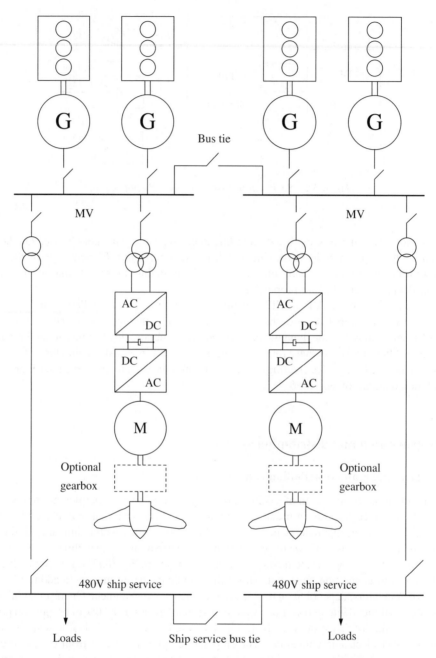

Fig. 6-57   *Electric power system of a ship (single line) with ABS-R2 redundancy (single-phase diagram)*

typically employed in navy ships. Four generators supply the load in parallel, in such a way that if one or more generator fails with the help of bus ties, the rest of the generators can still supply the load. The physical location of the generators is at different fire zones of the ship, also providing redundant power generation and distribution with physical separation.

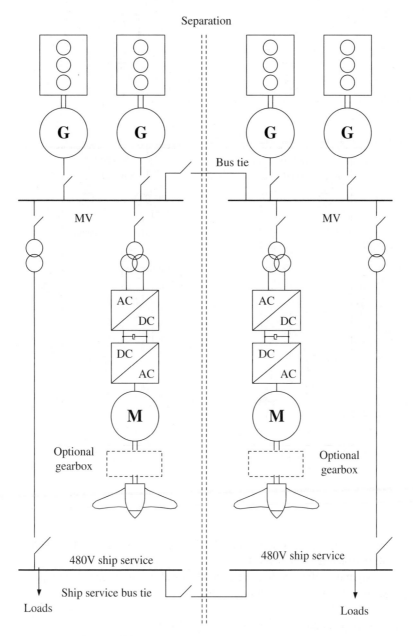

Fig. 6-58   *Electric power system of a ship (single line) with ABS-R2 redundancy and physical separation (single-phase diagram)*

Voltage generation and distribution is normally of three phases at 50 or 60Hz. Regarding the voltage levels employed at the distribution system, it normally depends on the required total amount of power. As the power increases, in general it is also necessary to increase the operating voltage, owing to limitations to work with too high currents such as: thermal and mechanical stresses and losses at bus bars and generators, switchgear

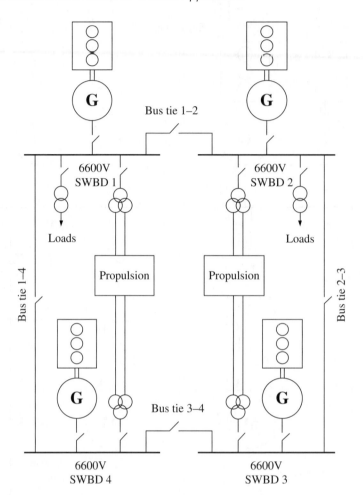

Fig. 6-59   *Electric power system of a ring bus in navy ships (single-phase diagram)*

capacities, and so on. Consequently, at higher installed power, the chosen working voltage is also higher in order to operate with reasonable current levels.

Attending to IEC low- and medium-voltage standard levels, the following indicative choices are commonly employed for the AC generation and distribution:

- 690V: For ships with generation capacity smaller than 4–5MW;
- 6.6kV: For ships with generation capacity, between 4–5 and 20MW;
- 11kV: For ships with generation capacity, between 20 and 70MW; and
- 13kV: For largest ship with generation capacity around 100MW.

For ship service lower voltage levels are used: 230/400V, although 440V distribution is also commonly used in ships.

It must be highlighted that not only the power level of the ship determines the voltage level. The fault or short-circuit current of the power electric system also influences the voltage choice. Thus, in particular ships where most of the loads are converter loads (VSDs for instance), which do not contribute to short-circuit current, the voltage choice can be different from the above-mentioned recommendations.

On the other hand, in ships where the ANSI standard applies, a wider range of voltages may be utilized: 120V, 208V, 230V, 240V, 380V, 450V, 480V, 600V, 690V, 2400V, 3300V, 4160V, 6600V, 11000V, and 13800V.

Finally, the power generation and distribution system is also composed of power cables and protective elements (fuses, circuit breakers, etc.), which are briefly described in the following sections.

6.5.2 Electric power generation

6.5.2.1 Prime mover

As mentioned before, the electric energy generation of the ship is typically made by prime movers such as diesel engines or gas turbines, which directly move synchronous generators. In general, the rotational speed of both generator and prime mover is controlled to a constant value, actuating on the prime mover. In this way, the generator is able to provide a constant frequency grid voltage which supplies the different loads of the ship, for instance the propulsion and the ship service loads. In addition, for the correct operation of these loads of the ship, it is necessary to maintain not only the frequency of the grid but also the amplitude. In general, this is achieved by actuating on the synchronous generator, more specifically on its excitation circuit, as will be explained later. Although it is possible to find different prime mover possibilities in ships and vessels, in this section only the diesel engine based generation is described.

The working principle of diesel engines is based on the transformation of chemical energy taken from fuels into mechanical energy at the output shaft. The energy conversion process is carried out in two steps. First, by means of combustion reactions of the fuel, chemical energy is converted into thermal energy. Second, this thermal energy is converted into rotational mechanical energy.

The theoretical cycle of a diesel engine consists of air inlet, compression, combustion and expansion, and finally exhaust. The ideal pressure and volume cycle is depicted in Fig. 6-60.

The actual cycle inside the engine can be performed in four or two strokes. Two-stroke diesel engines in general can achieve higher power and are lighter compared to four-stroke engines. On the other hand, four-stroke engines can operate efficiently at higher speeds than two-stroke engines [13]. Each type of engine has

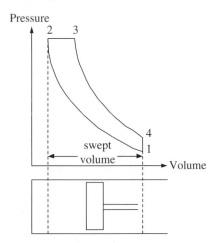

Fig. 6-60 *Theoretical cycle of a diesel engine*

found its suitable scenario in the context of ships. Fig. 6-61 shows an example of diesel engines offered by Man Diesel company.

A stroke is defined as the distance traveled by the piston between the top and the bottom. Fig. 6-62 shows a schematic of the four-stroke cycle of a diesel engine. It is composed of:

- (A) Compression stroke: Closing the inlet valve, the piston goes back up the cylinder compressing the fuel/air mixture. Just before the piston reaches the top of its compression stroke a spark plug emits a spark to combust the fuel/air mixture.

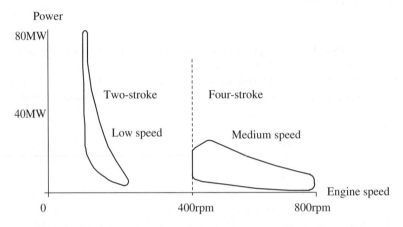

Fig. 6-61   *Available diesel engine characteristics of Man Diesel company [14]*

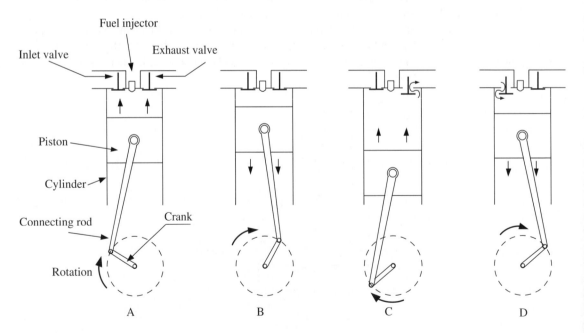

Fig. 6-62   *Illustration of a four-stroke cycle of a diesel engine [13]*

- (B) Power stroke: The piston is forced down by the pressure wave of the combustion of the fuel/air mixture. The power of the engine is generated from this stroke.
- (C) Exhaust stroke: The exhaust valve is opened and the piston travels back up expelling the exhaust gases through the exhaust valve. At the top of this stroke the exhaust valve is closed.
- (D) Intake stroke: The inlet valve is opened and the fuel/air mixture is drawn in as the piston travels down.

Generally, engines have two or more cylinders acting in concert with each other to generate the engine power.

Typical values of achieved efficiencies of diesel engines are shown in Fig. 6-63. In general, the efficiency depends on the load level. In order to optimize the efficiency under load variations, it is quite common to split the total generation into different generation units, managing the required power generation demanded by the sequential starting and stopping of units. This fact is also useful for a redundancy perspective, in order to obtain fault tolerance.

Another point to note is that the effective power, P_e, developed by the engine is proportional to the rotational speed and the mean effective pressure, p_e, achieved in the cylinder:

$$P_e = c \cdot p_e \cdot n \tag{6.44}$$

being c a constant that depends on the dimensions of the cylinder, the number of cylinders, and number of strokes (two or four). As noted in previous sections and described in subsequent sections, in general diesel generators are controlled to rotate at constant speed, producing constant frequency with synchronous generators. This is achieved (to put it very simply) by injecting more or less fuel into the diesel engine, thus obtaining different effective pressures and matching the power generation with the power demanded.

A simplified model of a diesel engine that describes the torque dynamic as a function of the diesel fuel rate input is shown in Fig. 6-64. The dynamics of the driver that converts the required fuel flow into actual fuel flow

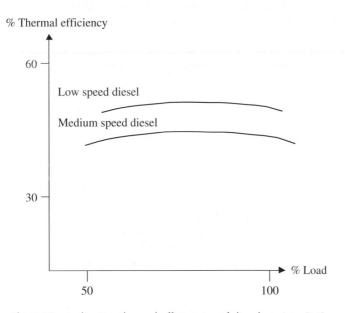

Fig. 6-63 *Indicative thermal efficiencies of diesel engines [14]*

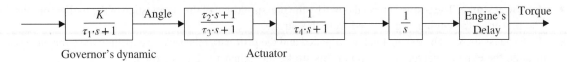

Fig. 6-64 illustration: blocks labeled $\dfrac{K}{\tau_1 \cdot s + 1}$ (Governor's dynamic) — Angle — $\dfrac{\tau_2 \cdot s + 1}{\tau_3 \cdot s + 1}$ — $\dfrac{1}{\tau_4 \cdot s + 1}$ (Actuator) — $\dfrac{1}{s}$ — Engine's Delay — Torque

Fig. 6-64 *Equivalent model of the diesel engine [15]*

ϕ is modeled as a first-order system with gain K and constant time τ_1. Then, the actuator and the engine are modeled by a succession of two poles, one zero, one integrator, and a delay (as shown in the block diagram). The output torque interacts with the torque of the synchronous generator at the shaft. Later sections show the speed control employed for the diesel engine.

The engine's time delay is modeled using Laplace transform as:

$$e^{-T_d \cdot s} \tag{6.45}$$

being the delay T_d, dependent of the engine's rotating speed and the number of cylinders (bigger rotating speed produces lower delay and bigger number of cylinders also produces lower delay).

This diesel model is considered accurate enough when the objective is to model the speed control dynamics interacting with the synchronous generator. Efficiency or temperature conditions cannot be modeled with this simplified approach.

6.5.2.2   Generator

As noted in the previous section, the prime movers, which are very often diesel engines in a ship, are coupled to electric generators. In general, the generator is a wound rotor synchronous generator whose stator three-phase winding creates the electric grid. The DC excitation of the rotor is supplied externally by a specifically dedicated converter. In general the total power generation is divided into several generator units.

The voltage and power levels of the generator are the same as the ones described for the distribution power grid. The frequency of the generated power can be either 50 or 60Hz. As seen in previous chapters, the relation between the rotational speed of the generator and the output frequency is provided by the pole pair number p, by the well-known expression:

$$f = \frac{p \cdot rotational\ speed}{60} \tag{6.46}$$

Thus, typical configurations of p, f, and speed are covered in Table 6-4; however, different options are also possible.

The simplest equivalent electric circuit of the generator per phase is composed of an induced electromotive force, E_f, the armature resistance, R_a, and the synchronous resistance, X_s. Fig. 6-65 illustrates the equivalent circuit. In most cases, R_a can be neglected and only E_f and X_s used.

Ideally, the induced electromotive force, E_f, depends on the speed and the DC current, I_f, of the magnetizing winding:

$$E_f = k \cdot I_f \cdot n \tag{6.47}$$

Table 6-4   *Typical generator configurations (typically it is also possible to find its equivalent with 50Hz)*

f (Hz)	p	Speed (rev/min)
60	3	1200
60	4	900
60	8	450
60	10	360

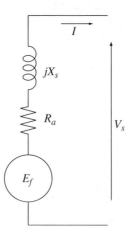

Fig. 6-65   *Single-phase equivalent circuit of the synchronous generator*

Modern generators are equipped with brushless excitation, in order to avoid maintenance due to brushes and slip-rings. The excitation is in the same shaft as the generator. The brushless excitation machine is an inverse synchronous machine with DC magnetization of the stator, a rotating three-phase winding, and a rotating diode rectifier.

Added to the excitation winding, at the rotor a damping winding is also included. This winding is to damp stator and rotor electromagnetic transients. The winding consists of copper bars located at the outer part of the rotor short-circuited with copper rings at both ends. A schematic representation is given in Fig. 6-66.

From the electric equivalent circuit of the synchronous generator represented in Fig. 6-65, the general phasor diagram is depicted in Fig. 6-67. The output stator voltage depends on the output stator current and the induced voltage E_f.

Thus, the single-phase active and reactive power exchange by the machine to the grid, neglecting the armature resistance, is given by the following equations:

$$P_s = \frac{E_f \cdot V_s}{X_s} \cdot \sin \delta \tag{6.48}$$

$$Q_s = V_s \cdot \frac{V_s - E_f \cdot \cos \delta}{X_s} \tag{6.49}$$

Hence, from these pair of equations it can be seen that E_f must be controlled, in order to maintain the grid voltage constant (which is V_s). In addition, the choice of E_f must be carefully done in order to share equal P_s and

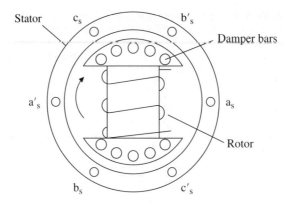

Fig. 6-66   *Schematic illustration of a cross-section of a synchronous generator*

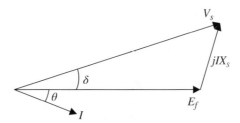

Fig. 6-67   *Synchronous generator phasor diagram neglecting armature resistance*

Q_s of all the generators working in parallel. To this end, special control strategies are described in the next subsection. From the modeling point of view, the exciter and generator model can each be described by a first-order system.

6.5.2.3   *Frequency and voltage control*

As noted before, the electrical energy generation of a ship requires a frequency and voltage amplitude control, as onshore electric grids do. Since the grid in general is created by the parallel connection of several generators, special attention must be paid to the frequency and voltage regulators employed, for balanced load sharing among generators. This section describes two commonly used solutions for these both controls. First of all, the frequency control is analyzed.

Since the electric energy generation of the ship, in general, is carried out by several generators connected in parallel, the frequency control strategy implemented must take into account the presence of more than one generator, which could be different from the simplest case where generation is only carried out by one generator. As mentioned before and described by (6.47), the frequency of the generated voltage by the synchronous generators is basically controlled by the rotational speed. Thus, imposing a desired speed, the corresponding frequency is obtained according to (6.46). Generally, the speed is controlled by actuating on the diesel engine, by means of its governor allowing injection of more or less fuel. An increase or decrease of the demanded load power will affect the frequency or speed equilibrium, requiring a change at the fuel

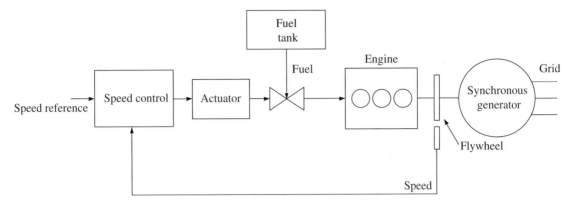

Fig. 6-68 *Schematic representation of the speed control of a diesel engine and its governor [2]*

entrance of the diesel engine. Fig. 6-68 illustrates a schematic block diagram of speed control of a diesel engine and its governor [2]. Note that each generator and diesel engine unit requires its own governor for speed control. Moreover, load sharing between generators must also be guaranteed in such a way that all the power demanded of the load is provided by the generators equally. In this way, all generators work at equal operating conditions, avoiding overloading or unequal stresses of generation units.

Different control philosophies exist for speed and frequency regulation of generators that operate in parallel and must share a load. Probably the most popular ones can be highlighted as "speed droop" and "isochronous load sharing". The first control mode does not require communication between generation units, while the second mode requires sending information between generation units, by high-speed bus communication or hardwired, in order to ensure load sharing. In ships, both speed droop and isochronous load sharing control modes can be found; however, for simplicity, this section only describes the first method.

Speed droop simply imposes that the steady-state frequency will drop proportionally with the load power. Without some form of droop, engine-speed control would be unstable in most cases. With speed droop, the reference frequency of the diesel engine is decreased with increasing delivered active power. The mathematical expression which describes droop is [16]:

$$\% \, Droop = \frac{no \, load \, frequency - full\text{-}load \, frequency}{full\text{-}load \, frequency} \times 100 \tag{6.50}$$

In general, a minimum of 2.5% droop is needed to ensure stability of the droop speed governor. Typical droop percent is between 3 and 5%. Once the droop is defined, the frequency is set as a line with slope equal to the droop, as the example shown in Fig. 6-69 where a 5% droop is selected with no load frequency of 52.5Hz and 1MW of rated power [17].

Thus, once the speed droop characteristic (frequency droop characteristic) is defined, the control block diagram of the speed droop control can be examined, as in Fig. 6-70. All the generation units must be controlled according to this control block philosophy. Note that, as commented before, there is no need of communication between different units. If all the generation units are equal and the droop is defined equally in all of them, the load sharing will also be equal. Thus, the combination of PI controllers with droop (which can be interpreted as a steady-state offset) achieves desired load sharing. A practical speed governor in general has a certain dead band or tolerance band, in order to avoid oscillations to any load changes.

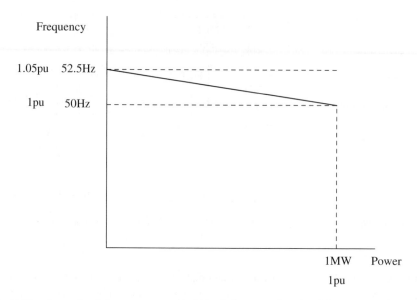

Fig. 6-69   *Example of droop curve with 5% of droop and no load frequency of 52.5Hz*

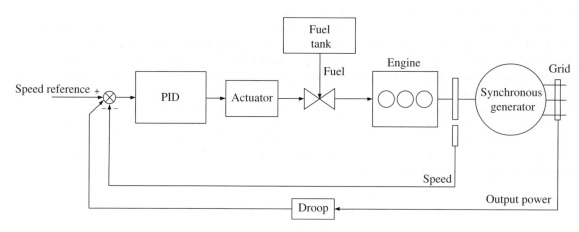

Fig. 6-70   *Speed droop control block diagram [2]*

To maintain the voltage amplitude regulation of the grid, it is possible to follow several control philosophies; however, here only one is described, which is similar to the control used for the frequency [2]. The amplitude voltage of the grid can be altered, as deduced from (6.47), by modifying the rotational speed and the DC field current of the synchronous generators. Since the speed is imposed by the frequency control, the voltage control is carried out by actuation of the field DC current. In this case again, as there may be several generators operating in parallel, voltage regulation can also be performed by droop control. This implies that, to the voltage sensing, a voltage variation is added which varies proportional to the load. This droop control philosophy results in a simple sharing of the reactive load of the grid (kVAR), among generators, which is necessary for the equal loading of generators in terms of current. Thus, for several generators connected in parallel, each one uses a voltage droop characteristic, as the example shown in Fig. 6-71.

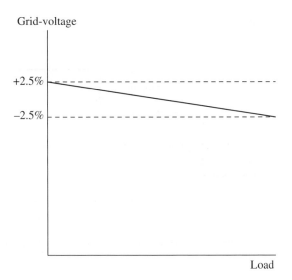

Fig. 6-71   *Voltage droop characteristic*

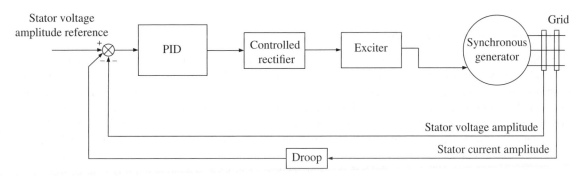

Fig. 6-72   *Voltage droop control block diagram [2]*

Then, the control block diagram implemented is shown in Fig. 6-72. The terminal voltage amplitude of the stator is sensed and the control tries to maintain the set point, while the droop signal compensation adds a voltage offset proportional to the load. The droop is in this case calculated from the current amplitude of the stator current, which is an image of the reactive power (kVAR). Thus, by changing the E_f value by means of the excitation circuit, as deduced from (6.49), the output reactive power can be controlled. Note that in modifying E_f the active power is also modified, as seen in (6.48). This means that both frequency and voltage loops are coupled, requiring a careful tuning of both control loops. In the context of an onshore grid system control, where many more generators are used and of more rated power, special stabilization loops are typically added (power system stabilizers) in order to avoid stability problems.

There are several other possibilities for frequency and voltage control in ships. One of them, for instance, uses a master generator with PI regulators and controls the frequency and the voltage amplitude. Then, the rest of the generators (slaves) are controlled with active and reactive power references taken from the master generator. This set-up also ensures equal load sharing, but communication between units is also required.

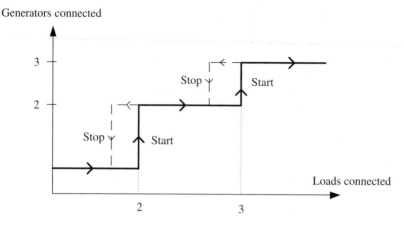

Fig. 6-73   *Number of engines connected as function of the load [18]*

Finally, it must be said that synchronization of new generators being added to the grid system must be carried out. However, this is beyond the scope of the chapter.

6.5.2.4   *Power management system*

The ship incorporates a power management system (PMS) which starts and stops engines according to the current load of the ship at any given moment. Thus, depending on the traveling conditions or simply the load conditions of the ship, it is more efficient to fit the number of diesel engine units to the actual demanding load. Fig. 6-73 illustrates a typical connection and disconnection of engine strategy. When the system detects that the load is above certain predefined limits, new generator units are enabled. In order to avoid instabilities in the system, hysteresis bands are employed for engine enabling and disabling. In general, the conditions for connection and disconnection of different generation units must be fulfilled for a pre-specified period, in order to avoid transient instabilities as well. Note that this PMS is useful for making engines operate at their more efficient ranges and allows them to not utilize part of the electric equipment when it is not required, thus reducing the possibility of system failure.

In addition, note also that generator units may need several dozens of seconds to start up, synchronize, and share load, which means that the PMS must manage this time to avoid dangerous overloads of the power system causing blackout.

The PMS also limits the power to heavy loads when there is no power available.

Thus, the typical main tasks of a PMS are [18]:

- automatic load dependent enabling and disabling of generation units;
- manual enabling and disabling of generation units;
- fault dependent enabling and disabling of generation units, in cases of under-frequency and/or under-voltage;
- start of generation units in case of blackout;
- determination and selection of the enabling and disabling sequence of generation units;
- start and supervision of the automatic synchronization of alternators and bus tie breakers;
- balanced and unbalanced load application and sharing between generation units;

- handling and limiting of heavy consumers;
- tripping of non-essential consumers; and
- bus tie and breaker monitoring and control.

6.5.3 Power cables

Ships with larger dimensions need long distribution power cables to transmit the generated energy to the loads. Depending on the ship's characteristics, the distance between the generator units and the most distant loads, for instance the propellers, can go from some few meters up to more than 100 meters. Especially for those cases where the length of the cables is more significant, careful attention must be paid to the electric characteristics of the distribution power cable.

Power cables are in general made up of some characteristic components, as illustrated in Fig. 6-74 [19]. The conductor typically is made of copper or aluminum and is responsible for carrying the electricity. The conductor may be solid or made up of various strands twisted together. The strand can be concentric, compressed, compacted, segmental, or annular to achieve the desired properties of flexibility, diameter, and current density. Depending on the nature of the built cable and the necessities of the application, the cable can be made of one conductor (mono-phase cables), three (three-phase cables), four, and so on. The insulation is a dielectric material layer (with a high-voltage breakdown strength) whose purpose is to provide insulation between conductors' phases and ground. Typical insulators are PVC (polyvinyl chloride), PE (polyethylene), XLPE (cross-linked polyethylene), and EPR (ethylene propylene rubber). The shield is a metal coating that works by confining the dielectric field to the insulation. Without proper shielding, the electrical stress can cause deterioration of the insulation and danger of electric shock. The shield can be made of multiple layers. The armor or sheath is a layer of heavy-duty material employed to protect the components of the cable from the external environment.

In order to derive the electrical characteristics of a power cable, it is possible to find in specialized literature several modeling possibilities, for instance pi circuits, pi circuits in series, Bergeron's model, J. Marti models, and Idempotent models. These models were originally developed to describe the electric behavior of power lines and power cables, with a length of several kilometers. However, the distribution power cable of a ship is not in the range of kilometers in length, but it is in the range of meters in length. Owing to this, the electric

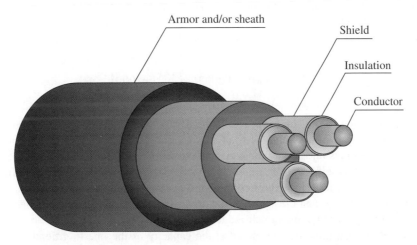

Fig. 6-74 *General representation of a three-phase power cable*

model of a ship's power cable is simplified typically to an equivalent series resistance, a self-inductance, and a capacitance to ground. All these elements are "parasitic elements" which would be better to be reduced; however, they are always present.

Thus, first of all, the DC resistance of a phase conductor is given by:

$$R_{DC} = \frac{\rho}{S}(\Omega/m) \tag{6.51}$$

with S being the effective area of the conductor and ρ the resistivity of the conductor at $20°$ of temperature (ρ_{cu} = $1.724 \times 10^{-8}\,\Omega m$ and $\rho_{al} = 2.830 \times 10^{-8}\,\Omega m$). Note that the resistance is given in function of the distance in m. The effect of the temperature in the resistivity of the conductor also affects the resistance of the conductor. The electrical resistivity of the materials in function of the temperature, in general, is linearly approximated as:

$$\rho_T = \rho_{20°}[1 + \alpha(T - 20°)](\Omega \cdot m) \tag{6.52}$$

with α the temperature coefficient of resistivity ($\alpha = 0.0039\ per\ °C$ for both copper and aluminum) and T the temperature of the test. However, because of the skin effect, a conductor offers a greater resistance to the flow of alternating current than it does to direct current. Thus, at higher frequencies most of the current travels along the skin of the conductor. Consequently, the AC resistance of the conductor is different from the DC resistance and can be calculated as:

$$R_{AC} = \frac{2\pi\rho\int_0^a J^2 r\,dr}{I^2}(\Omega/m) \tag{6.53}$$

with a being the conductor radius, I the current flowing through it, and J the current density in function of r and the skin effect. Therefore, the DC resistance depends on the area of the conductor, while the AC resistance requires that the current density be well defined starting from the skin of the conductor (depends on the frequency). Alternative to this expression, in a simplified approach, the magnitude of the resistance increase is often expressed as an "AC/DC ratio" and is provided tabulated based on previously calculated values.

However, owing to the effect of the electromagnetic field created by the circulating current through the conductor, a self-inductance is generated that can be described as [19]:

$$L = 0.00005 + 0.0002Ln\left(\frac{D}{a}\right)(mH/m) \tag{6.54}$$

with a again being the radius of the conductor and D the distance between conductor axes. It is important to highlight that this expression is only valid for triangular spatial disposition of the conductors. If the spatial disposition of the conductors is linear, the self-inductance equation will be different. There are more accurate alternative approaches to this self-inductance modeling method that consider the effect of different factors or phenomena such as the frequency. However, in ship applications it is not really important since the frequency is moderately short (50 or 60Hz) and the length of the cables is in the range of meters rather than kilometers, resulting in a not very significant value of self-inductance.

Finally, another effect that must be considered is the capacity of the line to ground, which is typically represented by a capacitor between the conductor and ground. This capacitor represents the capacitive

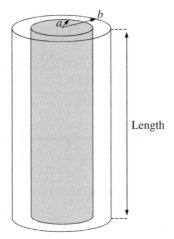

Length

Fig. 6-75   *Geometrical approximation of the physical form of the cable*

behavior between the conductor and the shield, which is normally connected to ground. Thus, the difference of voltage between the conductor and ground generates a capacitive effect that can be described by:

$$C = \frac{2\pi\varepsilon_0\varepsilon_r}{Ln(b/a)}(F/m) \tag{6.55}$$

being ε_r the relative permittivity of the insulating material and ε_0 permittivity or dielectric constant in the vacuum ($\varepsilon_0 = 8.854 \times 10^{-12}$ *F/m*). The parameter a again is the radius of the conductor and b the external radius of the insulation. This expression is valid only when the cylindrical shape geometry of conductor and insulation are both concentric, as illustrated in Fig. 6-75.

Hence, for an XLPE insulation:

$$C = \frac{\varepsilon_r}{18000\,Ln(b/a)}(\mu F/m) \tag{6.56}$$

being $\varepsilon_r = 2.5$ according to IEC 60287 for XLPE. Hence, the total equivalent R-L-C values of a given cable can be calculated by multiplying the length of the cable by (6.53), (6.54), and (6.56). The equivalent model of the three-phase power cable is represented in Fig. 6-76. It should be noted that if the cable is not very long the R, L, and C values of the cable will not be significant and will not produce significant losses or voltage droop.

On the other hand, when the voltage droop generated by the cable is going to be calculated analytically by hand (and not, for instance, by complex computer-based methods), typically the capacitance to ground is neglected assuming an equivalent R–L electric circuit for the cable model, as shown in Fig. 6-77.

The voltage droop, assuming loads of high power factors, is typically approximately calculated as a function of the consuming current [4]:

$$V_{droop} = I(R\cos\theta + X_L\sin\theta) \tag{6.57}$$

This expression is valid for $R << X_L$ for a wide range of power factors.

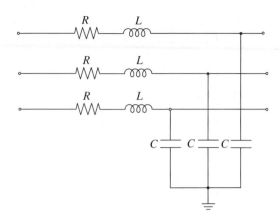

Fig. 6-76 *Equivalent simplified electric model of the three-phase power cable*

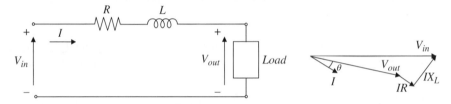

Fig. 6-77 *Voltage droop representation due to cable impedance (neglecting capacitive effect) [4]*

6.5.4 Fault protection

Short circuits or faults are events that can occur in electric systems and therefore also in electric generation and distribution systems of ships. These short circuits or faults (called only faults in this section for simplicity) in general can produce abnormal increase of currents at the system that can cause damage to the equipment or even injury to people. Consequently, protective elements are typically incorporated in the electric system of the ship, located at strategic places, in order to detect and fix the fault reducing the short-circuit time and therefore also potential damage. In addition, a very important aspect taken into account in fault clearance is the ability to obtain selectivity, which means that only the faulty branches of the electric system are disconnected during the fault, so the ship can operate with the non-affected branches of the electric system increasing their availability. Hence, it can be affirmed that the overall function of the protection system is to detect and isolate faults quickly and accurately in order to minimize the effects of disturbances.

Therefore, this section analyses in a generic way how the distribution system of the ship is protected against faults. It must be mentioned that short circuits occur when a live conductor touches another live conductor or ground in a grounded system. This situation can be caused by accident or by different factors, for instance insulation deterioration in cables (due to vibrations, corrosion, over-voltages, aging, etc.).

6.5.4.1 Grounded or ungrounded electric system

The electric system of a ship can be designed grounded or ungrounded. By a grounded system, it is understood that the neutrals of generators and wye windings of transformers are connected to ground. In an ungrounded system, the neutrals are not electrically connected to ground.

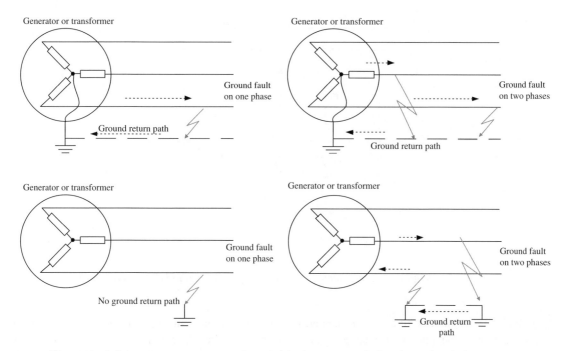

Fig. 6-78   *Schematic representation of ground faults, in grounded and ungrounded systems [4]*

One major difference of selecting a grounded or ungrounded system is the possibility to obtain a single-fault tolerant system with an ungrounded choice [4]. As illustrated in Fig. 6-78, when a single ground fault occurs in a grounded system, the short-circuit current appears, owing to the returning path found by the current through ground. However, in an ungrounded system, a single ground fault is not sufficient to create a short-circuit current. Two lines must be faulted to ground to produce the short-circuit current.

This is an important advantage in ships and makes ungrounded electric system configurations the preferred choice. However, owing to safety reasons, it is necessary to detect single ground faults in order to correct the problem caused the fault. Fig. 6-79 illustrates one possible detection schema, based on lamps connected to the low-voltage distribution grid. As can be seen, if for instance a grounded faults occurs in phase C, the lamp of phase C will turn dark.

6.5.4.2   Types of faults

As mentioned before, short circuits and faults can occur in a ship just as they occur onshore. There are only five types of possible faults [4]:

- L-G: One phase touches ground. Note that this is not a fault in ungrounded grids.
- L-L: Two phases touch together.
- L-L-G: Two phases touch together and ground.
- L-L-L: All three phases touch together.
- L-L-L-G: All three phases and ground touch together.

Note that the list above is provided in order of probability of occurrence. The most probable fault involves only one phase and ground, while the less probable fault involves all the phases and ground. Often, if the fault

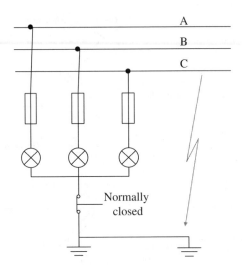

Fig. 6-79   *Single ground fault detection, in low-voltage ungrounded distribution systems*

lasts a significant amount of time, it can happen that the overcurrent and over-voltages produced by the fault cause a fault in other phases, producing a chain fault which falls into the extreme case of an L-L-L-G fault.

All the faults described can be produced by direct touching or by equivalent impedance, depending on the causes creating the fault. Faults can be produced in cables, in the terminals, or in connection boards of transformers, generators, motors, converters, or even service loads. It is also typical in power electronic converters, when a failure (damage, fault, etc.) occurs and, for instance, one of its semiconductors remains in short circuit, it produces a fault seen by the entire grid system. Some possible fault situations are simplified and illustrated in Fig. 6-80. Depending on where the fault occurs and under what conditions, the faulty overcurrent and over-voltage will be of different amplitude.

Figure 6-81 illustrates an exponentially decaying fault current. Typically, a short circuit generates an initial high current amplitude peak at the affected phase (or phases), which is damped exponentially until a stationary sinusoidal short-circuit current is reached (can be zero depending on the fault type, location, and system configuration). The initial peak current can go more than 10 times above the rated current, while the damping transient time can be of several cycles, depending on the electric system configuration and characteristics. Commonly, the high initial peak currents are employed for fault detection and actuation for protection. The protection system tries to quickly disconnect the faulty branches, not reaching the steady-state current of Fig. 6-81, protecting the system from the high currents as soon as possible. Also depending on the fault and location, it can happen that the initial peak current presents not so high an amplitude, which will bring more difficulties for detection (and therefore protection). Voltage anomaly detection (sensing of an uncommon high or low voltage) is also employed for fault detection and protection.

6.5.4.3   *Fault current sources*

Factors of the electric system which influence the fault current may be summarized as:

- Transient impedance of the generators: Bigger or lower values affect;
- Length of the cables, owing to their equivalent impedance; and
- Transformer's leakage and equivalent impedances.

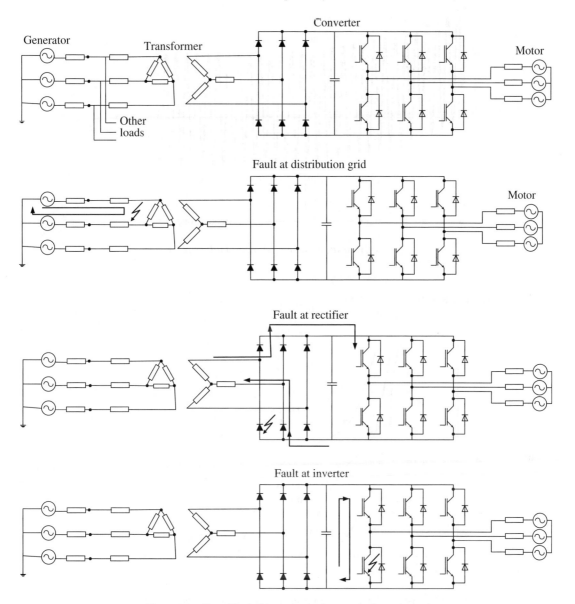

Fig. 6-80   *Simplified illustrations of some fault situations*

6.5.4.4   *Most important protective elements*

This section provides a brief overview of the most commonly used protective elements in the electric power systems of ships. Although it is possible to find a large number of different protections in the market, this section shows only the most representative ones, so the reader can have a general idea of their working principle and main characteristics.

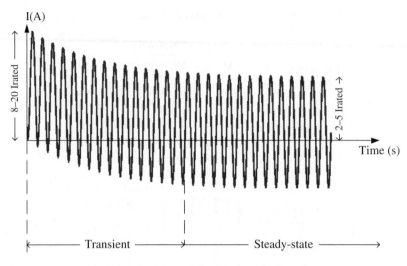

Fig. 6-81   *Exponentially decaying fault current*

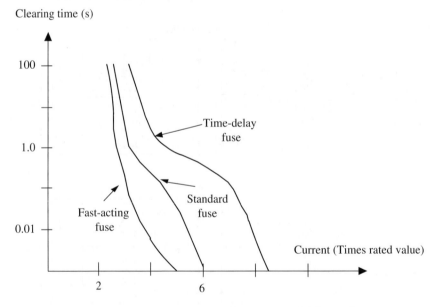

Fig. 6-82   *Fault current versus clearing time of different fuses*

Fuses   A fuse is a ceramic cartridge with a fuse link and a heat absorbing material. The fuse link is in general made of copper or silver, while its design depends mainly on its current, voltage, and time ratings. The heat-absorbing material employed to quench the arc is normally silica sand. Fuses are characterized by their current versus average clearing time curves. There are several types of fuses, oriented to different applications. Fig. 6-82 illustrates the characteristic fault current versus clearing time curves of different fuses [4]. It is seen that the fuse clears at fault currents greater than twice the rated current. Then, depending on the fuse type and fault current, the time delay clearing the fault is different. Thus, for instance, fast-acting fuses are useful for protecting sensitive circuits in high currents (e.g. power electronic circuits). Conversely, time-delay fuses can be appropriate for protecting elements which typically experience a high inrush or transient currents, and the fuse must not view their presence as a fault. These elements can be motors, transformers, and capacitor banks.

The fuse's fault current versus clearing time characteristic can be chosen from a variety of options of different manufacturers.

Circuit breaker   A circuit breaker is mainly composed of a contactor, a quenching chamber, and a tripping device. Circuit breakers normally incorporate a thermal-magnetic tripping device but can also incorporate an electronic one. In order to increase the rated voltage of the circuit breaker, multiple poles can be connected in series.

Hence, circuit breakers open and close the electric circuit automatically when an overcurrent is detected, or manually when required by the user. When it must be opened automatically due to a disturbance in the system, it presents similar characteristics to fault current versus clearing time as fuses, which means that higher overcurrents are interrupted in shorter time. The manual trip of the breaker enables one to open the circuit when needed, even when there is no fault. The circuit breaker is a more expensive option than a fuse for protecting a circuit. However, it enables one to open the circuit manually, and replacement is not needed when it clears a fault.

6.5.4.5   *Protection coordination*

Once the electric system is defined, or at least at a preliminary stage, it is necessary to proceed with a study which predicts how the electric system of the ship behaves under different types of faults. It is fundamental and crucial that the protection against faults is as quick as possible, to avoid any damage to the system or humans. Thus, a good system protection will be characterized by fast fault detection and will disconnect only the faulty branches, keeping the remaining safe parts of the electric system operating and avoiding a dangerous blackout in a ship. Note that a blackout can cause a total loss of control of the ship and, therefore, in dangerous situations can result in collisions or the ship running aground. Nowadays, the use of programmable protection devices is usual.

Hence, in general the protection system design requires the following tasks under faults:

- Fault current and voltage analysis, ideally performed under simulation-based studies in order to know the behavior of voltages and currents of the electric system, under different types and locations of faults. Alternatively to the simulation study, simplified current analysis based on simplified analytical equations is also typically used [4], [20].
- Definition of circuit breaker and fuse protection at strategic locations of the system, with their corresponding ampere and time delay interruption.
- Definition of the protection coordination of the system, to ensure good selectivity or disconnection of only faulty branches of the system. Note that for that purpose strategic locations of voltage and current sensors may be required as well.

An appropriate protection design scheme is a trade-off between a general desire to disconnect faulty equipment with minimum time delay and the need for disconnecting as few parts of the electric system as possible. This is achieved by protective device coordination studies. Fig. 6-83 illustrates a general procedure for designing the protection and coordination system, starting with a preliminary design of the electric system, which must include all the predefined generators, distribution electric system, propulsive drives, and loads, with their

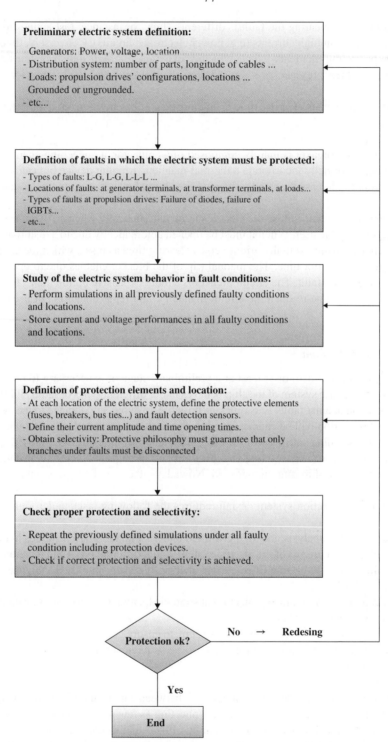

Fig. 6-83 *Schematic procedure for protection definition and coordination*

corresponding computer-based simulation models. A set of different simulations is carried out, which shows the behavior of the fault currents and voltages at different strategic places of the system, under different types of faults and locations of faults. The more situations simulated, the more guarantees there will be that a proper protection strategy has been designed.

Therefore, once the simulations are performed and the voltage and currents of the different tests are stored, a detailed study must be carried out selecting the protective fuses and breakers locations, with their corresponding strategic locations and ampere and time delay characteristics. For breakers and bus tie disconnection, voltage and currents sensors may also be needed at strategic places of the electric system. Finally, once the protective elements together with the corresponding coordination strategy are selected, the simulations at different faults must be repeated again, confirming good selectivity and system protection. If the results obtained are not satisfactory at any stage, redesign of some elements of the protective philosophy or even of the electric system itself may be necessary, always in order to provide good availability of the ship's operation in the event of a failure.

When we speak about proper protection device coordination, we understand that when faults occur, for instance far away from the generation or, near the load, a protection device near the fault (therefore far away from the generation) must interrupt the fault. For that purpose, in general the protection devices located near loads present a lower interrupting current than protection devices near generation. Such coordination philosophy avoids an upstream feeder from losing power before the downstream feeder. Thus, this is obtained by selecting fuses and circuit breakers of lower current versus trip time, as we move closer to loads in the electric system of the ship. A graphical illustration of different possible situations is show in Fig. 6-84 [12]. Figure 6-84 (a) shows a fault far away from the generation in a motor load and the coordinated fault design enables us to interrupt the fault by a protection device located close by. In a similar way, in Fig. 6-84(b), a fault in a transformer is shown which must be protected upstream and downstream, since the current fault can come from both sides of the transformer (through the motor load and through the generation part).

On the other hand, a different situation can be distinguished when the fault occurs in a bus and the ship presents different buses. Fig. 6-85 shows that the coordinated fault design enables one to isolate the fault by protection devices located close by (including bus ties), avoiding the need to interrupt half of the ship's electric system (and thus increasing availability of the ship).

Note that the protective element must consider also heavy natural current transients, such as the inrush current of transformers. As is known, when closing the breaker of a transformer's primary side connecting it to the

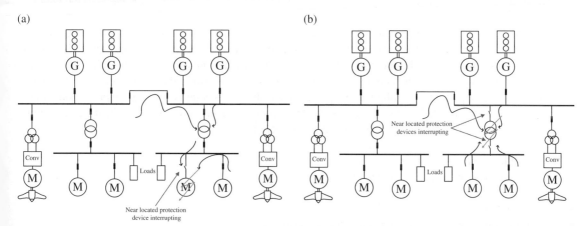

Fig. 6-84 *Faulty situation examples on a ship's electric system, showing that the coordinated fault design enables one to interrupt the faults by protection devices located close by: (a) fault far away from the generation in a motor load and (b) fault in a transformer which must be protected by upstream and downstream [12]*

(a)   (b)

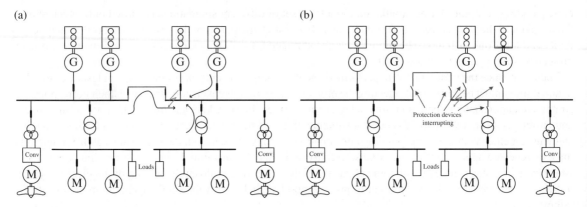

Fig. 6-85 *Faulty situation example on a bus, showing that the coordinated fault design enables one to isolate the fault by protection devices located close by, avoiding the need to interrupt half of the ship's electric system: (a) fault in a bus and (b) fault isolation by several sides of the electric system [12]*

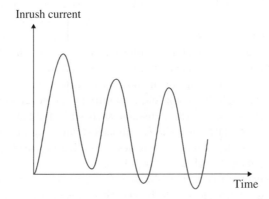

Fig. 6-86 *Magnetizing inrush current of the primary side of a transformer after connection to the grid*

electric grid, the primary side experiences a high current transient, called an inrush current. No matter how the secondary side is connected and even being at no load, a magnetizing inrush current, as depicted in Fig. 6-86, is always present. Consequently, the protective device design must take care of this fact in such a way that the inrush currents of transformers must not make the protection actuate, allowing the connection of transformers.

Finally, it is also important to highlight that, apart from the above protection coordination philosophy, many other protections are included along the ship's electric circuit elements, such as switchboards, generators, transformers, converters, and motors. Some of these protections are:

- under-voltage and over-voltage
- under-frequency and over-frequency
- overcurrents
- ground faults
- synchronizing check for feeder or generator connections
- thermal overloads.

Some of these protections may actuate as an alarm or tripping.

6.5.5 Harmonic distortion and voltage droop analyses

6.5.5.1 Voltage droop analysis

The voltage created by the generator and controlled by the automatic voltage regulator at the generator's terminal may be different from the voltage seen by the loads after distribution through the ship. Between the loads and the constant regulated AC bus voltage, there are typically several meters of AC cables and transformers. If nothing special is done (and in the absence of harmonic filters that, as is explained in a later section, may be used), the natural equivalent circuit of both cables and transformers are a combination of an inductance and a resistance. The values of the resistance and inductance are typically provided by the manufacturers, or in the case of the cables can also be calculated with analytic formulas that can be found in Section 6.5.3. A simplified representation is given in Fig. 6-87. These impedances produce a voltage droop along the electric system which is dependent on the amount of consuming load current. Thus, the bigger the load current, the lower the voltage seen by the loads. The resistive part in addition produces power losses that should always be minimized when looking to create an efficient design.

On the other hand, some loads, especially power electronic converters of thrusters or other loads, may not work properly at highest speed-power operating points if the input AC voltage is significantly reduced. Consequently, a voltage droop analysis is normally carried out by means of computer-based simulation, which studies the behavior of this fact under different operating conditions and tries to ensure the proper operation of the loads independent of the load consumption profiles.

In addition, voltage transients must also be studied in detail. During the energizing of transformers that are suddenly connected, or start-up of directly connected motors, a transitory high current is demanded for the generators. This fact provokes an unavoidable voltage amplitude transient in the electric system, which normally classification bodies require to be kept within certain limits. The automatic voltage regulator, which tries to maintain this voltage, is generally not able to totally eliminate this brief phenomenon; therefore, special attention must be paid to its control in order to not provoke amplification of these sudden voltage variations. In addition, capacitor banks are often also located at key points of the electric system in order to increase the voltage amplitude of some critical loads or simply to enable the increase of the equivalent power factor of all the loads.

6.5.5.2 Harmonic distortion analysis

The presence of loads that consume current harmonics in an electric system can produce voltage distortions in the electric system, which affect the performance of the entire system. Typically, loads with diode-based rectifiers are the most significant ones to cause this phenomenon. In onshore grids that in general tend to be

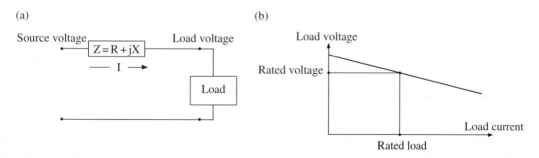

Fig. 6-87 *(a) Simplified equivalent impedance circuit between the constant controlled AC bus and the load, (b) Drooping voltage characteristic in function of the load current [2], [4]*

stronger than the electric grids of ships, the presence of these harmonics may not be so problematic. However, in the relatively weaker grids of ships, this fact must be especially taken into consideration. In general, the most important problems caused by the presence of harmonics are [2], [4]:

- Potential malfunctioning of electronic equipment, which is designed to operate in a non-polluted context. Power electronic equipment and low power loads can operate non-properly due to the presence of harmonics.
- Increase of losses of transformers, generators, and motors directly connected to the distorted grid voltage produce overheating and accelerate the deterioration of such equipment.
- The increase of power consumption in losses due to the presence of harmonics can generate deterioration of the insulation of all the equipment, increasing also the probability of short circuits and faults.

Hence, the presence of harmonics is determined by several characteristics of the electric system of the ship. Typically, attempts are made at the design stage to identify and avoid problems like this in order to meet the requirements of the classification bodies. The amount of harmonic distortion in the electric system is mainly influenced by the following elements [2], [4]:

- Converter topology of heavy loads: As seen in previous sections, the electric drives associated with a ship's thrusters are the most important loads in many ships. Owing to this, converter topologies employed in the electric drives of thrusters are the most important source of harmonics in a ship's electrical systems. Nowadays, voltage source converters and their associated rectifiers are the most commonly adopted configurations. It is necessary to find a compromise between a large amount of pulse rectification (reduced harmonic distortion) and the cost, weight, and space of its corresponding transformer. Alternatively, as seen in previous subsections, especially in not very high-power ships where the distribution grid does not reach high voltage levels, it is possible to use AFE rectifiers instead of diode rectifiers. In this way, the increase of investment accompanied to the AFE is justified by the avoidance of needing the transformer and by obtaining better harmonic distortion.
- Generator design: The injected current harmonics are propagated through the electric system, including the generators. The main impedance seen by the harmonics through the generator is the sub-transient imped-ance of the generator. Thus, the designed value for this impedance is normally a compromise of being able to mitigate harmonic effects against obtaining an efficient and compact generator design, as well as achiev-ing a reasonable short-circuit current level.
- Transformer design: The leakage inductance of the transformer directly actuates as a low-pass filter (LPF) for the harmonics present in the electric system. A big transformer's leakage inductance improves the har-monic distortion level. However, a compromise must be found in order to not cause excessive voltage droop at the electric system as well.
- AC cables: The equivalent impedance of the AC cables used to distribute the electric energy in large ships can be significant. In general, the resistive part is more significant than their inductive part. However, it could happen in large ships that the inductive part has a significant importance, resulting in a factor which also influences the harmonic distortion of the ship's electrics, in combination with the rest of the elements.

Thus, the electric system configured with the combination of all these elements can make the harmonic dis-tortion level different. In order to predict at the design stage the voltage harmonic content level of the system, simulation-based analysis is carried out with equivalent electric models of all the equipment of the electric system. In addition, classification societies also typically impose THD levels that must be fulfilled in order to avoid problematic situations that may arise maybe not at the beginning but after some time of operation due to deteriorations in equipment described earlier in this section.

Finally, when the design of the electric system reveals an unacceptable harmonic distortion level and it is not possible to reduce it because there is no margin to change elements or any other reason, the solution to reduce the harmonic distortion needs the inclusion of harmonic filters.

There are different types of harmonic filters. Probably the most well-known ones are the passive filters and the active filters. Passive harmonic filters are composed of combinations of *RLC* elements, which are tuned to damp fixed chosen frequencies. Alternatively, more sophisticated and complex active filters which are based on power electronic circuits can filter a group of harmonics present in the system. Passive filters, owing to their simplicity, are the most extended harmonic filters, and because of this they are briefly described next.

Passive filters are typically connected in parallel to the electric grid and tuned to one specific frequency which is to be eliminated from the system [2], [4]. Thus, as depicted in Fig. 6-88, the harmonic filter is typically located near the harmonic source (power electronic converter) in such a way that the harmonics are not propagated throughout the rest of the electric system. A common configuration of parallel passive filter is given in Fig. 6-88. As advanced before, it is composed of a series connection of a resistance, an inductance, and a capacitance.

The complex form of the impedance of the filter can be represented as:

$$Z = R + j\left(L\omega - \frac{1}{C\omega}\right) \tag{6.58}$$

The filter is tuned to the frequency that makes the inductive and capacitive impedances equal, which is:

$$\omega = \sqrt{\frac{1}{LC}} \tag{6.59}$$

Thus, the resistance is the impedance which limits the current when the circuit is in resonance at the tuned frequency. If there are several dominant polluting harmonics (for instance fifth and seventh), it is possible to use different filters in parallel, each one tuned to one harmonic's frequency. Fig. 6-89(a) illustrates each component's impedance variation (resistance, inductance, and capacitance), as well as the total impedance of the filter. It will be noticed that the impedance of the resistance does not vary with the frequency, while the resonance frequency occurs when both impedances of the capacitance and inductance are equal. Note also that in the *RLC* circuit there are two independent parameters, *L* and *C*, which give infinite combinations to obtain a resonance at any specific frequency.

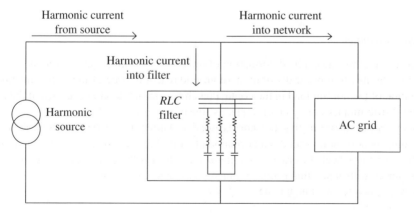

Fig. 6-88   *Schematic representation of the filtering action*

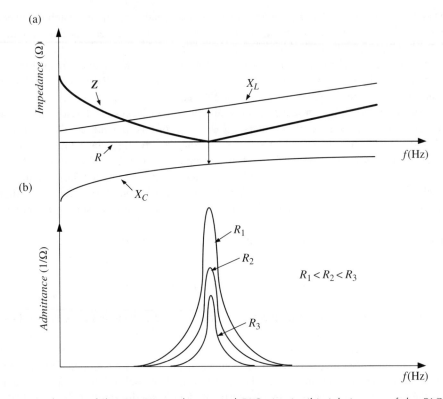

Fig. 6-89 *(a) Impedances of the R-L-C impedances and RLC circuit, (b) Admittance of the RLC circuit with different R values*

On the other hand, the quality of a filter (Q_f) determines the sharpness of tuning. A passive filter can be either a high or a low Q_f type. The former is sharply tuned to one of the lower harmonic frequencies (e.g. the fifth) and a typical value is between 30 and 60. The low Q_f filter, typically in the region of between 0.5 and 5, has low impedance over a wide range of frequencies. Thus, Q is defined as the ratio between the impedance of the inductance (or the capacitance) and the resistor at the resonance frequency:

$$Q_f = \frac{1}{R}\sqrt{\frac{L}{C}} \tag{6.60}$$

Therefore, it is possible to see in Fig. 6-89(b), representing admittance of the filter (inverse of the impedance), how the quality Q_f of a filter is bigger the smaller the resistor, which means also that it is more sharply tuned to the resonance frequency.

Consequently, it can be said that the filter design must find a compromise between the following interrelated aspects:

- power losses at fundamental frequency;
- reactive power generated at fundamental frequency;
- the quality factor (Q_f);
- resonance frequency of the filter; and
- the nominal voltage and current of the RLC components.

Thus, if more than one passive filter is connected to the electric system, aiming to mitigate several polluting harmonics, the overall filters effect must also be examined. In addition, adding filters to the system will modify the harmonic contents of the electric system, which often yields a complex iterative approach before the final design of the filters is achieved.

6.6   Computer-based simulation example

6.6.1   Ship under study

This section provides one example of ship propulsion modeling and simulation. As an illustrative example, the fictitious ship is composed of only one propeller as, for instance, the electrical layout shown in Fig. 6-90. The main characteristics of the ship itself, which are needed for propulsion modeling purposes, are as studied in previous subsections and are provided in Table 6-5.

It can be seen that the propeller is 3.35MW, rotating at a speed range of between 0 and 143rpm (propeller shaft). The rated ship's speed is 15.5knots.

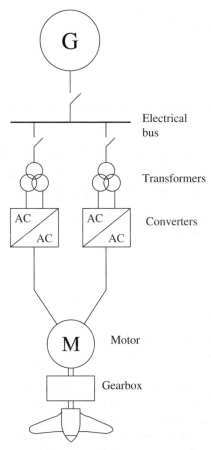

Two parallel converters supplying a
double-winding synchronous motor

Fig. 6-90   *Invented electric ship propulsion system, composed of only one propeller driven by a unique motor*

Table 6-5 *Main characteristics of the fictitious simulated ships for propulsion modeling*

V		
	Rated ship's speed	15.5knots
w		
	Wake fraction coefficient	0.322
t		
	Thrust deduction factor	0.045
	Rated power of propeller	3350kW
	Rated torque of propeller	191.94kNm
	(referred to the propeller shaft)	
	Rotational speed of propeller	0–143rpm
	(referred to the propeller shaft)	
	Gearbox ratio	7
	(between shaft of propeller and shaft of motor)	
D		3.4m
	Diameter of propeller	
M		
	(Mass of the ship)	3850000kg
Ma		
	(Added mass)	962500kg
I		
	Inertia	232.8kgm^2
Ct		6920

The characteristic η_o, k_f, and k_t coefficient curves in function of J of the simulated propellers are shown in Fig. 6-91. With a gearbox ratio of 7, it gives a rotational speed of the motor driving the propellers of about 0–1000 rpm. The total mass of the ship $(M + M_a)$ is supposed to be 4,812,500kg.

With this, the power and torque curves in function of the rotational speed of the propeller and in function of the ship speed are presented in Fig. 6-92. Note that, depending on the speed required for the ship, the rotational speed required is different as well as the torque produced by the propeller. These torque and speed values are obtained by substituting the values of this specific ship and propeller characteristics (Table 6-5) in the model block diagram of Fig. 6-28 supposing it to be on the open sea without external forces.

6.6.2   Electric propulsion drive

As introduced in Fig. 6-90, the electric propulsion drive is composed of two parallel converters supplying a unique double winding wound rotor synchronous machine. The supplying circuitry of the field is not shown in the figure. A more-detailed converter configuration is given in Fig. 6-93. Each stator of the synchronous machine is supplied by a wye–delta transformer, a 12-pulse diode rectifier, and the three-level NPC converter. The crowbar circuit for each capacitor of the DC bus is not shown, for simplicity. The middle point of the DC bus of the 3L-NPC converter is not connected to the middle point of the 12-pulse rectifier. This is commonly done in this way to ensure the voltage balancing of the DC bus capacitor by control and modulation rather than by the rectifier itself. This solution based on balancing the modulation voltage normally provides better performance than the very dynamically slow correction provided by the passive 12-pulse rectifier. The rated output AC voltage of the converter is calculated to provide around 3.3k$V_{LL \text{ (RMS)}}$. This choice of medium voltage

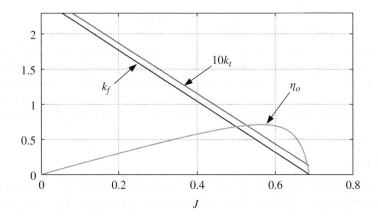

Fig. 6-91 η_o, k_f, and k_t coefficient curves in function of J of the simulated propellers

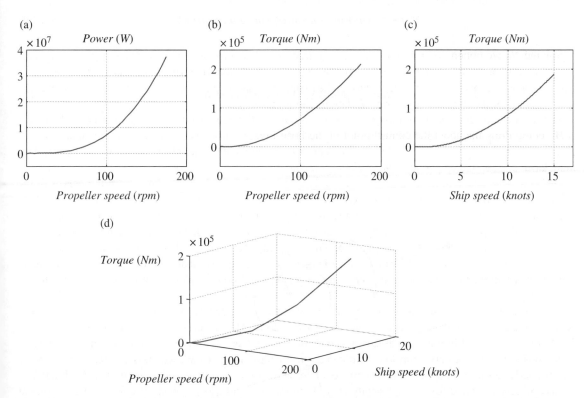

Fig. 6-92 Torque and power curves of the ship: (a) power in function of the propeller's rotational speed (at variable V), (b) torque in function of the propeller's rotational speed (at variable V), (c) torque in function of the ship's speed (at variable rotational speed), and (d) three-dimensional torque curve

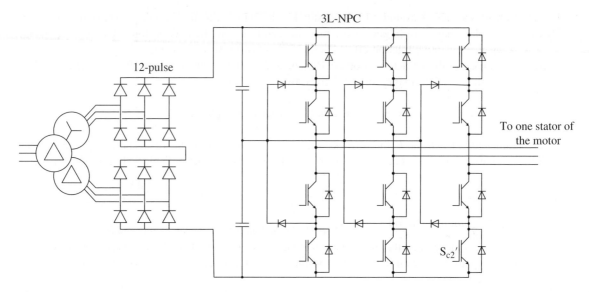

Fig. 6-93   *One converter configuration, for each stator of the synchronous machine, composed of a wye–delta transformer, a 12-pulse diode rectifier, and a three-level NPC converter*

is due to the high-power motor design which is 3.35MW. For this purpose, the 3L-NPC converter is very suitable, providing as it does a good quality of output AC voltage without requiring very high-voltage semiconductors and also applying lower voltage steps at the motor terminals than a two-level converter does, which is a great benefit for the motor's long-term insulation deterioration.

The main characteristics of the motor are shown in Table 6-6. Speed and torques are referred to the motor shaft, considering that the gearbox ration is 7, as described above. The electric dynamic equivalent model is given in Chapter 3. Both stator windings have been considered ideally equivalent, showing equal equivalent inductance and resistance parameters.

Once the parameters of the motor are known, it is possible to obtain the maximum capability curves of the machine by means of the procedure described in Chapter 3. These maximum achievable magnitude curves are shown in Fig. 6-94. The x axis in all of them represents the rotational speed in rd/sec of the motor shaft (the propeller's shaft speed must be divided by the gearbox ratio 7). The curves have been evaluated above the rated speed value, up to the flux-weakening region. Together with the torque and powers, the load torques and powers have also been drawn, which means the opposite torque and power that the motor must create at steady state at every speed to move the ship. These are the torque and power values of Fig. 6-92, referred to the motor shaft.

As the system has been dimensioned, the maximum torque demanded by the ship, which is around 30kNm, is given at higher speed than the rated torque, around 128rd/sec. This means that the rated torque of the motor 32kNm is not reached in principle at steady state, but could be useful as a reserve for dynamic or transient conditions. Note that at maximum load torque and power (around 3.84MW, which is greater than the rated power), and even at lower speeds, the machine operates at rated stator voltage and currents. The machine must be accordingly designed for this. At steady state, both stator d and q component of currents achieve equal values, owing to the idealized and symmetry model of the machine being considered. The difference between the d and q inductances of the motor makes it necessary to apply an MTPA (maximum torque per ampere) strategy, as described in Chapter 3, generating negative d stator current components in both stators, to achieve

Table 6-6 *Main characteristics of the wound rotor synchronous machines employed for the propellers*

Rated power	3.35MW
	(can reach 3.84MW at 127.5rd/sec)
Rated speed	1000rpm
	(can reach 127.5rd/sec)
Rated torque	32kNm
	(30kNm at 127.5rd/sec)
Rated frequency	50Hz
Rated voltage	3.3kV
Rated current	355.5A_{RMS} for each stator
Pole pairs	3
R_s	20.6mΩ
$L_{d1} = L_{d2}$	4.5mH
$L_{q1} = L_{q2}$	6.5mH
$L_{si1} = L_{si2}$	112.37μH
Field rotor flux created by excitation	4.8Wb$_{RMS}$

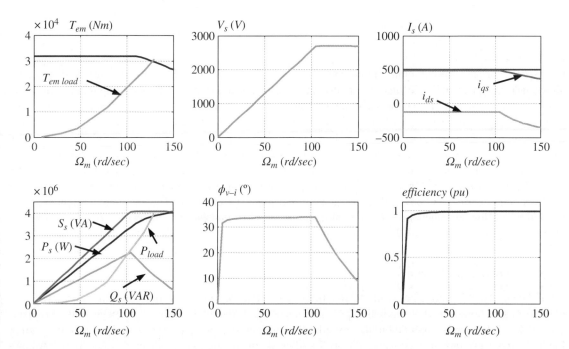

Fig. 6-94 *Maximum capability curves of the 3.35MW wound rotor synchronous propeller motor (only in motor mode operation), together with torque and power curves of the ship representing the load connected to the propeller*

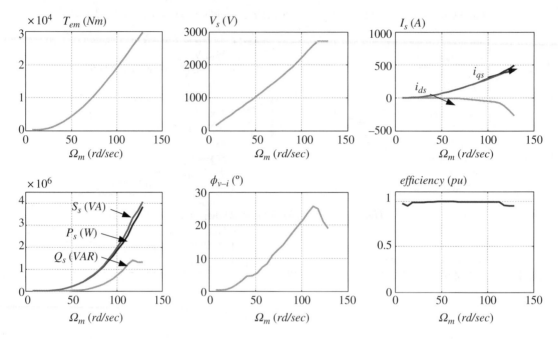

Fig. 6-95 *Operating point curves of the 3.35MW wound rotor synchronous propeller motor when driving the ship's load*

the required torque with the minimum current amplitude. Note that when the stator flux weakening begins, above the rated speed, the performance of both d and q stator currents of both stators vary their behavior, changing also the variation in the stator reactive power as well as the phase shift between the stator voltages and currents. Note that the efficiency of the machine is very high at higher torque values. During the operation of the system, it has been supposed that the field excitation has been maintained constant, obtaining a constant rotor flux value. It must be highlighted that the final real magnitudes in which the motor work are not the maximum capability curves but curves that depend on the torque-speed load curve. These curves can be easily calculated and are shown in Fig. 6-95. As can be seen, the low torques demanded at lower speeds make a lower consumption of current and power than the maximum available to the motor. The maximum current and voltage limits are simultaneously only reached at the point of maximum torque, 30kNm.

6.6.3   Simulation performance

Once the electric drive, together with the electric propulsive motor and converter configurations, have been defined, this section shows the simulation performance of the ship model, when it is propelled with this drive. The model of the ship is explained in the preceding sections and graphically represented in block diagram of Fig. 6-28. Now it is implemented using Matlab-Simulink simulation blocks, as depicted in Fig. 6-96. It follows exactly the same philosophy as described before. The entrance is the speed of the propeller's shaft n, while by calculating J, it achieves values of force F and torque T and finally the speed of the ship V. As external force (F_{ext}), a sinusoidal entrance has been simulated of 5kNm and 5mHz, while all the resistive forces have been merged into only one for simplicity, described by $C_t = 6920$ coefficient, as shown before in Table 6-5.

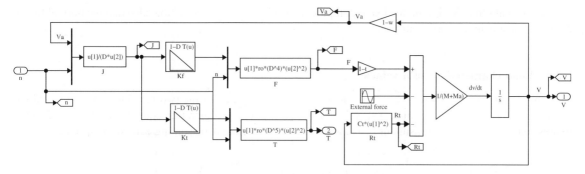

Fig. 6-96 *Block diagram of the ship model in Matlab-Simulink*

This simulation block model mainly provides the torque, T, to the mechanical shaft of the propeller, which will be contra-posed to the torque generated by the propulsive motor T_{em}, providing the speed by means of the well-known differential equation:

$$T_{em} - T = I\dot{\Omega} \tag{6.61}$$

Hence, the entire Matlab-Simulink block diagram, including the propulsive electric drive of the ship shown in Fig. 6-90, together with the model of the ship shown in Fig. 6-96, is represented in the overall block diagram Fig. 6-97, including all power electronics and control of the system as described in Section 6.4.1 and Chapter 3. The converters are fed by their corresponding 12-pulse rectifiers and delta–wye double secondary winding transformer, while the i_{ds} and i_{qs} current control of both stator machines are independently performed. The double stator wound field synchronous machine is an own model system, as presented in Chapter 3, while the excitation circuit is within the model of the machine itself. Note again that the block diagram representing the model of the ship receives as input the speed of the propeller's shaft, while it provides the torque, T, which must be contained in (6.61), which is implemented within the machine model itself.

Thus, the behavior of the most interesting magnitudes of the ship when accelerating from 2 to 4m/s of linear speed are served in Fig. 6-98. It will be noted that to achieve the increase of the ship's speed it is necessary that the rotational speed experiences an increase also as illustrated in Fig. 6-98(b). In that case, the J coefficient suffers a transitory modification, as seen in Fig. 6-98(c), while the torque and forces shown in Fig. 6-98(d) and (e) also experience an increase to address the speed increase requirement. Finally, the efficiency η_o depicted in Fig. 6-98(f) shows how when reaching the steady state again takes an equivalent value to the one that it had before the speed increase. It must be highlighted that the strong oscillatory behavior shown by all the magnitudes corresponds to the severe external force imposed (5kNm and 5mHz), trying to emulate severe sea conditions as ships advance on the open sea.

After this, in a different set of simulation graphics, the magnitudes closer to the electric drive are shown in Fig. 6-99 and Fig. 6-100. At time instant 1180 approximately of the simulation performed, and corresponding with Fig. 6-98, stator currents, rotational motor shaft's speed, the i_{ds} and i_{qs} currents, stator voltage references, electromagnetic torque created by the motor, and mechanical power also created by the motor are shown in Fig. 6-99. With regards to the currents, only one set of stator currents is shown. The second stator current set would be exactly equal in this idealized simulated case in fundamental components at steady state, while the only difference would be the ripple or harmonic behavior, which in this case, owing to the parameters employed, is significantly low. The i_{ds} current presents a negative value approximately equal to 6A, to achieve the MTPA behavior. Note that the shown operating point is quite far away from the rated operating point, providing reasonably slow current, voltage, and torque–power values. Regarding the grid current behavior, phase a currents at

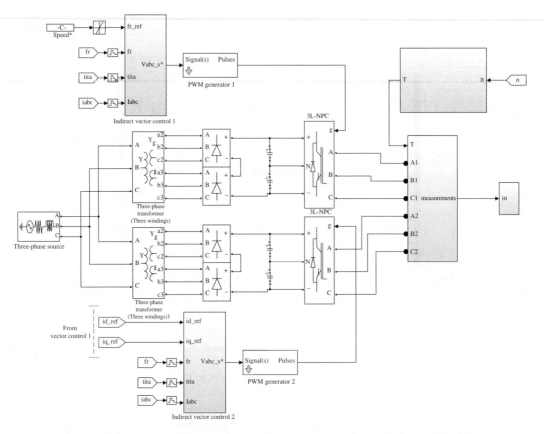

Fig. 6-97 *Block diagram of the entire propulsive system together with the model of the ship*

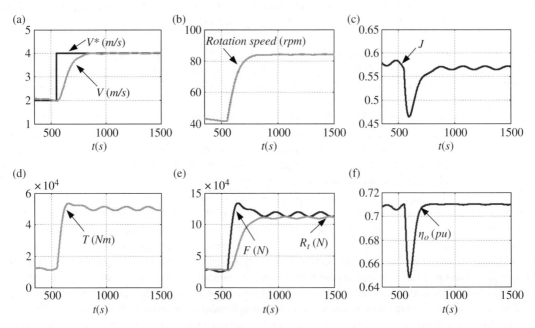

Fig. 6-98 *Most interesting magnitudes of the ship when accelerating from 2 to 4m/s of linear speed. An external sinusoidal force (F_{ext}) of 5kNm and 5mHz has been simulated (a) Linear speed of the ship, (b) Rotational speed of the propeller, (c) Advance coefficient J, (d) Torque, (e) Linear forces, (f) propeller open-water efficiency*

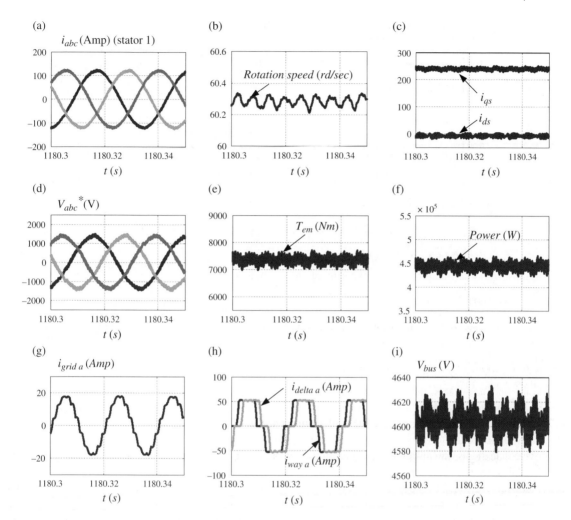

Fig. 6-99 *Most interesting magnitudes of the ship's propulsion motor, corresponding to simulation of Fig. 6-98, at time instant 1180 approximately*

the input of one of the transformers and at the output of both secondaries of the same transformer are shown. It is seen that the currents coming from the grid in Fig. 6-99(g), owing to the presence of the 12-pulse rectifiers, achieve a reasonably good quality. This is achieved by the shift angle of current supplied by the delta–wye secondary of the transformer as depicted in Fig. 6-99(h) and by employing a considerably high impedance at transformers and grid. Then, the total DC bus voltage of this converter supplying one of the stators is shown in Fig. 6-99(i).

To conclude, Fig. 6-100 illustrates one zoom of some of the above mentioned variables where the ripple of these variables can be distinguished under detail.

If the ship were composed of more propellers or even bow propellers, enabling complicated maneuvers, the model needed to be simulated would be much more complicated. This simulation example is included here only as a first step to understand the complex behavior of a ship's propulsion.

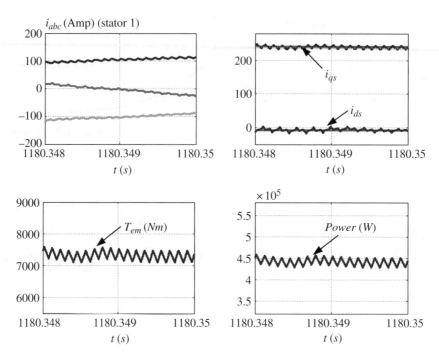

Fig. 6-100 *Zoom of the most interesting magnitudes of the ship's propulsion motor, corresponding to the simulation of Fig. 6-99*

6.7   Design and dimensioning of the electric system

The overall design of the ship's electric configuration is not an easy task. Every single element (generator, power cables, motors, etc.) or even parameter (loss resistance, equivalent impedance, rated rotating speed, etc.) can affect the overall behavior of the electric system. Fig. 6-101 roughly orientates how the system design can be approached step by step. Obviously, every step or task carried out in the flow diagram of Fig. 6-101 requires further analysis and calculations, with detailed data and parameters of the elements involved being taken into account. This diagram only seeks to provide a general idea of the main issues that must be considered. For a detailed electric ship's layout design, many hours of experienced engineering are necessary, ideally under close cooperation with many other people involved with the ship's design, and not necessarily engineers. Therefore, first of all in general, the ship itself is designed based on: ambient conditions of operation (waves, wind, etc.), operation necessities, hydrodynamics of the ship, speed requirements, redundancy requirements, auxiliary equipment's, and so on. Then, once the ship design is finalized, the electric loads are designed, where commonly propulsion is the most important one. Then, an iterative process is conducted, as depicted in Fig. 6-101, where aspects such as power, network voltage, number of sub-networks, type of drives (6p, 12p, AFE, etc.), and generator characteristics are defined and globally evaluated. Often, the type of energy generation (diesel, gas, high, medium, or low speed, etc.) is already imposed by the concept of the ship itself, and therefore the design is conducted to the initially established main ideas. As shown in the power flow diagram of Fig. 6-83, when protection analysis is being carried out, many simulations and studies are also necessary. Moreover, this protection analysis could also be considered a part of the overall ship's electric layout design, shown in Fig. 6-101. Note, for instance, that the generators' design characteristics and parameters affect issues such as circuit current level, the impedance to the THD of the entire system, and the

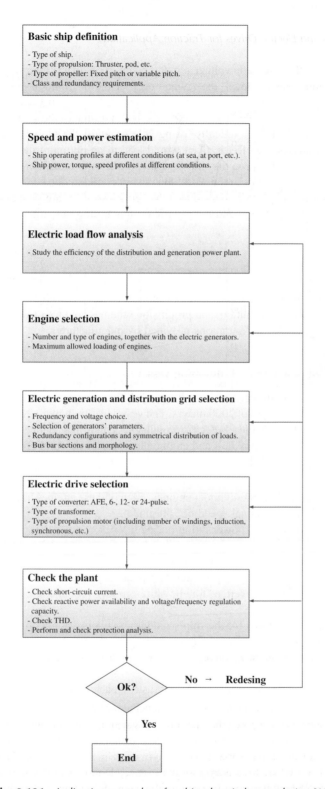

Fig. 6-101 *Indicative procedure for ship electric layout design [18]*

reactive power available to the automatic voltage regulator, and so on. In a similar way, the transformer design also affects the short-circuit current level, the voltage drop of the distribution grid, the THD, the volume, weight, cost, etc.

Thus, this iterative design process could even affect the initial design of the ship itself, owing to the necessities of the electric power plant. Theoretically, being able to adjust the global design of the ship itself in accordance to the necessities of the electric plant would lead to an optimum design. However, in practice, it depends on the actual relation between the ship designer, the electric integrator, and the shipyard. It is possible to find different working realities in the world, but a very common one consists of: definition of a fixed ship design and then the electric plant is adapted to it. Nevertheless, there are several important companies in the world that integrate the ship design and the electric design at the same time.

6.8   Real examples

This section shows some representative commercial examples of vessels with electric propulsion of different manufacturers. The generation, distribution, and propulsion schemas are provided, together with some details and discussion about the electric configurations provided by manufacturers.

6.8.1   Simon Stevin fall pipe and rock dumping vessel

This type of vessel, known as a fall pipe and rock dumping ship, is designed to cover channels, cable systems, and gas pipelines at a maximum depth of 2000 meters. The vessel was built by La Naval and the owner is Jan de Nul. It is equipped with an advanced diesel-electric propulsion system. The main electrical generation plant for the propulsion system and the auxiliary hull and engine plants are made up of five main generator sets comprising the following equipment [21]:

- five MAN Diesel, model 32/40 nine-cylinder engines. Each of these engines has a power rating of 4500kW (480kW per cylinder) and a nominal speed of 750rpm;
- five main generators, each with a power rating of 5625kVA;
- 6.6kV main switchboard;
- two 2000kVA transformers for power distribution;
- four 4100kVA transformers for the ship's rudder propellers;
- INGEDRIVE MV frequency converters for the ship's rudder propellers, each with a power rating of 3350kW;
- four asynchronous electric motors for the ship's rudder propellers, each with a power rating of 3350kW;
- four 2500kVA transformers for the ship's thrusters;
- four INGEDRIVE MV frequency converters for the ship's thrusters, each with a power output of 2000kW; and
- four asynchronous electric motors for the ship's thrusters, each with a power output of 2000kW.

All the electric and electronic equipment is supplied by Ingeteam. The schema of the electric configuration is illustrated in Fig. 6-102.

The Simon Stevin propulsion plant is made up of two main azimuth propellers. These are fixed pitch. Each has a power output of 3350kW. It also incorporates two retractable azimuth thrusters, to facilitate the maneuverability. These fixed-pitch bow thrusters each have a diameter of 2500mm and a power output of 2000kW. It is also equipped with two thrusters with dismountable tunnels. These fixed-pitch thrusters each have a diameter of 3000mm and a power output of 2000kW.

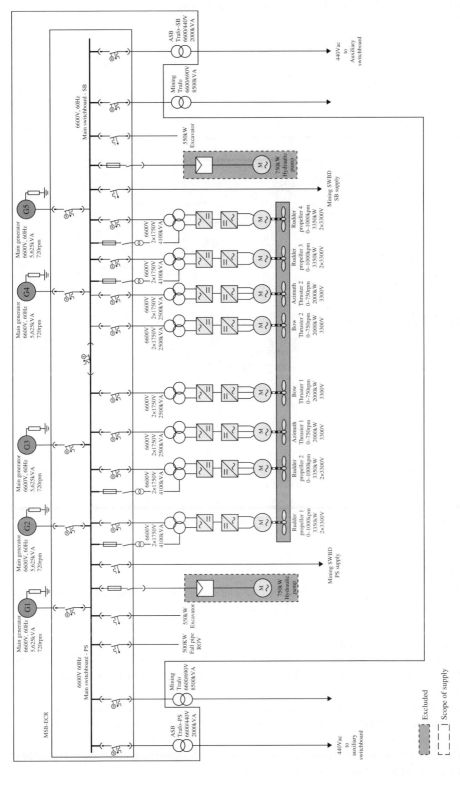

Fig. 6-102 *Electric configuration of Simon Stevin fall pipe and rock dumping vessel [21] (Source: Ingeteam. Reproduced with permission of Ingeteam)*

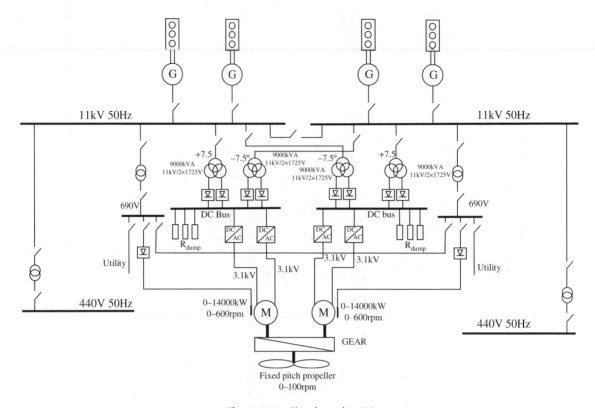

Fig. 6-103 *Shuttle tanker [2]*

In addition, the Simon Stevin is equipped with an emergency genset, with a power output of 345kW at 1800rpm. This in turn drives a 450kVA, 60Hz alternator. The vessel is also equipped with an auxiliary genset, with a power output of 1539kW at 1800 rpm. The Simon Stevin is fitted with a class II DP system (Dynapos AM/AT R compliant).

6.8.2 Shuttle tanker

In this section, the configuration of a shuttle tanker developed by ABB is presented, as depicted in Fig. 6-103. The main characteristics are summarized as [2]:

- Four generators of 8MW in redundancy;
- Owing to the high power installation, 11kV 60Hz distribution electric system with redundancy;
- Single screw propulsion, commanded by gearbox and two 14MW brushless synchronous motors with double stator of 3.1kV;
- Each stator of the motor is variable speed supplied by an ACS6000 DTC controlled converter;
- The converters are supplied in 24-pulse configuration by their corresponding transformers and diode rectifiers;
- For each DC bus of the converter, breaking resistors being able to dissipate dozens of MJ in less than one minute;
- The ship service grid is distributed at 440V.

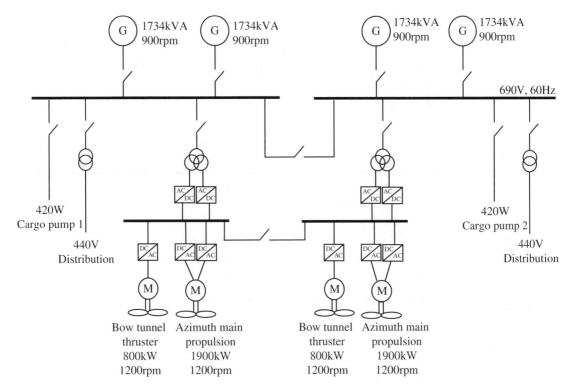

Fig. 6-104   *Platform supply vessel, with low voltage electric system configuration [22]*

6.8.3   Field support vessels

Field support vessels can be used for a wide range of applications, for instance platform supply vessels (PSV), ROV support vessels, multipurpose vessels, anchor handlers (AHTS), etc. Their power requirements for station keeping, sailing, and loads vary widely. Fig. 6-104 is an example of a platform supply vessel with low voltage electric configuration (690V) of Siemens. Its main characteristics can be summarized as:

- The generation is composed of four generators of 1734kVA.
- There are two main Azimuth main propellers of 1900kW with double stator motors.
- There are also two bow tunnel thrusters of 800kW.
- It also incorporates a special redundancy at the DC side of the converters supplying the propellers.
- The ship service is distributed at 440V.

On the other hand, a large anchor handler such as the one shown in Fig. 6-105 often incorporate hybrid mechanical and electrical propulsion. The hybrid solution allows the efficient use of the diesel engines with the electrical drives as booster drives in bollard pull condition, while electric drives are used for transit and station-keeping purposes [2]. Its main characteristics can be summarized as:

- The generation is composed of three main generators of 2000kVA.
- There are two main propellers of 5850kW, driven by hybrid electrical and mechanical propulsion each. Each propeller is driven by a main diesel engine of 4000kW and one VSD controlled motor of 1800kW.

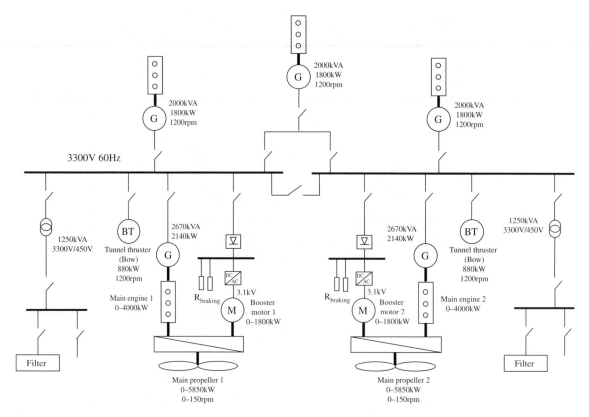

Fig. 6-105   *Large anchor handler, with hybrid mechanical and electrical propulsion system [2]*

- In addition, there are also two tunnel bow thrusters of 880 kW, directly connected to the 3.3kV distribution grid without drive.
- Note that the choice of a 3kV distribution voltage level, together with the small power electronic drives' power (2 × 1800 kW) compared to the total generated power, means the transformer can be eliminated, reducing considerably weight, space, and cost. In turn, the quality of the grid is not so severely degraded, since the source of current harmonics due to the converter effect is not so determinant. However, for more susceptible, low-power loads, harmonic filters are employed.
- The ship service is distributed at 450V.

6.8.4   Cruise liner

To end with, this section illustrates a large cruise liner, as depicted in Fig. 6-106. It can be seen that, in general, there are no significant differences with the electrical configurations of previous examples. The generation consists of four main diesel engines rotating four synchronous generators, creating an 11kV and 50Hz electric grid. Safety and onboard comfort for passenger vessels are very important issues, implying strong levels of redundancies in the propulsion plant, as can be seen in Fig. 6-106. The main propellers are twin screw (redundant) and gearless. For cruise liners geared transmission is also often applied as well as pods. The supply

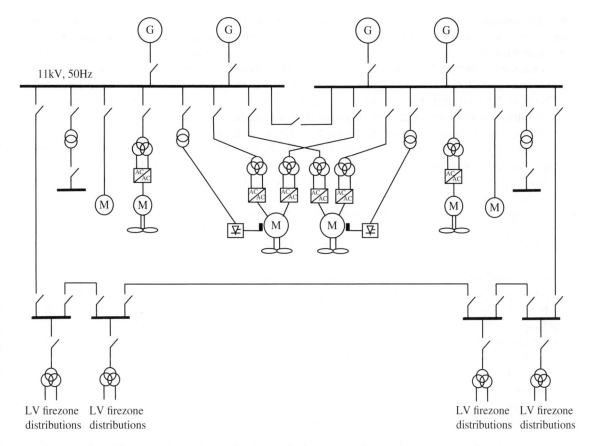

Fig. 6-106 *Configuration (redundant) of a large cruise liner, twin screw, gear less [18]*

converters are 24-pulse configurations with synchronous machines. There are also two auxiliary low-power thrusters, and many loads distributed along different locations of the ship. The protection system of fuses and circuit breakers must be carefully chosen to allow good availability in case of faults.

6.9 Dynamic positioning (DP)

According to the American Bureau of Shipping (ABS), DP is a vessel or unit which automatically maintains its position (fixed location or predetermined track) exclusively by means of thruster force. This can be achieved in several ways, including:

* installing tunnel thrusters added to the main screw;
* installing azimuth or podded thrusters, which have the ability to produce thrust at any direction.

Reference sensors, wind sensors, motor sensors, gyro-compasses, and so on provide the computer system with the information to gauge the vessel's position. Within the computer system, mathematical models of the vessel describe the effect of the wind and current drag to the vessel. This information, combined with the sensor information, allows the computer system to calculate the necessary steering angle and thruster forces for each thruster of the vessel.

This positioning philosophy allows operating at sea, where alternatives such as mooring or anchoring are not possible because of deep water, congestion at sea bottom due to the presence of pipelines for instance and so on. Therefore, DP guarantees a vessel will hold its place against wind, waves, and current by thrusters which are commanded by computers taking signals from satellites, sonar, and compasses.

DP is employed in vessels needing the following applications [23]:

- servicing aids to navigation
- cable-laying
- crane vessels
- cruise ships
- diving support
- dredging
- drilling
- landing platform docks
- maritime research
- mine sweepers
- pipe-laying
- platform supply
- rock-dumping
- sea launch
- shuttle tankers
- surveying.

Compared to other alternative types of positioning systems, such as anchoring and jack-up barges, DP is nowadays commonly used, mainly because it allows excellent maneuverability that is not dependent on water depth or limited by an obstructed seabed, and it also enables a quick dynamic set-up. Obviously, these advantages are achieved by means of a more complex and costly investment in a thruster and control system for the vessel. In addition, DP also must deal with hazards such as running off position due to system failure or blackouts.

On the other hand, it must be mentioned that it is possible to consider six degrees of freedom in a ship. Three of them involve translation: surge (forward/astern), sway (starboard/port), and heave (up/down). While the other three degrees of freedom involve rotation: roll (rotation about surge axis), pitch (rotation about sway axis), and yaw (rotation about heave axis). DP is concerned primarily with control of the ship in the horizontal plane (i.e. surge, sway, and yaw). An illustrative example showing these degrees of freedom is provided in Fig. 6-107.

As mentioned before, in order to dynamically control the position of the ship it is necessary to use position reference systems and sensors [10]. Two typical position reference systems are:

- Global navigation satellite systems (GNSS): The most typical one used is the Navstar GPS, which is a US satellite navigation system with world coverage.
- Hydroacoustic position reference systems: Locating several transponders at fixed positions on the seabed and one or several mounted under the hull, the position of the vessel can be measured.

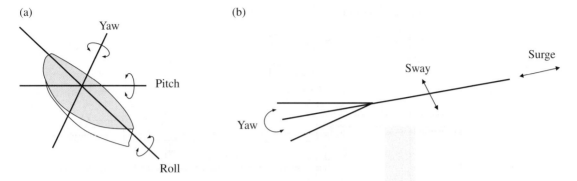

Fig. 6-107 *(a) Rotation roll, pitch, and yaw angles in a ship, (b) Example of surge, sway, and yaw control of a ship*

While sensor systems can comprise:

- gyrocompasses, which measure the heading of the vessel;
- vertical reference unit (VRU), which measures the vessel's heave, roll, and pitch motions;
- an inertial measurement unit (IMU) typically contains gyros and accelerometers in three axes that can be used to measure the body-fixed accelerations in all the six degrees of freedom of the ship;
- wind sensors;
- draft sensors;
- environmental sensors, such as wave or current sensors.

With regards to the control system, although at the beginning proportional integral derivative (PID) controllers were used, nowadays modern controllers use mathematical models based often on Kalman filters, which describe the hydrodynamic and aerodynamic behavior of the ship. The Kalman model is recursively updated with the information provided by sensors and position reference systems. The simplified block diagram of a DP control system is shown in Fig. 6-108.

On the other hand, to standardize the response to failures in a vessel, typically the owner decides which type of classification attends to the positioning system installed [10]. Three examples of organizations that have rules for the classification of DP systems are Det Norske Veritas (DNV), American Bureau of Shipping (ABS), and Lloyd's. A DP system must achieve a sufficiently reliable positing keeping capability. Depending on the consequence of a loss of position-keeping capability, the reliability of the system would be different. Thus, the requirements of classification societies—International Marine Organization (IMO), DNV, Lloyd's, ABS, and so on—have been grouped into three different equipment classes [10]:

- **Class 1:** Loss of position can occur in the event of a single fault, which means that the DP system does not need to be redundant.
- **Class 2:** Loss of position cannot occur in the event of a single fault in any active element of the system. Consequently, any active component of the DP control must have redundancy, for instance two independent computers, three gyrocompasses, etc.
- **Class 3:** Equal requirement as DP of class 2, with additional redundancy requirement in technical design and physical arrangement.

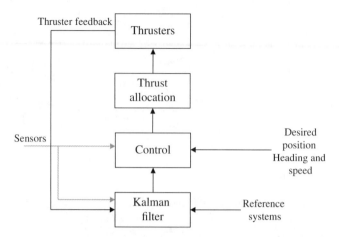

Fig. 6-108 *Simplified control block diagram of the DP (Source: https://en.wikipedia.org/wiki/File:Control-Kalman.svg. Used under CC- BY- SA 3.0 http://creativecommons.org/licenses/by-sa/3.0)*

6.10   Historical evolution

This section tries to show in a synthesized way the historical evolution of ship propulsion. The beginning occurred a very long time ago. Since its early origins, ship propulsion has experienced a massive evolution over the centuries. Table 6-7 illustrates the most remarkable events related to the evolution of ship propulsion. This table is not intended to be a rigorous and exhaustive historical analysis. Instead, it means to give the reader a wide perspective of the most significant stages of ship propulsion development. Hence, it is possible to observe that, until the 18th century, the main energy source was mainly the wind (with sails) and prior to that also human force was typically used (oars). But later, the development of steam turbines in the 19th century meant that sail ships and steam-turbine propelled ships (and combinations of both) worked alongside each other.

Then, in the 1920s, owing to the strong competition between transatlantic passenger lines, turbo-electric propulsion became popular (steam-turbines supplying synchronous electric motors moving the screw shafts). The rotational speed of the shaft was given by the electric frequency of the generators, and each generator could run one or more propulsion motors. In the 1950s, the development of highly efficient diesel engines dominated, prompting the disappearance of electric and steam-turbine propulsion.

However, in the 1980s, again thanks to the advances of power electronics, electric propulsion was introduced with VSDs, being mainly diesel engines as prime movers. This propulsion philosophy has dominated until today, as has been discussed in this chapter.

One of the first diesel-propelled ships was the *Vandal* launched in 1903 in Saint Petersburgh. This vessel was composed of three diesel engines, each with a generator of 87 kW and 500V [24]. Each generator at the same time moved its corresponding propulsion electric motor of 75 kW directly coupled to the propeller's shaft. The control of the speed of each propeller was performed with a "controller," which acted on the excitation current of the generator.

Some years later, motivated by the strong competition between transatlantic passenger lines, Normandie implemented four turboelectric steam turbines. Thus, oil was burnt to make steam that turned turbines to generate electricity to power electric motors that revolved propellers. That is, steam from the boilers turned turbines which generated the electricity. There were two sets of four three-bladed propellers each (eight propellers in total).

Table 6-7 *Synthesized evolution of ship propulsion [2], [3]*

Year	Development
Thousands of years BC	Canoes
1800BC–200BC	Successions of civilizations: Egyptian ships, Phoenician ships, Roman ships, etc.
	Combination of sails and oars
	One, two, and three banks of oars
200 AD	Viking ships:
	• More sophisticated ships in terms of: stability, speed, maneuverability
1400	Caravels:
	• Christopher Columbus: arrival to America (1492)
	• Beginning of regular overseas journeys
	• Only sails
1606	Jerónimo de Ayanz y Beaumont maybe patents the first steam turbine and designs the first submarine. Many other individuals worked on steam power before.
1707	Probably one of the first experimental applications of a steam turbine installed on a paddle wheel boat by Denis Papin
1770	James Watt builds the first modern steam turbine
1776	Beginning of the steam-turbine era in ship propulsion.
	Claude-François-Dorothée, Marquis de Jouffroy d'Abbans, first uses the steam turbine for ship propulsion
1791	John Barber proposes one of the original ideas of the gas turbine
1802	Symington tests the ship with steam-turbine propulsion actuating to paddle wheels
1818	Savannah US navy ship is built
	• complements paddle wheels moved by steam turbines and sails
	• the world's first transatlantic steamship service
1837 and 1843	Great Western and Great Britain ships are built, where I. K. Brunel employs screw propulsion moved by steam turbines. After that screw propulsion moved by steam turbines begins to become popular
1880	It can be said that the screw propeller is well established
1884	Charles Parsons develops a new generation of steam turbines very suitable for ship propulsion
1888	Isaac Peral develops the first submarine with electric motor propulsion and batteries as energy source
1891	Steam turbine with electric generator: Charles Parsons develops a steam-turbine propulsion coupled to an electric generator
1892	Diesel patents its diesel engine
1897	Turbinia ship:
	• prototype, with steam-turbine propulsion
	• after that, further improvement and development of ships with steam-turbine propulsion is seen
1903	Diesel-electric ships:
	• introduction of diesel-electric ship propulsion
	• *Vandal*: three diesel engines coupled to three synchronous generators, supplying three motors driving the propeller
1905	Only diesel ships
	• first directly reversible diesel engine introduced into ship propulsion
	• delays the development of electric propulsion

(*continued overleaf*)

Table 6-7   (*continued*)

Year	Development
1935	Steam turbines driving directly synchronous machines coupled to the propellers: • rotational speed was typically constant • motivated by the strong competition between transatlantic passenger lines, reducing traveling times and increasing engine efficiency • transition from steam to diesel propulsion mainly justified by lower fuel consumption
1928	Frank Whittle patents a turbo-reactor for aircrafts • development of gas turbine
1938	Otto Hahn obtains nuclear power
Mid-20th century	Diesel propulsion again • introduction of efficient and economically favorable diesel engines • steam-turbine and electric propulsion begin to be displaced by diesel propulsion
1954	US navy develops the firsts ships and submarines with nuclear power
1960	The first DP system is developed, by using simple PID controllers
1967	General Electric develops a gas turbine for probably the first ship propelled by this turbine
1970s and 1980s	Development of VSDs, first with AC/DC rectifiers Development of VSDs, with AC/AC converters
1989	First Azipod prototype installation at "Seili" built by ABB
Early 1990s	Introduction of podded propulsion: • originally developed to improve the performance of icebreakers • found to present benefits for efficiency and maneuverability of other types of ships
Late 1990s and early 2000s	Wholesale introduction of VSC-based electric drive propulsion
2000–2016	Introduction of DC distribution networks Hybrid propulsions Energy storage systems introduced

Some decades later, one of the first modern ship propulsion systems was based on cyclo-converters and current-source converters or load commutated converters, such as the *Queen Elizabeth II* cruiser, whose electric propulsion system is shown in Fig. 6-109 and was renewed in the mid-1980s. Its main features were [25]: nine diesel generator units each being 10.5MW, 10kV, 60Hz, and 400 rpm. It incorporates two wound field synchronous machines each being 44MW, 0–144 rpm, and 50-pole. It uses six-pulse rectifiers and six-pulse load commutated inverters systems for each motor. The propeller is variable pitch. On the other hand, the propulsion speed range of the propeller is 72–144 rpm. The speed control is performed with inner loop of current control. When the frequency reaches 60 Hz, the machine is switched directly to being supplied from the 60 Hz grid, improving the efficiency of the system. The motor is started with the inverter up to 50% of rated speed at no load, and then the load torque is established by controlling the propeller pitch angle. Unity power factor is almost achieved by means of the machine's excitation.

Current-source converters were commonly used in the past, and still are today, to supply the highest power loads. Fig. 6-110 illustrates a detailed scheme of the topology with thyristor-based six-pulse rectifier and inverter and main voltage and current waveforms. The grid supplied current contains a considerable amount of harmonics and can be reduced by increasing the number of pulse configuration and transformer as occurred in diode rectifiers at voltage source converters. In fact, it operates in an equivalent way, as described in previous sections (see also Fig. 6-47). The DC current link is smoothed by an inductor. The inverter side works equivalently also as the rectifier side, by using the induced voltage of the motor to perform the commutations.

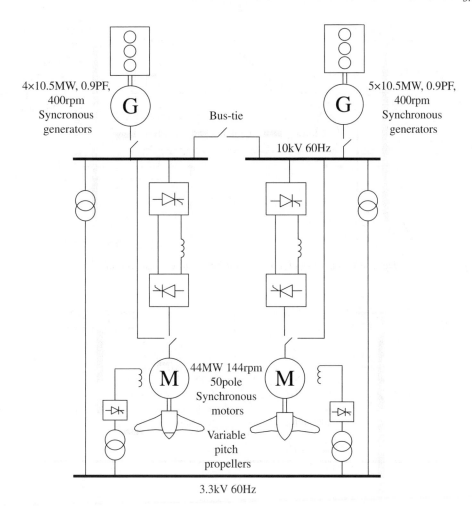

4×10.5MW, 0.9PF, 400rpm Syncronous generators

5×10.5MW, 0.9PF, 400rpm Syncronous generators

Bus-tie

10kV 60Hz

44MW 144rpm 50pole Synchronous motors

Variable pitch propellers

3.3kV 60Hz

Fig. 6-109 Queen Elizabeth II *diesel-electric propulsion system with current source converter and wound field synchronous machine (Source: [25]. Reproduced with permission of John Wiley & Sons)*

In terms of controllability, dynamic performance, and harmonic distortion, the current source inverter in principle is inferior to the voltage source inverters. However, some advantages also exist in this current source topology, which explains why some manufacturers still work with this converter philosophy.

As a different example of the 1980s, Fig. 6-111 shows the propulsion system of an icebreaker with diesel electric propulsion, with cyclo-converter and wound field synchronous machine. The main characteristics can be summarized as [25], [26]: three fixed-speed diesel engines and synchronous generators, producing a bus voltage of 4160V and 60Hz. Thirty-six thyristors and six-pulse blocking mode cyclo-converter. The machines are WFSM with brushless excitation of 8000HP, 12-pole, 0–180rpm, and 0–8Hz. It presents speed reversal capability but not regeneration. While vector control until rated speed and scalar control at field weakening mode (145–180 rpm) is utilized.

The cyclo-converter fabricates the output voltage from the input voltage by enabling conduction of the thyristors. Detailed circuit of the cyclo-converter is shown in Fig. 6-112. In general, the motor voltage quality is better than in a current-source converter, owing to a reduction of the harmonics. There is no speed limitation at

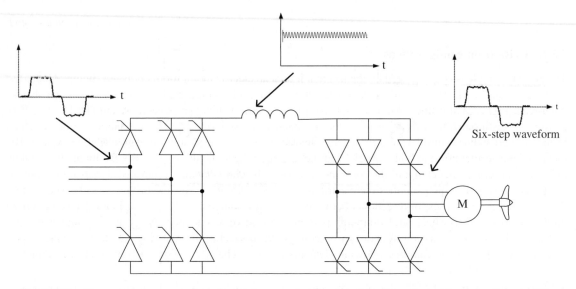

Six-step waveform

Fig. 6-110 *Current source converter, also known as a load commutated inverter, with main characteristics*

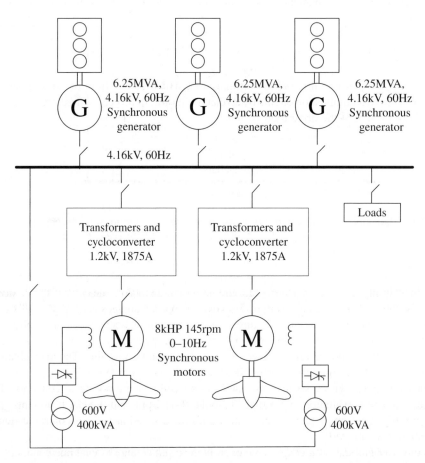

6.25MVA,
4.16kV, 60Hz
Synchronous
generator

6.25MVA,
4.16kV, 60Hz
Synchronous
generator

6.25MVA,
4.16kV, 60Hz
Synchronous
generator

4.16kV, 60Hz

Loads

Transformers and
cycloconverter
1.2kV, 1875A

Transformers and
cycloconverter
1.2kV, 1875A

8kHP 145rpm
0–10Hz
Synchronous
motors

600V
400kVA

600V
400kVA

Fig. 6-111 *Icebreaker diesel-electric propulsion with cyclo-converter and wound field synchronous machine, manufactured by Canadian GE [25], [26]*

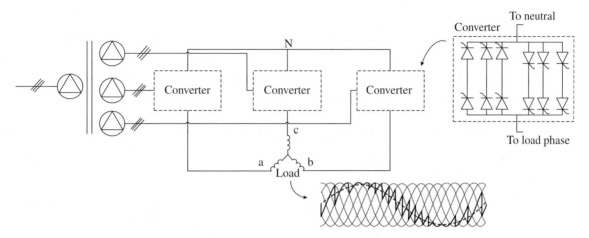

Fig. 6-112 *36 thyristors and six-pulse blocking mode cyclo-converter and fabricated waveform*

lower speeds, since the commutation is made with the supply voltage. This converter topology was commonly used for multi-megawatt drive applications.

6.11 New trends and future challenges

This section gives a summary of the most remarkable tendencies and future challenges regarding ship propulsion. In general terms, it is possible to say that in ship propulsion as in many other applications the continuously innovative tendencies are:

- improve performance: more accurate DP, improve comfort (noise, vibrations, etc.), increase maneuverability;
- improve the efficiency and/or fuel consumption;
- reduce cost;
- reduce volume and weight of distribution and propulsion systems;
- improve reliability and availability of electronic equipment as well as the lifetime of operation;
- introduction of energy storage systems, which enable one to reduce the fuel consumption among other benefits; and
- be able to adapt to different characteristics of ships (power, voltages, etc.) in an easy, quick, and efficient manner (versatility).

These general goals have produced the following remarkable tendencies:

6.11.1 DC distribution networks

There is clear tendency to use more and more DC, rather than AC, distribution networks, as has been seen in this chapter. In fact, at the time of writing, there are several developments in the world of low-power ships regarding DC distributions that seem very promising. Thus, mainly thanks to the development of power electronics and DC breakers and fuses, nowadays it is possible to efficiently and reliably go for DC distribution networks. As is schematically represented in Fig. 6-113, the idea is to locate the power converter at the generator's side (can

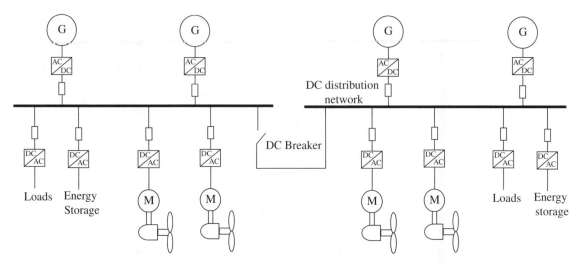

Fig. 6-113   *Simplified representation of DC distribution network for ship propulsion*

be diode-, thyristor-, or VSC-based), so the distribution through the different loads of the ship is done with DC rather than AC. With this idea, the main advantages that can be achieved can be summarized as:

- The DC distribution system can collect several DC-links coming from different sources/loads at different locations of the ship and distributes the electrical energy by means of a DC circuit. In this way, the main AC switchboards, the distributed rectifiers, and their correspondingly costly and bulky transformers are avoided.
- Energy efficiency can be increased and fuel consumption and polluting emissions can be reduced thanks to the operation of the diesel generators at variable speed according to the power demand of the ship's network, therefore tending to operate at optimum or most efficient operation points.
- The available space for cargo is increased, owing to the reduction of the size and weight of the electrical equipment, which provides higher flexibility when designing the electrical distribution system according to the ship's functionality.
- The integration of renewable energy systems and energy storage systems (solar energy generation, electrical-batteries-based storage systems, etc.) within the distribution system is, in principle, simplified.

However, there are some technological challenges that must be met to obtain a proper DC distribution network: safe and reliable protection systems (DC breakers, fuses, etc.), and proper functioning of the power electronic conversion systems.

6.11.2   Hybrid mechanical-electrical propulsion

Although not new (see Fig. 6-105), this concept of propulsion is becoming increasingly popular, especially in ships where the main propulsion is mechanical and there exists a need for high-power propulsion, but only in punctual moments. Clear examples of this context are, for instance:

- Warships, where the need for high speed exists for military maneuvers. In general, the propulsion is directly made with high-power gas turbines.

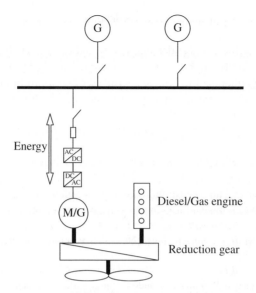

Fig. 6-114   *Hybrid shaft generator of Rolls-Royce*

- Tugboats and AHTS, where they are traditionally based on diesel mechanical propulsions, owing to the necessity of a high bollard pull. They typically incorporate big low-speed diesel motors to be able to produce that pull.

In general, these types of propulsion are very inefficient at low loads and they work at full load during short time intervals, since most of the time they are waiting, maneuvering, or slowly navigating. Therefore, this means that their principal propulsion system operates at low load.

Thus, in these cases the solution that is usually employed incorporates a reduced-power electric motor in the same shaft of the main propulsion or in an additional output of the gearbox. This electric motor allows one, for instance, to switch off the main diesel motor when it is not required or even reduce its power when operating because it can operate as a booster.

One more concept that is very similar to the previous one is also becoming very popular and will probably be more so in the future. It can be said that it is based on the same idea as the previously described hybrid mechanical-electrical propulsion, but in this case an additional functionality is provided to the electric motor, that is the electric motor and the associated electric equipment are able to operate as a generator, making it possible to supply the auxiliaries of the ship with the main diesel at variable speed. These systems have been used for many decades in large LNG cargo ships, but nowadays with the introduction of efficient, voltage-soured converters, the efficiency and versatility of this idea is significantly improved. One example of this system is depicted in Fig. 6-114.

6.11.3   Hybrid generation

There is another hybrid concept that is becoming increasingly popular. This concept is based on the traditional diesel-electric propulsion system which incorporates a battery-based energy storage system. The main objectives of including an efficient and cost-effective battery system are to:

- improve the efficiency of the diesel engine by eliminating the consumption variations due to sudden load changes (peak shaving);
- eliminate the necessity of connecting and disconnecting diesel engines in short transients, to meet demands for higher power;
- be used in PTH (power take home), which means arriving in port in full electric mode.

References

[1]   MAN Diesel & Turbo, Basic principles of ship propulsion. December 2011. Copenhagen, Denmark, https://marine.man.eu/docs/librariesprovider6/propeller-aftship/basic-principles-of-propulsion.pdf?sfvrsn=0.

[2]   Adnanes AK. *Maritime Electrical Installations and Diesel Electrical Propulsion*. Oslo, Norway: ABB, 2003.

[3]   de la Llana I. *Nuevo Sistema de Propulsión Naval* (New Ship Propulsion System). PhD thesis, University of the Basque Country, 2011.

[4]   Patel MR. *Shipboard Electrical Power Systems*. Boca Raton, FL: CRC Press, 2012.

[5]   Carlton J. *Marine Propellers and Propulsion*. London: Butterworth-Heinemann, 2007.

[6]   Bertram V. *Practical Ship Hydrodynamics*. London: Butterworth-Heinemann, 2000.

[7]   Karlsen AT. *On Modeling of a Ship Propulsion System for Control Purposes*. PhD thesis, Norwegian University of Science and Technology, 2012.

[8]   Smogeli OV. *Control of Marine Propellers: From normal to extreme conditions*. PhD thesis, Norwegian University of Science and Technology, 2011.

[9]   Izadi-Zamanabadi R. *A Ship Propulsion System as a Benchmark for Fault Tolerant Control*. Department of Control Engineering, Aalborg University, 1997.

[10] Sørensen AJ. *Marine Control Systems: Propulsion and motion control of ships and ocean structures*. Lecture at Norwegian University of Science and Technology, Department of Marine Technology, January 2011.

[11] Mashayekh S, Wang Z, Qi L, et al. Optimum sizing of energy storage for an electric ferry ship. *2012 IEEE Power and Energy Society General Meeting*, 22–26 July 2012, San Diego: 1–8.

[12] Hansen JF. *Electric Propulsion*. Lecture at Norwegian University of Science and Technology, Trondheim, 9 March 2010, ABB Group.

[13] El-Gohary MM, El-Sherif HA. *Ship Propulsion Systems*. Faculty of Engineering, Naval Architecture and Marine Engineering Department, Alexandrie University.

[14] *Two Stroke Low Speed Diesel Engines For Independent Power Producers and Captive Power Plans*. Copenhagen, Denmark: Man Diesel, 2009.

[15] SimPowerSystems Library from Matlab-Simulink.

[16] Ingebrigtsen Bø T. *Dynamic Model Predictive Control for Load Sharing in Electric Power Plants for Ships*. PhD thesis, Norwegian University of Science and Technology, Trondheim, 2012.

[17] Woodward (2011). *Governing Fundamentals and Power Management*, http://www.woodward.com/pdf/ic/26260.pdf.

[18] *Diesel Electric Propulsion Plants: A brief guideline how to design a diesel-electric propulsion plant*. Man Diesel Turbo, https://marine.man.eu/docs/librariesprovider6/marine-broschures/diesel-electric-drives-guideline.pdf?sfvrsn=0.

[19] Zubiaga M, Abad G, Aurtenetxea S, et al. *Energy Transmission and Grid Integration of AC Offshore Wind Farms*. Intech Open Acces, http://www.intechopen.com/books/energy-transmission-and-grid-integration-of-ac-offshore-wind-farms, Jan 2012. This is an open access book.

[20] Standard IEEE 45. *Recommended Practice for Electrical Installations on Shipboard*. DOI 10.1109/IEEESTD.2002.94134.

[21] La Naval. *Simon Stevin: An exclusive ship report*, http://www.lanaval.es/archivos/201205/simon-stevin_nb333.pdf?1, 2010.

[22] *Reference List: Offshore vessels: Low voltage propulsion*, https://w3.siemens.no/home/no/no/sector/industry/marine/Documents/Offshore-References%20V2012_02.pdf, 2012.

[23] http://en.wikipedia.org/wiki/Dynamic_positioning.

[24] Borrás R, Rodríguez R, Luaces M. Starting of the naval diesel-electric propulsion: *The Vandal. JMR* 2011; **3**(3): 3–16.

[25] Bose BK. *Power Electronics and Drives*. Oxford: Elsevier, 2006.

[26] Hill WA, Turton RA, Dugan RJ, Schwalm CL. Vector controlled cycloconverter drive for an icebreaker. *IEEE IAS Annual Meeting Conference Record*, 1986: 309–313.

7

Electric and hybrid vehicles

David Garrido and Gonzalo Abad

7.1 Introduction

This chapter introduces the reader to the main characteristics of electric and hybrid vehicles. These vehicles are propelled by electric motors, sometimes in combination with combustion engines, and draw their power from onboard energy sources.

Pure electric vehicles appear to have advantages over those powered by the internal combustion engine. They do not release pollutants at their point of use, they are quiet, the infrastructure needed for their implementation can be achieved using the existing one, the motors exhibit high-efficiency performance, and they can recover the kinetic energy by means of regenerative braking. On the other hand, limited range and charging times present major disadvantages for these vehicles users.

This chapter focuses on the physical laws acting on these road vehicles, on the different vehicle configurations, and on the pure electric vehicle and hybrid vehicle configurations and the subsystems that integrate them. Finally, battery management systems, some computer-based simulations, design examples, and future trends are presented and described.

7.2 Physical approach to the electric vehicle: Dynamic model

This section describes a longitudinal model of a road vehicle. The vehicle is modeled by its corresponding dynamic differential equations, which are derived from the physical laws describing its movement. Only longitudinal movement along a line is modeled, we do not consider the movement during curves. This model is

Power Electronics and Electric Drives for Traction Applications, First Edition. Edited by Gonzalo Abad.
© 2017 John Wiley & Sons, Ltd. Published 2017 by John Wiley & Sons, Ltd.

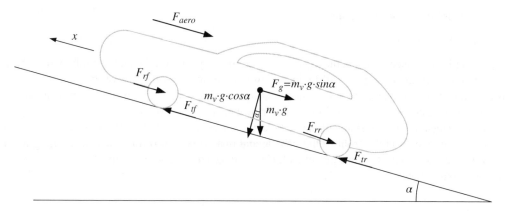

Fig. 7-1 *Forces affecting a vehicle*

considered accurate enough for studying the traction drive of the vehicle. Readers seeking to study a model which considers the dynamic movement during curves are referred to more specialized literature. The model that is presented in this section is useful for electric, hybrid, and combustion-engine vehicles. In other words, the dynamic equations developed in this model are useful no matter the nature of the traction motor it is. Much of the analysis and development of this section is based on the well-recognized reference [1] in the field of electric vehicles. On the other hand, as noted in Chapter 5, the models developed for the electric vehicle and for the train are similar. In fact, in essence the electric vehicle model follows almost the same modeling philosophy as the train model, and only a few, superficial, details are different.

Hence, the dynamic equation of a vehicle in the longitudinal axis taking into consideration the forces along the vehicle of Fig. 7-1 can be described by means of (7.1)

$$m_v \cdot \ddot{x} = \left(F_{tf} + F_{tr}\right) - \left(F_{rf} + F_{rr} + F_{aero} + F_g\right) \tag{7.1}$$

where:

- F_{tf} is the traction longitudinal force at the front tires generated by the prime mover.
- F_{tr} is the traction longitudinal force at the rear tires generated by the prime mover.
- F_{rf} is the force caused by the rolling resistance at the front tires.
- F_{rr} is the force caused by the rolling resistance at the rear tires.
- F_{aero} is the equivalent longitudinal aerodynamic force.
- F_g is the force produced by gravity.
- m_v is the vehicle mass.
- x is the longitudinal direction of motion.

It must be pointed out that F_{tf} is zero for a rear-wheel-drive vehicle, and F_{tr} is zero for a front-wheel-drive vehicle. Thus, the traction force created by the drive and transmitted through the drive wheels tends to move the vehicle forward. Then, the vehicle experiences some resistive forces, such as rolling resistance, aerodynamic force, and the force produced by gravity (when it moves uphill), which tends to oppose the movement. The following subsection provides more details of these forces.

7.2.1   Forces actuating on the vehicle

7.2.1.1   *Aerodynamic force*

The aerodynamic resistance which actuates on a vehicle moving forward depends mainly on two basic phenomena: viscous friction and pressure difference between the front and the rear of the vehicle [1].

Thus, viscous friction is caused by the air surrounding a vehicle's skin (which is moving at almost the vehicle's speed) moving at a different speed than the air further from the vehicle.

Conversely, depending on the shape of the vehicle, the forward movement of the vehicle creates a pressure difference between the air at the front of the vehicle and that at the rear of the vehicle, resulting in an aerodynamic force opposite to the forward movement. Hence, in general the aerodynamic force is approximated to the following expression:

$$F_{aero} = \frac{1}{2} \cdot \rho \cdot A_f \cdot C_d \cdot (\dot{x} + v_{wind})^2 \qquad (7.2)$$

with $\dot{x}$ being the vehicle's speed, A_f the vehicle's front area, ρ (1.225kg/m^3) the air density, and v_{wind} the wind speed (positive for a head wind and negative for a tail wind). C_d is the aerodynamic drag coefficient that can be estimated, for instance by experiments in wind tunnels. Its value depends on the shape of the vehicle and can go from 0.15 to 0.7 depending on the vehicle's design.

7.2.1.2   *Force produced by gravity*

The force produced by gravity depends on the mass of the vehicle and the shape of the non-horizontal road. Depending on whether the vehicle is moving uphill or downhill, this force opposes or assists forward motion respectively. Thus, as depicted in Fig. 7-1, the car which is moving uphill is affected by an opposing force to the movement, which can be modeled as:

$$F_g = m_v \cdot g \cdot \sin(\alpha) \qquad (7.3)$$

with g the acceleration due to gravity, m_v the mass of the vehicle, and α the inclination of the road. This force is often called grading resistance.

7.2.1.3   *Rolling resistance*

Rolling resistance is mainly caused by deformation of the tire with contact path, while the tire is rolling. This deformation generates an asymmetric distribution of ground reaction forces. Thus, when the tires are not rotating, the distribution of the normal load in the contact patch is symmetric. However, when the tires are rolling, the normal load distribution is not symmetric [1], as illustrated in Fig. 7-2.

It will be seen from the figure that the normal load is larger in the forward half of the contact path. Consequently, this phenomenon produces a normal load shifted forward, creating a moment opposite (resistant) to the rolling of the wheel. In order to maintain the wheel rolling, a force, F, acting on the center of the wheel is required, to balance this rolling moment. This equivalent force is called rolling resistance and is modeled as:

$$F = N \cdot C_r \qquad (7.4)$$

with C_r being the rolling resistance coefficient and N the normal load. If the vehicle is moving uphill or downhill, the normal road, N, must be substituted by its component perpendicular to the road surface, also taking into account whether the wheel is at the front or the rear of the vehicle (see Section 7.2.1.6 for more details).

The rolling resistance coefficient depends on many variables, the most influencing ones being probably vehicle speed, v, tire pressure, and road surface conditions. It can take values that go from 0.001–0.002 in wheels on rails to 0.013 in car tires on concrete or asphalt, to even 0.3 in wheels rolling on field. The influence of the vehicle speed to this coefficient often takes the shape depicted in Fig. 7-3.

This behavior depends on many aspects. In many cases, especially when the vehicle's speed remains moderate, the rolling coefficient C_r is often assumed to be constant, without loss of accuracy.

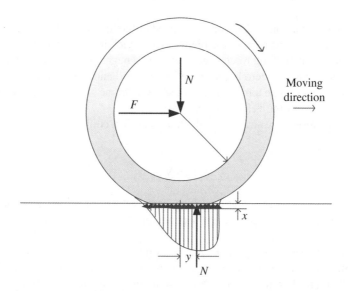

Fig. 7-2   *Normal load distribution on a tire on a hard road surface*

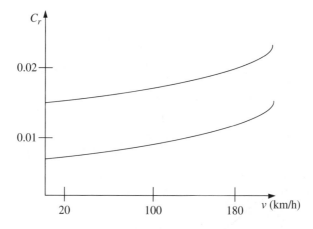

Fig. 7-3   *Rolling resistance coefficient, C_r, in function of the vehicle speed*

7.2.1.4   *Tire-ground adhesion*

In a vehicle, the traction effort generated by the drive wheels depends on the slip of the wheels on the ground. However, the adhesive capability between the tire and the ground is limited. This fact is significant when the vehicle runs on icy or soft roads, where the traction torque can cause significant slipping, limiting the performance of the vehicle. Thus, the traction force is created by the slip between the driving wheel and the vehicle linear speed. The slip, s, of a tire is defined as the relative difference of wheel speed and vehicle linear speed.

$$s = \frac{r \cdot \Omega - v}{\max(r \cdot \Omega, v)} \tag{7.5}$$

with v the linear vehicle speed along x, Ω the angular speed of the tire, and r the rolling effective radius of the free rolling tire. During traction, the term $r \cdot \Omega$ is bigger than v, yielding to slip values between 0 and 1. However, for braking operation, the braking force is generated by a negative slip, owing to a bigger value of term $r \cdot \Omega$ than v. Thus, the slip is equal to zero because the linear speed and the tire rolling speed are equal, which implies that there is no motor or brake torque on the wheel. At extreme slips, ±1, the wheels are either "locked" at zero speed or "spinning" with the vehicle at zero speed.

Therefore, experimental tests have shown that the traction effort corresponding to a certain tire slip can be fairly represented by:

$$F_t = N \cdot \mu \tag{7.6}$$

being F_t the traction effort of the corresponding drive wheel, N the normal load, and μ the traction (adhesive) effort coefficient. This traction effort coefficient depends on the slip and presents the shape as depicted in Fig. 7-4.

Note that in the slip range between region OA the traction force, F_t, is nearly proportional to the slip value. This region is the normal operation range. Conversely, in region AB, μ increases nonlinearly in function of the slip. This fact means that a further increase in wheel torque results in part of the tire traction force. Thus, the

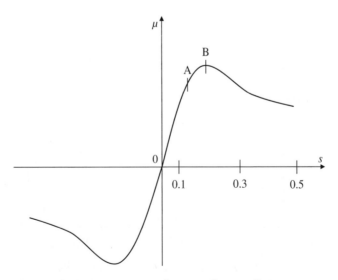

Fig. 7-4   *Traction (adhesive) effort coefficient*

peak traction effort is achieved around 0.15 or 0.2 slip. Finally, a higher increase on the slip beyond this region produces an unstable behavior of μ, decreasing its value and therefore the traction force. On the other hand, Fig. 7-5 shows some indicative traction effort coefficients, of different road surfaces.

Analytical treatment of these traction effort coefficients is possible by employing the Pacejka tire model [2]. It is described by the so-called magic formula, given by:

$$\mu(s) = D \cdot \sin[C \cdot a\tan(B \cdot s - E(B \cdot s - a\tan(B \cdot s)))] \tag{7.7}$$

being the parameters B, C, D, and E, the stiffness, shape, peak, and curvature factors respectively. Experimental tests have shown that these coefficients can take the values shown in Table 7-1, providing a good approximation.

7.2.1.5 Wheel's dynamic equation

In a similar way as we developed the linear motion equations of the vehicle body, it is necessary to derive a wheel's dynamic equations as well. Hence, by considering the torques applied to a wheel as can be seen in Fig. 7-6., the equation yields:

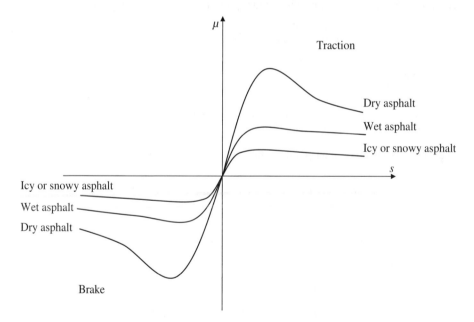

Fig. 7-5 *Traction (adhesive) effort coefficients on different road surfaces*

Table 7-1 *Parameters of the Pacejka tire model*

	B	C	D	E
Dry	10	1.9	1	0.97
Wet	12	2.3	0.82	1
Snow	5	2	0.3	1
Ice	4	2	0.1	1

$$T_{em} - T_b - T_t = J \cdot \dot{\Omega} \tag{7.8}$$

where T_{em} is the torque produced by the driving motor (only present at driving wheels), T_b is the brake torque applied to each wheel due to the effect of the brakes, and T_t is the reaction torque of each wheel due to the effect of the traction longitudinal force, being r the effective radius of the wheel and T_t is calculated by multiplying r by F_t. J is the inertia of the rotating wheel, plus the equivalent inertia of the drive motor and transmission system. Note that the rolling resistance is modeled as force and therefore is not considered here in this model equation.

7.2.1.6   *Normal forces*

For computation of the rolling resistances and the traction longitudinal forces, it is necessary to derive the normal forces. The static load distribution of the vehicle mainly depends on the vehicle geometry, the slope of the road, and the acceleration of the vehicle. Hence, the normal force distribution on the tires can be determined by assuming that the net pitch torque on the vehicle is zero, which means that by summing the moments about contact point of the rear and front tires, they must be equal to zero. For simpler computation of these moments, Fig. 7-7 is prepared [1]. In the figure, aerodynamic forces and rolling resistances are omitted since they are usually neglected when calculating normal tire forces.

Thus, by computing the moment about the contact point of the rear tire [1]:

$$N_f \cdot l + m_v \cdot \ddot{x} \cdot h + m_v \cdot g \cdot h \cdot \sin(\alpha) - m_v \cdot g \cdot b \cdot \cos(\alpha) = 0 \tag{7.9}$$

Solving N_f, we obtain:

$$N_f = \frac{m_v}{l} \cdot [g \cdot b \cdot \cos(\alpha) - g \cdot h \cdot \sin(\alpha) - \ddot{x} \cdot h] \tag{7.10}$$

By doing the same about the contact point of the front tire:

$$N_r \cdot l - m_v \cdot \ddot{x} \cdot h - m_v \cdot g \cdot h \cdot \sin(\alpha) - m_v \cdot g \cdot a \cdot \cos(\alpha) = 0 \tag{7.11}$$

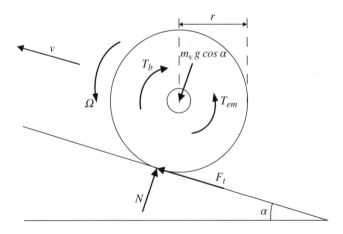

Fig. 7-6   *Wheel's dynamic under the influence of corresponding torques and forces*

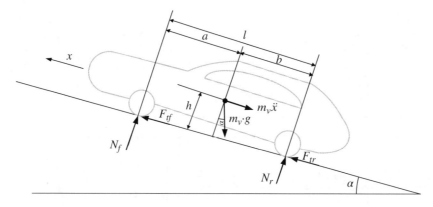

Fig. 7-7   *Calculation of normal tire forces*

Solving N_r, we obtain:

$$N_r = \frac{m_v}{l} \cdot [g \cdot a \cdot \cos(\alpha) + g \cdot h \cdot \sin(\alpha) + \ddot{x} \cdot h] \qquad (7.12)$$

Note that at the resulting normal load expressions there are some static terms and some others dependent on the acceleration of the vehicle. Thus, when the vehicle is accelerating, the normal load is transferred to the rear wheels, while during braking it is transferred to the front wheels. Note also that when the vehicle speed is constant (zero acceleration) and there is no slope to the road the normal force yields:

$$N_f = \frac{m_v \cdot g \cdot b}{l} \qquad (7.13)$$

$$N_r = \frac{m_v \cdot g \cdot a}{l} \qquad (7.14)$$

And the addition of both normal forces results in:

$$N_f + N_r = \frac{m_v \cdot g \cdot (b + a)}{l} = m_v \cdot g \qquad (7.15)$$

7.2.2   Model block diagram of a four-wheel drive car

Once the most important mathematical expressions of the vehicle model have been described, this section groups all of them to form the overall vehicle model, represented in Fig. 7-8. Thus, considering the linear motion equation (7.1) and the rotating equation (7.8) applied to the front and rear wheels, the overall model block diagram is depicted in Fig. 7-8.

Note that it has been assumed that the rear wheels and the front wheels are driving wheels, both driven by two different torques (T_{em_f} and T_{em_r}). For each front and rear axle, both front and rear wheels are considered identical ($r_f = r_r = r$), so the adhesion coefficient and efficient radio has been also considered identical.

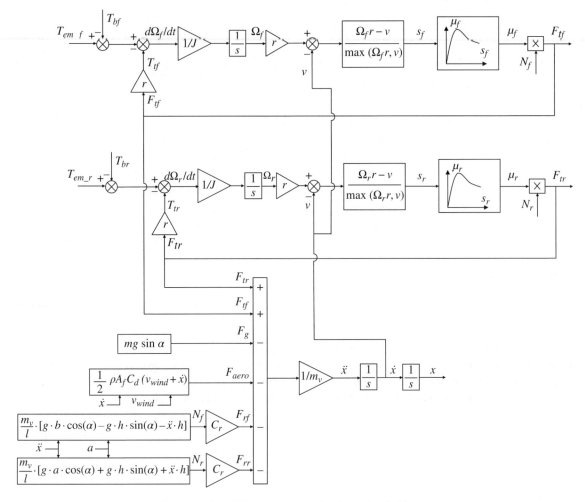

Fig. 7-8   *Model block diagram of a four-wheel drive car*

In addition, the inertias J_f and J_r of the front and rear wheels, also considered identical ($J_f = J_r = J$), are the sum of two wheel inertias, the shaft inertia, the transmission system inertia, and the motor drive inertia.

7.2.3   Model block diagram of a two-wheel drive car

In an alternative way to the model of a four-wheel drive car presented in the previous subsection, it is probably more common to find vehicles with only two driving wheels. Hence, this section shows the model block diagram of a car where only the front wheels are the driving wheels. Fig. 7-9 shows the car configuration and Fig. 7-10 shows the model block diagram. It is seen that, in this case, the traction longitudinal force at the rear wheels is zero (non-driving wheels), consequently compared with the previous model block diagram of Fig. 7-8 its complexity is significantly reduced since the rear wheel model is omitted. It can also be noted

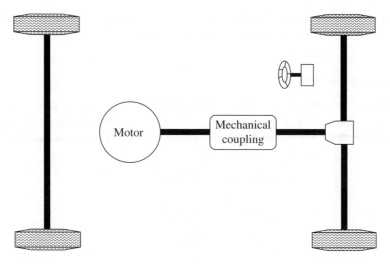

Fig. 7-9   *Wheel-tire-motor assembling inertia of a two-wheel drive car*

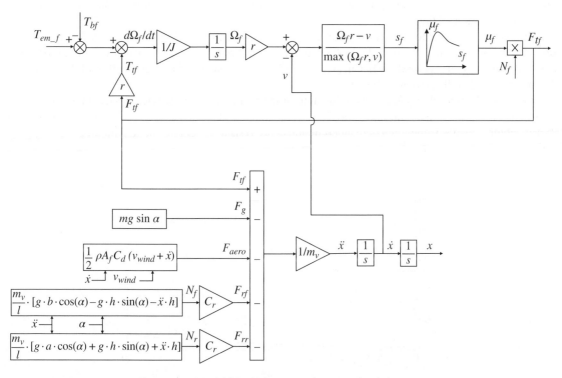

Fig. 7-10   *Model block diagram of a two-wheel drive car*

that the normal force, N_r, is still needed to compute the rolling resistance F_{rr}. Note also that the normal forces N_r and N_f maintain the same value as derived in Section 7.2.1.6.

7.2.4 Load characteristic of a vehicle

This subsection shows how the load of a vehicle behaves in function of the two variables: the speed and the slope [1]. Hence, assuming a constant speed of vehicle navigation (acceleration equal to zero), from (7.1) the expression that provides the resultant resistance load of the vehicle is (assuming zero wind speed):

$$
\begin{aligned}
Resistance = {}& m_v \cdot g \cdot \sin(\alpha) + \frac{1}{2} \cdot \rho \cdot A_f \cdot C_d \cdot V^2 \\
& + \frac{m_v}{l}(g \cdot b \cdot \cos(\alpha) - g \cdot h \cdot \sin(\alpha)) \cdot C_r \\
& + \frac{m_v}{l}(g \cdot a \cdot \cos(\alpha) - g \cdot h \cdot \sin(\alpha)) \cdot C_r
\end{aligned}
\tag{7.16}
$$

This corresponds with one term for the grade resistance, one term for the aerodynamic resistance, and two terms for the rolling resistances. This expression gives the actual force that the traction motor of the vehicle must provide, to reach a constant speed. Note that there are two independent variables for a given vehicle, as mentioned above: the speed of the vehicle, V, and the slope, α. If we consider a constant angle slope for a given vehicle, this expression results in a constant term plus a term dependent on the speed squared. Hence, if we represent different resistance curves in function of the vehicle speed, for different slope angles, we obtain the graphical illustration of Fig. 7-11.

Therefore, it is seen that for a given speed an increase on the demanded speed requires a quadratic increase of force. On the other hand, if we include the typical maximum capability curves of an electric motor (synchronous and asynchronous, as we show in Chapters 2 and 3) that is producing the traction effort of the vehicle, we obtain the graphical illustration of Fig. 7-12.

In this case, the typical torque curves in function of the rotational speed provided in Chapter 2 and 3 have been transformed into linear variables, producing equivalent curve shapes. Note also that the torque the electric motor provides is the force demanded by the curve resistance, only if the maximum capability curve of the motor is above the corresponding resistance curve.

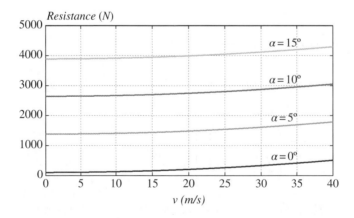

Fig. 7-11 *Resistance of a car in function of the grade and the vehicle's speed*

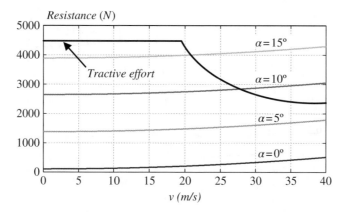

Fig. 7-12   *Resistance of a car in function of the grade and the vehicle's speed and tractive effort of an electric motor*

Table 7-2   *Main characteristics of the simulated fictitious vehicle*

Front wheels-tire assembling inertia	$J_f = 1\text{kgm}^2$
Rear wheels-tire assembling inertia	$J_r = 1\text{kgm}^2$
Wheel's radio	$r = 0.29\text{m}$
Mass of the vehicle	$m_v = 1418 + 75.5\text{kg}$
	(mass of a passenger = 75.5kg)
a	0.96m
b	1.65m
h	0.6m
Equivalent front area	$A_f = 1.37\text{m}^2$
Drag coefficient	$C_d = 0.3$
Rolling resistance coefficient	$C_r = 0.007$

7.2.5   Simulation performance

This section shows some simulation experiments based on the dynamic model of vehicles described in previous sections. Different simulation conditions are emulated, obtaining different vehicle performances and behaviors. The main characteristics of the simulated fictitious vehicle are listed in Table 7-2.

It is assumed that there is one passenger driving the vehicle with a mass of 75.5kg. The following subsections illustrate different operation conditions of the vehicle.

7.2.5.1   *Vehicle with four driving wheels on a dry surface*

The first simulation experiment shows the vehicle performance of a vehicle with the characteristics listed in Table 7-2, where the four wheels are driving wheels. It is supposed that the road surface is dry and is modeled according to the coefficients and formulas (7.6) and (7.7). The resulting adhesion coefficient is illustrated in Fig. 7-13.

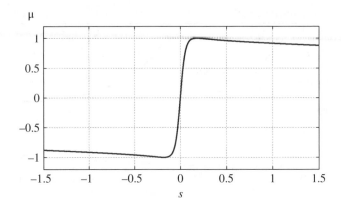

Fig. 7-13   *Adhesion curve µ(s) on a dry road*

The most important features of the experiment are summarized as:

- The wind speed is supposed to be zero for the modeled aerodynamic force.
- The four wheels of the vehicle are assumed to be identical, which means that they are modeled with equal inertias, equal adhesion coefficients, and equal effective radiuses.
- The slope of the road is $\alpha = 2°$ uphill.
- The traction torque of the vehicle is ideally created (the electric drive has not been included).
- At the driving wheels, the following torques are applied:
  - $T_{em} - T_b = 200$Nm for all front and rear wheels. The vehicle accelerates with constant traction torque. The breaking torque is set to $T_b = 0$ in all the wheels.
  - After 30sec the torques are set to $T_{em} - T_b = 0$Nm for all front and rear wheels, which means that the vehicle slows downs due to the effect of the rolling resistances, aerodynamic resistance, and the grade resistance.
  - Finally, at the last 10sec of the experiment the torques are set to $T_{em} - T_b = -200$Nm, which means that the vehicle decelerates more consistently with the strong effect of the brakes ($T_{em} = 0$ and $T_b = -200$Nm).

Thus in Fig. 7-14 and Fig. 7-15, the vehicle's most interesting characteristics are presented. In Fig. 7-14 the whole experiment is shown, while in Fig. 7-15 only detail of what occurs around time instant 31sec is depicted. It will be seen that the vehicle advances 400m during the whole simulation. The linear speed reaches a maximum point of 15m/sec. The acceleration is positive during the traction process, while it is negative during the braking process. Note also that deceleration is much higher when the braking torque is applied to the wheels. During the traction process, it is possible to distinguish a positive slip at the front and rear wheels, producing also positive adhesion coefficients, producing positive traction forces and torques. However, when there is no traction or braking torque applied to the wheels, the slip is zero, producing zero adhesion coefficients, and also producing zero traction forces at the vehicle. Finally, when a braking torque is applied at the front and rear wheels, the slips become negative, producing negative adhesion coefficients, also producing negative traction torques and forces (braking torque and forces). It will be seen that, although normal forces (N_f and N_r) are of different magnitude (owing to the geometry of the vehicle) at the front and rear wheels, the resulting traction forces (T_{tf} and T_{tr}) are nearly equal since the slips of the front and rear wheels take different values. Finally, the traction power produced by the driving motor takes a peak value of around 21kW, while the braking force produced by the brakes takes a peak of around −15kW.

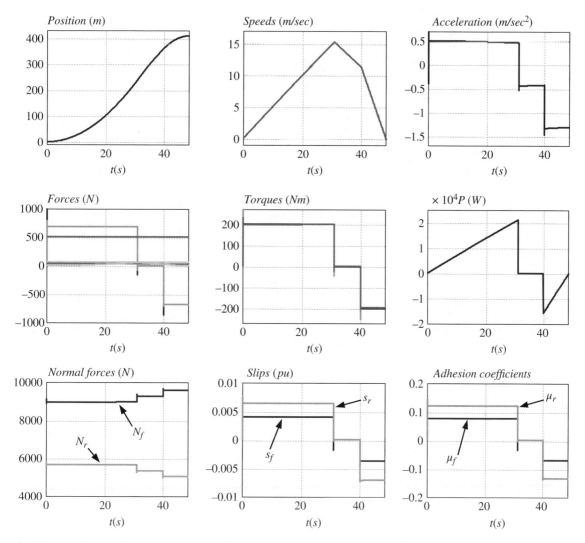

Fig. 7-14   *Vehicle performance with four driving wheels and on a dry surface. The traction and braking torques are set to ±200Nm*

7.2.5.2   *Vehicle with four driving wheels on an icy surface*

This experiment is equal to the previous experiment, with the only difference being that the road surface is icy, modeled with coefficients of Table 7-1 and (7.7). Thus the adhesion coefficient in this case results as depicted in Fig. 7-16.

Thus, as in the previous experiment, a traction-resulting torque of $T_{em} - T_b = 200$Nm for all front and rear wheels is set. Fig. 7-17 shows the simulation performance of the vehicle. It is seen that the slip of the rear wheels becomes uncontrolled after only 1sec of simulation, owing to the low adhesion of the surface. Therefore, the adhesion coefficient is in its unstable area, producing a reduced amount of traction force.

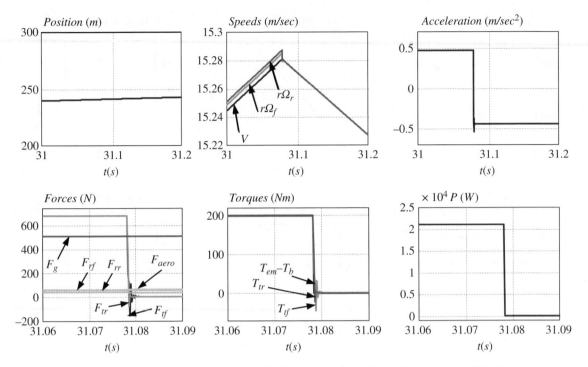

Fig. 7-15 *Vehicle performance with four driving wheels and on a dry surface. The traction and braking torques are set to ±200Nm. Zoom around 31sec*

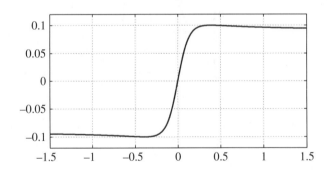

Fig. 7-16 *Adhesion curve μ(s) on an icy road*

Only the rear wheels tend to slip with an uncontrolled behavior because, as seen in the previous simulation, since they experience lower normal force they tend to work with higher slip and adhesion coefficient. In this case, the high input torque produces slip at the rear wheels, which reaches their unstable region, while the slip at the front wheels reaches the limit of the stable region.

On the other hand, if we reduce the traction torque from 200Nm to 100Nm, the slip of the rear wheels is maintained within its stable region, as can be seen in Fig. 7-18, producing an acceptable acceleration performance of the vehicle.

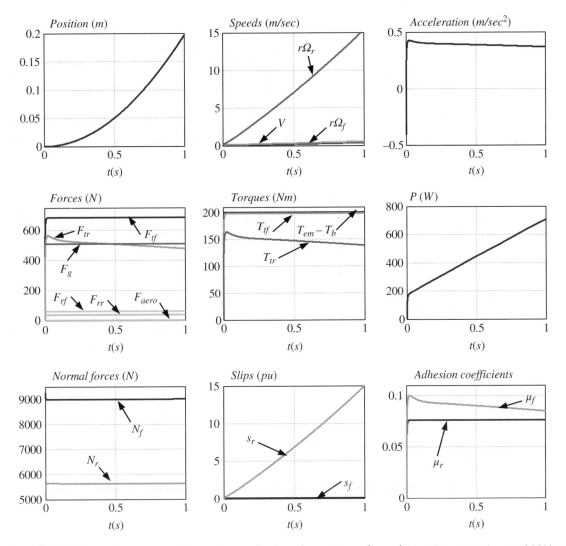

Fig. 7-17 *Vehicle performance with four driving wheels and on an icy surface. The traction torque is set to 200Nm. The slip of the rear wheels becomes uncontrolled.*

7.2.5.3 *Vehicle with two driving wheels on an dry surface*

In this section, a very similar result is presented to the previous cases, but in this case the vehicle simulated only presents driving wheels at the front axle. Thus, the parameters of the vehicle are identical to the previously simulated ones (Table 7-2). The driving torque applied by the drive to the wheels is 200Nm when accelerating and –200Nm when braking. Hence, the simulation performance of the most interesting variables is shown in Fig. 7-19. Compared with the previous simulations, the general behavior of the vehicle is very similar. In this case, the reached speed is much lower, since there are only two wheels generating torque at the vehicle.

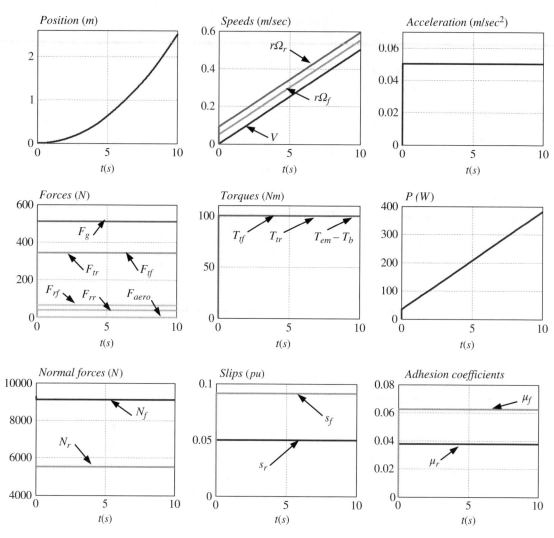

Fig. 7-18 *Vehicle performance with four driving wheels and on an icy surface. The traction torque is set to 100Nm. The slip of the rear wheels is maintained within the stable region*

7.2.6 Anti-slip control

In this section, a brief description is made of the anti-slip control philosophy normally carried out in vehicles. This anti-slip control enables the driver to avoid the uncontrolled slip of the wheels during the acceleration or braking process of the vehicle. The adhesion level is an important factor to efficiently accelerate or brake the electric vehicle. Hence, if the applied traction torque is greater than that allowed by the adhesion level at the wheel–road interface, then the uncontrolled wheel slip occurs exceeding a fixed (a previously defined) value that then produces a reduction of the traction forces. In an equivalent way, at the braking process, if the applied braking torque is too big, an uncontrolled slip will also occur, locking the wheels and therefore also minimizing the braking forces. In addition, uncontrolled slip of wheels is an undesired phenomenon which reduces the life

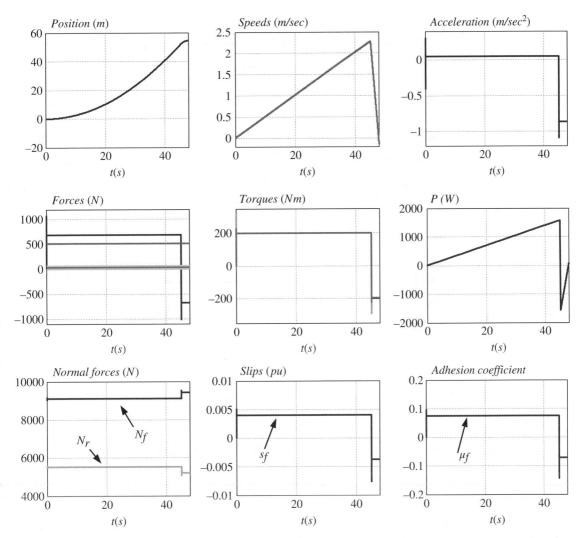

Fig. 7-19   *Vehicle performance with two driving wheels and on a dry surface. The traction and braking torques are set to ±200Nm; anti-slip control*

of mechanical parts. Vehicle manufacturers have developed many control techniques to avoid wheel slip. This section is not intended to describe any one of them in detail, owing to the subject's complexity. Instead, a rough general idea is provided about how the anti-slip control philosophy is carried out, so the reader can gain a basic knowledge of it.

Hence, to understand the influence of the slip ratio on the traction and braking forces, consider the adhesion characteristics of a dry and icy surface shown in Fig. 7-20. As was described in (7.6), the traction and braking forces result from the product of the normal forces and their adhesive characteristics. Therefore, as can be seen in Fig. 7-20, the magnitude of the tire force typically increases linearly with the slip for small slip ratios. It reaches a maximum of adhesion coefficient (therefore also of force) at a given slip depending on the characteristic. At slip ratios beyond this value, the magnitude of the tire forces decreases. Thus, if the driver presses

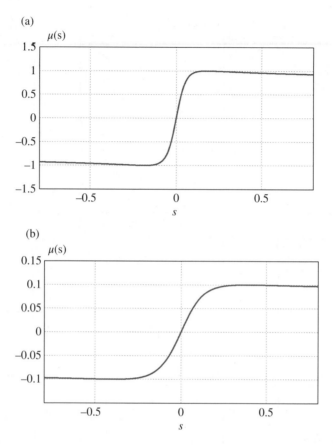

Fig. 7-20   *(a) Adhesion curve μ(s) on a dry road, (b) Adhesion curve μ(s) on an icy road*

the accelerator hard, the wheels will accelerate faster than the vehicle, resulting in a big slip value. Similarly, if the driver presses the brake pedal hard, the wheels will tend to slow down faster than the vehicle, resulting again in a big slip value. In both situations, the big slip value will produce reduced tire forces during the acceleration or deceleration process of the vehicle.

Therefore, the anti-slip control philosophy basically prevents excessive traction or braking torques from being applied to the wheels, in order to avoid the slip exceeding the optimum value. Hence, if the driver presses too hard on the accelerator or the brake, the anti-slip control will reduce the torque in order to keep the slip below the optimum value and maximize the traction or braking forces.

The following two subsections show some examples of traction and braking performances of a vehicle which help to understand the anti-slip control philosophy.

7.2.6.1   *Traction*

This section shows the performance of a specific vehicle during the acceleration process. The simulated vehicle characteristics are the ones presented in previous subsections and in Table 7-2 and it has two driving wheels at the front. The slope of the road is supposed to be 2°. The simulated road is dry and, therefore, the adhesion coefficient of tire/road is as depicted in Fig. 7-13. The most interesting magnitudes of the vehicle performance when it accelerates at constant traction torque of $T_{em} = 300\text{Nm}$ are shown in Fig. 7-21. It is possible to observe that the vehicle accelerates normally, under a small and controlled slip value. All the torque and forces of the vehicle behave in a controlled way. Therefore, it is possible to conclude that, with this traction torque value and on this surface, this vehicle does not need a special anti-slip control actuation.

However, if the vehicle is running on an icy surface as the one presented in Fig. 7-16, for instance, the same vehicle with the same traction torque of $T_{em} = 300\text{Nm}$ will slip, not being possible to perform an efficient acceleration performance. Owing to this, it is necessary to implement an anti-slip control. The basic idea of the anti-slip control philosophy can be understood with the help of Fig. 7-22. There, it will be seen that the traction torque demanded by the driver is supposed to be constant. However, when it is detected that the slip goes

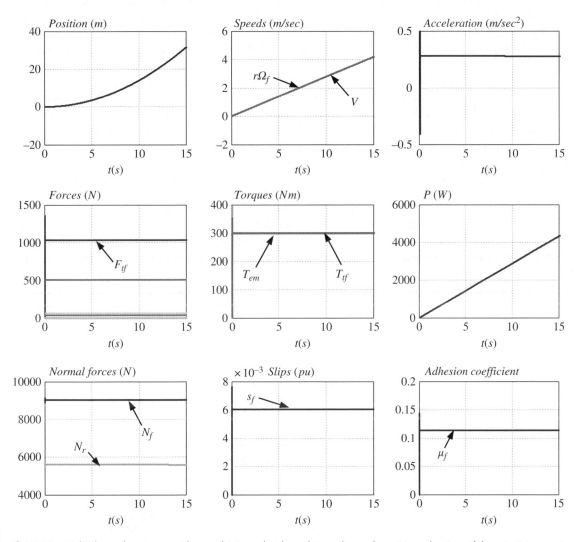

Fig. 7-21 *Vehicle performance with two driving wheels and on a dry surface. No reduction of the traction torque* T_{em} *is required to avoid uncontrolled slip; acceleration process*

beyond the optimum value reducing the traction forces, the anti-slip control progressively reduces the actual torque applied to the wheels (by the traction motor). Thanks to this, the speed of the wheel is reduced until the slip is within the linear region, increasing therefore the traction longitudinal force and producing a more effective acceleration. After that, once the slip has been reduced to its linear region, the anti-slip control enables the driver to again increase the torque progressively if the torque demanded by the driver is still bigger. Typically, if the road conditions do not change, the increase of the torque will make the driver lose the slip control once more, because it will become necessary again to decrease the torque. Thus, this process is repeated several times during the acceleration process, producing the typical torque profile depicted in Fig. 7-22. Therefore, when the actual torque is being reduced, it means that the slip has been lost, and the anti-slip control is trying to reduce the speed of the wheel, also reducing the slip. On the other hand, when the actual torque is being increased, it means that the slip has been recovered and the control is trying to progressively increase the torque to reach the level demanded by the user. The system will always try to reach the torque demanded by the driver, but if the slip takes too large values, the control reduces the actual torque applied to the wheels. Obviously, if the slip is estimated to be inside the linear region, the anti-slip control will not reduce the actual torque and the torque applied to the wheel will be the one demanded by the user, as occurs in the simulations presented in Fig. 7-21.

One practical problem of anti-slip control is the slip estimation. In general, the slip is not measured, in order to avoid complex and expensive speed sensors. Often, the only measurement available in the vehicle is the rotating speed of the four individual wheels. Algorithms based on sophisticated models typically used to estimate the slip are beyond the scope of this book. In addition, there are several different methods to produce the torque reduction/increase applied to the wheels, and these are not analyzed here.

Fig. 7-23 shows a simulation-based experiment of the same vehicle as the previous example, but in this case running on an icy surface (Fig. 7-16). An anti-slip control algorithm is implemented to prevent undesirable large slip ratios. It is possible to observe that when the slip takes a significant big value the adhesion coefficient, μ_f, takes a very low value and, therefore, also the traction force, F_{tf}. If the traction effort is lower, the acceleration of the car is slower and therefore the slip continues increasing.

Note also that, in this example, for illustrative purposes the slip has been allowed to reach values of up to 1.65, producing a significant reduction of the adhesion coefficient. It would be possible to adjust the moment in which the torque is reduced to a lower value of slip than in this example.

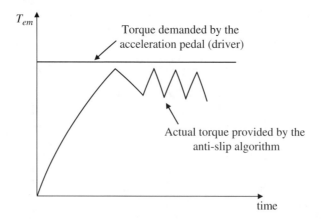

Fig. 7-22   *Traction torque* T$_{em}$ *provided by the driving wheels with anti-slip algorithm operating*

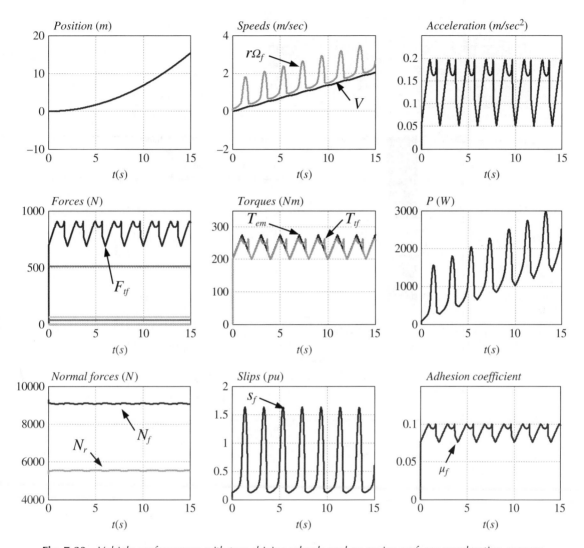

Fig. 7-23 *Vehicle performance with two driving wheels and on an icy surface; acceleration process*

In contrast, comparing this experiment with the previous one, it is possible to see that on the icy surface the car is able to move less distance (15m against 30m, in 15sec), since the vehicle on the icy surface is obliged to accelerate with lower traction torques to prevent the uncontrolled slips, resulting also in lower traction forces.

7.2.6.2 *Braking*

The performance of the vehicle during the braking process is similar to the acceleration process described in the previous subsection. The braking of a vehicle is normally made [1] by the combination of the mechanical

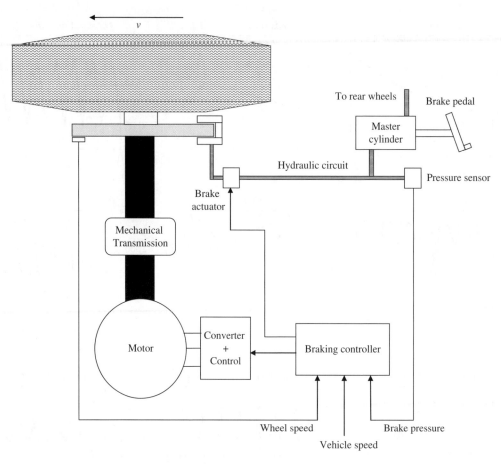

Fig. 7-24 *Regenerative braking integrated with conventional hydraulic braking system [3]*

friction brake and the regenerative electric brake (developing a negative electromagnetic torque by the electric motor). Both braking methods are normally combined to try to obtain basically the combination of two goals: recover the maximum kinetic energy of the vehicle and distribute the total braking forces on the front and rear axles, so as to achieve an appropriate braking. Thus, the braking process results in a compromise between the driver's comfort and the amount of energy recovered. Fig. 7-24 illustrates a schematic block diagram of a regenerative braking integrated with a conventional hydraulic braking system. Note that a controller coordinates the effect of both braking solutions.

It is possible to find different braking philosophies in electric vehicles, basically grouped as: series braking with optimal braking feel, series braking with optimal energy recovery, and parallel braking [1]. In this section, only a short description of the parallel braking is provided. In general, the parallel brake employs simultaneous electrical regenerative brake and mechanical friction brake. Often, the regenerative brake is applied only to the front wheels, while the mechanical brake presents a fixed ratio distribution between the front wheels and rear wheels, as depicted in Fig. 7-25.

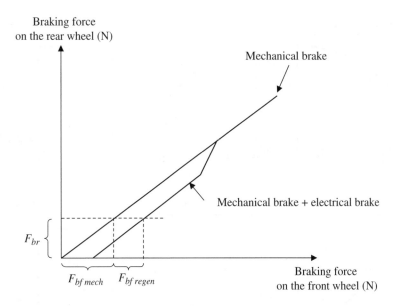

Fig. 7-25 *Braking forces distribution between the rear and front wheels with parallel braking strategy*

It is possible to see in Fig. 7-26 how the regenerative brake adds additional force to the mechanical brake at the front wheels. On the other hand, the mechanical braking forces at both front and rear axles are proportional to the force pressure at the pedal, developed by the user. In a similar way, the regenerative force is also in function of the pressure at the pedal. In addition, in general, when the vehicle deceleration demand is high, only the mechanical brake is used, maintaining the braking balance. Then, below this high deceleration demand, both mechanical and regenerative brakes are commanded. Finally, when the deceleration command is less than a minimum value, only regenerative braking is employed.

As mentioned before, this is just an example of different possible braking strategies. Series braking philosophies are braking alternatives that can be found in specialized literature. Alternatively, the anti-lock braking system (ABS) is a feature that electric vehicles and hybrid vehicles often incorporate. The ABS system prevents the vehicle's wheels from locking, when a strong and sudden braking maneuver is demanded by the driver. Thanks to the ABS system's ability to prevent the wheels locking, more-effective, faster, and safer braking of the vehicle is possible.

The ABS typically distributes the braking force between the regenerative torque and the electronically controlled mechanical brake. Thus, depending on the pressure imposed on the brake pedal by the driver, there is one controller that determines the overall braking torques at the front and rear wheels and how it is distributed between the regenerative torque and the mechanical brake.

Fig. 7-26 illustrates the braking process using an ABS control algorithm (anti-slip) of the car simulated in the previous subsection driven on an icy surface. As with the acceleration process, the vehicle brakes with a torque profile determined by the action of the anti-slip algorithm. In order to avoid an excessive slip ratio during deceleration, the braking torque is increased and decreased regularly, producing a traction (negative) force, F_{tf}. Thanks to the anti-slip control, the vehicle is able to brake in an efficient way. Note that, as mentioned during the traction process, it is possible to reduce the peak slip reached during braking, by reducing the braking torque quicker than is done in this example.

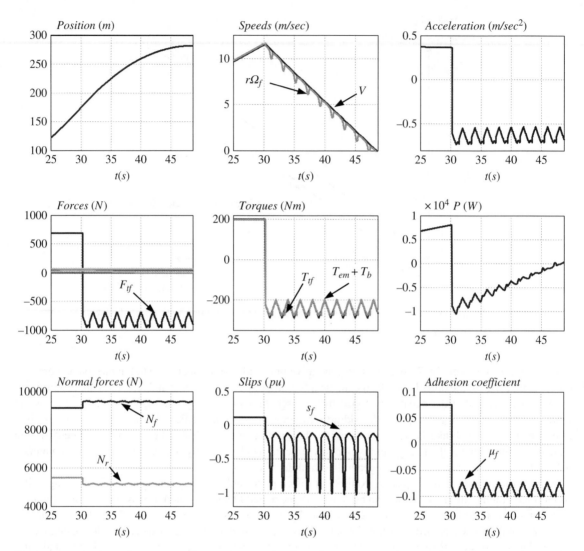

Fig. 7-26 *Vehicle performance with two driving wheels and on an icy surface. Braking process with anti-slip control algorithm*

7.3   Electric vehicle configurations

As in the internal combustion vehicle case, there are a lot of possible electric vehicle configurations. The pursuit of the optimum power train for each scenario (city car, medium-range car, off-road cars, etc.) remains a challenge. Electric vehicles also have the drawback of having a limited range, so selecting the appropriate drive train configuration which yields better efficiency is required.

7.3.1   Drive train configurations

The electric vehicle is characterized by the fact that it uses an electric motor for traction and chemical batteries as energy sources. Alternatively to the chemical batteries, fuel cells, ultracapacitors, and flywheels can also be employed. Fig. 7-27 illustrates a general scheme of an electric vehicle propulsion system [1], [4], [5]. It is possible to distinguish the driving wheels, the mechanical transmission, the electric motor, the power electronic converter, the controller, and the energy source with its corresponding energy refueling system. By means of the accelerator, the brake pedals, and the steering wheel, the driver provides the inputs for driving the vehicle. From the pedal inputs, the controller generates the corresponding controlled signals for the power electronic converter, which supplies the electric motor and therefore moves the driving wheels. The energy flow can be, in most cases, bi-directional between the energy source and the electric motor, since batteries, ultracapacitors, and flywheels are commonly prepared to receive regenerated energy at a certain stage.

The mechanical transmission system can be configured in several different ways in electric vehicles [1]. Basically, it is possible to find the following elements in the mechanical transmission of an electric vehicle:

- Differential: Is the mechanical device that permits the wheels to be driven at different speeds, when they move along curves. In general, it consists of a simple planetary gear train.
- Clutch: The clutch is the mechanical device that connects and disconnects the power transmission, from driving shaft to wheels.
- Gearbox: Is a mechanical device that has a set of gear ratios, in order to adapt the speed/torque ratios from the driving shaft to the wheels. The gearbox can be multispeed or fixed speed.

Fig. 7-28 illustrates one possible drive train configuration, which comes from the simple replacement of a conventional car's internal combustion engine by an electric motor. In this case, therefore, the clutch, gearbox, and differential are present at the drive train.

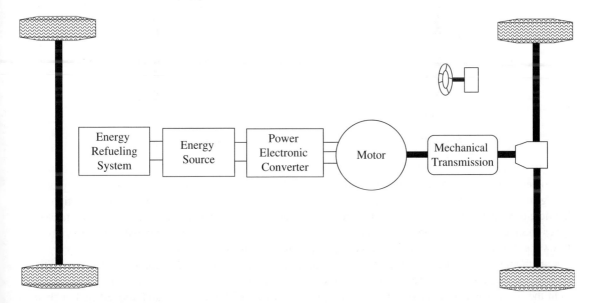

Fig. 7-27   *General scheme of an electric vehicle propulsion system*

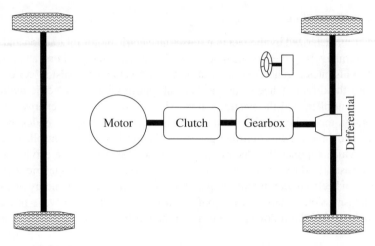

Fig. 7-28   *First possible configuration of an electric vehicle [1]*

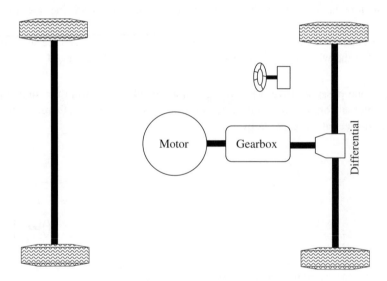

Fig. 7-29   *Second possible configuration of an electric vehicle [1]*

Another possible configuration is shown in Fig. 7-29. In this case, the multispeed gearbox is substituted by fixed gearing, avoiding the necessity of a clutch. This fact can be achieved by an appropriate selection of the electric motor characteristics. Compared to the previous configuration, the motor must be prepared to work at higher speeds, providing higher torques at these higher speeds. This solution, compared to the previous one, allows the dimensions, weight, and complexity of the drive train to be reduced.

In an attempt to further reduce the size and weight of the mechanical transmission system, it is possible to integrate the fixed gearbox, differential, and motor at the axle of the driving wheel, as depicted in Fig. 7-30.

On the other hand, some drive train configurations do away with the differential altogether. For them, a motor is needed to drive each wheel, so each wheel can operate at different speeds when running on curves. Thus, a possible configuration is depicted at Fig. 7-31, where the differential is eliminated and one motor and fixed gearing is utilized for each wheel.

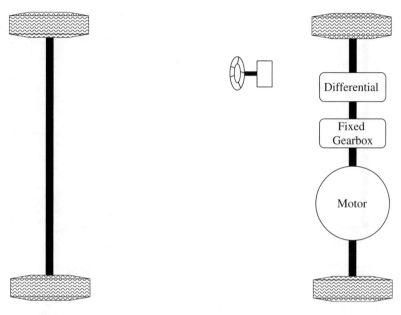

Fig. 7-30   *Third possible configuration of an electric vehicle [1]*

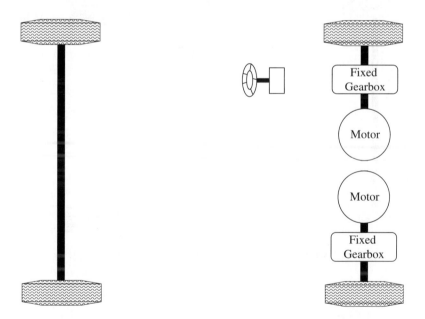

Fig. 7-31   *Fourth possible configuration of an electric vehicle [1]*

Using the same arrangement philosophy, it is possible to find the in-wheel concept, which locates the traction motors within the wheel. This concept enables the weight and size of the drive train to be further reduced, but it requires a more sophisticated motor design. Fig. 7-32 illustrates a schematic of this configuration, which includes also a thin fixed gearbox to match torques and speeds of the electric motor and the wheels.

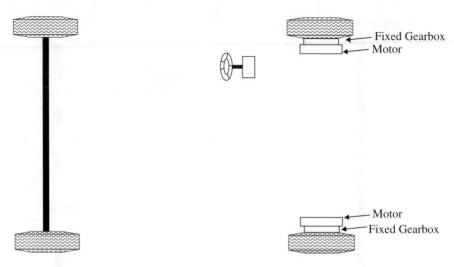

Fig. 7-32   *Fifth possible configuration of an electric vehicle [1]*

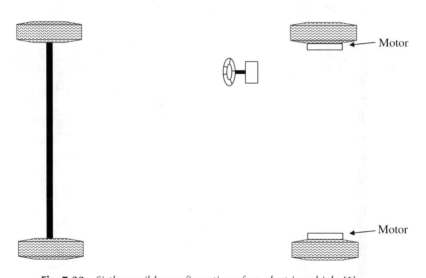

Fig. 7-33   *Sixth possible configuration of an electric vehicle [1]*

Finally, Fig. 7-33 illustrates a configuration where the gearbox is eliminated from the drive train, making the electric motor directly in charge of driving the wheels. In this way, the motor operates at the same speed and torques as the wheels of the vehicle. For that purpose, the electric motor in general must be specially designed, a typical configuration being the synchronous multi-pole motor. In addition, the motor must also be prepared to be able to provide the required high torques at the start-up and during acceleration of the vehicle.

As will be noted, all the drive train configurations shown present a single axle driven configuration.

7.4   Hybrid electric vehicle configurations

A hybrid electric vehicle (HEV) combines the internal combustion engine power with electric power. Thus, it uses both internal combustion engine and electric motor for propulsion of the vehicle. There are two basic configurations of HEV: series and parallel. In a series HEV the internal combustion engine and the electric motor are connected in series, the electric motor being the one that provides power to the wheels. Conversely, in a parallel HEV, both the internal combustion engine and electric motor are connected in parallel, making it possible to provide power to the wheels in parallel. As will be seen below, apart from these two basic configurations, there are other types of HEV configurations.

Thus, in an HEV the fuel and the internal combustion engine are the main sources of energy. On top of this, the electric motor allows the vehicle's overall efficiency to be increased by adjusting to the optimum operating point of the internal combustion engine, and it recovers energy from regenerative braking, storing it in the batteries.

It must be noted that the internal combustion engine could be replaced by other types of power sources, such as fuel cells. In a similar way, an HEV's batteries can be replaced by ultracapacitors or flywheels, or by a combination of both.

7.4.1   Series hybrid electric vehicle

The series HEV schematic is depicted in Fig. 7-34. In this configuration [4], the fuel tank and engine are the main energy sources. The engine powers an electric generator which converts the mechanical energy produced by the engine into electric energy. These elements allow making the engine function at its optimum operation point, improving its overall system efficiency and therefore reducing fuel consumption. Then, the energy provided by the engine-generator set and/or the batteries is supplied to the electric motor, producing propulsion for the wheels. The electric traction motor can work as a motor or as a generator and also in forward or reverse modes. When braking in regenerative mode, the energy is stored in the batteries. For that purpose, the battery pack must be sized accordingly. In addition, a battery charger may also be needed to charge the batteries by a wall plug-in from the electric grid.

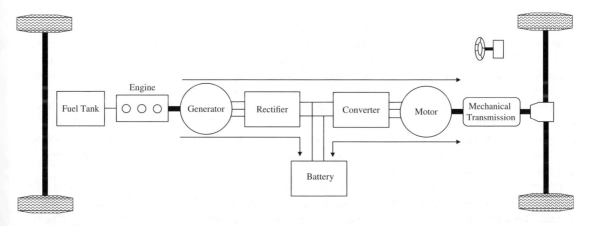

Fig. 7-34   *Schematic of a series HEV (Source: [4]. Reproduced with permission of John Wiley & Sons)*

Therefore, the series HEV can work at different modes depending on the operating conditions demanded [4]:

- **Battery-only mode:** If the vehicle power demand is low and can be satisfied by the energy stored at the batteries, the engine is turned off and the vehicle is propelled solely by the batteries.
- **Engine-only mode:** In this situation, the vehicle energy comes only from the engine and the batteries are not charged or discharged. This happens typically at moderately high power demands, when the vehicle cruises, for instance on freeways.
- **Hybrid mode:** At the highest power demands, the traction power is provided by both the engine and the batteries.
- **Engine traction and battery charging mode:** When the engine is working and the state of charge (SOC) of the batteries is low, the engine propels the vehicle and it charges the batteries at the same time.
- **Regenerative braking mode:** The electric motor works as a generator and it charges the batteries.
- **Battery charging mode:** When the engine provides only energy to charge the batteries (the traction motor does not receive energy).
- **Hybrid battery charging mode:** When the batteries are charged by both the engine and the electric motor working in regenerative braking.

On the other hand, it can be highlighted that this HEV configuration is advantageous, since it can make the engine operate continuously at maximum efficiency, because it can freely rotate at different speeds, not directly depending on the travelling speed. In addition, the usage of the electric motor does not need multi-gear transmission, simplifying the mechanical transmission. Conversely, the main disadvantages of this HEV configuration are mainly the increase of losses due to the energy conversion through the engine and motor, and the cost and weight added by the generator.

In addition, it must be remarked that, depending on the design, it is also possible to find ultracapacitors connected in parallel to the battery pack, in order to reduce the stress on the batteries due to high current transient demands.

Finally, Fig. 7-35 illustrates an alternative series HEV configuration, such as the wheel hub motor. It will be seen that there is one traction motor installed at each wheel, allowing the mechanical transmission to be simplified, by means of a more complex control philosophy of the four wheels independently.

7.4.1.1   *Plug-in hybrid electric vehicles (PHEVs)*

A PHEV is the combination of an electric vehicle and a HEV that can be recharged with an electric plug from the electric grid. PHEVs are often called range-extended electric vehicles (ReEVs) or extended range electric vehicles (EREVs). These names come from the idea that the vehicle predominantly operates as an electric vehicle and incorporates a diesel or gasoline tank, for running an extended range when the battery energy is ended. Compared to an HEV, the PHEV normally presents a larger battery pack, allowing it to work as a pure electric vehicle over greater distances. Note that during the extended range this vehicle concept can accept regenerative braking energy for charging the battery pack. PHEVs, just like HEVs, can present different configurations: series, parallel, or series-parallel. A general layout for a series configuration PHEV is depicted in Fig. 7-36. Compared with the series HEV configuration, a battery charger is included as well as a larger-sized battery pack. Additionally, it is also possible to incorporate a DC/DC converter between the battery pack and the converter, for adapting the voltage.

Basically, while driving when the battery is almost empty the range extender part (fuel-tank, engine, generator, and rectifier) starts to operate. Note that the rectifier needs to be a bi-directional converter to start the engine. This takes a few seconds and then the engine plus generator start to generate electric energy. The size, in power terms, of the range extender is determined by the energy management strategy of the vehicle.

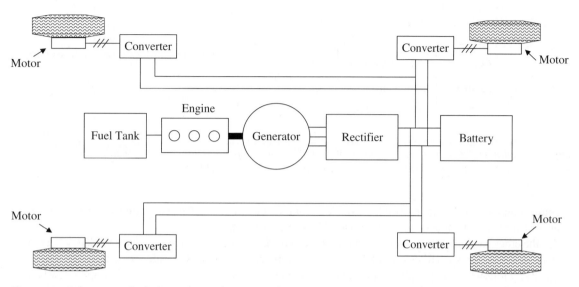

Fig. 7-35 *Schematic of a hub motor configuration of a series HEV (Source: [4]. Reproduced with permission of John Wiley & Sons)*

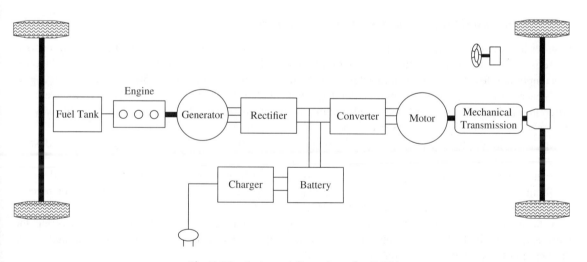

Fig. 7-36 *Series configuration of a PHEV*

A simplified version of the battery charger that typically incorporates the PHEV is depicted in Fig. 7-37. It includes a diode rectifier which converts the AC voltage from the grid into DC voltage. Then a DC/DC converter that can take different topologies adapts this DC voltage to the voltage of the battery pack. After that, often an additional DC/DC bi-directional converter is included, adapting the voltage of the battery pack to the high voltage of the DC bus of the vehicle. In Section 7.6, details of representative battery chargers are presented compared to this simple topology. It is shown how a power factor correction is often included, in order to mitigate the negative impact on the grid. In addition, safety requirements and standards often oblige

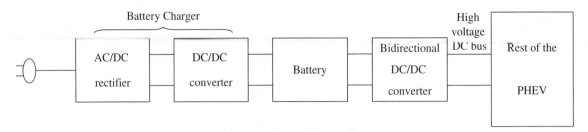

Fig. 7-37   *Simplified battery charger scheme of the PHEV. Often a power factor corrector can be included as well as isolation from the grid (Source: [5]. Reproduced with permission of John Wiley & Sons)*

manufacturers and designers to provide isolation from the electric grid. This isolation can be provided by a transformer at the AC side or even at the DC/DC converter itself.

On the other hand, it must be mentioned that some vehicle manufacturers have shown a tendency to convert already designed regular HEVs into PHEVs. Thus, in a practical way, it is possible to achieve this objective basically by following two possible options: one is to replace the existing battery pack with a larger new battery pack (normally from nickel metal hydride to lithium-ion) or another is to add an extra battery pack to the existing one. Hence, today's HEVs typically use battery packs with nickel metal hydride (NiMH), with a range between 1.2k and 2kWh. However, the converted PHEVs can have a range of between 7 and 16kWh [4].

7.4.1.2   *Series hybrid energy management control strategies*

In series hybrid vehicles, as well as in renewable energy microgrids, there is more than one source of energy. There is a suitable framework for the adoption of advanced control strategies flow of energy to reduce the consumption of fossil fuels. As mentioned previously, in series hybrid vehicles the internal combustion engine plus generator system, also named range extender, is not coupled to the tractor vehicle axle, so it can operate at different points of power (torque vs. speed) without being influenced by the vehicle's speed. In this way the internal combustion engine can operate in areas where its performance is greater.

Fig. 7-38 shows the efficiency maps of the two parts that form a range extender system: the combustion engine map and the electric generator plus power electronics map. Both systems are designed to match their efficiency peaks in the same region. In this specific case, they are designed to deliver 19kWs (90Nm @ 2000rpm) efficiently, as the combustion engine reaches 34% efficiency and the electric generator reaches 96%, achieving a total system efficiency of $\approx 33\%$.

As the internal combustion engine and the generator are in the same axle, if both efficiency maps overlap, the range extender overall efficiency is obtained, as shown in Fig. 7-39. It is also easy to obtain the optimum curve of operation for a given power demand.

Series hybrid vehicle energy flow control is usually based on the experience of developers, governed by rules imposed for different subsystems (battery, combustion engine, power inverter) to safeguard their integrity and try to make the vehicle as efficient as possible.

Given the nature of the subsystems that make up the series hybrid vehicle, whose behavior is rarely linear or can be represented by equations, fuzzy and heuristic controls appear more robust than other solutions to the nonlinearities of the plant to be controlled.

In the specific case of the series HEV, inputs as the state of battery charge, power demand (requested electric power), combustion engine power, temperatures, etc. are fuzzified defining the behavior of the algorithm power control based on the "windows" that developers can define. This behavior has been analyzed for parallel hybrid vehicles in several articles [6].

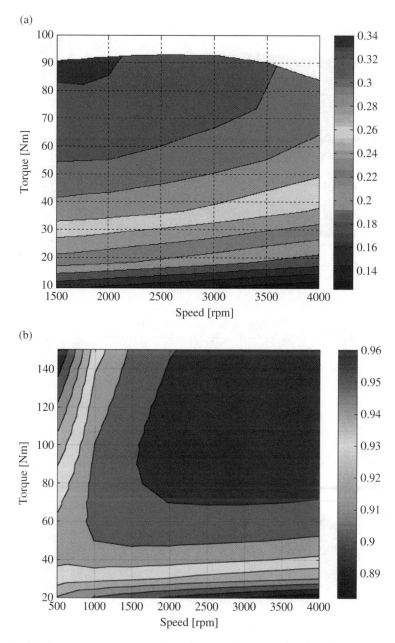

Fig. 7-38   *(a) Example of a range extender internal combustion engine efficiency map, (b) Example of a range extender inverter plus electric motor efficiency map*

Battery SOC hysteresis control example   This simple control is based on the switch-ON and switch-OFF the generator according to the SOC level of the traction battery. As the generator is not connected to the wheels axis, normally, the generator works in a very narrow area of its efficiency map. The power delivered by the generator is almost constant in this strategy. This allows the generator to work in its best efficiency map area,

(a)

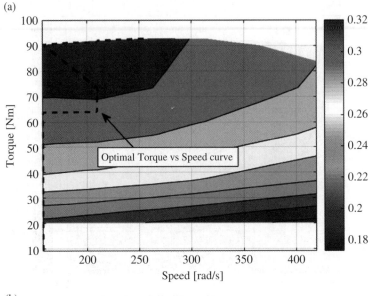

(b)

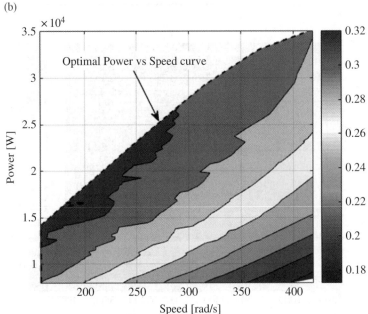

Fig. 7-39   *(a) Example of a range extender overall efficiency map (torque vs speed), (b) Example of a range extender overall efficiency map (power vs speed)*

always within the limits of the rules imposed by the battery charge, both current and temperature, as the charge could degrade the battery because of bad energy management.

In Fig. 7-40, a variant of hysteresis control can be seen. In the first stage the series hybrid vehicle is working as a pure electric vehicle till its battery SOC reaches a certain low SOC level. In that moment the range extender starts to operate at a constant power level, charging the battery to a medium SOC level. In the final stage, the

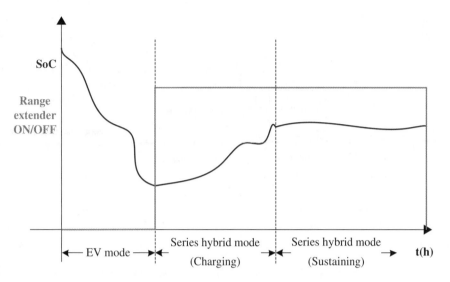

Fig. 7-40   *Battery SOC hysteresis control with charge sustaining mode*

range extender delivers the same energy that the traction system consumes, so there is no charge in the battery. Even though the range extender and the vehicle traction motor are power balanced, the range extender can operate in a more efficient point as its working speed is not defined by the vehicle's speed.

Within this range of strategies are also those that make use of the location of the vehicle, whether it is on the highway, approaching a city, or within the city itself [7]. In this way, vehicles can meet emissions legislation in cities like London, legislation that is being more extended to other areas of the world.

7.4.2   Parallel hybrid electric vehicle

The parallel HEV can be understood as a conventional internal-combustion-engine-powered vehicle assisted by an electric motor [4]. The schematic of the parallel HEV is depicted in Fig. 7-41. It will be seen that both the engine and the electric motor can deliver in parallel mechanical power to the wheels. The engine and electric motor are coupled together through a special mechanical coupling. Compared with the series HEV, the generator is not needed in this configuration, but a more complex mechanical coupling is mandatory. In this configuration, the electric motor can operate as a generator in regenerative braking or absorbing energy from the engine. As with the series HEV, the same operation modes with the same operation principles can be found in the parallel HEV [4]: battery-only mode, engine-only mode, hybrid mode, engine traction and battery charging mode, regenerative braking mode, battery charging mode, and finally hybrid battery charging mode.

The mechanical coupling of the parallel HEV shown in Fig. 7-41 is a device that can be a torque coupling or a speed coupling. The general representation of both devices is depicted in Fig. 7-42 [4]. One of the input shafts comes from the engine, while the other comes from the electric motor. Then the output of the coupling goes to the mechanical transmission.

Thus, the torque coupling ideally adds the input torques according to the following expression (losses are neglected) [4]:

$$T_{input1} \cdot k_1 + T_{input2} \cdot k_2 = T_{output} \tag{7.17}$$

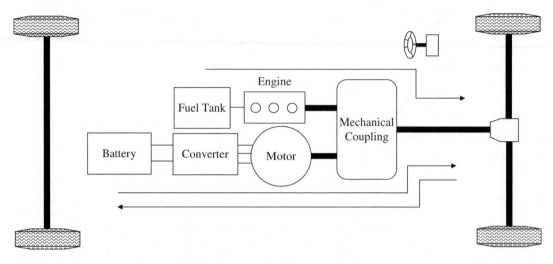

Fig. 7-41   *Schematic of a parallel HEV (Source: [4]. Reproduced with permission of John Wiley & Sons)*

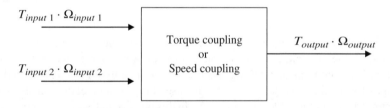

Fig. 7-42   *General representation of torque coupling and speed coupling devices*

While the speed relations are:

$$\frac{\Omega_{input1}}{k_1} = \frac{\Omega_{input2}}{k_2} = \Omega_{output} \qquad (7.18)$$

being k_1 and k_2 constants that depend on the torque coupling design. On the other hand, the speed coupling ideally adds the two input speeds according to:

$$\Omega_{input1} \cdot k_1 + \Omega_{input2} \cdot k_2 = \Omega_{output} \qquad (7.19)$$

with torque relations as:

$$\frac{T_{input1}}{k_1} = \frac{T_{input2}}{k_2} = T_{output} \qquad (7.20)$$

being again k_1 and k_2 constants that depend on the speed coupling design. Thus, it is possible to find different types of devices for each torque and speed coupling system. For instance, one commonly used mechanical speed coupling is the planetary gear. A basic planetary gear is composed of three different-sized gears: the sun gear, the planetary gear, and the ring gear. The sun gear is mounted on a shaft located at the center. The planetary gear, which is connected to the planet carrier, rotates around the sun gear. The ring gear encircles the sun and planetary gears. The ring gear presents internal teeth, while the planet and sun gears present external teeth. One conceptual representation of a planetary gear is depicted in Fig. 7-43.

Therefore, the speed relations between the different gears are determined by the number of teeth of each gear. These speeds are then the speeds of the shafts of the motor, engine, and final drive, which are connected to the planetary gear.

7.4.3   Series-parallel hybrid electric vehicle

It is possible to combine both the series and parallel architectures, producing the series-parallel HEV [4] shown in Fig. 7-44. It will be seen that the engine is connected to the mechanical transmission, so it can drive the wheels directly. This configuration is able to be operated as a series or parallel HEV. Therefore, the fuel consumption can be optimized choosing the right configuration for each working condition. However, the more

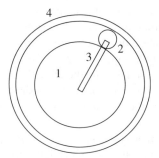

Fig. 7-43   *General representation of a planetary gear: 1 is the sun gear, 2 is the planetary gear, 3 is the planet carrier, and 4 is the ring gear (Source: [2]. Reproduced with permission of John Wiley & Sons)*

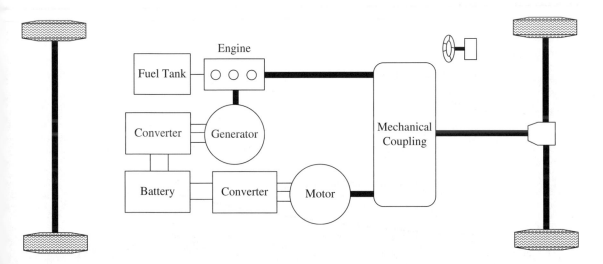

Fig. 7-44   *Schematic of a series-parallel HEV*

components needed makes this configuration more complex and expensive than the pure series or parallel basic configurations.

7.5   Variable speed drive of the electric vehicle

7.5.1   General view

In this section, a general view of the variable speed drive of the electric vehicle is provided. Some common features of the power converter, the electric motor, and the control strategies utilized for traction of the electric vehicle are described.

Thus, as is shown in Chapters 2 and 3 and discussed throughout this book, the maximum capability curve of the torque magnitude and power magnitude of the typical electric motor employed in electric vehicles is depicted in Fig. 7-45 (equivalent to Fig. 7-12). In addition to Fig. 7-12, Fig. 7-45 represents the load torque curve when the vehicle is at steady state. As is studied in earlier sections of this chapter, the load torque of the vehicle is increased with the speed. In general, the maximum capability torque of the motor is greater than the load torque of the vehicle. With this torque difference, the vehicle is able to accelerate and brake. At lower speeds, the torque difference between the motor maximum torque and the load torque is greater, therefore the achievable acceleration and deceleration is greater but the risk of slip is also increased. Conversely, above the base speed, the torque difference is not so high and therefore the achievable accelerations and decelerations are lower. The motor is designed or chosen according to the desired features (more or less torque, greater or lower base speed, etc.). Accordingly also, the cost and volume of the traction drive of the electric vehicle will be affected.

On the other hand, often in electric vehicles there is a mechanical gearbox which equalizes the speeds of the electric motor to the speeds required by the wheel. In general, the efficient rotational speed ratios of the motor are greater than the speeds of the wheels, so a gearbox reducing the speed is necessary. In fact, single-gear transmission and multi-gear transmission can be used, but nowadays probably a single-gear transmission is more common. Moreover, with the introduction of high pole number electric synchronous motors, the geared transmission can be avoided obtaining a direct drive design, as is shown in the configuration in Fig. 7-33.

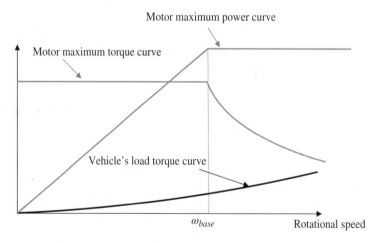

Fig. 7-45   *Electric motor characteristic and vehicle's load torque curve*

In addition, it is usual to represent the torque and speed operating points of the vehicle during given and categorized drive cycles. These operating points of traction torque vs speed are then represented as depicted in Fig. 7-46. In this example, the NEDC (New European Driving Cycle) urban drive cycle has been employed. This is a standard cycle defined by the United Nations Economic Commission for Europe (UNECE). It is

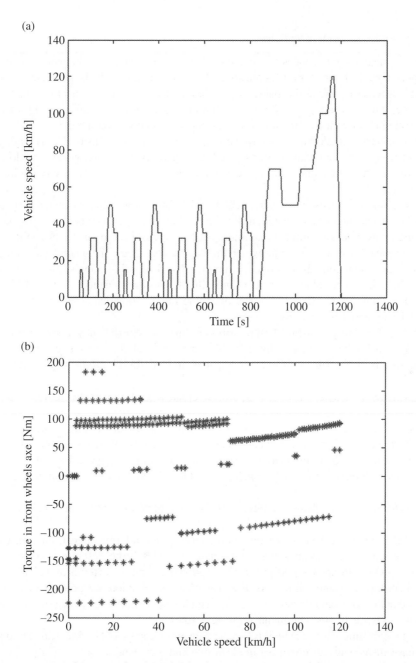

Fig. 7-46 *(a) NEDC driving cycle, (b) Wheel axle torque operating points running the NEDC cycle*

designed to assess the emission levels of car engines and fuel economy in passenger cars. Note that the traction torque strongly depends on the speed variations demands imposed by the drive cycle. The operating point representation shows in which region the wheel axle operates. Using this information and selecting the proper reduction gearbox help with choosing the appropriate electric motor.

7.5.2   Electric motors in electric vehicles

Electric motors are ideal traction devices for vehicles. They are quiet, operate at high efficiency, and have very good torque/speed characteristics for this application. Their maximum torque is available at very low speeds, and even at high speeds they are still capable of producing adequate torque at near constant power across a wide speed range, which means that a simple single-ratio transmission is sufficient. Furthermore, electric motors can operate in an overload condition for brief periods when it is required.

An appropriate electric vehicle traction motor should have high power density (kW/l), high specific power (kW/kg), high efficiency in all operating points, overload capability, low noise generation, low cost, and low maintenance. Also, it has to take into account each converter technology and necessary cooling system associated to each motor.

The first electric motors to be used in electric vehicles were DC motors. This choice made sense 30 years ago because it was the easiest motor to drive with a chopper; however, using this type of electric motor raises several issues: limited power density, limited maximum rotation speed, heat losses located in the rotor, and thus difficult to evacuate and brush deterioration (which implies maintenance). The second generation of motors for automotive use were induction (asynchronous) motors. The increasing sophistication in motor inverters allowed the use of these very robust, compact, and reliable (no maintenance) motors. Efficiency tops out at 82%, which falls rapidly to 75% over a wider operating range. Some cars use asynchronous motors currently, Renault Twizy and Tesla cars for example.

Then, synchronous motors provide the best performance in terms of efficiency. Permanent magnet synchronous machines (PMSMs) do not require power supply for the rotor, but present a potential risk of demagnetization at high speeds, when flux-weakening control is applied. In addition, these motors use rare earth magnets.

Finally, the switched reluctance machines use coils to create the rotor magnetic field that can be modulated electronically with a chopper, allowing an easy drive at high speed (field weakening).

Nowadays, there are electric vehicles on the market using these technologies. Peugeot vehicles and the Toyota Prius (both electric traction motor and electric generator) use PMSMs. At present, a disadvantage is the availability of rare earths in the market for this usage. On the other hand, Renault-Nissan and Bolloré use a wound rotor synchronous motor, which is larger and heavier and requires additional electronics, but not rare earths.

The main advantages and drawbacks of various motors suitable for electric vehicles can be summarized as follows.

It can be said that the asynchronous machine is the best option in terms of cost, manufacturing ease, and robustness. Conversely, the permanent synchronous motor provides the poorest results in these characteristics. However, permanent synchronous motors provide very good performance in terms of: power density (kW/l) or specific power (kW/kg), efficiency, and versatility to adapt to different torque-speed characteristics. Finally, it can also be said that the switched reluctance machine, which is sometimes used in electric vehicles can provide very good results in terms of cost and manufacturing easiness.

The following motor subsections are mainly based on the work carried out by Oak Ridge Laboratory [8], [9]. These works are well-recognized references in electric vehicles research area.

7.5.2.1   *Interior permanent magnet motors*

Nowadays, interior permanent magnet (IPM) motors with rare earth magnets are almost universally used for hybrid and electric vehicles because of their high power density, specific power, and constant power/speed ratio. In Chapter 3, IPM motors used in electric drives are described. Performance of these motors is optimized when the strongest possible magnets—i.e. RE neodymium-iron-boron (NdFeB) magnets—are used. However, at present, the possibility of the supply becoming limited and the cost of RE magnets becoming very high could make IPM motors unavailable or too expensive, as mentioned before.

An IPM motor is a hybrid that uses both reluctance torque and magnetic torque to improve efficiency and torque. These motors are created by adding a small number of magnets inside the barriers of a synchronous reluctance machine. They have excellent torque, efficiency, and low torque ripple. They have now become the motor of choice for most HEV and electric vehicle applications.

IPM machines have high power density and maintain high efficiency over the entire drive cycle, except in the field-weakening speed range where there are losses in motor efficiency. This translates into a challenge to increase the constant power speed range without loss of efficiency. Other major issues are failure modes and the high cost of the motor. These machines are relatively expensive, owing to the cost of the magnets and rotor fabrication. In Fig. 7-47 the typical efficiency map of an IPM motor is shown. It has a great efficiency peak area and the performance over the whole torque vs speed map is greater than 85%, which makes this electric motor the best option from the performance point of view.

Note that the efficiency map is obtained by computing the efficiency of the motor, at every torque-speed point. This analysis shows that the machines in general present approximately constant efficiency regions. Consequently, some regions are more efficient than others. As described in Section 7.4.1.2, in general it is desirable to operate the motor at the maximum efficiency region in order to reduce energy consumption. For that purpose, a global and complex design is often suggested that tries to operate the different elements of the system at higher efficiency regions. Note also that the efficiency maps could also be obtained in other applications described in this book, helping to perform efficient and global designs. However, they are not so extensively employed as yet in electric and hybrid vehicle applications.

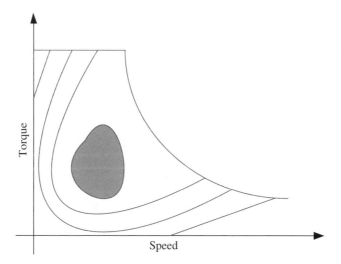

Fig. 7-47   *Typical efficiency map for an IPM motor. The grey area represents the most efficient working area (about 96% in this motor technology)*

7.5.2.2 *Induction motors*

Induction motors (IMs) have lower power density compared with IPM motors but also cost less. They are robust and have a medium constant power speed ratio (CPSR). Being a mature technology, they are reliable but have little margin for improvement. Most manufacturers consider IMs the first choice if IPM motors are not available.

The IM was invented more than 100 years ago and is the most widely used type of electric motor. Mostly because of its ability to run directly from an AC voltage source without an inverter, it has been widely accepted for constant-speed applications. IMs have the advantage of being the most reliable; they require low maintenance, have a high starting torque, and are widely manufactured and utilized in the industry today. These machines offer robust construction, good CPSR, low cost, and excellent peak torque capability.

The only ways to increase the power density of an IM used for vehicle traction are to use die-cast copper rotors for superior performance over the aluminum die-cast rotor and to increase the speed. Aluminum has only 56% of the conductivity of copper, which leads to an inferior performance when used in the rotor of an IM. In a first-pass analysis of a 50kW aluminum rotor IM, losses were 4% higher and power/torque densities 5% lower than the equivalent copper rotor motor [10]. Induction machine speed goes to 12,000rpm and even 15,000rpm at maximum vehicle speeds. This use of high motor speeds always results in smaller, lightweight traction motors, but it requires a high-ratio gearbox that also has a mass and losses.

The most difficult problem to deal with when using an AC IM is to extract the heat generated by the rotor conductors. The use of the lower-resistance copper over aluminum can be beneficial in two ways. The first is to reduce the ohmic losses in the rotor conductors, thereby reducing the heat that must be extracted to achieve high power and torque densities. The second possible advantage of using copper rather than aluminum in the rotor conductors is that the cross-section of the conductors can be reduced for the same ohmic losses, owing to the lower resistivity of copper compared to aluminum so that higher rotor magnetizing flux can be permitted, which can improve vehicle traction by increasing the starting torque. In Fig. 7-48 the efficiency map of an IM is represented. The peak efficiency is less than in PMSMs, but it also has a good performance over the required working area for an electric vehicle with a lower manufacturing cost.

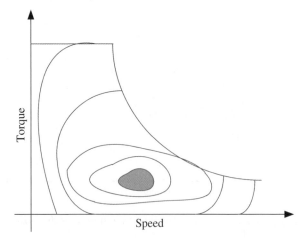

Fig. 7-48 *Typical efficiency map for an IM. The grey area represents the most efficient working area (about 87% in this motor technology)*

7.5.2.3 *Switched reluctance motors*

Switched reluctance motors (SRMs) use a doubly salient structure with toothed poles on the rotor and stator. Each set of coils is energized to attract a rotor pole in sequence so it acts much like a stepper motor. With current technology, SRMs have inherently high torque ripple. In addition, the high radial forces can create excessive noise levels if not carefully designed. These machines are best suited to high-speed applications where ripple is not an issue.

Unlike most other motor technologies, both the rotor and stator of the SRM comprise salient teeth such that torque is produced by the tendency of its rotor to move to a position where the inductance of an excited stator winding is maximized and reluctance is minimized. This condition generally occurs when the corresponding stator tooth is fully aligned with a rotor tooth. The non-steady-state manner in which torque is produced in the SRM introduces the requirement of a sophisticated control algorithm which, for optimal operation, requires current and position feedback. In addition to non-steady-state operation, the SRM often operates with the rotor and stator iron in saturation, increasing the difficulty of optimal control and making the machine very difficult to accurately model without the aid of computer processing and modeling techniques.

The absence of PM material, copper, aluminum, or other artifacts in the rotor greatly reduces the requirement of mechanical retention needed to counteract centrifugal and tangential forces. This causes the SRM to be especially well suited for rugged applications or high-speed applications wherein high power density is desired. As there are no conductors in the rotor, only a low amount of heat is generated therein, and most of the heat is generated in the stator, which is easily accessible in regard to thermal management. In addition to having low material costs, the simplicity of the SRM facilitates low manufacturing costs as well.

Having much lower material and manufacturing costs, the SRM presents a competitive alternative to the PMSM. Although the power density and efficiency of the PMSM will probably not be surpassed, the SRM comes close to matching these characteristics. Various comparison studies have shown that the efficiency and power density of the SRM and AC induction machine with copper rotor bars are roughly equivalent.

In regard to mass-transportation vehicle propulsion, the primary problems with SRM technology are the torque ripple and acoustic noise that are associated with the fundamental manner in which torque is produced. There are various methods to reduce torque ripple and acoustic noise, but they often bring about important sacrifices of efficiency and/or power density.

The most significant problem with the conventional SRM is that it cannot be driven with the conventional three-phase power inverter. Although the volt-amp requirement of the SRM converter for a given power rating is typically somewhat higher than that of the conventional drive system, the layout of this inverter is such that the risk of catastrophic DC rail-to-rail failure is eliminated. Fig. 7-49 shows the SRM motor efficiency map. Note that the peak efficiency is at high speeds, so the design of the power trains carrying this technology should work with higher gearbox ratios. This also avoids the torque ripple at lower vehicle speeds.

7.5.3 Power electronic converter

A conventional two-level voltage-source inverter is used to drive the traction electric motor. The inverter can be configured mainly according to two basic topologies, as depicted in Fig. 7-50. One consists of a traditional battery-powered two-level converter and the other incorporates a DC/DC converter between the battery and the two-level converter. The main advantages of the configuration that incorporates the DC/DC converter can be highlighted as:

- It keeps the DC-link voltage at the inverter independent from the SOC of the batteries.
- It minimizes the stress of the inverter controlling the DC-link voltage, without increasing the battery cost and size.
- It permits manufacturers to design the inverter and battery packs separately.

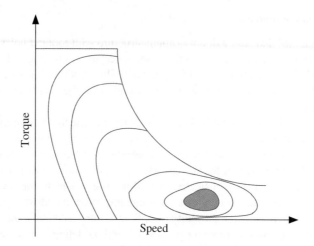

Fig. 7-49   *Typical efficiency map for an SRM. The grey area represents the most efficient working area (about 92% in this motor technology)*

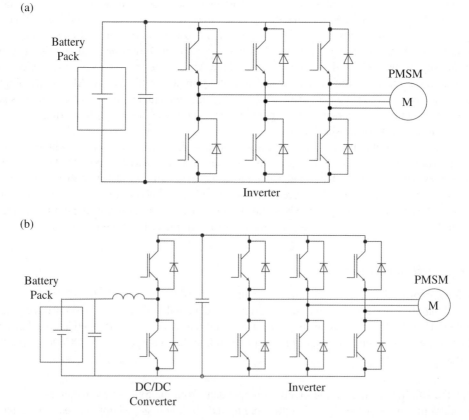

Fig. 7-50   *Power electronic configuration possibilities: (a) traditional PWM battery-powered inverter and (b) DC/DC converter between battery and DC-link*

However, its main disadvantages are:

- It increases power losses due to the DC/DC converter.
- It increases system complexity and cost, although this topology requires smaller batteries.

Note that the battery pack must be carefully dimensioned in such a way that the system is able to work as a motor and generator, in all working conditions without the necessity of a crowbar or resistive energy-dissipating element.

Thus, this topology in general operates in such a way that when motor speed is low, the inverter voltage of the DC-link is imposed by the battery. Then, above the motor base speed, the DC/DC converter is controlled to supply the inverter with system rated voltage. Finally, for the intermediate range, the system voltage dynamically changes in accordance with the machine mechanical speed. Thanks to this variable DC-link voltage philosophy, the overall drive performance can be greatly improved by reducing the voltage and current distortions as well as increasing the efficiencies.

In this case (Fig. 7-50), the converter presented employs insulated-gate bipolar transistors (IGBTs) as controlled switches and a PMSM as a traction motor.

When the battery pack voltage is lower than 100V, the inverter is normally composed of metal-oxide-semiconductor field-effect transistors (MOSFETs). To achieve high powers these MOSFETs are parallelized. This configuration is used in small vehicles, such as airport caddies or electric towing vehicles. The switching frequencies swing from 8kHz up to 30kHz in the state-of-the-art models which use SiC switches and ZVS (zero voltage switching) topologies.

On the other hand, it is possible to combine different energy sources in the electric vehicle, as illustrated in Fig. 7-51, in an attempt to optimize the energy source dimensioning. Thus, the idea is to combine the energy storage of the batteries, for instance, with ultracapacitors and fuel cells. For a proper storage combination, each energy source requires its own DC/DC converter. Then, all the DC/DC converters are connected in parallel to the DC-link, supplying simultaneously the inverter feeding the electric motor. Thus, the ultracapacitors enable inductor current ripples to be reduced and global efficiency to be improved, while the fuel cells allow higher efficiency values of the overall vehicle to be obtained.

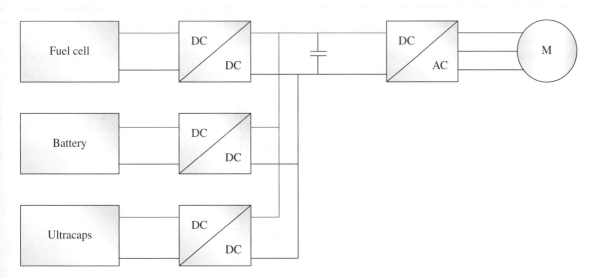

Fig. 7-51 *Alternative power electronic configuration combining different energy sources*

On the other hand, the bi-directional DC/DC converter that converts the DC voltage of the batteries to the voltage required by the inverter to operate is depicted in Fig. 7-52. It links the lower voltage of the batteries (V_1) to the higher voltage of the DC-link of the inverter (V_2). Note that in this case, for analysis purposes, a battery has been placed at V_2, but in reality this battery does not exist and the inverter supplying the motor is used instead. Thus, when the energy goes from the battery (V_1) to the inverter (V_2), the semiconductors that operate alternatively at constant switching frequency (f_{sw}) are D_1 and T_2. The relation between both DC voltages is given by the boost operation well-known equation:

$$V_2 = \frac{1}{1-\delta} V_1 \tag{7.21}$$

being δ the duty cycle related to the *On Time* of switch T_2:

$$\delta = (T_2 \; On \; Time) \cdot f_{sw} \tag{7.22}$$

Conversely, when the energy flows from the inverter (V_2) to the battery (V_1) during the regenerative brake operation, the semiconductors that operate are D_2 and T_1. In this case, the relation between both voltages is given by the buck operation well-known equation:

$$V_1 = \delta \cdot V_2 \tag{7.23}$$

with δ being the duty cycle related to the *On Time* of switch T_1:

$$\delta = (T_1 \; On \; Time) \cdot f_{sw} \tag{7.24}$$

As an example, this configuration is used in the 2010 Toyota Prius model. The nominal battery voltage (V_1) in this case is 201.6V and DC-link voltage (V_2) swings from 201.6V to 650V. The DC/DC converter has a nominal power of 27kW, with a 329µH (at 1kHz) inductor and a 750V DC, 2629µF DC-link capacitor. The switching frequency of the DC/DC converter is established at 5kHz. One advantage of using a DC/DC converter is related to the current ripple of the batteries. When the DC/DC converter is not used, the batteries are directly connected to the inverter, with the capacitor in the middle, making a current ripple appear in the batteries. This current ripple depends on factors such as: switching frequency, amount of current flowing through the motor, capacitance value and its parasitic impedance, and so on. When the DC/DC converter is used, the current ripple through the batteries is minimized.

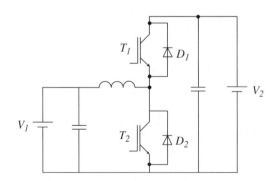

Fig. 7-52   *Non-isolated bi-directional DC/DC converter*

7.5.3.1 Inverter control

Traction inverters for controlling three-phase motors have a flexible control-software architecture that allows customization of the vehicle final integrator by loading software application modules. These application modules communicate with the motor-control core and implement the interface to the higher-level controls or respond directly to the driver inputs and outputs for stand-alone systems, as depicted in Fig. 7-53. The application module can be as simple as a CAN communication layer or as complex as to provide a complete electric vehicle control. They also offer several layers of protection to prevent safety-critical conditions.

Both operation modules are based on finite state machines (FSMs). The application FSM takes care of human–machine interface (HMI) tasks, for instance vehicle proper switch-ON and inverter disable when charging. These FSMs check vehicle status and compare it with the driver's inputs (via CAN frames or analogic signals) to work properly, as shown in Fig. 7-54.

On the other hand, motor control module handles electric power control FSMs. For example, in this module the pre-charge of the inverter and also the motor control are performed.

The motor control is typically performed by a typical torque vector control, which is shown in Fig. 7-55 (and the same as described in Chapters 2 and 3).

Besides the motor control itself, this FSM also has the battery protection submodule. The battery is protected from overcharge/overdischarge by means of ramps. Fig. 7-56 shows how this protection is performed. Depending on battery SOC, the inverter allows power flowing to/from the battery. If the SOC is reaching a low SOC level, the acceleration ramp acts and the power flowing from the battery is limited, and the same happens when the battery is nearly fully charged and regeneration is limited by regeneration ramp.

7.6   Battery chargers in electric vehicles

7.6.1   Introduction

A battery charger is a device that is composed of several power electronic circuits used to convert AC electrical energy into DC with an appropriate voltage level to charge the battery.

There are many possible battery charger topologies available. Safety requirements and standards often include isolation from the grid of the charger, which is typically done by means of using a transformer that can be located directly at the AC side or, also, within the DC/DC converter if it is present. The efficiency,

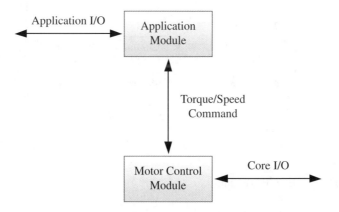

Fig. 7-53 *Application module and control core*

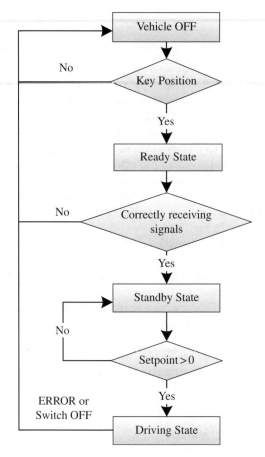

Fig. 7-54 *Vehicle switch-ON FSM example*

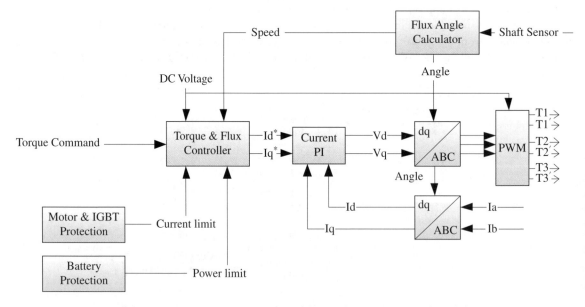

Fig. 7-55 *Motor torque control implemented in motor control module*

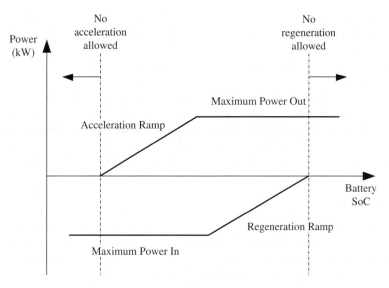

Fig. 7-56 *Battery protection graph*

charging speed, and cost of the chargers are important issues to be considered when designing commercial vehicles. In connection to this, the charging time is a characteristic that can range from minutes to several hours. However it is possible to affirm that there are four possible charging algorithms for batteries: constant voltage, constant current, constant voltage and current and pulse charging.

Thus, it is possible to classify battery chargers as off-board and on-board types, with uni- or bi-directional power flow [11]. Bi-directional power flow capability enables energy transmission to the grid, by means of a more complex power circuit.

In addition to this, it is possible to distinguish three charging power levels. The main characteristics of each charging power level are [11]:

- Level 1: With voltages of 120V AC (US) and 230V AC (EU), the charger (one phase) is typically located on-board and charging is performed at home or at the office. The expected power-energy levels are:
  - 1.4kW and 5–15kWh with charging time 4–11 hours.
  - 1.9kW and 16–50kWh with charging time 11–36 hours.
- Level 2: With voltages of 240V AC (US) and 400V AC (EU), the charger (one or three phases) is typically located on-board and charging is performed at private or public outlets. The expected power-energy levels are:
  - 4kW and 5–15kWh with charging time 1–4 hours.
  - 8kW and 16–30kWh with charging time 2–6 hours.
  - 19.2kW and 3–50kWh with charging time 2–3 hours.
- Level 3: With voltages of 208 to 600V AC or V DC, the charger (three phases) is typically located off-board and charging is performed at a filling station. The expected power-energy levels are:
  - 50kW and 20–50kWh with charging time 0.4–1 hour.
  - 100kW and 20–50kWh with charging time 0.2–0.5 hours.

Most of the charges of electric vehicles are level 1, which is the slowest method of charging, when the vehicle is left in the garage overnight, for instance. On the other hand, level 2 is the primary method for dedicated private and public facilities. It may need a dedicated power charger and a powerful connection installation (up to 19.2kW). Level 3 charging is the fastest and is normally placed in highway rest areas and city refueling points.

Note that the power, energy, and voltage ranges shown may in the future evolve to different values from those presented in this section.

Finally, it must be mentioned that it is possible to find on the market some other battery charger philosophies, such as contactless inductive chargers. These are not covered in this book, but here we should say that they can be classified into two major groups: conductive chargers and inductive chargers. The conductive chargers use metal-to-metal contact, and inductive charging of electric vehicles is based on a magnetic contactless power transfer. There are already a number of contactless charging units on the market of many countries.

7.6.2 General structure

There are basically two power conversion stages required to charge the battery using grid electricity: one is the AC/DC rectification and the other is the DC/DC conversion. Each of these stages can be formed with many different passive and active component combinations (inductors, capacitors, and semiconductor switches).

The general structure of a battery charger is depicted in Fig. 7-57. It is composed of an AC/DC rectifier connected to the AC grid, a power factor corrector (PFC), to improve the quality of the exchanged currents with the grid, and finally a DC/DC converter with galvanic isolation. It must be mentioned that the isolation can also be included at the input stage, rather than at the DC/DC conversion stage.

For a uni-directional battery charger, the AC/DC rectifier at the input is normally a diode rectifier. For a bi-directional battery charger, the AC/DC rectifier is more sophisticated as will be shown, including controlled switches as well as diodes, to enable the injection of energy into the grid. When the bi-directional battery charger is implemented, the PFC is not normally needed, since the input rectifier actuates itself as a PFC. Finally, the input filter is also often of a different nature, depending on the type of input rectifier that has been adopted.

The common nominal battery voltage levels in electric vehicles that are in the market are between 300–400V. These high terminal voltages allow smaller cabling size and considerably decrease the current ratings of active and passive devices for a given power level. Owing to high battery voltage and a 120V/240V grid connection, a boost rectification stage is preferred over a buck rectification stage to prevent an unnecessary high conversion ratio between the DC-link and the battery terminals.

7.6.3 Power factor corrector

PFC uni-directional chargers only transmit power from the grid to the vehicle battery and operate with almost unity input power factor.

The diode rectifier of Fig. 7-57 is a simple converter structure, which was specified in Chapter 1. A possible power factor correction stage is depicted in Fig. 7-58 [4]. It is basically employed to correct the power factor but also helps to reduce the contents of current harmonic exchanged with the grid. This example of PFC is

Fig. 7-57 *Basic electric vehicle battery charger structure (Source: [4]. Reproduced with permission of John Wiley & Sons)*

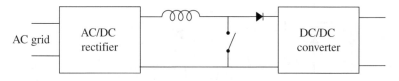

Fig. 7-58 PFC stage

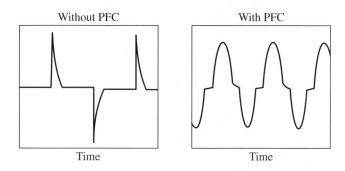

Fig. 7-59 Simulated current exchange with the grid, without and with PFC

Fig. 7-60 Typical current mode waveform

composed of an inductor, a controlled switch, and a diode. Mainly it is a boost converter. Once again, it is supposed that the input rectifier is a diode bridge.

In Fig. 7-59, the improvement of the PFC is represented by means of the exchanged current with the grid. It is possible to see how the current becomes more sinusoidal, also reducing the current peak and therefore the stress of the semiconductors. This stage also helps to increase the voltage level of the DC bus.

Therefore, although there are different control philosophies for these PFC circuits, a typical make operates the controlled switch in pulse-width modulation (PWM) form, obtaining an average current waveform through the inductor, as shown in Fig. 7-60.

Fig. 7-61 shows an interleaving proposal, an alternative topology. The main advantages of this topology are decreased rectifier AC input and DC output current ripple for the same switching frequency compared to a conventional AC/DC boost converter. Reduced input current ripple decreases the required switching frequency to meet a current total harmonic distortion (THD) level required by the utility. Reduced output ripple

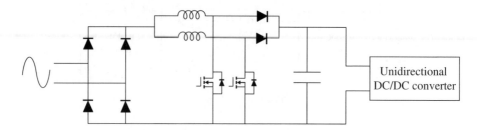

Fig. 7-61   *Interleaved uni-directional charger topology example showing components [11]*

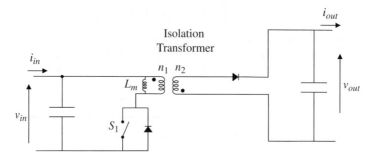

Fig. 7-62   *Flyback DC/DC converter including galvanic isolation*

also results in decreased high-frequency DC-link capacitor ripple requirement. It is based on two parallel-connected boost converters, operating 180° phase shifted.

7.6.4   DC/DC converters

Often DC/DC converters incorporate a high-frequency transformer. There are many possible DC/DC converter configurations. Some are bi-directional and some are not. There are some essential design considerations for vehicle chargers that can be grouped as:

- light weight
- high efficiency
- small volume
- low electromagnetic interference
- low current ripple at the battery
- the step-up function of the converter.

This section shows some of the potentially most employed DC/DC converter topologies in battery chargers. The first topology presented is the flyback converter with galvanic isolation, depicted in Fig. 7-62. In all the converters listed, the controlled switches are commutated in PWM mode. This converter is a buck-boost converter, whose principle is based on the transfer of energy from the magnetizing inductance of the transformer, L_m, and the output capacitor, C_{out}. The energy transferred through the input is stored at L_m during the ON-state of the controlled switch. During this time interval, the output capacitor imposes the output voltage providing the required current. During the time that the controlled switch is in the OFF-state, the magnetizing inductance and the input provide energy to the output. Thus, the transfer of energy is uni-directional, from the input to the output. The flyback converter is probably one of the simplest DC/DC converter topologies. It must be mentioned that the magnetizing inductance of the transformer is in charge of storing considerably large amounts of energy, necessitating a considerably high dimension core.

There is also a fairly simple DC/DC converter topology which is called the forward converter and is depicted in Fig. 7-63. It is a step-down converter, whose most complex element is a transformer with three windings. During the ON-state of the controlled switch, the energy is transferred from the input to the output and to the L_1 inductance. While at OFF-state, L_1 transfers energy to the output. The number of turns n_1, n_2, and n_3 affects the time interval in which the L_m inductance operates. Here again, the transfer of energy is uni-directional, from the input to the output.

There are many other DC/DC converter topologies that are more complex and can provide better performances. Some of these topologies are depicted in Fig. 7-64. It can be said that the configurations with more controlled switches can achieve less stress on the semiconductors. In addition, configurations with transformers that are magnetized bi-directionally are also advantageous, taking into account that there is no need for a demagnetization circuit. It must be mentioned that, apart from those presented here, there are alternative control philosophies for the switches of these converter configurations. In specialized literature also, it is possible to find alternative converter topologies.

7.6.5   Bi-directional battery charger

As mentioned before, there are different charger approaches from the ones that we have seen in previous sections. A quite obvious one is the bi-directional battery charger that, for instance, allows the transfer of energy stored in the vehicle battery pack to the home. One converter configuration can be obtained by combining a single-phase, two-level AC/DC converter with a dual-active bridge DC/DC converter, as depicted in Fig. 7-65. In this case, the PFC is not necessary since the AC/DC converter can control the power factor with a sinusoidal current at any power factor. Besides, the input rectifier can also take different structures for single-phase and three-phase inputs, as depicted in Fig. 7-66.

These configurations of bi-directional battery chargers are not so common nowadays, but in the future if the market share of electric vehicles grows substantially they may become more common.

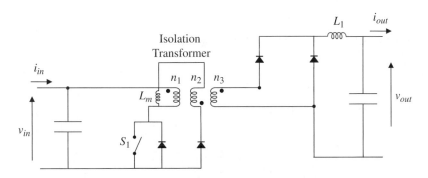

Fig. 7-63   *Forward DC/DC converter*

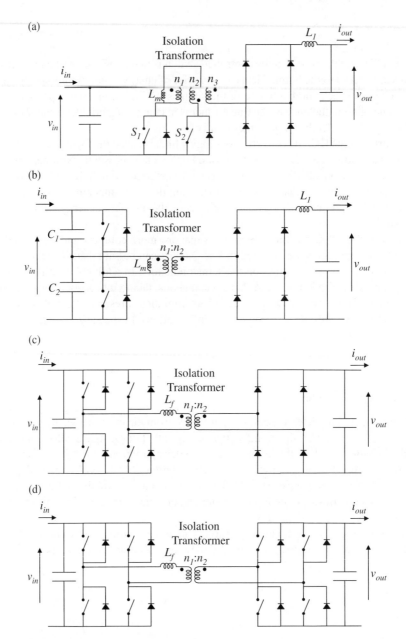

Fig. 7-64 (a) Push-pull DC/DC converter, (b) half-bridge DC/DC converter, (c) single active bridge DC/DC converter, and (d) dual-active bridge DC/DC converter

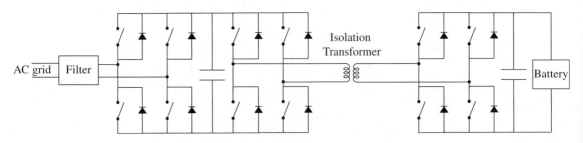

Fig. 7-65 Bi-directional battery charger

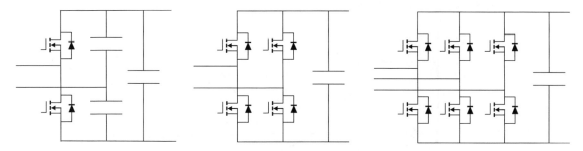

Fig. 7-66 *Input rectifier structures for bi-directional chargers*

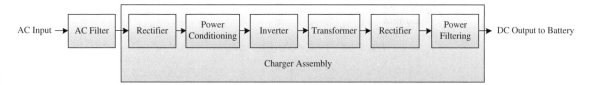

Fig. 7-67 *2013 Nissan LEAF on-board charger power stages*

7.6.6 Example: 2013 Nissan LEAF charger

The 2013 Nissan Leaf vehicle includes a 6.6kW onboard charger, accepting AC power in standard forms (120V@60Hz and 240V@50Hz). It also accepts fast charging (DC charging) from a separate port via an off-board charger. The charger meets the SAE J1772 standard, which regulates charger functionality and impact on the grid power, mainly regarding THD issues.

The 2013 Nissan Leaf on-board charger weighs 16.3kg and has a volume of approximately 11.1l giving specific power and power density figures of 0.4kW/kg and 0.6kW/l respectively.

The block diagram in Fig. 7-67 describes the various stages of the charger. After the external line filter, the AC input is protected with fuses and additional filtering (including common mode) is also incorporated. There is a pre-charge stage to avoid high in-rush currents charging the charger's inner capacitors. The first stage is a conventional rectifier with four diodes. The next stage, power conditioning, is a boosting PFC stage to minimize the impact on grid power quality caused by the harmonics. The third stage of the charger is a half-bridge inverter which drives the primary coil of one large isolation transformer, which has two secondary windings. Output from the two secondary windings is fed to two full-bridge rectifiers, whose outputs are placed in series. Finally, the rectifier output is fed through a power filtering before passing through the output connector that connects to the battery.

As the battery voltage is not constant and varies with the SOC, the output voltage of the charge must vary accordingly. The charger output voltage range goes from 300 to 400V.

The reported [9] total efficiency of the charger is about 87% for operation at a power level of 3.3kW and 120V and about 91.5% for operation at 6.6kW and 240V.

7.6.7 Battery charging strategies

Battery charging strategies depend on the type of batteries themselves. There are four major methods to charge an electric vehicle's battery: on-board charging, DC fast charging, inductive charging, and battery fast replace, which is not a charging method itself but can be an option, although it is beyond the scope of this chapter.

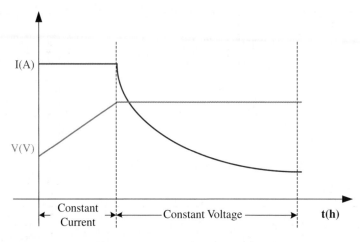

Fig. 7-68   *CCCV charging strategy*

Besides subjects like connector compatibility, payment for electricity and so on, the most important issue when charging a battery is the charging strategy below each charging method.

On-board charging, DC fast charging, and inductive charging have common objectives to meet: to have galvanic isolation between the infrastructure and vehicle, wake up of the battery management system (BMS) during charging period, communication between battery and charger, protection against overcharging, cooling of the battery, and so on. Some of these objectives are shared with the BMS.

7.6.7.1   *Constant current constant voltage (CCCV) charging strategy*

CCCV charging as illustrated in Figure 7-68, is the most used and safest method for charging batteries. On-board chargers normally use this strategy. The battery is charged with a constant current until a certain voltage is reached. Then, the battery voltage is kept constant while the current goes down. Charging will be stopped when the current reaches a certain threshold. When constant voltage is reached, it takes a long time until the battery is fully charged. This strategy is the best choice when the charging time is not important, like at home overnight or during an eight-hour work period in the same place. This charging strategy powers units of 2 and 5kW, which means a current/charge ratio of around C/8 for the battery pack. This ratio is quite low to heat up the battery, so the charge is more efficient and the battery cells and BMS electronics experience less stress.

7.6.7.2   *Fast-charging strategy*

A fast-charging strategy has to be controlled from the BMS. The target of this strategy is to charge the battery from 20% of the SOC to 80% within minutes. The operational SOC range (20–80%) is due to battery resistance increase in the SOC ranges from 0 to 20% and from 80 to 100%, which implies higher losses. A charge rate of 1.2C will fill up the battery from 20 to 80% in 30 minutes. Charge currents of 40A up to 100A are used depending on the size of the battery pack. The charging infrastructure must be able to provide the power, and that means a power of 33kW to charge a vehicle of 28kWh, for instance in Fig. 7-69 it is possible to observe the charging peak current of 1.2C and the SOC swinging from 20 to 80%. For a 28kWh battery pack, with 350V bus voltage, this 1.2C means a 96A constant current for 30 minutes, so the internal battery cables and busbars have to be designed to bear this current.

7.7 Energy storage systems in electric vehicles

The essential part of an electric car energy storage system is the battery. There are other energy storage systems to be used in electric cars, such as ultracapacitors and fuel cells, but nowadays the most promising storage systems are made up of batteries or hybrid systems (e.g. batteries plus ultracapacitors). Batteries are made up of various elements. These include cells, which are connected together in series or parallel strings to achieve the desired voltage and capacity. A cell is a closed power source, in which energy is stored chemically. This energy is released by internal chemical reactions as a flow of electrons through an external circuit.

A cell can be either primary (single-use) or secondary (rechargeable), which is the one used in electric vehicles. A cell is a device that converts the chemical energy contained in its active materials directly into electric energy by means of electrochemical oxidation/reduction (redox) reactions. A cell comprises a number of positive (cathode) and negative (anode) charged plates immersed in an electrolyte that produces an electrical charge by means of an electrochemical reaction. These cells have three types of construction: prismatic, cylindrical, and pouch (Fig. 7-70). From the electric traction perspective these types of construction or packages have to have good mechanical stability, electrical conductivity, and thermal conductivity.

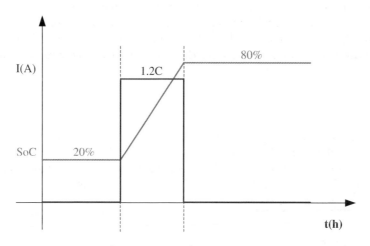

Fig. 7-69 *Fast-charging strategy*

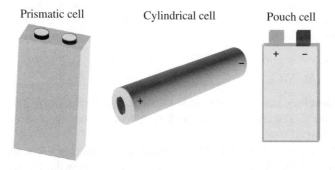

Fig. 7-70 *Prismatic cell (Calb), cylindrical cell, and pouch cell (A123)*

7.7.1 Battery cell chemistries for electric vehicles

The main categories that the vehicle battery research has focused on are: energy, power, lifespan, safety, and cost. The energy stored in a battery determines the electric drive range, and since the available space is limited in vehicles, vehicle makers focus on energy density (Wh/l) or specific energy (Wh/kg). Different cell chemistries have different energy densities, and between them lithium-ion cells have considerably greater energy density than previously used ones, lead-acid-based and nickel-based chemistries basically. This makes them particularly suitable for automotive applications. They are also considered safer, less toxic, and are more highly energy efficient with a significantly longer cycle life.

7.7.1.1 Lead acid (Pb)

Composition: Lead-acid batteries are composed of a lead-dioxide cathode, a sponge metallic lead anode and a sulfuric-acid solution electrolyte.

This chemistry is used in starter batteries for internal combustion engine vehicles because it has good discharge power capacity. However, it is heavy and has poor energy density. It is a popular, low-cost secondary battery, has a good high rate performance, moderately good low and high-temperature performance, easy state-of-charge indication, and good charge retention for intermittent charge applications. Cell components can be easily recycled. Because of the irreversible physical changes in the electrodes, failure occurs between several hundred and 2,000 cycles. The main drawbacks of these batteries are their comparatively low energy density, long charge time, and need for careful maintenance. They have been used in earlier generations of electric and hybrid vehicles.

7.7.1.2 Nickel cadmium (NiCd)

Composition: These cells use nickel hydroxide $Ni(OH)_2$ for the cathode, cadmium as the anode, and alkaline potassium hydroxide (KOH) as the electrolyte.

It was used in the beginning of space exploration. It has a long cycle life, good low temperature, and high-rate performance capability. As drawbacks, it has memory effect and cadmium is highly toxic, so the use of these batteries is banned for electric vehicles.

7.7.1.3 Nickel metal hydride (NiMH)

Composition: These cells use nickel hydroxide $Ni(OH)_2$ for the cathode. Hydrogen is used as the active element in a hydrogen-absorbing anode. The electrolyte is alkaline, usually potassium hydroxide.

These cells have higher capacity than nickel cadmium ones, and a rapid recharge capability, long cycle life, and long shelf life in any SOC. The drawbacks of this technology are poor charge retention, memory effect, cost, and environmental problems.

7.7.1.4 Zebra sodium (Na-NiCl₂)

Composition: This technology utilizes molten sodium chloroaluminate $(Na-NiCl_2)$ as an electrolyte, which has a melting point of 160°C. The anode is molten sodium and the cathode is nickel in the discharged state and nickel chloride in the charged state.

The major drawback of Zebra batteries is that they need to be charging when they are not in use in order to be ready to use when needed. If shut down, the reheating process lasts a whole day and then another 6–8 hours for a full charge. It is also inefficient as it consumes energy when not in use.

7.7.1.5   *Lithium-ion battery*

Composition: Lithium-ion (Lithium-ion) battery cells use a carbon based anode, although there are some exceptions, lithium titanate anodes for example. Various compounds can be used for the cathode, each of which offers different characteristics and electrochemical performance. The electrolyte is usually a lithium salt dissolved in a non-aqueous inorganic solvent.

The energy of these batteries is limited by the specific capacity of the electrodes. The superiority of Li-on batteries have been demonstrated over other types of batteries supplying greater discharge power for faster acceleration and higher energy density for increased all-electric range in electric vehicles. However, some issues, including cell life (calendar life and cycle lifetime), cost, and safety need to be dealt with. One important issue with Lithium-ion batteries is the need to equalize each cell charge to balance out the total charge among the cells in a more precise way compared to other chemistries.

The term lithium-ion does not specifically correspond to a particular battery chemistry. Some of the chemistries beyond this term are: lithium cobalt oxide ($LiCoO_2$), lithium manganese oxide spinel ($LiMn_2O_4$), lithium nickel cobalt manganese ($LiNi_xCo_yMn_zO_2$), lithium titanate oxide ($Li_4Ti_5O_{12}$), and lithium iron phosphate ($LiFePO_4$). Although each type of Lithium-ion cell has some advantages, the $LiFePO_4$ cathode is the most used technology in electromobility applications with increased safety and stability features. Its failure due to overcharging does not emit too much heat. However, it has lower cell voltages compared to other cathodes and hence many of these have to be connected in series requiring more balancing control.

7.7.1.6   *Comparison of battery technologies*

As a summary, battery technologies are compared in Table 7-3. As NiCd is banned for electric vehicles, it is not shown in the table.

7.7.2   Battery pack

Battery packs are built by assembling cells in series (to increase voltage) and parallel (to increase capacity, Ah), building modules, and then the final battery pack. The total energy of the pack is the total voltage multiplied by the total capacity. The cell voltage for a given type of battery is more or less constant (e.g. lithium-ion cells are approximately 3–4V), while the capacity will vary based on the cell's design and size. The pack also needs components to ensure the inner cells are performing at their best:

- BMS: an assembly of circuit boards that monitors the cells (e.g. temperature, voltage) and the whole pack to determine state of health and SOC. It ensures the safety of the pack and interfaces with the vehicle electronics and charger. Its cost increases with the number of cells it has to monitor. One of the major tasks of the BMS is temperature control, maintaining the battery pack at its optimum temperature to extend its life. BMS systems are explained in more detail in Section 7.8.
- Power distribution and safety devices: shunts, fuses, contactors, and safety disconnect. Wiring harnesses, power and control wiring, and connectors are made to connect cells and modules. Connectors to the vehicle electronics have to pass automotive standards and sometimes require high ingress protection (IP) levels, which increases cost.
- Internal cell support: made of plastic and/or metal, it holds the cells together to the correct compression levels and allows a module assembly process. When the pack is liquid cooled, the cell support is more complex as it acts as a cooling matrix, for example it has a network of grooves for the coolant to circulate around.

Table 7-3 Comparison of battery technologies

Battery type	Energy density (Wh/l)	Specific energy (Wh/kg)	Specific power (W/kg)	Life Calendar	Life Cycle	Discharge power (C-rate)	Cell voltage (V)	Temperature range (°C)	Safety	Cost
Lead Acid	60–75	30–40	100–200	Low	Low	6–10	2	−40 + 60	Proven	Very Low
NiMH	140–300	50–80	100–500	Good	Good	15	1.2	−30 + 60	Proven	Moderate
Zebra	160	90	150	Very good	Good	—	2.58	270–350	Moderate	High
$LiCoO_2$	450–490	170–185	1000	Low	Poor	1	3.6	−20 + 60	Low	High
$LiFePO_4$	130–240	80–125	5000–6000	Good	Good	5–50	3.2	−20 + 60	Excellert	Low
$LiNi_xCo_yMn_zO_2$	270–365	150–190	—	Moderate	Poor	1–40	3.7	−20 + 60	Moderate	Moderate
$Li_4Ti_5O_{12}$	118–200	65–100	—	Good	Very good	10–20	2.5	−50 + 75	Excellent	Very High
$LiMn_2O_4$	280	90–110	1800	Moderate	Poor	3–5	3.8	−20 + 50	Moderate	Moderate

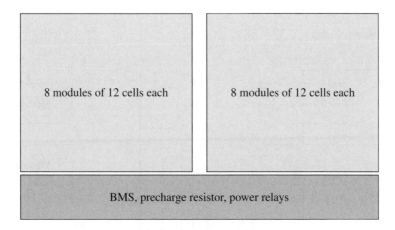

Fig. 7-71   *Azure van battery pack distribution*

7.7.2.1   *Example of a battery pack*

This subsection presents, as an example, Azure vehicle's battery cells characteristics, which are Johnson Controls-Saft cells. They present the following electrical characteristics (Source: Johnson Controls-Saft): The nominal voltage of the cell is 3.6V with a specific power of 794W/kg and power density of 1667W/dm^3. They present an average capacity of C/3 after charge to 4V/cell of 41Ah. They present voltage limits of 4V in charge and 2.7V in discharge. The maximum continuous current limit is 150A and maximum peak current (during 30sec) of 300A.

From a pack perspective, this is divided into three different sections, as depicted in Fig. 7-71, two of them dedicated to battery modules, independent from each other, and the third one dedicated to a central BMS and protection.

The details of the battery pack architecture are shown in Fig. 7-72. Two identical battery strings comprise the battery pack. In the figure, only four modules are shown, but there are actually eight. In the middle of each battery string there is a safety relay. This relay switches off when the battery envelope is removed, assuring that the present voltage is low when the battery pack is manipulated.

In the left part of the Fig. 7-72 there are positive relays. One of them is connected to the positive terminal of the battery string and to the battery pack output terminal. The other one connects the same points but with a pre-charge resistance. The pre-charge resistance commitment is:

- When the battery pack connects with an inverter, this one has a capacitor at its input. If the connection is done without any pre-charge, the current is very high, owing to voltage difference between both points.
- If two or more battery strings are present in a battery pack, if a voltage difference exists between them, the current may be very high. To limit this current, a pre-charge resistor is used.

To measure the current each battery string has a shunt resistance in the negative terminal. The voltage is measured in the external terminals of the battery pack. In this way, the BMS has two measurements: the sum of each cell voltage and the pack voltage. With these measurements, any difference between bus voltage and battery voltage can be analyzed in order to activate the pre-charge contactor.

Finally, the battery pack incorporates another safety system. If someone removes the external connector, all battery power relays remain open so that the battery pack can be manipulated safely.

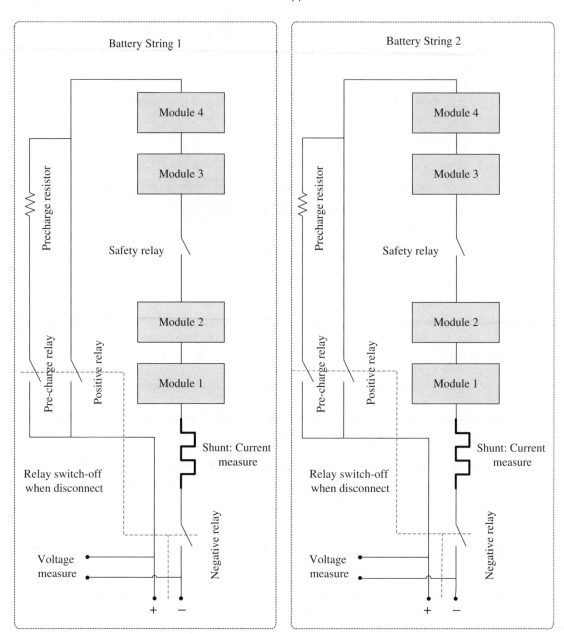

Fig. 7-72   *Battery pack architecture*

7.8   Battery management systems (BMS)

The BMS is the device that is responsible for the correct management of the cells of a battery pack. The principal issues of the BMS are the assurance of the security limits of the individual cells and the management of the energy of each cell.

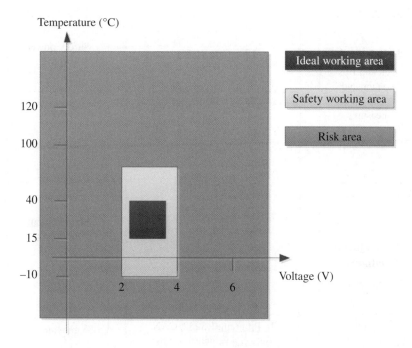

Fig. 7-73 *Safe operation area (SOA) of lithium-ion cells*

The major responsibility of the BMS is to assure that all the lithium-ion cells are inside the safe operation area (SOA) presented in Fig. 7-73. Typical security regions regarding voltage and temperature are presented in Fig. 7-73. If the cell works inside the ideal working area, it would reach the manufacturer's cycle and performance claim. If it works outside of the ideal working area but inside the SOA, it would degrade its performance during cycles. Finally, if a cell enters the risk area, the cell could be damaged and no longer operate. For this issue voltage, current, and temperature measurements are important to guarantee lithium-ion cell security.

For the control and management of the three main parameters (voltage, current, and temperature), the BMS may have different functionalities, as presented in Fig. 7-74. The BMS of a battery pack has three main functionalities: measurement, communication, and action.

Action is the most important functionality of the BMS, implementing three main tasks:

- **Security:** Security circuits allow the battery to be disconnected from the application when a security hazard is presented by a static or electromechanical actuator.
- **Balancing:** The balancing system is responsible for maximizing battery life and improving its performance.
- **Others:** Battery thermal management systems (BTMS), charger interconnectivity, and so on.

Measurement functionality is another important function of the BMS, with two different measurement variables: variables for protection circuits and auxiliary variables.

- **Protection measurements:** Voltage, current, and temperature are the most important variables and should be controlled by the BMS.

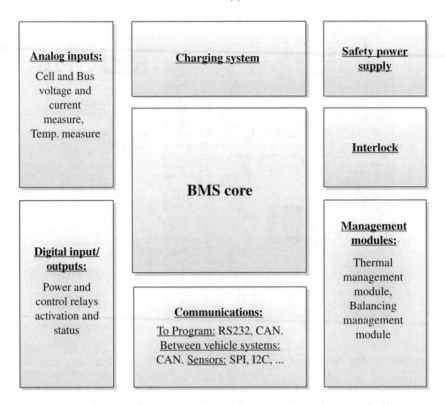

Fig. 7-74   *BMS configuration and complexity inside an electric vehicle [12]*

- **Auxiliary measurements:** Auxiliary measurements are useful for the user. The typical auxiliary functions are related to the battery state. The state of charge (SOC), state of health (SOH), state of function (SOF), and state of life (SOL) are the most common auxiliary measurements.

Interaction with the human and the rest of the vehicle is the last functionality of the BMS, although this functionality is not necessary for the correct operation of the battery. CAN interface is the most common communication protocol for battery packs, sharing information between the BMS and the vehicle or the user and making functionalities as displaying information, data logging, and so on.

A BMS structure of a large lithium-ion battery pack can be very complex. Different BMS architectures present state-of-the-art available options to deal with the whole battery management problem. Different architectures are presented in Table 7-4.

Centralized topologies present high connection complexities. They are not modular and are limited to the number of series connected cells. They have the advantage that if one cell is disconnected all the other cells continue under control.

Distributed topologies reduce connection complexity thanks to advance communication protocols between slave and master BMS. Individual structures permit total flexibility in cell series connection. Daisy chain connection reduces connection difficulty between contiguous slave BMSs. However, a communication connection failure can lead to a battery control loss. Parallel connection permits greater control during possible communication failures. Different architecture connections are presented in Fig. 7-75.

Table 7-4 *Main BMS architecture and structure topologies*

Architecture	Structure	Connection
Centralized	Individual	Star
Distributed	Individual	Daisy chain
	Modular	Daisy chain
	Modular	Parallel

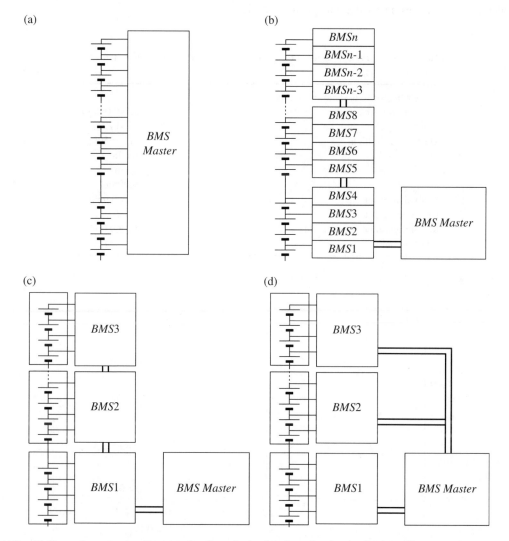

Fig. 7-75 *BMS topology connection: (a) distributed, (b) distributed individually daisy chain, (c) distributed modular daisy chain, and (d) distributed modular parallel*

7.9   Computer-based simulation example

7.9.1   Vehicle study

In this section, the developed physical model of the electric vehicle is utilized, combined with the model of a traction electric drive. The fictitious electric motor simulated presents two traction wheels at the front. The main characteristics of the simulated vehicle are listed in Table 7-5.

    Therefore, depending on the linear speed at which the vehicle is running, as well as the slope of the road, the opposite linear force to the movement and the equivalent torque that the driving motor must provide are graphically illustrated in Fig. 7-76. Note that for the transformation from linear speed to rotating speed of the motor the gearbox ration, the front wheel's radio must be considered. On the other hand, the road is assumed to be dry and modeled by the adhesion coefficient specified in Fig. 7-13.

Table 7-5 *Main characteristics of the simulated fictitious vehicle with two traction wheels*

Front wheels-tire assembling inertia (referred to the motor's shaft)	$J_f = 0.08 \text{kgm}^2$
Wheel's radio	$r = 0.29\text{m}$
Mass of the vehicle	$m_v = 550\text{kg}$
a	0.64m
b	1.1m
h	0.095m
Equivalent front area	$A_f = 0.959\text{m}^2$
Drag coefficient	$C_d = 0.21$
Rolling resistance coefficient	$C_r = 0.0049$
Gearbox ratio	9

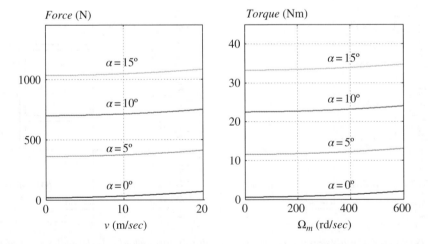

Fig. 7-76 *Resistance of the vehicle and equivalent torque provided by the motor (referred to the motor's shaft), at different speeds and slopes*

Table 7-6   *Main characteristics of the driving IM*

Rated power	13.1kW
Rated stator voltage (rms-LL)	28V
Rated stator frequency	125Hz
Pole pairs	2
Rated mechanical speed	3500rpm
L_m	3.7e–4H
$L_{\sigma s}$	1.79e–5H
$L_{\sigma r}$	1.79e–5H
R_r	3.98e–3Ω
R_s	2.3e–3Ω
Other characteristics	
Switching frequency of two-level power converter	8kHz

As a traction motor, an IM is employed with the characteristics shown in Table 7-6. It is seen that the stator-rated voltage is only 28V, in order to require not a high number of battery cells placed in series. The relatively highly rated stator frequency of 125Hz, combined with its two-pole pairs, enables a considerable rated speed of 3500rpm to be reached.

With these motor characteristics and its electric parameters, it is possible to obtain the maximum capability curves of the most interesting electrical magnitudes of the motor (see the procedure described in Chapter 2). These curves are provided in Fig. 7-77. In all cases, the maximum magnitudes are graphically represented by function of the mechanical rotating speed of the motor's shaft (Ω_m). Together with the maximum capability curve of the motor's torque (rated torque is 35.74Nm), the equivalent resistance torque of the vehicle is represented for four different slopes of the road. It will be noticed that the smaller the slope's inclination, the smaller the required torque at steady state and; therefore, the remaining torque up to the maximum can be employed for accelerating the vehicle faster. It can be seen that with slope $\alpha = 15°$, the steady-state torque needed is very close to the maximum torque that the motor can provide.

By comparison, the resulting rated rotor flux of this motor is relatively small (0.03Wb), owing to the fact that the rated stator voltage is relatively small and the rated speed or stator frequency is relatively high. Continuing with the analysis, it is possible to see that the current that the motor must handle is relatively high, to reach the 13.1kW rated voltage at the considerably lower stator voltage. Then also, since the rotor flux takes such considerably low values, consequently ω_r takes considerably high values, as depicted in the figure. Finally, this motor does not present very spectacular efficiency ratios, or the angle between the stator voltage and the current (power factor angle). Accordingly, for these motor characteristics, the appropriate two-level converter is selected and the required batteries.

7.9.2   Simulation

Once the vehicle's model and the motor characteristics are defined, the vehicle's dynamic performance is obtained by means of a computer-based simulation implementation in Matlab-Simulink. In order to control the vehicle, the vector control strategy of the induction machine is utilized as described in Chapter 2, and is discussed briefly in this chapter (anti-slip control is not necessary in this case). The torque is the reference

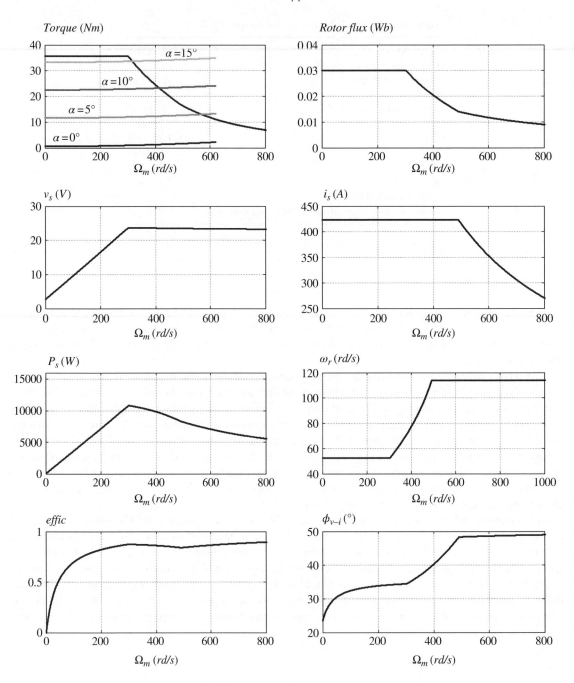

Fig. 7-77   *Maximum capability curves of the employed induction machine for vehicle traction*

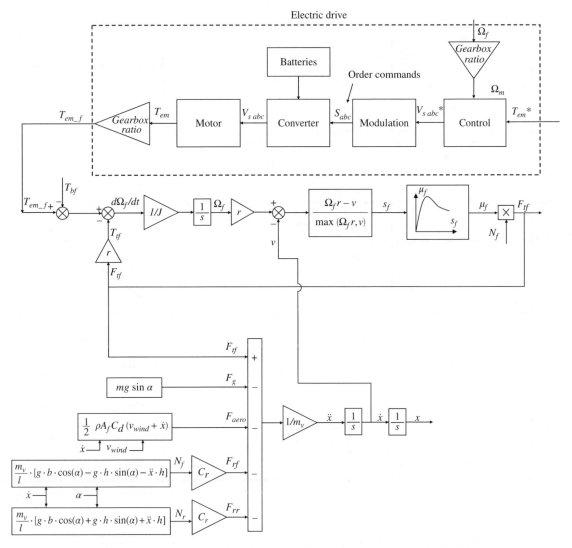

Fig. 7-78 *Model block diagram of a two-wheel driving car with its drive for the front wheels*

commanded by the user. The schematic block diagram of the vehicle model with its corresponding electric drive is depicted in Fig. 7-78.

In this simulated example, only an acceleration process is studied, where the traction torque is set to the maximum (35.74Nm). It is supposed that the road inclination is $\alpha = 0°$. Consequently, after building up all the models described and detailed, the simulation performance during the acceleration is provided in Fig. 7-79. It can be seen that in 4sec the vehicle can reach a linear speed of approximately 7m/sec, with a linear acceleration of approximately 1.7m/sec^2. Under these conditions, the opposition forces are very small compared with what the traction motor can generate (approximately 1000N). Thus, the equivalent torques at the wheels are around 300Nm, producing a traction power of almost 8kW at second four of the simulation.

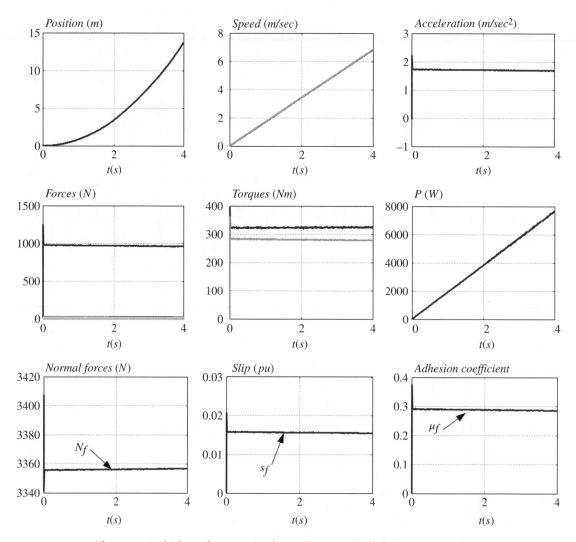

Fig. 7-79   *Vehicle performance with two driving wheels and on a dry surface*

Finally, Fig. 7-80 illustrates the most interesting electric characteristics of the motor during the last 60msec of the acceleration process. It is possible to note that the acceleration is performed at a rated torque of 36Nm (motor's shaft). For this purpose, the vector control imposes a rated flux of 0.03Wb. Since the torque and flux are the rated values, the i_{ds} and i_{qs} currents are also its corresponding rated values. Since the speed is not the rated value, the stator voltage is not the rated value either.

7.9.3   Quasi-static simulations

The drive cycle simulation of longitudinal vehicle models is an important tool for the design and analysis of power trains. Lateral forces are normally beyond the scope of this analysis. Vehicle simulation is usually done at sampling times in the order of seconds or tenths of seconds. Those times would seem like an eternity for a

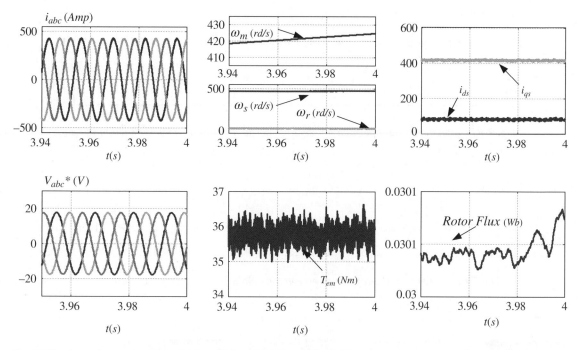

Fig. 7-80 *Most interesting magnitudes of the electric vehicle's driving motor, corresponding to the simulation depicted in Fig. 7-79, at time 3.94sec (speeds and torques referred to the motor side)*

power conversion simulation with device details, as required sampling times for such conversions would be in the order of micro- or nanoseconds.

There are several tools for such simulations on the market, and these tools mainly use two different methods of simulation: forward dynamic simulation and quasi-static simulation.

Forward dynamic simulation: It is capable of describing the dynamic behavior of a system in great detail, but suffers from long simulation times. In these simulations, subsystem models are defined as actual real components (with the same inputs and outputs) and used in rapid prototyping strategies to develop accurate controls. Car manufacturers and engineers use them in hardware-in-the-loop platforms to validate their behavior before implementing them in the final functional prototype. Fig. 7-81 shows a schematic description of a forward dynamic simulation flow.

Quasi-static simulation: These simulations are very fast as the vehicle system models are described as performance look-up tables and static equations instead of differential equations. Given a drive cycle, velocity and acceleration are used to calculate the required torques and speeds backward through the driveline. This is done in order to calculate the required energy input to the system, making the vehicle follow the prescribed velocity profile. Fig. 7-82 shows a schematic description of a quasi-static simulation flow. Note that there is no feedback as this simulation avoids using proportional integral derivatives or other closed-loop control algorithms.

The vehicle's motor/engine only operates in the first and fourth quadrants (positive/negative torque and positive speed).

There are several simulators, such as Advisor, Autonomie, or QSS-Toolbox (Fig. 7-83), that use quasi-static philosophy to calculate power train subsystems requirements. These simulators have example models of the power train subsystems to build the desired vehicle topology model.

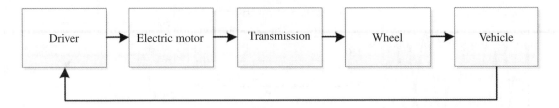

Fig. 7-81   *Schematic description of an electric vehicle forward dynamic simulation model*

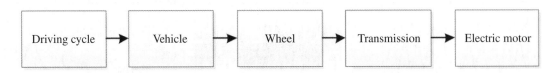

Fig. 7-82   *Schematic description of an electric vehicle quasi-static simulation model*

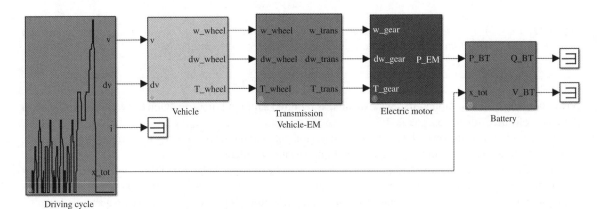

Fig. 7-83   *Electric vehicle model using QSS-Toolbox*

The problem comes when an accurate subsystem quasi-static model is needed to validate that it fulfills the driving cycle requirements. Automotive manufacturers usually do not publish details about the design, functionality, and operation of electric vehicle/HEV technologies (even the CAN protocol of each vehicle is encrypted). For example, single-value power ratings for motors and inverters are often published, but they do not include information about the power capability throughout the full operation range, the duration for which this power can be maintained, the efficiency throughout the operation region, and many other important characteristics that are needed to develop accurate quasi-static models.

Using the vehicle parameters shown in Table 5.5, the quasi-static model of the electric vehicle (Fig. 7-83) has been simulated over a standardized New European Driving Cycle (NEDC). For a given electric machine, the simulation analysis reveals that the gearbox ratio value of 9 is not appropriate to fulfill the NEDC (Fig. 7-84).

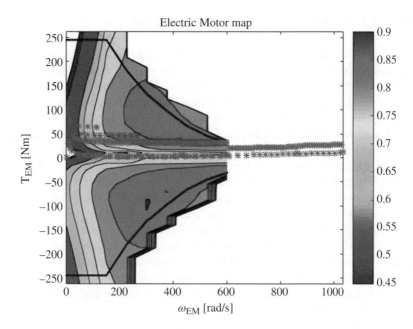

Fig. 7-84 *Motor axle working points over the Electric Motor map with a gearbox ratio of 9*

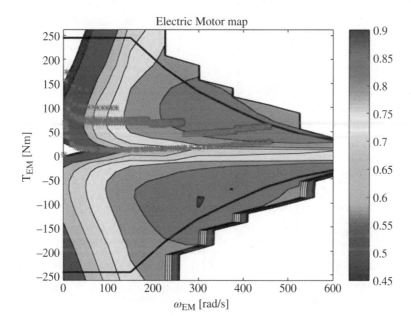

Fig. 7-85 *Motor axle working points over the Electric Motor map with a gearbox ratio of 4*

If the gear ratio is changed to a 1:4 relation, all the working points are inside the motor map and below the maximum thermal torque boundary, as can be seen in Fig. 7-85. This means that this 37.5kW maximum rated machine is suitable for the NEDC.

7.10   Electric vehicle design example: Battery pack design

In Section 7.9, we present how to dimension the power requirements of the electric motor and inverter using longitudinal force models of the vehicle. In fact, some of the quasi-static models presented are also valid for electric battery design, as they give information about the energy needed to complete a cycle and the power peaks along the cycle mentioned. Even so, a very good estimation of the battery pack can be achieved following the Fig. 7-86 flowchart. First of all, the electrical power demanded by the vehicle has to be generated. To do this, the battery pack has to match the motor/inverter DC-link's voltage, which gives the total cell numbers in a string, and also match with the current demand, which gives the total cell string number or the capacity per cell. With this information, the battery pack estimation can be done. Adding the battery SOC swing information, namely the percentage of the total battery energy used in a typical cycle, the battery pack estimation is much more accurate. Finally, using a continuous iteration method, different battery technologies are taken into account regarding their cost, safety, thermal condition needs, and so on.

Once the battery pack forming cell is selected, the cell data collection process has to be performed using various standard tests in order to validate this chemistry for the selected vehicle. After these tests, information about the cell behavior is used to program the BMS and to build the battery pack enclosure assuring good thermal performance. Finally, some vehicle tests are conducted to validate the whole system. Obviously, this is a simplification of the process and each of the steps can have months of engineering tasks. The simplified procedure is depicted in Fig. 7-86.

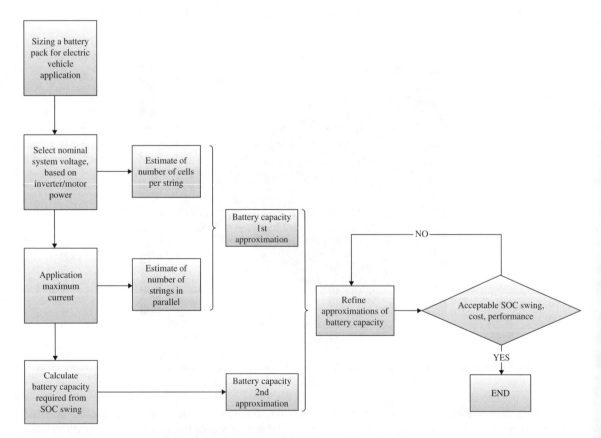

Fig. 7-86   *Battery pack design flowchart [13]*

7.11   Real examples

This section discusses some market examples of electric and HEVs. Nowadays, it is possible to find a wide range of possibilities in the real market. However, this section only gives a glimpse of only a few examples. This section is divided into two parts, first some electric vehicle examples are examined, followed by a few representative hybrid vehicles. The reader must be aware that this technology is continuously under development. Consequently, it is expected that what is described in this section will be improved by future developments in the years to come.

7.11.1   Electric vehicle examples

Some authors distinguish basically two market tendencies in electric vehicles. One is focused on short operating ranges with low battery weight, for instance oriented to city traffic. The other is focused on long operating ranges, oriented to the high performance market. Thus, Table 7-7 gives some representative examples of electric vehicles that can be found in the market, along with some relevant data about them, such as battery type, voltage, amount of energy storage, power, and range. Note that there is a chance that some of the information may in some way be inaccurate, so the survey has only a qualitative value. As can be noted in the table, most of the vehicles have a range between 100 and 150 kilometers, enough for commuters but insufficient for long journeys. In fact, this range is far from that achieved by internal combustion engine vehicles, which can reach a range of 800–1000 kilometers. Apart from that, charging times swing from three to eight hours to reach the full charge of the battery. Meanwhile, hybrid vehicles are being introduced in the market with great success, as is described in Section 7.11.2.

7.11.2   Hybrid electric vehicle examples

Nowadays the path to the full electrification of a vehicle goes through hybridization. This hybridization can be larger, in battery size, depending on the target market. For example, as can be seen in Table 7-8, commuters can use a Chevy Volt vehicle daily using the electric energy only and use the engine only when they go further distances (range extender concept). There are other hybridizations which allow smoother performances of both electric (less power and battery SOC swing) and combustion parts, improving the vehicle's overall performance.

The HEV example that is briefly described here is the Toyota Prius as this is one example of a successful hybrid vehicle that has been sold all around the world. There are several generations of Prius vehicles on the market. The schematic representation of the power train of the Toyota Prius is given in Fig. 7-87. It is, indeed, a special configuration of power train that differs from the most common HEV architectures seen in previous sections.

There is a special arrangement between the electric motor, the electric generator, and the engine. It is a planetary gear linking the three motors with three shafts. Thus, the engine is coupled to the carrier of the planetary gear. Then, the electric generator is connected to the sun gear. Finally, the ring gear is connected to the electric motor, which is the final drive of the front wheels. In this way, the vehicle speed is determined by the electric motor with the ring gear. On the other hand, the speed of the generator and the engine can be controlled in such a way that the engine speed is in its optimum efficiency operation point, providing a certain torque. As was described when modeling the vehicles, the vehicle speed and the forces in opposition to the movement impose motor torque. Therefore, for established torques of the engine and motor, the generator must provide a torque that exactly balances the other two torques. On the other hand, the battery pack is charged by the generator and/ or by the traction motor when it operates during regenerative braking.

Table 7-7 *Some representative electric vehicle models in the market*

Model	Market release	Body type	Battery type	Battery voltage (V)	Battery energy (kWh)	Charging rate (kW)	Motor power (kW)	EPA range (km)
BMW i3	2014	Sedan	Lithium-ion	360	22	6.6	125	130
Chevrolet Spark electric vehicle	2013	Coupé	LiFePO4	360	19	3.3	64	132
Fiat 500e	2013	Sedan	NiMnCo lithium-ion	364	24	6.6	83	140
Ford Focus Electric	2012	Sedan	Lithium-ion	325	23	6.6	107	122
Kia Soul electric vehicle	2014	Sedan	Lithium-ion		27	6.6	81.4	150
Mercedes B-Class Electric Drive	2014	Sedan	LiNiO$_2$	360	28	10	132	137
Mitsubishi i-MiEV	2009	Sedan	Lithium-ion: Yuasa	330	16	3.3	47	100
Nissan LEAF	2012	Sedan	Laminated lithium-ion battery	360	24	6.6	80	135
Smart Electric Drive	2013	Coupé	Lithium-ion: Deutsche ACCUmotive	300	17	3.3	55	110
Volkswagen E-Golf	2014	Sedan	Lithium-ion battery: Panasonic	323	24	7.2	85	134
Tesla Model S	2012	Sedan	Lithium-ion: Panasonic 18650	400	85	10	225	426

Table 7-8 *Some representative plug-in hybrid vehicle models on the market*

Hybrid vehicle model	Year	Hybridization type	Battery energy (kWh)	Traction maximum electric rated power (kW)	Traction maximum combustion rated power (kW)	EPA electric range (km)
Hyundai Sonata	2015	Plug-in hybrid: Parallel	10	50	115	35
Audi A3 E-Tron	2015	Plug-in hybrid: Combination	9	75	150	48
Toyota Prius	2012	Plug-in hybrid: Combination	4	60	73	18
Honda Accord	2015	Plug-in hybrid: Combination	7	124	146 (combined power)	20
Ford Fusion Energi	2015	Plug-in hybrid: Combination	7	88	105	32
Chevy Volt	2011	Plug-in hybrid: Series (range extender)	17	111	63	61

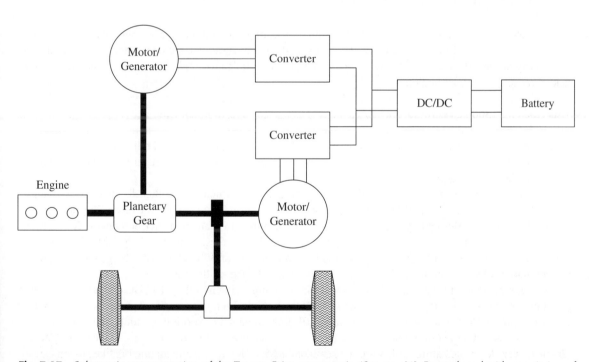

Fig. 7-87 *Schematic representation of the Toyota Prius power train (Source: [4]. Reproduced with permission of John Wiley & Sons)*

A great benchmarking of the 2010 Toyota Prius model can be found in the public report carried out by the Oak Ridge Laboratory [14].

7.12   Historical evolution

This section presents a brief historical evolution of electric vehicles and hybrid vehicles. Table 7-9 shows a short summary of the most important events in the history of electric vehicles development. The table is not exhaustive and only presents those events deemed the most important in the evolution of the electric vehicle. It will be noticed that electric vehicles were invented around 1839, about 50 or 60 years earlier than cars with gasoline motors. Prior to the 1830s, the fundamentals of transportation were only through steam power, basically because the electromagnetic laws were not sufficiently developed to be able to develop the electric motors. However, accompanied by the development of rechargeable batteries, by the end of the 19th century and beginning of the 20th century, the electric vehicle became fairly widely used. Nevertheless, by the 1920s, the internal combustion engine proved a more attraction option for powering vehicles, becoming predominant for the rest of the 20th century. At the same time, the electric vehicles started to resurge in the 1960s and 1970s, primarily owing to environmental concerns about emissions and later because of the energy crises of the 1970s and the increase of gasoline prices. Since then, significant efforts has been made to develop both electric and HEV technology.

7.13   New trends and future challenges

Electric vehicles are here to stay. The path from internal combustion engine cars to pure electric vehicles is clear but it has some hurdles to overcome. High cost is the major disadvantage compared with internal combustion engines and the main reason why electric vehicles have failed to break through, alongside range and a lack of recharging infrastructure.

In 2013, the International Energy Agency (IEA) estimated cost-parity could be reached in 2020, with battery costs reaching $300 per kilowatt hour of capacity. Battery costs have fallen from above $1000 per kilowatt hour in 2007 to around $410 in 2014, a 14% annual reduction. Costs for market-leading firms have fallen by 8% per year, reaching $300 per kilowatt hour in 2014 [18]. We estimate prices will fall further to around $230 per kilowatt hour in 2017/18, the crossover point where electric cars become cheaper depends on electricity costs, vehicle taxes, and prices at the pump.

As mentioned before, range anxiety and lack of recharging infrastructure are other disadvantages. Even so, new carbon-based batteries should resolve this problem in the near future. As Power Japan Plus and Kyushu University claim [19], this battery in going to be cheap to manufacture, safe, and environmentally friendly, and could massively improve the range and charging times of electric cars, as it can charge 20× faster than its Lithium-ion counterpart. This new battery would have a long life, of 3000 charge/discharge cycles, and will also discharge fully without the risk of short-circuiting and damaging the battery.

In power electronics, research and development focus on improving inverters, DC/DC converters, and chargers. For instance, the US Department of Energy's (DOE) Vehicle Technologies Program sets targets to reduce the cost of power electronics and electric motors [20]. Researchers are working to reduce the dimensions of inverters and DC/DC converters, and to reduce part count by integrating functionality and reduce cost. Today's vehicle power electronics utilize silicon-based semiconductors. However, semiconductors like silicon carbide (SiC) and gallium nitride (GaN) are more efficient and can withstand higher temperatures than silicon components. The ability to operate at higher temperatures and higher switching frequencies can decrease system

Table 7-9 *Synthesized evolution of electric and hybrid vehicles [15], [16], [17]*

Year	Historic event
1839	Non-rechargeable battery-powered electric vehicle developed in the UK. This concept of vehicle could not compete with the steam-powered vehicles already in use at that time
1859	Gaston Planté invents the first lead acid battery
1870	In the UK, an electric vehicle with a sophisticated electric motor is developed. It is able to reach a maximum speed of 13km/h. Its performance is still inferior to steam-powered vehicles' performance
1895	Charles Duryea registers the first American patent for a gasoline vehicle. One year later Henry Ford sells his first quadri-cycle of gasoline
1897	The company London Electric Car inaugurates a service of electric vehicles powered by lead acid batteries. The power of the electric motor is approximately of 2kW. The battery has 40 cells and its range is approximately 80km
1898	Dr Ferdinand Porsche in Germany probably develops the first HEV. This concept of vehicle is not extensively established until the end of 20th century
1899	Electric Company is created in the USA. It is the first company dedicated to the mass manufacture of electric cars. In 1904, this company produces 2000 electric taxis, for the cities of New York, Chicago, and Boston
1900	Approximately, 50 plants exist worldwide, producing around 4000 vehicles based on one of three technologies: steam motors (40% approximately), electric motors (38% approximately), and gasoline motors (22% approximately)
1912	This is probably the year of major diffusion of the electric vehicle, with technological advances improving its performance. For instance, the model fabricated in Chicago by the Woods company reaches speeds of up to 60km/h
1920	The performances of vehicles with internal combustion engines become superior to those of electric vehicles. Henry Ford works on the gasoline engine, and the electric vehicle begins to disappear. Ironically, the main market for rechargeable batteries was adapted for starting internal combustion engines
1960s	Electric vehicles start to resurge, basically because of environmental concerns about emissions of the internal combustion engine vehicles. The major internal combustion engine vehicle manufacturers, General Motors (GM) and Ford, become involved in electric vehicle research and development. However, the technology is not mature enough to produce commercially viable electric vehicles satisfying the demands of users accustomed to vehicles powered by internal combustion engines
1970s	Gasoline prices increase owing to various energy crises and the Arab oil embargo (1973). A desire to break dependency on oil together with environmental concerns relights interest in the electric vehicle
1976	In the USA, Congress enacts Public Law 94–413, the Electric and Hybrid Vehicle Research, Development and Demonstration Act of 1976. Since this year, many public and private initiatives have been created to ensure that the electric vehicle becomes a reality
1980s–1990s	Technological developments of power semiconductors, and microprocessors enable industry to obtain better and more efficient converters for electric vehicles. This increases the performance of electric motors. In addition, zero-emission vehicles are also promoted
1990s	Manufacturers realize that their significant efforts on electric vehicle technologies are hindered by unsuitable battery technologies. Some auto industries start to develop HEVs to overcome the battery and range problem of pure electric vehicles
1997	The first modern hybrid electric car, the Toyota Prius, is sold in Japan. After that, the hybrid electric car is launched onto the market
2000	General Motors issues a recall of its 450 Gen 1 EV1s

(*continued overleaf*)

Table 7-9   (continued)

Year	Historic event
2003	Tesla begins development of the Tesla Roadster
2008	"All the pieces of the puzzle are in place for making a mass-production vehicle in the near future: battery range, optimized energy consumption, and performance and driving pleasure" Carlos Ghosn (Nissan-Renault). Mass production of Renault Fluence, Kangoo, Zoe, and Twizy electric cars begins
2009	Lithium-ion batteries cost $650 per kWh according to Deutsche Bank
2010	Lithium-ion batteries cost $450 per kWh according to Deutsche Bank. Chevy Volt is sold to customers for $41,000
2013	Ford, Mercedes, Toyota, Chevrolet, Fiat, and other major carmakers have at least one fully electric vehicle

costs by reducing thermal management requirements and passive components dimension. There are ongoing improvements in several areas, including device packaging, innovative power module designs, high-temperature capacitors, and new inverter designs. Chargers will probably be the focus of research in the field of power electronics as they will have to manage the vehicle to grid (V2G) paradigm and facilitate fast-charging stations in the coming years. Research is also underway to deal with challenges such as interaction with the grid and the minimization of charging losses when generating over 20kW. Also wireless charging has long been a wish of electric vehicle designers. The big challenge is that it's been less efficient than using a direct-wired connection.

Electric machine research is highly dependent on rare-earth materials. Neodymium and Dysprosium in high-strength magnets are essential for compact, high torque drive motors that tolerate high vehicle operating temperatures. Critical supply risk and rising price trend motivate research into finding alternative permanent magnets (PMs). According to researchers [21] these PM materials must:

- achieve superiority for elevated temperature (150–200°C) operation to minimize motor cooling needs;
- remain competitive at room temperature with current high magnetic energy density materials to conserve weight, space, and valuable materials;
- minimize or eliminate use of scarce rare-earth materials or develop rare-earth-free magnet alloys.

References

[1] Ehsani M, Gao Y, Gay SE, Emadi A. *Modern Electric, Hybrid Electric, and Fuel Cell Vehicles: Fundamentals, theory, and design*. Taylor & Francis, 2004.

[2] Pacejka HB. *Tire and Vehicle Dynamics*. London: Butterworth-Heinemann, 2006.

[3] Fush AE. *Hybrid Vehicles and the Future of Personal Transportation*. Upper Saddle River, NJ: CRC Press, 2009

[4] Mi C, Masrur A, Gao DW. *Hybrid Electric Vehicles: Principles and Applications with Practical Perspectives*. Chichester: John Wiley & Sons, 2011.

[5] Abu-Rub H, Malinowski M, Al-Haddad K. *Power Electronics for Renewable Energy Systems, Transportation and Industrial Applications*. Chichester: John Wiley & Sons, 2014.

[6] Paschero M, Storti GL, Rizzi A, Mascioli FMF. Implementation of a fuzzy control system for a parallel hybrid vehicle powertrain on CompactRio. *Int J Comput Theory Eng* 2013; **5**(2): 273–278.

[7]   Agostoni S, Cheli F, Mapelli F, Tarsitano D. Plug-in hybrid electrical commercial vehicle: Energy flow control strategies. In: *Advanced Microsystems for Automotive Applications*, G. Meyer, ed. Heidelberg, Germany: Springer, 2012: 131–144.

[8]   Miller JM, Manager P. *Electrical and Electronics Systems Research Division Oak Ridge National Laboratory: Annual Progress Report for the Power Electronics and Electric Motors Program*, http://info.ornl.gov/sites/publications/files/Pub46377.pdf, 2013.

[9]   Ozpineci B, Manager P. *Electrical and Electronics Systems Research Division Oak Ridge National Laboratory*. Annual Progress Report for the Power Electronics and Electric Motors Program, November 2014.

[10] http://www.coppermotor.com/wp-content/uploads/2013/08/Techno-Frontier-2013-MBurwell-ICA-EV-Traction-Motor-Comparison-v1.8-Eng1.pdf.

[11] Yilmaz M, Krein PT, Review of battery charger topologies, charging power levels, and infrastructure for plug-in electric and hybrid vehicles. *IEEE Trans Power Electron* 2013; **28**(5).

[12] Lu L, Han X, Li J, et al. A review on the key issues for lithium-ion battery management in electric vehicles. *J. Power Sources* 2013; **226**(March): 272–288.

[13] Ball R, Keers N, Alexander M, Bower E. *Mobile Energy Resources in Grids of Electricity*, http://www.transport-research.info/sites/default/files/project/documents/20140203_154622_76425_Deliverable_2.1_Modelling_Electric_Storage_devices_for_Electric_Vehicles.pdf 2012.

[14] Olszewski M, Burress TA, Campbell SL, et al. *Evaluation of the 2010 Toyota Prius Hybrid Synergy Drive System 2011*. Oak Ridge, TN: UT-Batelle/US Department of Energy.

[15] Husain I. *Electric and Hybrid Vehicles: Design fundamentals*. Upper Saddle River, NJ: CRC Press, 2010.

[16] López FT. *El Vehículo Eléctrico: Tecnología, desarrollo y perspectiva de futuro*. Mcgraw-Hill/Interamericana de España, 1997.

[17] Larminie J, Lowry J. *Electric Vehicle Technology Explained*. Chichester: John Wiley & Sons, 2012.

[18] Nykvist B, Nilsson M. Rapidly falling costs of battery packs for electric vehicles. *Nat Clim Chang* 2015; **5**(4): 329–332.

[19] Power Japan Plus. Power Japan Plus: Balancing the energy storage equation, https://www.youtube.com/watch?t=91&v=OJwZ9uEpJOo.

[20] http://energy.gov/sites/prod/files/2014/03/f10/pl004_davis_joint_plenary_2011_o.pdf.

[21] Marlino L, Anderson I. *EV Everywhere Grand Challenge Electric Drive Status and Targets*, http://energy.gov/sites/prod/files/2014/03/f8/7a_marlino_ed.pdf, 2012.

8

Elevators

Ana Escalada and Gonzalo Abad

8.1 Introduction

In this chapter, elevators from an electric and electronic point of view are studied. Hence, first of all, a general description of the elevator is carried out, providing insight about relevant aspects such as: elevator's definition, their classification, their technical parameters, their main components, and a study of the elevation system. After that, as with the previous three chapters, the physical aspects of the elevator are described. This allows us to model the elevator dynamics enabling further analysis for design, dimensioning, control, and many other purposes.

Then, the electric core of the elevator is described (i.e. the electric drive in charge of controlling the movement of the elevator). Details of the power electronic conversion system are provided, together with the electric machine, the control strategy, the brake, and an exercise showing how the electric drive can be dimensioned, and then a computer-based simulation example is investigated. Finally, some specifics about elevators are briefly introduced: standards and norms, door opening and closing mechanisms, rescue systems, and traffic. To conclude, as with other chapters, a summary of the technology employed by the most representative elevator manufacturers is carried out, together with a short history of the evolution of the elevator as well as a discussion of future trends.

8.2 General description

8.2.1 Definition

An elevator (in some parts of the world, the word "lift" is used) in general is defined as an equipment oriented to elevate different loads, including cars for passenger transport, to two or more different levels and conducted

Power Electronics and Electric Drives for Traction Applications, First Edition. Edited by Gonzalo Abad.
© 2017 John Wiley & Sons, Ltd. Published 2017 by John Wiley & Sons, Ltd.

into guides disposed with an inclination of between 0° and 15° in respect of the vertical. Thus, as some illustrative examples, Fig. 8-1 shows different aspects of elevators.

There are many mechanical and electrical elements that make up elevators. However, as depicted in Fig. 8-2, it can be said that the principal components of an electric elevator are:

- Car: cabin where transport of persons or other elements is carried out;
- Motor: this is the traction element in charge of transporting the car, using the mechanical arrangement of sheave-ropes and counterweight;
- Guides: metallic profiles located through the enclosure in order to guide the car;
- Counterweight: its function is to equilibrate the load of the car and it is joined to it by ropes.

Later in this chapter more details are provided and more elements are described.

8.2.2 Classification

Elevators can be classified according to various characteristics; however, probably the one most employed one classifies elevators in function of the traction method employed, which leads to different design principles, construction, and installation:

- Electric elevators, where the movement is governed by electric machines.
- Hydraulic elevators, where the movement is governed by hydraulic actuators.

In this chapter, only electric elevators are analyzed. However, it must be mentioned that hydraulic elevators are commonly used nowadays, mainly because they do not need any counterweight and the space occupied in the shaft is minimal.

In parallel, depending on the location of the traction motor in the elevator itself, it is possible to distinguish three major electric elevator configurations, which are briefly described in the following subsections.

8.2.2.1 *Elevator with machine at the top of the shaft*

This is probably the most employed configuration in recent years in so-called developed countries. Basically, it is characterized by having the electric motor located at the top of the elevator. Technically, it is generally agreed that overall it is the simplest configuration, since the resulting elevator is robust. In addition, out of the three configurations discussed here, it is the most convenient for high-speed elevators. On the other hand, it presents architectonic disadvantages since it needs a machine room to be included in the building. Fig. 8-3 shows a schematic example of this configuration. It must be noted that, nowadays, in most modern buildings this configuration is being displaced by the machine-room-less concept elevator that is described in Section 8.2.2.3.

8.2.2.2 *Elevator with machine beside the bottom of the shaft*

In this type of electric elevator, the traction motor is located at the bottom. Technically, this configuration results in an elevator that is more complex than the previous one and therefore also more expensive. In general, compared with the previous configuration, it requires additional deflector sheaves and longer ropes, which is a disadvantage. These items assure the adherence of the ropes to the sheaves. Because the machine is at the bottom of the shaft, as is the drive sheave, correct adherence to rope configuration guidelines must be guaranteed to ensure the ropes do not slip. Fig. 8-4 shows an illustrative example of this configuration.

(a)

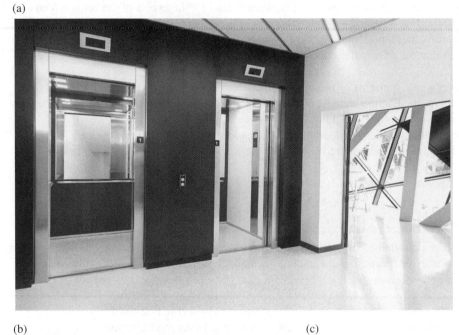

(b) (c)

Fig. 8-1 *Different elevator illustrations: (a) modern elevator, (b) inside of a car, and (c) schematic structure of a conventional elevator (Source: ORONA. Reproduced with permission of ORONA)*

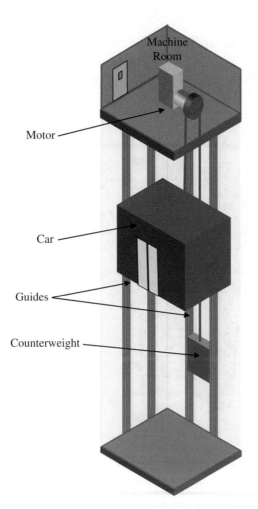

Fig. 8-2 *Schematic of an elevator with machine room, showing its principal components*

8.2.2.3 Machine-room-less elevator

This configuration is the most recent one and is characterized by the elimination of the machine room. And slowly but surely it is being incorporated into the regulations of many modern countries. In essence, this configuration does not bring any advantage to the user or to the installer, because installation and maintenance are more complex than for the other two configurations, but it brings advantages to the architect because it eliminates the machine room. Fig. 8-5 shows an illustrative example of this elevator configuration.

8.2.2.4 Hydraulic elevator

Hydraulic elevators also exist, although in advanced countries they have been displaced by electric elevators. The traction force is generated by a hydraulic piston and does not need a counterweight. It is adequate for moderately high loads and for speeds no greater than 1 m/sec. One of its main disadvantages, compared to the electric elevator concept, is energy efficiency. The hydraulic elevator can consume 3–4 times more

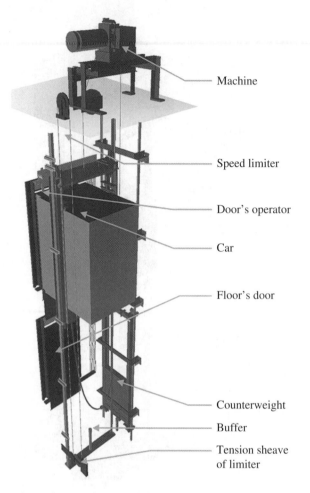

Machine

Speed limiter

Door's operator

Car

Floor's door

Counterweight

Buffer

Tension sheave
of limiter

Fig. 8-3 *Electric elevator with machine at the top of the shaft (Source: ORONA. Reproduced with permission of ORONA)*

energy than the equivalent electric elevator. Fig. 8-6 shows an illustrative example of this elevator configuration.

8.2.3   Technical parameters

The most important technical parameters defining elevators are: the rated load Q (kg) and the rated speed v (m/sec). Commonly, the speed of the elevator is constant, but nowadays there are control techniques being introduced that adapt the speed of the elevator to the specific load of the moment in an attempt to reduce power consumption and so improve energy efficiency. Therefore, the elevator and its installation are designed, constructed, and carried out according to these main technical parameters. In general, in elevators for passengers, the rated load is related to the maximum amount of admissible passengers (4, 6, etc.). The average weight of a passenger is normally approximated to 75kg in any calculation of elevator load. There are a huge variety of

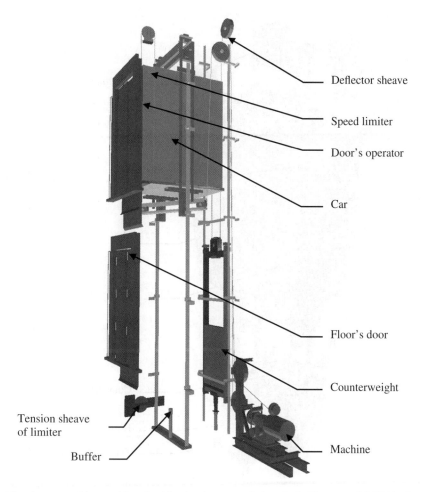

Deflector sheave

Speed limiter

Door's operator

Car

Floor's door

Counterweight

Tension sheave
of limiter

Buffer

Machine

Fig. 8-4   *Electric elevator with machine beside the bottom of the shaft (Source: ORONA. Reproduced with permission of ORONA)*

elevators of different rated loads and speed characteristics on the market. Table 8-1 lists examples of the standard elevators of the manufacturer ORONA.

In addition, some other more specific technical parameters used include:

- maximum journey with number and position of the stops;
- dimensions of the elevator's cavity, car, and machine room;
- grid voltage, number of journeys per hour, and load factor;
- control system;
- doors of the floor and car;
- number of elevators and its position in the building;
- ambient conditions.

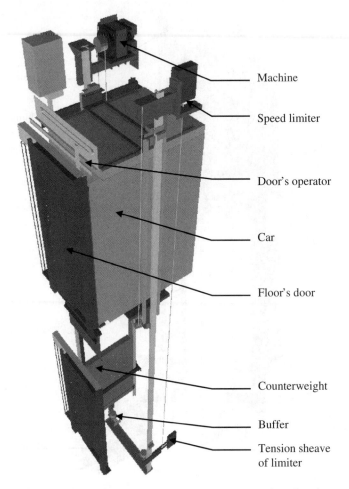

Machine

Speed limiter

Door's operator

Car

Floor's door

Counterweight

Buffer

Tension sheave
of limiter

Fig. 8-5   *Machine-room-less electric elevator (Source: ORONA. Reproduced with permission of ORONA)*

8.2.4   Components of an elevator installation

Section 8.2.1 briefly presents the most important elements comprising a modern elevator. The following subsections delve deeper into this question, providing more information about the different elements of the elevator needed for its correct functioning.

8.2.4.1   Elevation system

From the user's point of view, the main objective of an elevator is to transport vertically passengers or loads. These elements travel inside the car. The electric machine or motor starts moving when the elevator is required to go to a certain level. It moves the drive sheave in one or other direction depending on whether the elevator must go up or down. It must be mentioned that there are different traction methods, but the most common ones are based on sheaves and traction ropes. Alternatively, there are some experimental prototypes which employ

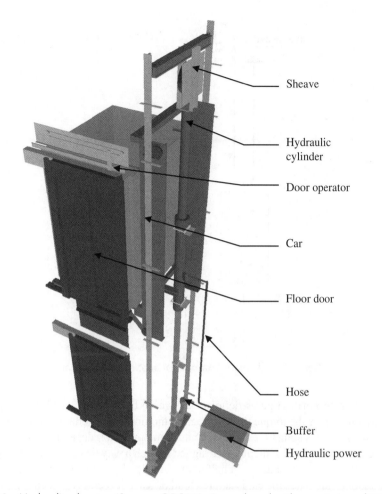

Sheave

Hydraulic cylinder

Door operator

Car

Floor door

Hose

Buffer

Hydraulic power

Fig. 8-6 *Hydraulic elevator (Source: ORONA. Reproduced with permission of ORONA)*

Table 8-1 *Example showing the technical parameters of a typical elevator manufacturer*

Type of installation	Residential				Hoist				Hospital
Number of passengers	4	6	8	13	—	—	—	—	—
Rated load (kg)	320	450	630	1000	1000	2000	3000	5000	1600
Rated speed (m/sec)	0.63	1	1.6	2.5	0.63	0.36	0.5	0.25	1

linear motors to move the car, not using drive sheaves but still needing sheaves to join the car to the counterweight.

Therefore, the car is suspended by means of the traction ropes, which pass through the drive sheave and finish at the counterweight. The counterweight travels the opposite way to the car, and its function consists of compensating the weight of the car and part of the rated load. In Fig. 8-7 a schematic example of an elevation

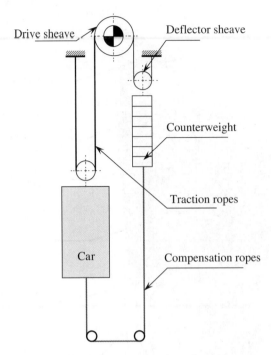

Fig. 8-7   *Schematic example of an elevation system*

system is represented. Note that in this figure a deflector sheave has been included with the elevation system. This is often added in order to reduce the torque that the traction motor must generate for moving the car. Later, we show that the deflector sheave is sometimes included and sometimes not, depending on the elevator's design.

On the other hand, in general the installation is designed in such a way that the counterweight is equal to the mass of the car plus half of the rated load. Under half of the rated load, it is said that the load is equilibrated. The most unbalanced situations are when the car is at rated load or when it is empty, which is half of the rated load. During the entire journey, the car and the counterweight are conducted by their respective guides. Fig. 8-8 shows an illustrative example of a counterweight.

In order to detect the position of the car, usually permanent magnets are located at the cavity of the elevator, so the car can detect its location by electromagnetic sensors. For a more accurate positioning, modern installations also use incremental encoders which facilitate the traction movement, as is described below. Thus, the encoder is used by the control system of the vertical traction system of the elevator, while the positioning system based on magnets is additionally and redundantly employed to know where the car is located by the maneuver. At the maneuver, which is the central control element governing all the parts of the elevator, all the information of the different elements of the elevator is centralized and managed.

Hence, the elevation system in an elevator is a group comprising a car, a counterweight, and ropes. Nowadays, it is possible to find four main types of ropes or cables used by elevator manufacturers:

- Conventional steel rope: It is formed by steel threads, weaved with cord to form around a metallic heart.
- Covered rope: This is similar to conventional rope, but with the external surface covered by a thermoplastic material.
- Synthetic rope: Composed by synthetic fibers and externally covered to increase adherence to the sheave.
- Flat rope: It is a tape, internally constituted by steel threads and externally coated.

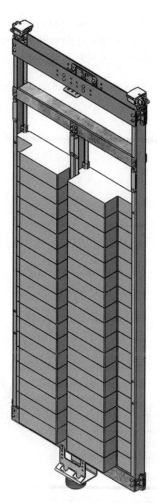

Fig. 8-8 *Illustrative example of a counterweight (Source: ORONA. Reproduced with permission of ORONA)*

Nowadays, it can be said that there is a move away from using steel ropes. Instead, the other three rope types are more often employed, because they increase the lifetime of the ropes and allow the use of drive sheaves of increasingly reduced diameters. This last fact is advantageous, since the larger the diameter of the sheave, the higher the torque that the motor must realize to move a given load. This means, in general, a larger traction motor and more power losses at it (note that the higher the torque, often the larger is the necessary current and therefore the larger the joule power losses, although it depends on the resulting resistance). It must be emphasized that there is a tendency to reduce the volume and dimensions of traction elements while at the same time trying to be as efficient as possible. In order to satisfy these objectives, it is desirable to employ a drive sheave with a small diameter that guarantees the life expectancy of the traction rope and the safety of the installation.

Finally, in relation to the compensation ropes, their use depends mainly on two factors: the rise and the type of traction rope employed. In this matter, each manufacturer follows its own philosophy. For instance,

ORONA's synthetic ropes, up to height of 75m, do not need compensation ropes. For rises higher than 75m, synthetic ropes are not used, because of the risk of elongation of the ropes. For these longer rises, covered ropes are employed. When covered ropes are used, the compensation rope is approximately necessary with heights higher than 30m.

With regard to the car, in general each manufacturer has its own car design, but we can distinguish the following main elements that compose the car:

- chassis: metallic structure forming the car's skeleton;
- floor platform;
- door and its opening mechanism;
- guides support;
- union of the ropes at the top.

Fig. 8-9 shows examples of two different types of chassis.

Fig. 8-9 *Two examples of chassis (Source: ORONA. Reproduced with permission of ORONA)*

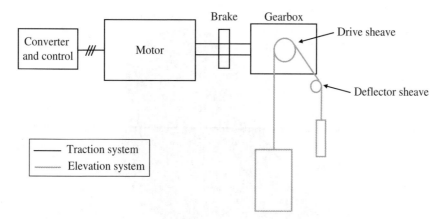

Fig. 8-10 *Simplified illustration of a traction system with a deflector sheave*

8.2.4.2 Traction system

The traction system of an elevator is normally made up of an electric machine and electric drive, coupled to an electromagnetic brake and a gearbox (the tendency nowadays is to eliminate the gearbox), and a drive sheave. If the car and the counterweight are not at a minimum distance, it is necessary to install a deflector sheave as introduced earlier and illustrated in Fig. 8-10. Safety norms oblige one to apply the brake directly to the drive sheave. This implies that modern traction systems are formed by compact developments of motor, brake, encoder, and drive sheave mounted solidly to the same shaft.

Therefore, as briefly discussed earlier, in electric elevators it is possible to find machines with a gearbox and gearless machines. When a gearbox is utilized to adapt the torques and speeds of the machine to those of the drive sheave, basically there are two main gearbox technologies: worm gearboxes and planetary gearboxes.

Thus, Fig. 8-11 illustrates an example of a machine with a worm gearbox. These machines typically have a low mechanical efficiency (normally > 70%), reduced noise, and reasonably reduced cost. They are useful for low and medium speeds (up to 2m/sec).

Planetary gearboxes in general are more complex than worm gearboxes, have a better mechanical efficiency, have an improved noise behavior compared to the worm gearbox but they are more expensive.

In general, geared machines are associated with induction electric machines, which are not able to be designed to operate efficiently at low speeds and high torques and therefore need a gearbox to adapt these two variables to the torques and speeds of the drive sheave. On the contrary, the development of most sophisticated and complex multi-pole permanent based synchronous motors and their inclusion in the elevator application enables one to adapt efficient designs of electric motors coupled directly to the drive sheaves, without the need of a gearbox adapting torque and rotational speeds. Hence, Fig. 8-12 shows some examples of gearless machines. In general, their performance is the most efficient (around 90% of efficiency), they eliminate the maintenance required for gearboxes, they are the quietest, they do not have any speed limitations, and the cost is similar to that of the geared versions. In addition, one of the main improvements connected to the gearless concept is the reduction of space obtained compared to its equivalent geared versions, because the need for the machine room is eliminated.

8.2.4.3 Doors and accesses

The doors and access ways of the elevator fulfill several functions. One of them consists of permitting access to the car when it is at floor level. Access must be secure and free from the risk of passengers or maintenance

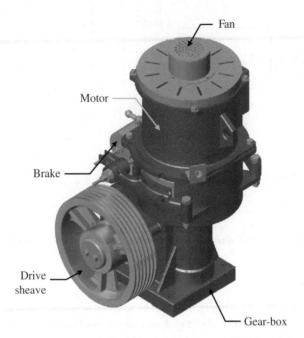

Fan

Motor

Brake

Drive
sheave

Gear-box

Fig. 8-11 *Machine with worm gearbox (Source: ORONA. Reproduced with permission of ORONA)*

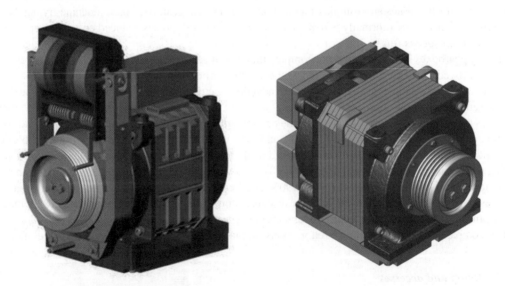

Fig. 8-12 *Gearless examples of machines based on multi-pole permanent magnetic synchronous machines (PMSMs) (Source: ORONA. Reproduced with permission of ORONA)*

Fig. 8-13   *Illustration of doors for elevators*

operators becoming trapped. In the event of an emergency or during maintenance, access ways must also allow access to the shaft. In some cases, the doors must be fireproof, in the sense of helping to reduce the spread of flames in the event of fire.

In normal operating conditions, the floor doors can only be opened when the car is at or near floor level. With regards to the car's doors, the system in charge of opening and closing the doors is located in the car itself and enables the opening of the floor doors when the car is level with the doors. There exist many different types of doors, the simplest and maybe the oldest ones work manually, but the most typical ones work automatically. With the automatic ones, we can find lateral aperture doors, central aperture doors, and so on. Finally, Fig. 8-13 illustrates one example of doors for elevators.

8.2.4.4   *Guides*

With regards to the guides, their main function is to guide the car and the counterweight through their displacement. They normally are designed to minimize the horizontal movements and possible oscillations due to passenger movement within the car, since they reduce the comfort as well as the lifetime of the elements. In addition also, they are in charge of maintaining the car and counterweight after the activation of the parachute.

In general, the guides are laminated profiles of standard T geometry that can be fabricated with different materials and tolerances. It is important to highlight that the mounting process of the guides is crucial to obtain a good comfort inside the elevator. Fig. 8-14 shows an illustrative example of the guide arrangement of an elevator. It is possible to see that they require adjustable supports in order to adapt to any irregularities of the building.

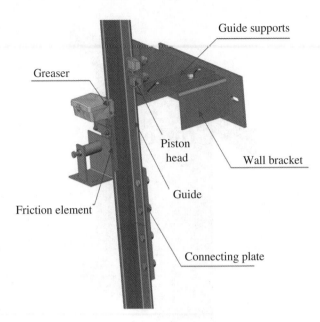

Fig. 8-14   *Illustration of the guides*

8.2.4.5   *Safety elements*

For obvious reasons, safety in elevators is a key issue. Thus, elevators incorporate an over-speed protection, to protect the elevator and the passengers in the event of the elevator falling or over-speed. This protection is purely mechanical. It stops the car by using friction against the guides, in case an over-speed is produced. This mechanism is normally located at the lowest part of the car. In simple terms, it is possible to say that it is composed of four elements:

- Speed limiter: detects the over-speed and generates at the parachute the necessary actuating force.
- Parachute: stops the car, by means of the friction effort produced against the guides when activated.
- Limiter rope: transmits the actuating effort to the parachute.
- Tension weight: is in charge of maintaining the necessary tension at the rope.

Fig. 8-15 illustrates a graphical example of a protection system against over-speed.

In the following lines, some more details are provided of some of the mentioned elements. Hence, the first element described is the speed limiter. This is a mechanical device designed to suddenly block when a given speed is exceeded. There are speed limiters which work only in one direction and others that work in both directions (going up and down). This element consists of a sheave moved by a rope fixed to the car. Thus, when the limiter is blocked the rope pulls the parachute, activating it.

With regards to the parachute, it operates stopping the car or counterweight, trapping the guides like a pincer. The parachute, as mentioned before, is activated by the speed limiter. There is one parachute at each guide. All parachutes are joined together by a rigid bar. The parachute can stop the car instantaneously or it can do so gradually, using controlled deceleration.

Finally, in the unlikely case that the car falls to the pit, there is another safety element located at the pit: the shock absorbers. They stop the car or counterweight absorbing the kinetic energy.

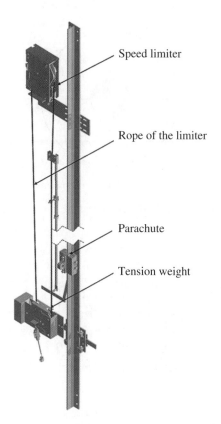

Speed limiter

Rope of the limiter

Parachute

Tension weight

Fig. 8-15 *Protection system against over-speed (Source: ORONA. Reproduced with permission of ORONA)*

8.2.4.6   *Control system*

The elevator also incorporates a high-level control system, which supervises and controls all the elements of the elevator. It centralizes sensor and actions of elements such as:

- Vertical transport: car positioning and electric drive control;
- Door movement: accesses control and control of the door's operation;
- Signaling pushbuttons: from floor, from car, registers, etc.;
- Service: type of traffic;
- Safety: different norm requirements, firefighters, etc.;
- Communications: monitoring, to other elevators, tele-service, etc.

8.2.5   Elevation system

In this subsection, deeper analysis about the elevation system is provided, which has been briefly introduced in Section 8.2.4.1. This short introduction to the elevation system and different roping arrangements allows us to study the physical aspects of the elevator in a subsequent section. Thus, as described before, the drive sheave is used to transmit the power between the machine and the ropes of the elevator. The transmission force

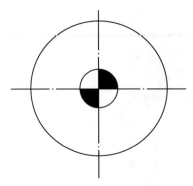

Fig. 8-16   *Simplified representation of a drive sheave*

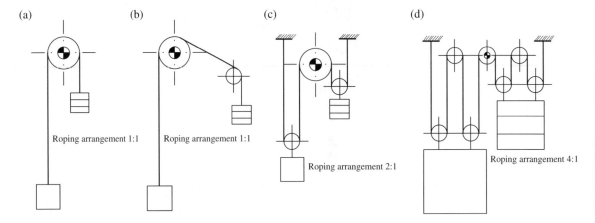

Fig. 8-17   *(a) Simple 1:1 roping arrangement, (b) simple 1:1 roping arrangement, with deflector sheave, (c) multiple 2:1 roping arrangement, and (d) multiple 4:1 roping arrangement*

is transmitted by friction between the ropes and the drive sheave. Note that the ropes must always be able to ensure adherence (a certain level of friction) and avoid uncontrolled slip. Fig. 8-16 illustrates the simplified representation of a drive sheave that will be used in subsequent schemas and has been already shown before.

Fig. 8-17 graphically illustrates different types of roping arrangements. It can be noted that with 1:1 roping the number of ropes attached to the car or counterweight is 1. With 2:1 roping arrangement, the number of ropes attached to the car or counterweight is 2, and the same holds with 4:1 and so on.

As will be later studied in detail, a higher-number roping arrangement, as opposed to the simple 1:1 roping arrangement, allows the electric machine to operate at higher rotational speeds, for equal linear speed of the car. One advantage is that the electric machine uses only a fraction of the weight to move, which is proportional to the roping arrangement relation. However, one disadvantage is that with a high-number roping arrangement, the installation complexity increases, since more sheaves and ropes are required.

Therefore, Fig. 8-17(a) shows the simplest configuration 1:1 roping, with diameter of the drive sheave equal to the distance between the car and the counterweight. In Fig. 8-17(b), a deflector sheave has been installed, mainly because the distance between the car and the counterweight is great. In cases when this distance is still greater and at the same time the rated load is also increased, the roping arrangement ratio is also increased, sharing the weights of car and counterweight between several ropes, as illustrated in Fig. 8-17(c) and (d). In such cases, as is discussed below, the forces of the ropes are reduced. In addition, note that the extreme points of the ropes are tied up to the building joists. Note also that roping arrangements higher than 1:1, from a different point of view, can be simply adopted in order to increase the linear speed of the car and to reduce the torque requirement of the electric machine. Note that the required power to move a given load in the car at a constant linear speed is independent of the roping arrangement. Mainly, by increasing the roping arrangement, the torque that the drive sheave must produce is diminished, but the rotational speed is increased.

Hence, in general, for a given elevator, depending on the load, weight of the car, and the route, a diameter of drive is fixed and then the required number of ropes is calculated guaranteeing the adherence between sheave and ropes. Normally, the diameter of the drive sheave is fixed at an initial state together with the type of ropes, in order to ensure the minimum lifetime of an installation. Mainly, the diameter of the drive sheave is determined by the diameter of the rope. Indicatively, a ratio of 40:1 is approximately employed for metallic ropes. Thus, Fig. 8-18 shows a variety of different roping arrangements for elevators.

In general the ropes are wrapped over the drive sheave in grooves. The weight of both the car and the counterweight ensures the seating of the ropes in the groove and also the traction. Thus, the pinching action of the grooves on the rope creates the necessary friction. Various types of grooving are used for different loads and traction requirements. In general, the sharper the undercut angle, the greater the angle obtained. This fact is graphically illustrated in Fig. 8-19. A deeper analysis reveals that several manufacturers improve the friction action by using several techniques, for instance by including polyurethane between the grooves and ropes [1].

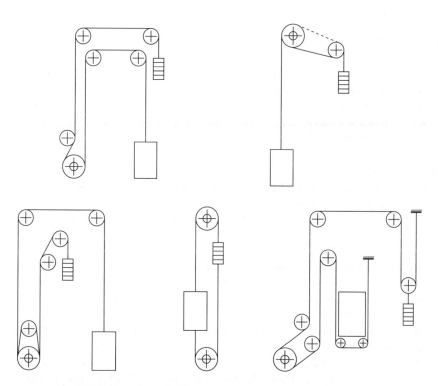

Fig. 8-18   *A variety of different roping arrangements for elevators*

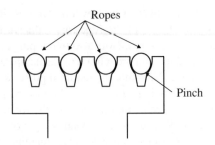

Fig. 8-19   *Undercut sheave groove*

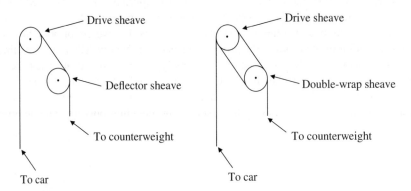

Fig. 8-20   *Single-wrap arrangement simplified schema and double-wrap arrangement simplified schema*

In addition, for elevators with high speed, the ropes are double-wrapped through the drive sheave in order to obtain an embrace angle between rope and sheave that ensures proper adherence. Fig. 8-20 graphically illustrates the difference between single- and double-wrap arrangements.

Therefore, the design of a new elevator, and the corresponding roping arrangement, results from the detailed analysis of different factors that can be summarized as:

- The diameter of the drive sheave is defined by the type of rope and its characteristics.
- Each elevator manufacturer employs different traction ropes, but it can be said that a common one is a metallic rope of diameters: 8, 10, and 12mm.
- The groove shapes and the number of ropes are fixed in order to guarantee the adherence between the rope and the sheave.
- The roping arrangement is chosen in an attempt to find a compromise. Roping arrangements with high ratios are good to reduce the required torque of the motor and therefore its volume. However, if the roping arrangement ratio is too high, the installation is more complex and also more space is required.
- Deflector sheaves are necessary, depending on the distance between the drive sheave and the point where the rope is connected to the car.
- A compensation rope is included for traveling distances larger than a fixed limit and also depending on the weight of the traction cables.
- Finally, the result must be coordinated with the electric drive technology employed to govern the movement of the drive sheave, as is described in later sections: motor technology and electromechanical characteristics, gearbox ratio or gear-less, power electronic converter, control strategy, and so on.

This subsection concludes with some hints for choosing the suspensions system or roping arrangement of a specific elevator. First of all, it must be said that the elimination of the gearbox is desirable, mainly to reduce the required space by eliminating the machine room and also to avoid the maintenance and cost associated with the gearbox. However, by eliminating the gearbox, the torque that the motor must generate increases considerably, since it will need to provide the same defined power at a lower rotating speed. In order to design traction motors optimized in efficiency and of reduced volume, it is desirable to reduce the traction torque required. Therefore, it is quite typical to include roping arrangements of 2:1 or even greater, enabling the reduction of the torque demanded at the drive sheave and therefore at the traction motor.

However, the reader can easily observe that high roping arrangements demand greater space at the cavity of the elevator in order to include the necessary sheave arrangements and ropes. Consequently, it becomes crucial to find a good technological balance between the roping arrangement and the traction electrical machine.

Thus, in general, the principal advantages of the gearless elevator are:

- The closed-loop electric drive allows better performance to be obtained and greater ride comfort.
- Since the machine rotates at the same speed required by the elevator itself, the high-speed rotating elements are avoided, improving significantly the acoustic noise levels against geared elevators.
- It is possible to obtain better efficiency in the energy consumption of the elevator, since normally the electric machines employed (PMSMs) are very efficient.
- The reduced space required for the drive reduces the architectural impact on the building.

8.3 Physical approach

This section analyses the physical equations of the elevators that permit us to develop its model. The derived physical equations are differential equations which are obtained under several assumptions and simplifications [1], [2]. The main purpose of this section is not to obtain a detailed model of the elevator, describing complex behaviors or performances. Instead, having gained a general perspective of the forces and factors which describe the movement of the elevator above allows us to attempt to understand the dimensions and design of the electric drive which controls it.

Hence, first of all the model equations of the simplest elevator with a 1:1 roping arrangement are derived. After that, the same is carried out with an elevator of more complicated physics, i.e. with a 2:1 roping arrangement.

8.3.1 Model equations of elevators with a 1:1 roping arrangement

This section derives the model differential equations describing the torques, forces, and movement of the most simple elevator configuration: a 1:1 roping arrangement. A simplified representation of the elevation system is presented in Fig. 8-21. The following is assumed:

- Only one rope is considered rotating with the drive sheave, whose mass is neglected and also is supposed rigid (non-elastic).
- The mass of the car, m_1, and the mass of the counterweight, m_2, is supposed to only move vertically.
- The inertia of the machine (and gearbox, if there is any) and sheave is very small in comparison to the mass of the car and counterweight. Therefore, only the car and counterweight are considered as elements contributing to the inertia, with T_s being the torque applied to the sheave by action of the electric machine, $\dot{\theta}_s$

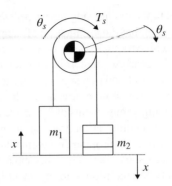

Fig. 8-21   *Ideal model of an elevator with a 1:1 roping arrangement*

the rotational speed of the sheave, θ_s the angular position of the sheave, and x the vertical displacement of the car and counterweight.

The mass of the car is noted as:

$$m_1 = M + Q \tag{8.1}$$

While the mass of the counterweight:

$$m_2 = M + \psi \cdot Q_{max} \tag{8.2}$$

with M being the mass of the car with no load, Q the load mass, and ψ the fraction of the maximum load in pu.

Note that the load may change from one journey to other, thus the counterweight is chosen in order to minimize the energy consumption of the elevator, in the most common expected cycles of operation. In general, the ψ parameter can take values of between 0.45 and 0.5. In our model, we will consider that $\psi = 0.5$.

Hence, the model differential equations of the elevator can be described with the help of Fig. 8-22 as:

$$F_1 - m_1 g = m_1 \ddot{x} \tag{8.3}$$

$$m_2 g - F_2 = m_2 \ddot{x} \tag{8.4}$$

$$T_s + (F_2 - F_1)R = J\ddot{\theta}_s \tag{8.5}$$

with J being the inertia of the sheave and R the radius of the sheave. Note that the weight of the cables or friction of ropes and counterweight with the guides has not been considered in this simplified model approach. Thus, rearranging (8.3)–(8.5) into a unique equation:

$$T_s = (m_1 - m_2)gR + (m_1 + m_2)R\ddot{x} + J\ddot{\theta}_s \tag{8.6}$$

assuming that $x = R\theta_s$, the last expression yields:

$$T_s = (m_1 - m_2)gR + (m_1 + m_2)R^2\ddot{\theta}_s + J\ddot{\theta}_s \tag{8.7}$$

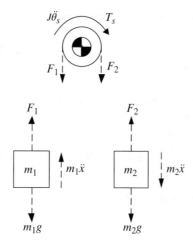

Fig. 8-22 *Force decomposition of the elevator with a 1:1 roping arrangement*

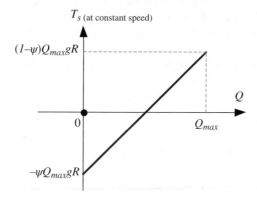

Fig. 8-23 *Torque provided at the sheave at constant speed, depending on the load of the car Q, with a 1:1 roping arrangement*

which means that the sheave must function with a constant term which does not depend on the acceleration $(m_1 - m_2)gR$ and a term which does depend on the acceleration: $(m_1 + m_2)R^2\ddot{\theta} + J\ddot{\theta}$. When the elevator is operating at constant speed $\ddot{\theta}_s = 0$, the torque that the sheave must provide is:

$$(m_1 - m_2)gR \tag{8.8}$$

If we substitute (8.1) and (8.2), this torque yields as function of the load:

$$(Q - \psi Q_{max})gR \tag{8.9}$$

This means that, depending on the load of each journey, the torque of the sheave at constant speed will be between a maximum $(1 - \psi)Q_{max}gR$ and a minimum $\psi Q_{max}gR$, as illustrated in Fig. 8-23.

On the other hand, from (8.7), it is possible to obtain the movement equation in function of the load of the elevator:

$$T_s = (Q - \psi Q_{max})gR + (2MR^2 + QR^2 + Q_{max}\psi R^2 + J)\ddot{\theta}_s \qquad (8.10)$$

The maximum torque that the sheave must provide is at the acceleration moment when going up at maximum load $Q = \psi Q_{max}$ (considering $\psi = 0.5$):

$$T_{s\_max} = \frac{1}{2}Q_{max}gR + \left(2MR^2 + \frac{3}{2}Q_{max}R^2 + J\right)\ddot{\theta}_s \qquad (8.11)$$

This expression is useful for design purposes of the electric drive of the elevator. Similarly, the power at the sheave can be described as:

$$P = T_s\dot{\theta}_s \qquad (8.12)$$

Hence, this physical approach shows that by paying attention to (8.10) the torque that the electric machine must apply at the sheave T_s finds in opposition the torque produced by the car and counterweight and the inertia of the sheave itself. Fig. 8-24 illustrates this fact, in which T_s must be controlled by the electric machine drive all the time, in order to obtain the desired movement of the elevator in terms of position, speed, acceleration, etc.

Finally, from (8.7) is possible to obtain the model equation of the elevator as:

$$\ddot{\theta}_s = \frac{T_s + (m_2 - m_1)gR}{(m_1 + m_2)R^2 + J} \qquad (8.13)$$

While in function of the load:

$$\ddot{\theta}_s = \frac{T_s + (\psi Q_{max} - Q)gR}{(2M + \psi Q_{max} + Q)R^2 + J} \qquad (8.14)$$

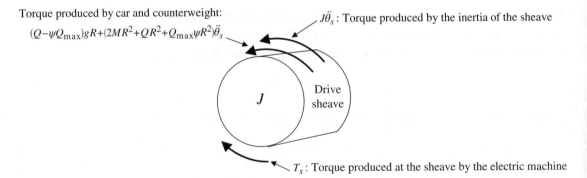

Torque produced by car and counterweight:
$(Q - \psi Q_{max})gR + (2MR^2 + QR^2 + Q_{max}\psi R^2)\ddot{\theta}_s$

$J\ddot{\theta}_s$: Torque produced by the inertia of the sheave

J

Drive sheave

T_s : Torque produced at the sheave by the electric machine

Fig. 8-24   *Resulting torques at the drive sheave*

(a)

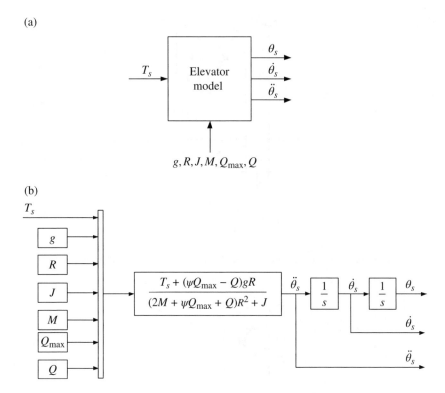

Fig. 8-25 *Model block diagram of the elevator with a 1:1 roping arrangement for computer-based simulation: (a) general block and (b) detailed model*

This expression is the model of the elevator that can be represented in block diagram, as illustrated in Fig. 8-25 for computer-based simulation purposes. The input of the elevator model is the torque applied to the sheave, T_s. The parameters, which are supposed constant during the whole journey, are: g, R, J, M, Q_{max}, and Q, while the outputs are the angular position θ_s, speed $\dot{\theta}_s$, and acceleration of the sheave $\ddot{\theta}_s$. Note that simply multiplying by the radius of the sheave it is also possible to obtain the linear position, speeds, and accelerations as outputs.

8.3.2 Model equations of elevators with a 2:1 roping arrangement

In this section, an equal procedure to derive the model of an elevator with a 2:1 roping arrangement is carried out. The same assumptions and notations as before are considered, in order to obtain a set of simplified model differential equations. Thus, in this case Fig. 8-26 illustrates the simplified schema of the elevator.

Therefore, with the help of the force decomposition shown in Fig. 8-27, the following equations can be derived:

$$F_1 + F_2 - m_1 g = m_1 \ddot{x} \tag{8.15}$$

$$m_2 g - F_3 - F_4 = m_2 \ddot{x} \tag{8.16}$$

$$T_s + (F_3 - F_2)R = J\ddot{\theta}_s \tag{8.17}$$

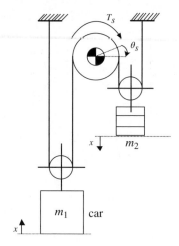

Fig. 8-26   *Ideal model of an elevator with a 2:1 roping arrangement*

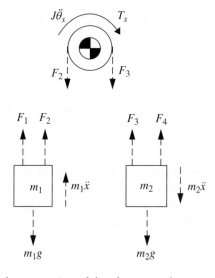

Fig. 8-27   *Force decomposition of the elevator with a 2:1 roping arrangement*

Knowing that $F_1 = F_2$ and $F_3 = F_4$, it is possible to convert the last three expressions into only one:

$$T_s = \frac{1}{2}(m_1 - m_2)gR + \frac{1}{2}(m_1 + m_2)R\ddot{x} + J\ddot{\theta}_s \qquad (8.18)$$

Assuming that $x = \frac{1}{2}R\theta_S$, the last expression yields:

$$T_s = \frac{1}{2}(m_1 - m_2)gR + \frac{1}{4}(m_1 + m_2)R^2\ddot{\theta}_s + J\ddot{\theta}_s \qquad (8.19)$$

As occurs with a 1:1 roping arrangement, the sheave must handle with a constant term which does not depend on the acceleration. $\frac{1}{2}(m_1-m_2)gR$ and a term which does depend on the acceleration: $\frac{1}{4}(m_1+m_2)R^2\ddot{\theta}+J\ddot{\theta}$. When the elevator is operating at constant speed $\ddot{\theta}_s=0$, the torque that the sheave must provide is:

$$\frac{1}{2}(m_1-m_2)gR \tag{8.20}$$

It should be noted that a 2:1 roping arrangement, compared to a 1:1 arrangement, reduces the torque to one-half required at the drive sheave owing to forces experienced in the car and at the counterweight. As has been mentioned before, this reduction of torque requirement is accompanied by an increase in the complexity of the installation. In addition, the relation between the linear displacement and the rotational angle is reduced to $R/2$, which is favorable to reduce the gearbox requirement, in case that exists.

Hence, as we did in the previous subsection, if we substitute (8.1) and (8.2), the torque expression (8.20) yields as a function of the load:

$$\frac{1}{2}(Q-\psi Q_{\max})gR \tag{8.21}$$

which means that, depending on the load of each travel, the torque of the sheave at constant speed will be between a maximum $\frac{1}{2}(1-\psi)Q_{\max}gR$ and a minimum $\frac{1}{2}\psi Q_{\max}gR$, as illustrated in Fig. 8-28.

On the other hand, from (8.19), the torque equation in function of the load yields:

$$T_s=\frac{1}{2}(Q-\psi Q_{\max})gR+\frac{1}{4}(2MR^2+QR^2+Q_{\max}\psi R^2+4J)\ddot{\theta}_s \tag{8.22}$$

The maximum torque that the sheave must provide is at the acceleration moment when going up at maximum load $Q=\psi Q_{\max}$ (considering $\psi=0.5$):

$$T_{s\_\max}=\frac{1}{4}Q_{\max}gR+\frac{1}{4}\left(2MR^2+\frac{3}{2}Q_{\max}R^2+4J\right)\ddot{\theta}_s \tag{8.23}$$

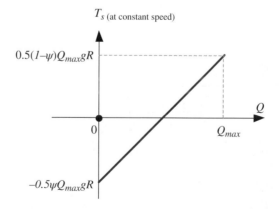

Fig. 8-28 *Torque provided at the sheave at constant speed, depending on the load of the car, Q, with a 2:1 roping arrangement*

Finally, the model equation of the elevator from (8.19) is:

$$\ddot{\theta}_s = \frac{T_s + \frac{1}{2}(m_2 - m_1)gR}{\frac{1}{4}(m_1 + m_2)R^2 + J} \tag{8.24}$$

while in function of the load:

$$\ddot{\theta}_s = \frac{T_s + \frac{1}{2}(\psi Q_{max} - Q)gR}{\frac{1}{4}(2M + \psi Q_{max} + Q)R^2 + J} \tag{8.25}$$

This expression is the model of the elevator that can be represented in the block diagram, as illustrated in Fig. 8-29 for computer-based simulation purposes.

Finally, the reader can simply deduce the model equations of higher roping arrangements by extrapolating the procedures that have been followed. Table 8-2 gives a short summary of some interesting model equations of different elevator configurations.

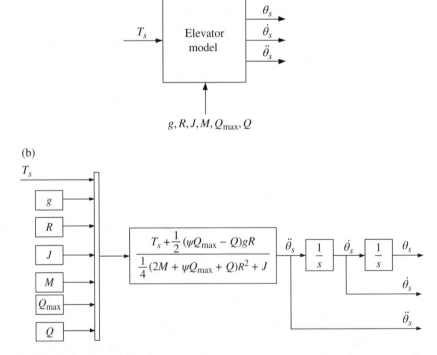

Fig. 8-29   *Model block diagram of the elevator with a 2:1 roping arrangement for computer-based simulation: (a) general block and (b) detailed model*

Table 8-2 *Some interesting model characteristics, for different elevator configurations*

Roping arrangement	Torque at constant speed	Maximum torque at constant acceleration with ($\psi = 0.5$)	Model differential equation
1:1	$(Q - \psi Q_{max})gR$	$T_{s\_max} = \frac{1}{2}Q_{max}gR + \left(2MR^2 + \frac{3}{2}Q_{max}R^2 + J\right)\ddot{\theta}_s$	$\ddot{\theta}_s = \dfrac{T_s + (\psi Q_{max} - Q)gR}{(2M + \psi Q_{max} + Q)R^2 + J}$
2:1	$\frac{1}{2}(Q - \psi Q_{max})gR$	$T_{s\_max} = \frac{1}{4}Q_{max}gR + \frac{1}{4}\left(2MR^2 + \frac{3}{2}Q_{max}R^2 + 4J\right)\ddot{\theta}_s$	$\ddot{\theta}_s = \dfrac{T_s + \frac{1}{2}(\psi Q_{max} - Q)gR}{\frac{1}{4}(2M + \psi Q_{max} + Q)R^2 + J}$
4:1	$\frac{1}{4}(Q - \psi Q_{max})gR$	$T_{s\_max} = \frac{1}{8}Q_{max}gR + \frac{1}{8}\left(2MR^2 + \frac{3}{2}Q_{max}R^2 + 8J\right)\ddot{\theta}_s$	$\ddot{\theta}_s = \dfrac{T_s + \frac{1}{4}(\psi Q_{max} - Q)gR}{\frac{1}{8}(2M + \psi Q_{max} + Q)R^2 + J}$

In general terms, the roping arrangements more commonly employed are 1:1 and 2:1. Bigger factors are typically employed for elevators with rated loads greater than 4000 kg, where the shaft for the elevator is bigger.

Finally, it must be noted that a more accurate model approach would need to include losses in friction between the ropes and the sheave, which depend on the nature of the rope, the mass of the car, and the counterweight. Manufacturers have found that losses due to this friction phenomenon can considerably affect the dimensioning of the drive, being a significant part of the consumed power. However, this is beyond the scope of this chapter.

8.4 Electric drive

This section presents the electric drive of an elevator. First of all, a general view of the control system of an elevator is described, providing the electric drive configuration of modern and common elevators. Then a view of the control strategy, the electric machine, and the power electronic converter supplying the machine is provided, together with the electric brake. Finally, a common elevator dimensioning is detailed, showing details of the functioning of the elevator under different operating conditions.

8.4.1 General scheme and general specifications

8.4.1.1 Introduction

Fig. 8-30 shows a simplified general schema of an elevator with a 1:1 roping arrangement. Only the most important elements have been included, showing the schematic communication or interaction between them. It is possible to distinguish the following elements:

- The elevation system that is composed of the car, ropes, and counterweight.
- The elevation system is governed by the electric drive, which is composed of the machine, the converter, and its control. Often, in elevator contexts as commented before, when we refer to the machine we are considering the set: electric motor, drive sheave, brake, encoder, and, in the case of a geared machine, the gearbox and the flywheel as well.
- The maneuver decides the speed reference for the electric drive, relating to processing the floor calls received and the position of the car. The maneuver is the elevator's central element, normally implemented by software, which controls the car's movement and answers passengers' calls from each floor and the car.

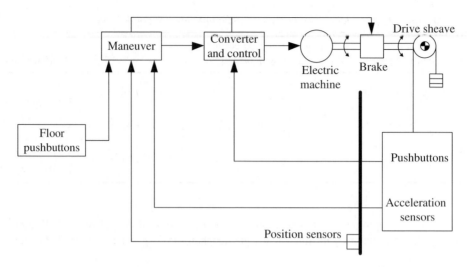

Fig. 8-30   *General communication schema of the different parts of an elevator*

8.4.1.2   *General specifications*

The electric drive of the elevator must guarantee some minimum requirements that can be briefly summarized as:

- **Starts up and stops:** The elevator can make a huge number of journeys throughout a single day (even up to 240 start ups in one hour). This particular demand affects the thermal behavior of the machine and the converter, so it is important to take it into account at the dimensioning and design stages of the drive.
- **Accuracy at the stop:** When the car reaches the requested floor, an accuracy that typically can go from 1 to 2 mm is required at the stop.
- **Electric brake:** As is looked at in more specific detail below, the electric drive of the elevator must operate at the four quadrants of torque and speed. Therefore, it must be prepared to operate as a motor or as a generator. For this purpose, different power converter configurations allowing this requirement are described below.
- **Brake:** Once the car reaches the requested floor controlled by the electric drive, an additional brake guarantees the blocking state of the car. More details about this brake are provided in Section 8.4.5.
- **Ride comfort:** The movement of the car must be smooth, with acceleration and jerk values limited. The jerk value, which is defined as the time derivative of the acceleration, is particularly important because it is closely related to the comfort concept. However, the comfort concept is subjective to each individual. Each person can perceive it in a different way. The comfort concept in an elevator is complex, and includes characteristics such as waiting time, stopping accuracy, noises inside and outside the car and in the building, acceleration, jerk, vibrations, and so on. It is not simple to obtain an estimation of comfort, but often it is given using numerical measurements of these mentioned characteristics. It is important to highlight that there are two significant sources of vibrations in an elevator that must be specially taken into account:
  - Vibrations caused by the electric machine: The rotational speed of the machine can cause vibrations of between 5 and 150Hz that can be transmitted to the car itself or even to the building. To mitigate this effect, action such as obtaining a good coupling and balance between the mechanical axis of the different rotating elements and obtaining a proper mechanical isolation between the machine itself and the floor where it is located should be taken.

○ Vibrations caused by the movement of the chassis: The chassis of the car is centered on some guides and at the same time the car is suspended with some dampers inside the chassis. While the car is moving, all the irregularities of the guides produce vibrations that are transmitted to the car. A good design of the dampers will mitigate this effect, which typically can cause vibrations of between 1 and 10Hz.

It has been found that different ranges of frequencies in noise and vibrations have different effects on humans. For instance, frequencies of below 0.5Hz produce dizziness, frequencies of around 5Hz can disturb human actions such as writing, frequencies of between 10 and 20Hz disturb the speaking action, and so on. Owing to these reasons, there are several standards associated with an elevator's operating quality that try to unify definitions, tests and measurements to ensure ride comfort, but these are beyond the scope of this chapter.

8.4.1.3   Trajectory generation

The elevator is speed controlled by the electric drive. The speed reference must be created by the trajectory generator, taking into account that the jerk (a derivative of the acceleration) must be also kept within certain required values, ensuring certain comfort requirements. An illustrative example of linear speed, acceleration, and jerk trajectories of an elevator, generated by the trajectory generator, are shown in Fig. 8-31. Typically we can find five different zones in the trajectory of an elevator:

- **Acceleration:** Once the brake is deactivated, the car accelerates from zero speed up to the traveling speed, which can be typically 1m/sec. Depending on how the speed reference is generated during this acceleration (ramp or other trajectories and time reaching the final value), the acceleration and jerk magnitudes take different values.
- **Journey:** During the journey, or ride, the car is moving at a rated constant speed. This stage is normally the longest of the trajectory.

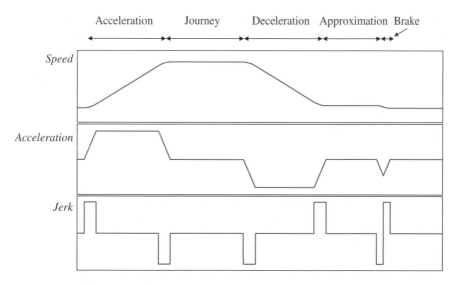

Fig. 8-31   *Linear speed, acceleration, and jerk trajectories of an elevator*

- **Deceleration:** The car's speed reduces from the journey speed up to the floor approximation speed, which can be typically between 50 and 80mm/sec.
- **Approximation:** The car approaches the required floor. Since there is not a position control, this approach is made at a reduced speed. This is also useful to guarantee the comfort of the journey.
- **Break:** Final deceleration before the car is stopped by the brake.

It will be noticed that the trajectory generator, only by creating the speed reference with certain speed values and variations under certain time ranges, indirectly imposes the acceleration and jerk as desired. Therefore, when creating the speed reference at the design stage, it is necessary to consider the rest of the relevant magnitudes that must also be controlled.

On the other hand, this example has shown linear velocities, accelerations, and jerks of the car. Note that the machine is creating this movement, but in rotational magnitudes. To convert from linear to rotational magnitudes, the diameter of the drive sheave (and the gearbox if there is one) must be considered.

Providing more details of the trajectory, Fig. 8-32 shows the accelerating process of the elevator. In order to achieve a comfortable ride, the constant jerk is applied typically to reach the maximum constant acceleration. During this time, the acceleration is a ramp and the speed is a parabolic. The time needed to reach the constant acceleration at constant jerk is:

$$t_j = \frac{a_{max}}{jerk} \tag{8.26}$$

During this time at constant jerk, the speed increase is given by:

$$\Delta v_j = \frac{1}{2} jerk \cdot t_j^2 \tag{8.27}$$

The speed increase at constant acceleration Δv_a is given by the rest of the total desired speed increase by two acceleration times at constant jerk, Δv_j:

$$\Delta v_a = \Delta v_t - \Delta v_j \cdot 2 \tag{8.28}$$

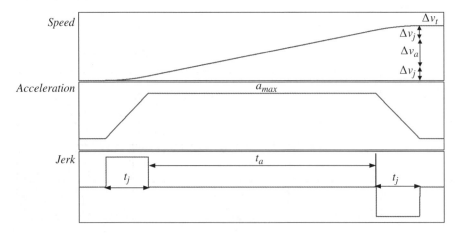

Fig. 8-32   *Zoom of the trajectories of an elevator showing the acceleration*

Consequently, the time accelerating at constant accelerations is:

$$t_a = \frac{\Delta v_a}{a_{max}} \tag{8.29}$$

This results in a total acceleration time:

$$t_{total} = t_a + t_j \cdot 2 \tag{8.30}$$

Therefore, the speed reference that will be then generated for the drive will be the composition of linear ramps with parabolic connections. Providing ramp and constant accelerations, suitable for being created by the electric drive of the elevator and meeting the comfort requirements for passengers.

By means of a positioning system which is often based on magnets, the maneuver detects the approach to the floor and imposes the approximation speed. In order to guarantee accuracy to within a few millimeters, during the installation of the elevator, the distances of acceleration and deceleration are adjusted as well. In addition, often the maneuver incorporates a position control in case a deviation from the floor of more than approximately 3mm is detected.

8.4.1.4 *Classic electric drive configuration*

There are several drive configurations for elevators available on the market, but this section describes a typical one. Fig. 8-33 shows an illustrative example driving an elevator with a 1:1 roping arrangement, but it can be extended to any other roping arrangement. It is possible to distinguish the power electronic converter supplying the AC machine. The typical converter configuration is composed of a uni-directional three-phase rectifier that ensures a constant DC bus voltage and allows the flow of energy from the grid to the machine. An insulated-gate bipolar transistor (IGBT), two-level, three-phase voltage source converter directly supplies the stator of the machine. A crowbar element allows energy to be dissipated from the machine when it operates as a generator. The sufficient stability of the DC bus voltage under steady-state, or even transient, operating conditions is proportionated by an appropriate amount of DC bus capacitors. The typical machine topologies are squirrel cage induction machines and permanent magnet synchronous machines. Depending on the machine design, it may be necessary to use a gearbox between the machine and the drive sheave which adapt the operating speed of the machine and the elevator.

The control strategy associated with this converter and machine topology nowadays typically ensures an accurate and high-dynamic-performance closed-loop speed control. For this purpose, a torque control based on a vector control technique with its associated modulation is typically employed, as described in Chapters 2 and 3. The speed reference is created by the trajectory generation block, ensuring the acceleration and jerk requirements. The drive is ultimately commanded by the maneuver, which decides the target floor based on the calls it receives from the different floors and from inside the car.

On the other hand, the machine of the drive works at four quadrants of torque versus speed operation. Depending on whether the elevator is going up or down and depending also on the load of the car, the drive can work at different torque-speed operation. Thus, for instance for an elevator with a 1:1 roping arrangement, in previous sections it has been seen that the torque can be described by the following model equation (8.10):

$$T_s = (Q - \psi Q_{max})gR + (2MR^2 + QR^2 + Q_{max}\psi R^2 + J)\ddot{\theta}_s \tag{8.31}$$

At steady state when the elevator operates at constant speed, the acceleration is zero and therefore the torque applied to the sheave by the electric machine is only the term: $T_s = (Q - \psi Q_{max})gR$, which means that

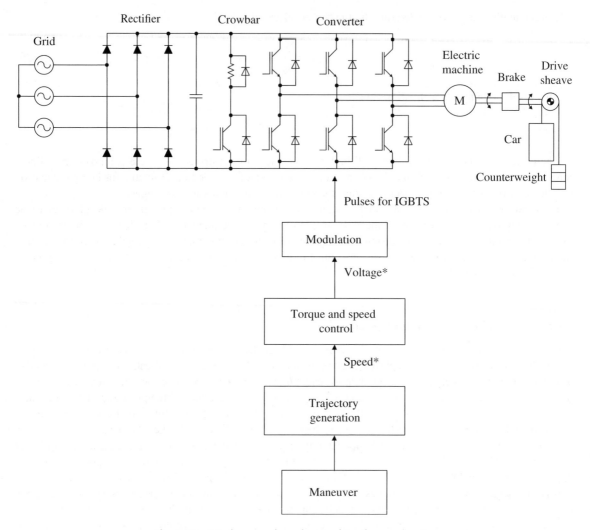

Fig. 8-33   *Modern gearless electric drive for an elevator*

depending on the load of the car, the torque can be positive ($Q > \psi Q_{max}$) or negative ($Q < \psi Q_{max}$), as illustrated in Fig. 8-34, no matter whether the speed is positive or negative.

This fact, excluding the acceleration and deceleration regimes, shows that the drive works in the four quadrants as illustrated in Fig. 8-35. Therefore, the drive must be prepared to dissipate the energy of the brake operation in quadrants 2 and 4. For that purpose, as mentioned before, a crowbar circuit can be included at the DC bus.

8.4.2   Control strategy

This section gives details of the control strategy employed at the elevators. More specifically, the electric machine, by means of the power electronic converter, controls the elevator according to comfort and safety recommendations and meets the grid code requirements.

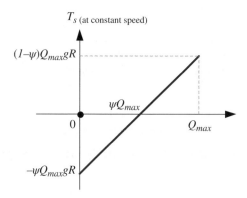

Fig. 8-34 *Torque provided at the sheave at constant speed, depending on the load of the car, Q, with a 1:1 roping arrangement*

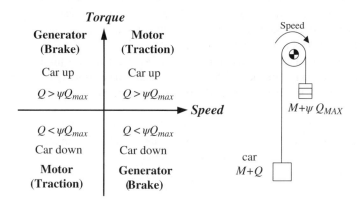

Fig. 8-35 *Four quadrants of operation of the drive at constant speed, for an elevator with a 1:1 roping arrangement*

Hence, as is shown in Fig. 8-33 and now again illustrated in a slightly different way in Fig. 8-36, the control strategy of the elevator receives the speed reference from the trajectory generation. The speed regulator, in general, can be a simple PI controller which generates the torque reference. As described in [3], for instance, it is possible to increase the performance of the speed regulator by including load estimators or observers, which compensate the load torque seen by the electric machine. For simplicity's sake, in Fig. 8-36, this sophisticated load compensation method based on estimations is not considered.

After that, once the final torque reference has been created, the torque control generates the required voltage reference for the modulator and this generates the conduction orders of the IGBTs of the converter.

The torque control, in general, is vector controlled, as seen in Chapter 2 and 3, depending on the electric machine being used: induction machine or PMSM. Thus, if an induction machine is being used, the vector control block diagram employed would be equivalent to the one studied in Chapter 2. Conversely, if a synchronous machine is used, the vector control strategy employed for this machine is described in Chapter 3. Note that in any of the cases it has been considered a sensorless control, since in both control of machines a position sensor is typically incorporated. In elevators, depending on the load of the car, when the brake

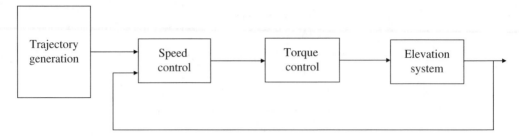

Fig. 8-36   *Schematic block diagram representation showing the main blocks of the control strategy employed at the elevator*

is deactivated, the drive can be required to start up going up or down, at motor or generator operation, from zero speed with variable torque load (can be any value from minus rated to rated). These conditions (always start up at zero speed and with variable torque that can be very high) are very severe and challenging for a proper sensorless control strategy, which, as seen in Chapter 2 and 3, the performance of the drive can be deteriorated when using sensorless control at near zero speed. For to this reason, amongst others, manufacturers generally still include the position sensors at the control stage.

On the other hand, the typical sample period in which the vector control is implemented is coordinated and synchronized with the switching frequency of the converter and can be around 8kHz. This is typically selected in an attempt to find a balance between reduction of power losses of converter and machine, small current and torque ripples, absence of noise for the user, etc.

Finally, it can be said that using the torque vector control at the elevator allows a fast torque control dynamic to be obtained that ranges in the order milliseconds, thus achieving accuracy in the elevator motion control and fulfilling the comfort and safety requirements of this product.

8.4.3   Electric machine

This section provides a general view of the machines that can be employed at the electric drive of the elevator, when this is vector controlled. There are several alternatives which do not use a torque control for the electric machine, but these are not studied in this section. Later sections provide a broad view of different and more primitive controls of the elevator.

Hence, closed-loop torque-controlled elevators, as has been already mentioned, normally use induction machines and PMSMs, which are typically brushless AC machines. In general, it is desirable to design the machine to perform the required operation demanded by the specific elevator where it is going to be used. The power for elevators used in residential buildings (transportation of up to 13 passengers) is around 3kW for six passengers and 7kW for 13 passengers. In a similar way, for public segment transport, for instance hospital or office elevators (transportation of up to 21 passengers), the power of the machines can reach 20kW.

The rated stator voltage of the machine is normally equal to or slightly lower than the grid voltage, which is typically 400V. This is because, with a three-phase diode rectifier and two-level converter with a space vector modulation or sinusoidal pulse-width modulation (PWM) with third-harmonic injection, assuming no losses and ideal behavior, the maximum AC voltage achievable on the machine side is equal to the AC voltage at the grid side. This fact is graphically represented in Fig. 8-37.

With regard to asynchronous machines, in some countries they are still the required machine technology. Not, for instance, in Europe, where they are not able to fulfil the safety norms, because with this

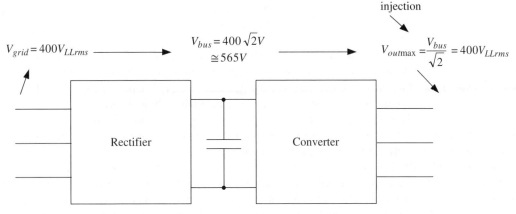

With space vector modulation or
sinusoidal PWM with third harmonic
injection

$V_{grid} = 400V_{LLrms}$

$V_{bus} = 400\sqrt{2}V$
$\cong 565V$

$V_{outmax} = \dfrac{V_{bus}}{\sqrt{2}} = 400V_{LLrms}$

Rectifier

Converter

Fig. 8-37   *Achievable maximum voltage on the machine side*

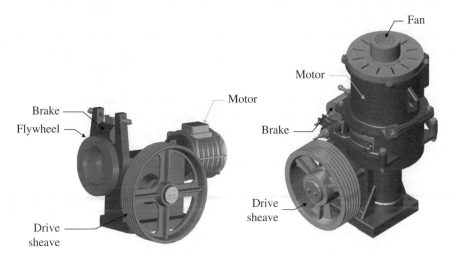

Fan

Motor

Brake

Flywheel

Motor

Brake

Drive
sheave

Drive
sheave

Fig. 8-38   *Geared induction machine examples for elevators. Gearboxes are integrated in the armor and therefore are not visible. One of their axles corresponds to the sheave axle (Source: ORONA. Reproduced with permission of ORONA)*

technology that needs the gearbox the brake is not applied at the drive sheave axis. Thus, with these induction machines, the standard frequencies typically are 50 and 60Hz, with pole pairs of 1, 2, or 3. Bigger pole pair numbers result in a machine that is big and not as efficient. These configurations produce rated machine speed rotations too high for the drive sheave of an elevator:

$$n_{synchronism} = \frac{60 \cdot frequency}{pole\,pairs} \tag{8.32}$$

Therefore, often a gearbox is used, in order to balance the speed of the rotation of the machine and sheave, as illustrated in Fig. 8-38.

On the other hand, the required gearbox ration normally is provided by:

$$r_{gearbox} = \frac{machine\_speed}{sheave\_speed}$$

(8.33)

The rotation speeds of induction machines with gearboxes are in the range of 25–60 rpm, depending on the roping arrangement and the gearbox ratio employed. In general, it must be highlighted that in most advanced countries elevators based on induction machines are disappearing, mainly because of their low efficiency, the presence of the gearbox, the required maintenance necessity, and their inability to meet safety norms. It can be affirmed that in advanced countries this machine concept is competitive in terms of cost. In fact, there is a clear tendency in public bodies to provide funding support for modernizing existing elevators based on induction machines using the more efficient PMSM-based elevators.

On the other hand, if PMSMs are used, it is possible to use efficient machine designs, with a high number of pole pairs being possible to adapt the speed of the machine and the drive sheave of the elevator. Thanks to this, the gearbox can be eliminated, obtaining a significant space reduction for the installation. Of course, this reduction is accompanied by an increased complexity and cost, but from a drive perspective no complexity is added. Thus, the versatility that can be achieved with a PMSM design in terms of electric and mechanic characteristics means it is possible to obtain very efficient machines that are more than sufficient to meet the needs of specific elevators.

Thus, for instance, Fig. 8-39 illustrates a PMSM-based direct drive of an elevator. In this case, the designs can present pole pair numbers that can range from 1 to 12. Nowadays, 10-pole pair number is a very common value; however, special attention must be paid to this parameter since with an attempt to choose a very high pole pair number in general the electric frequency also must be high and therefore the magnetic power losses are increased. Connected to this, the rated speeds that can be reached range from approximately 100 to 400rpm. The most important characteristics of one example of PMSM oriented to an elevator application are provided in Table 8-3. Its efficiency rounds 90% and obtains an optimum use of magnetic materials. It is a machine for a residential elevator, for more than 20 passengers, and a rise of up to 130m.

Therefore, in summary, typically PMSMs are designed in ranges of power of between 3 and 20kW. For instance, for an elevator of eight passengers and 1m/sec, the required machine power is 4.7kW and the approximated volume of the machine is $125 \times 300 \times 300$mm^3. For that purpose, radial PMSMs are probably the most employed ones, but axial flux PMSMs can be also used in some cases. The magnets are often set in the rotor

Fig. 8-39 *Example of gearless PMSM evolution for an elevator (Source: ORONA. Reproduced with permission of ORONA)*

Table 8-3 *PMSM parameters oriented to elevator application*

CE Orona		Made in EU Lancor	
COD. 8040233/01		G-02 Competitive LP100	
Art.-Nr.: 433208300155	Ser. No.: 21419382		
Motortype: MSIP-300.20-20		Connections: 人	
Pn	3.4 Kw	Duty-C/h	S5 40%-180
U	340 V	Nm	166 Rpm
Fn	27.7 Hz	Mn	195 Nm
In	7.1 A	Ma	341 Nm
Ia	12.3 A	Encl.	IP-21
Iso-class	F	Weight	157 Kg
Cooling	IC 00	Produced	11-2014

(Source: ORONA. Reproduced with permission of ORONA)

superficially but also inset magnets are commonly employed, in order to allow an easier construction of the machine, as well as an extension of the life of the machine.

With regards to the magnets, it is possible to employ: alnico, ceramics, rare-earths, samarium cobalt, and ferrites, which can provide different characteristics such as:

- different magnetizing levels, typically between 0.6T and 1.2T;
- optimum thermal behavior no higher than temperatures of 120 °C. This must be guaranteed in all the cycles of operation;
- dangerous zones of demagnetization (owing to temperature or effect of stator flux), designed to be located far from the normal operating regions of the elevator;
- cost and availability dependent on market.

With regards to the magnetic sheets, typically low losses sheets are employed with thicknesses of between 0.5 and 0.65mm. The coils are commonly constructed of copper, with thicknesses of around 1–2mm.

Then, with these materials the machines are designed, varying constructive parameters or degrees of freedom such as:

- dimensions, forms and number of teeth and slots;
- type of coils such as distributed or concentrated;
- number of turns;
- number of pole pairs;
- skew.

Thus, once the machine is designed the laminations stack is determined for construction of the magnetic sheets. In this way, the teeth and number of poles are fixed, together with their forms and inner and outer diameters. In order to improve passenger comfort in the car of the elevator, the torque ripple is minimized to levels lower than 5% of the rated torque, by applying skew or turning the magnets in respect of the teeth and therefore minimizing the magnet's reluctance.

Consequently, the resulting electromechanical characteristics of the designed machines are mainly:

- rated voltages in range of 100–400V, depending on the electric requirements, even lower voltages are also possible in cases where in the supply of energy batteries are involved;
- power between 3 and 20kW as commented before;
- range of torques between 100 and 1000Nm, depending on the roping arrangement and diameter of the sheave;
- rated frequencies between 10 and 100Hz;
- rotational speeds between 50 and 400rpm;
- efficiencies higher than 85%.

And these characteristics are used to achieve a well-balanced machine that fulfills the electromechanical demands of the technical requirements of a specific elevator, along with an efficiently designed elevator that is reliable, robust, and financially viable.

8.4.4   Power electronic converter

8.4.4.1   Standard configuration

One detailed example of a converter's configuration of a closed-loop controlled converter is illustrated in Fig. 8-40. It is a non-regenerative to grid system, since a three-phase diode rectifier is used at grid side. Owing to this, when the elevator is braking and the machine works as a generator, the crowbar circuit must be activated dissipating the generated energy at the crowbar resistance.

The pre-charge circuit of the DC bus capacitors is not shown in Fig. 8-40, but it is typically located on the DC side, rather than on AC side. In addition, one illustrative example of the input current and voltage to the diode rectifier is depicted in Fig. 8-41. It will be seen that the current waveform quality and power factor

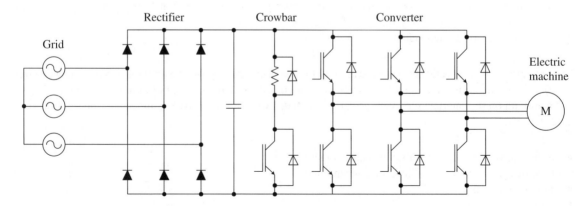

Fig. 8-40   *Typical power converter configuration of an elevator*

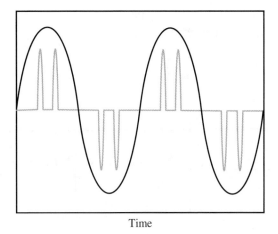

Time

Fig. 8-41   *Grid side voltage and current waveforms*

are very poor. In order to improve this quality it is possible to include inductances in the DC bus or AC input of the rectifier.

As mentioned before, assuming an ideal functioning of the rectifier, it imposes a low ripple DC bus voltage thanks to the DC bus capacitance. The voltage is given by:

$$V_{bus} = \sqrt{2}V_{grid\ LL\ RMS} \tag{8.34}$$

Thus, for a typical grid voltage of 400V, the mean DC bus voltage is around 565V. The ripple of the DC bus voltage depends on how the machine is supplied depending on the load trajectory conditions. A simple approximation for dimensioning the DC bus capacitance is to use 25–50uF, per $1 A_{RMS}$ of current at the machine side. However, since this value is important in terms of volume and the cost of the converter, a computer-based simulation is often used to choose a not too oversized value of capacitance.

The crowbar device, often also called a DC chopper in this context, is composed of a controlled switch, two diodes, and a resistance. The crowbar is activated automatically, when it detects a deviation of the DC bus voltage from the target value. That is when the machine actuates as a generator. Two automatic control philosophies are shown in Fig. 8-42 for the crowbar. In both the idea is to keep the DC bus voltage controlled near to the target autonomously, so as to not need to know when the elevator is regenerating and therefore avoiding communication or complex coordination between different parts of the converter. Thus, when using an ON/OFF-based control, the switching frequency of the crowbar IGBT is not controlled. On the contrary, if a PI-controller-based philosophy is adopted, it is possible to impose a constant switching cycle to the IGBT. In both cases, a specific dimensioning of the crowbar resistance is required that takes into account different regeneration powers owing to different loads at the elevator and based on how often the crowbar must work. For this dimensioning, a computer-based analysis is often carried out, finding and simulating the most unfavorable cases and obtaining the appropriate crowbar resistance characteristics.

8.4.4.2   *Innovative configurations*

This converter topology—a diode rectifier together with a two-level converter and a crowbar for regeneration—is the most extensively employed for modern elevator designs by manufacturers all around

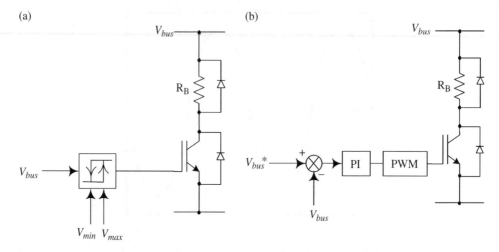

Fig. 8-42   *Crowbar device: (a) ON/OFF control and (b) linear control*

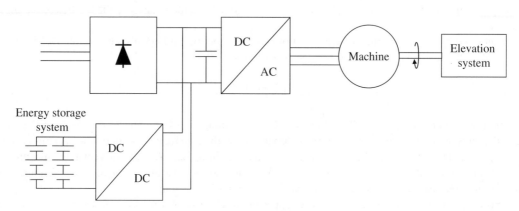

Fig. 8-43   *Elevator with electric drive that includes energy storage (input rectifier can be three phase or single phase, depending on how the entire system is dimensioned)*

the world. However, there is a tendency also to improve several aspects of the electric drive, including also innovation at the power electronic based conversion stage. Depending on the specific characteristics of the elevator, there are some other converter configurations that could be more attractive than the previously presented standard one. In the following, some few interesting innovative electric drive configurations are briefly described.

Thus, in an attempt to reduce the total energy consumption from the grid of the elevator, a first obvious solution could be based on the inclusion of a battery-based energy storage system in the drive, as depicted in Fig. 8-43. Thanks to the penetration of energy storage in many applications, as is discussed in previous chapters, battery performances and their characteristics have been improved significantly. In this possible solution the batteries are located on the DC side, through a DC/DC converter. The main idea is based on the fact that when the elevator is working in generator mode the regenerated power is stored at the battery instead of being dissipated in the resistance of the crowbar, as in a classic converter configuration. Therefore, if the elevator is required to work in motor mode the necessary energy can be taken from the stored energy at the batteries, thus reducing the energy consumption from the grid.

In principle, with this configuration the power contract from the grid operator can be reduced, since the energy storage permits to reduce the demands of power from the grid. This reduces the cost advantageously in many countries, since the electricity bill to the grid operator is based on the energy consumption and the power contract.

The necessity of the DC/DC converter is mainly due to the fact that, in general, the voltages of standard machines and battery-based storages do not match. Often, at least nowadays, an optimized and efficient battery-based energy storage system is less than 100V in DC, while the standard machines traditionally designed by the manufacturers are more likely to require much more than 100V of DC bus to operate. In addition, avoiding a costly and bulky transformer at the input, the AC grid connection also is always greater than 110V in AC, being also greater again than the efficient voltage of the storage system. Therefore, the DC/DC converter, among other functions, mainly matches the ideal DC voltage of the storage system to the grid and machine typical voltages; note also that the actual voltage of the batteries depends on the state of charge (SOC).

On the other hand, in principle, the inclusion of the energy storage system should eliminate the necessity of the crowbar for this configuration. However, in a practical real case, at least minimum-resistance-based dissipation should be included at some stage of the drive (batteries or main DC bus), because, depending on how the dimensioning of the batteries is made, the batteries may not be able to absorb all the regenerative power and/or energy in all situations, because, for instance, the SOC level is already high.

In addition, note that the batteries' efficiency, depending on what technology has been used and how the dimensioning has been carried out, is not always very high. Nowadays, it is typical to use batteries of 50% efficiency, which means that to obtain an amount of energy from them we need to charge them with twice that energy. Thus, this reality, together with the fact that the DC/DC converter also increases the losses at the conversion stage, obliges us to very carefully design the conversion and storage system in order to actually and effectively reduce the energy consumption from the electric grid.

Added to this, it must be remarked that the control system of the entire conversion system is more complex than in classic and standard conversion systems, since the energy management and control itself of the storage system must be carried out. Consequently, although ideally or theoretically it seems that this conversion configuration would be advantageous, it must be carefully studied and designed for the specific elevator for which it is going to be utilized. Note also that the inclusion of the batteries can require a considerable increase of space and modification of the mechanical structure to support their weight.

In a similar way, an alternative converter can be configured as depicted in Fig. 8-44. Note that, although this concept can be understood in many different manners, it can be said that it is similar to the electric drive of an electric vehicle seen in the previous chapter. Here, the idea is to reduce the maximum amount of contracted power from the grid operator, by connecting the batteries directly to the DC bus of the converter supplying the machine. As done in electric vehicles, it is also possible to connect a DC/DC converter between the batteries and the inverter, in order to adapt the DC voltage levels of the battery system and main DC bus of the inverter. In this way, the majority of the power and energy consumption of the elevator will be provided by the batteries. For that purpose, as noted for the previous configuration, the machine should be specially designed to match its rated voltage, to the low DC voltage of the efficient battery-based storage system. Nowadays, this depends on the battery technology employed, but the DC bus voltage should be probably less than 100V, resulting in a quite low-rated voltage machine and, therefore, the rated power of the machine also should not be very high, in order to avoid the machine working with very high and inefficient currents. Consequently, in principle, this configuration is attractive to low-powered elevators or those required to carry few passengers (i.e. in in residential buildings where the usage of the elevator is much lower than in hospitals or hotels) and, therefore, the energy consumption is very low as well. Note that in many residential houses the elevator is stopped most of the time of the day. On the other hand, the charger could be a simple and cheap diode-based rectifier and, in

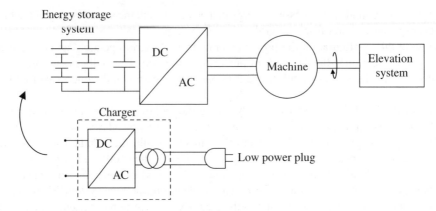

Fig. 8-44   *Elevator with electric drive that minimizes the input power contract*

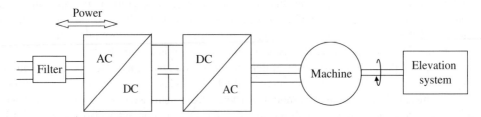

Fig. 8-45   *Elevator with regenerative electric drive to the grid*

general, since the DC bus voltage is relatively low, a transformer is necessary that steps down the AC grid voltage. As mentioned before, the charger is of very low power in order to reduce the grid power contract. For that purpose, again it is emphasized that the battery storage system must be accordingly and effectively designed. As in previous configurations, depending on how the battery system is dimensioned and the technology of the batteries is employed, it may be necessary to add a crowbar which enables the regenerative energy that the batteries are not able to absorb to dissipate. Note that batteries in general are not able to charge as quickly as they can discharge, (i.e. the discharge power is greater than the charge power). Finally, also, note that the variation of the voltages of the battery pack due to the SOC is directly seen by the DC bus and therefore by the machine.

Finally, the configuration depicted in Fig. 8-45 avoids the use of an energy storage system by using a regenerative active front end at the input. Thus, the regenerative energy created by the elevator is delivered to the grid. In this case, the complexity of the input converter is significantly increased, because it must be controlled by using the corresponding two-level converter, and an additional harmonic input filter must be employed, as described in Chapter 4. In addition, the grid operator must allow for regenerating energy back to the grid, which is not an obvious reality in most countries, at least nowadays. Moreover, nor is it a reality in most countries that the regenerative energy delivered to the grid will be paid for by the grid operator. Nevertheless, this regenerative energy could be employed by the installer of the elevator to supply auxiliary elements of the elevator, such as lighting, and therefore be able to reduce in some way the overall consumed energy of the elevator. In principle, note that this configuration does not allow reducing the power contract with the grid operator.

To conclude, it can be said that these innovative electric drive configurations are some promising options that could be suitable at least for some elevators with specific characteristics in the future. The reader can imagine alternative solutions by combinations of or small alterations to the three presented here. The appropriateness of each electric drive configuration at one specific moment depends on factors such as: the elevator's specifications, availability of technology that fits with those specifications, standards and norms, inertia or limitations of adaptation to new designs of the manufacturer.

8.4.5   Electric brake

The brakes of elevators are safety elements. The brake is an electromechanical system formed basically from one or more coils, which after being supplied overcome the effort produced by a spring. The schematic structure of the brake is graphically represented in Fig. 8-46. When the coil is not supplied, thrust springs press the armature disk against the brake disk. In this way, the brake disk is held between the friction pads. When the coil is supplied, owing to the magnetic force of the coil in the coil carrier, the armature disk is attracted against the spring force to the coil carrier. The brake is realized and the brake disk can rotate freely [4]. Therefore, when the brake is energized it does not brake, while if the brake is not energized, it brakes, stopping the elevator.

It will be noticed that the brake normally actuates on the drive sheave, therefore braking the whole elevation system. Hence, the brake normally is commanded by the maneuver. Before the elevator ride begins, the maneuver checks whether the brake is activated. If it is correct, it energizes the electric motor of the elevator and after that energizes the brakes, enabling the movement of the car. On the other hand, when the maneuver detects arrival at a floor, it activates zero speed reference to the drive of the elevator and immediately ceases supplying the brake, stopping therefore safely the elevator at the corresponding floor. Conversely, when any kind of failure is detected at the elevator (over-speed, over-acceleration, uncontrolled movements, etc.), the maneuver immediately de-energizes the brake, stopping the drive sheave and therefore the elevator itself.

In addition, it must be highlighted that the norm requires that the brake be at least divided into two separate parts in such a way that both brakes can be activated independently and each one is dimensioned to be able to stop the car independently if the other fails.

The elevator norms, in general, dictate that the brake must be dimensioned in such a way that it is able to stop the car with its rated load. In this way, in elevators the range of required braking torques goes from 100 to 600Nm approximately, depending on the individual elevator's characteristics.

Nowadays, the most commonly developed brakes are disk brakes, which can be square or circular. In the past, some other brakes, such as drum brakes, were also common.

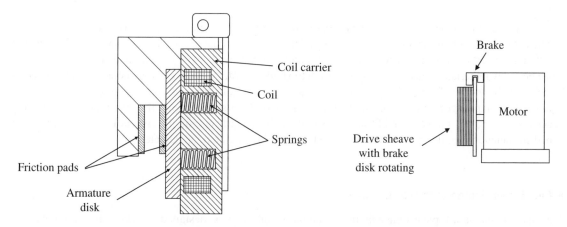

Fig. 8-46   *Schematic structure of the electric brake and its location*

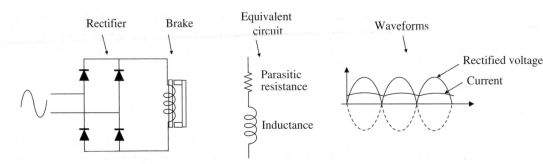

Fig. 8-47   *Supplying module of the brake, equivalent electric circuit, and voltage and current waveforms through the coil*

On the other hand, the coil of the brake must be DC supplied. Thus, the brake normally incorporates an external module, which consists of a simple diode rectifier, enabling supplying the coil in DC from the AC grid supply. The supplying module, the equivalent electric circuit, and voltage and current waveforms through the coil are represented in Fig. 8-47. The equivalent electric circuit mainly is an inductance accompanied with a small parasitic resistance. The elevated inductance of the coil, which rounds values of 50H, enables an almost constant current through the brake, when typically supplied from grids of 230V AC. Note that it is not necessary to add capacitance to the electric circuit, in order to obtain almost constant DC current, and therefore almost constant braking force. In general terms, the power needed to energize the brake when the elevator is moving, which is indicatively around 50 to 100W, for brakes of around 400 Nm. However, these values depend on the brake design.

With regards to the electromagnetic design of the brake, briefly, it is possible to say that unlike electric machines, the magnetic core is not laminated but is a solid magnetic core. This is mainly due to the fact that a constant magnetic field is required. In addition, a solid magnetic core against a laminated one simplifies the manufacturing process, allowing good constructive tolerances to be obtained. Thus, the force generated by the brake can be represented as:

$$F = \frac{N^2 \cdot i}{d^2} \tag{8.35}$$

where F is the force in newtons, N is the number of turns, i is the current through the coil, and d is the airgap distance. Often, the airgap is around 0.3–0.4 mm and the number of copper turns between 500 and 700. In order to obtain large forces, it is necessary to achieve an intelligent balance between an airgap that is not too small so it can be easily materialized and a current that is not too high, in order to reduce energy consumption. This must be achieved by means of cheap designs, which are also optimized in terms of volume, weight, and reliability. Another important aspect that affects comfort is the noise of the brakes. Finally, the time typically required to deactivate the brakes is lower than 300msec. As in the coil's electric circuit, the current DC level is not established instantaneously; the brake force also needs time to be liberated, as depicted in Fig. 8-48.

8.4.6   Dimensioning of the electric drive

This section shows a typical dimensioning example of an elevator designed for residential buildings [5]. Although the dimensioning of the elevator and the drive itself can be according to different criteria, the

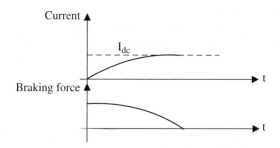

Fig. 8-48 *Current and braking forces when the brake is energized*

illustrative procedure discussed in this section leads to a deeper understanding of the concepts studied up to now in this chapter. It is an exercise of design, in which each time it is necessary to make a decision that decision is taken according to a practical point of view. It can happen that in the context of a real company which is designing an elevator, the decisions taken could go in a different direction, influenced by various tendencies restrictions of historical dependencies on some of the elements or design criteria of the elevator.

Hence, the first design specifications are:

- Maximum load: $Q_{max} = 800$kg (10 passengers)
- Speed of the car: $v = 1.6$m/sec
- Mass of the car: $M = 1650$kg

These are typical values of commercial elevators. In order to do a gearless design, often the roping arrangement is chosen as 2:1, which helps to include a relation of $s = 2$ between the linear speed of the car and the rotational speed of the electric machine. In addition, it also allows reducing the load torque seen by the electric drive (see Table 8-2). However, the mechanical complexity of the installation is increased with higher roping arrangements, as mentioned in previous sections.

On the other hand, the radius of the drive sheave is assumed to be:

$$R = 0.1 \text{m}.$$

As described in Section 8.2.5, the diameter of the sheave is determined by the technology of rope employed. Typically, for a conventional traction rope, the ratio between the rope and the sheave diameters should be around 40:1. With other types of rope technology, this ratio can be different, but always ensuring a correct adherence between the rope-sheave along the estimated lifetime of the elevator. With this philosophy, elevator manufacturers have put a great deal of effort into reducing the diameters of sheaves. Achieving this will reduce the traction torque produced by the motor at the sheave, at a given elevator's load. Nowadays, as a matter of example, the manufacturer ORONA has developed sheave diameters of around 115–130mm, for a reduced size of ropes with a plastic covering. For metallic ropes with a diameter of between 8 and 10mm, it employs diameters of sheave in the range of 320–400mm.

Consequently, with this information, the rated rotational speed of the machine is obtained:

$$\Omega_m = \frac{v \cdot s}{R} = \frac{1.6 \cdot 2}{0.1} = 32 rd/s \tag{8.36}$$

Another important characteristic that must be selected is the jerk and accelerating values of the elevator. As depicted in Fig. 8-31, in which the trajectories of speed, acceleration, and jerk are introduced, it can be seen that a good behavior from the comfort point of view is to operate at a constant jerk. Thus, the jerk and constant acceleration (maximum) values selected for a comfortable journey are:

- Jerk: $jerk = 2.5 \text{m/sec}^3$
- Acceleration: $a = 0.8 \text{m/sec}^2$

Thus the time needed to reach the constant acceleration at constant jerk is:

$$t_j = \frac{a}{jerk} = \frac{0.8}{2.5} = 0.32s \tag{8.37}$$

During this time at constant jerk, the speed increase is given by:

$$\Delta v_j = \frac{1}{2} jerk \cdot t_j^2 = \frac{1}{2} 2.5 \cdot 0.32^2 = 0.128s \tag{8.38}$$

The speed increase at constant accelerations is given by:

$$\Delta v_a = 1.6 - 0.128 \cdot 2 = 1.344 m/s \tag{8.39}$$

Consequently, the time accelerating at constant acceleration is:

$$t_a = \frac{\Delta v_a}{a} = \frac{1.344}{0.8} = 1.68s \tag{8.40}$$

This results in a total acceleration time:

$$t_{total} = 1.68 + 0.32 \cdot 2 = 2.32s \tag{8.41}$$

Once the speed and accelerations are fixed, it is possible to study the torques. First of all, the mass of the counterweight must be selected. By using (8.2) and choosing $\psi = 0.5$, we obtain:

$$m_2 = M + \psi \cdot Q_{max} = 1650 + 0.5 \cdot 800 = 2050 kg \tag{8.42}$$

Thus, by means of (8.21), the torque that the drive must apply at constant speed and at the most unfavorable case, i.e. at maximum load ($Q_{max} = 800$kg), is:

$$T_{em\ rated} = \frac{1}{2}(Q - \psi Q_{max})gR = \frac{1}{2}(800 - 0.5 \cdot 800) \cdot 9.8 \cdot 0.1 = 196Nm \tag{8.43}$$

This torque can be associated with the rated torque of the machine. However, it is important to note that the machine can be required to apply higher torques when accelerating. In the most unfavorable case of the maximum load and accelerating from zero speed, the required torque will be at its highest. This torque can be calculated by means of (8.23), assuming that the inertia of the sheave is $J = 0.4 \text{kg} \cdot \text{m}^2$ and the maximum acceleration is as chosen $a = 0.8 \text{m/sec}^2$:

$$T_{smax} = \frac{1}{4}Q_{max}gR + \frac{1}{4}\left(2MR^2 + \frac{3}{2}Q_{max}R^2 + 4J\right)\frac{a}{R}s$$

$$= 196 + \frac{1}{4}\left(2 \cdot 1650 \cdot 0.1^2 + \frac{3}{2}800 \cdot 0.1^2 + 4 \cdot 0.4\right)\frac{0.8}{0.1}2 \tag{8.44}$$

$$= 382.4 Nm$$

The resulting maximum torque which will be required in the most unfavorable case is almost two times greater than the rated torque, demanded also at maximum load. The difference is that the maximum torque will be required for a short time during one journey (2.32 sec), while the rated torque will be applied for much longer. Therefore, from the machine design perspective, an efficient design would be a machine of rated torque 196Nm, which is able to go punctually to 382.4Nm, which would produce a very competitive design. This means that the dimensions of the machine in terms of power could be:

$$P_{rated} = T_{em\,rated}\Omega_{m\,rated} = 196 \cdot 32 = 6.272\,kW \tag{8.45}$$

$$P_{max} = T_{em\,max}\Omega_{em\,rated} = 382.4 \cdot 32 = 12.24\,kW\,\text{(Punctually required)} \tag{8.46}$$

Thus, for that purpose the PMSM in Table 8-4 is selected. It fits reasonably well with the specifications of this elevator. Note that, thanks to the multi-pole design, it is possible to adapt the natural speed of rotation of the machine to the elevator requirement, obtaining a gearless installation.

It is supposed that the grid where the drive is going to be connected is of 400V; therefore the rated voltage of the machine has been chosen slightly lower, in order to be able to go to the maximum torques without the necessity of a transformer, which should always be avoided in order to reduce space, weight, losses, and cost. Thus, the rectifier and two-level converter arrangement, with a sinusoidal PWM, with third-harmonic injection, is able to generate approximately 400V at the machine terminals (see Fig. 8-37). Note that if the grid voltage is subjected to a certain variation (for instance ±10%) the machine may still have to have less rated voltage. For that purpose, the converter and rectifier also must be accordingly designed, choosing the semiconductors and elements which can work at 400V on the AC side and with 11.45A of rated current on the motor side.

Therefore, the maximum capability curves of this specific machine can be derived as studied in Chapter 3, resulting in the curves shown in Fig. 8-49.

Table 8-4 *Characteristics and parameters of the interior PMSM used for the elevator*

P_{rated}	6.675kW
$T_{em\,rated}$	197Nm
Ω_{rated}	32rd/sec
V_{rated} (LL-RMS)	345.4V
I_{rated} (RMS)	11.45A
p (pole pair)	10
f_{rated}	51Hz
R_s	1.015Ω
L_d	13.5mH
L_q	15mH
ψ_r (phase-peak)	0.8036Wb

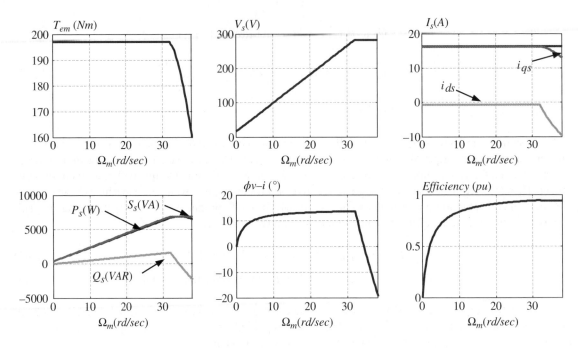

Fig. 8-49 *Maximum capability curves of the 6.6kW PMSM (only in motor mode operation)*

Therefore, the machine, depending on the load conditions at rated constant elevating speed at steady state, will work at different electric magnitude conditions, as illustrated in Fig. 8-50. Thus the torque and powers can be either positive or negative as well as the current module, depending on the loading conditions. However, since the speed is constant and rated, the required supplying voltage will always also be near the rated voltage. The phase shift between the stator voltage and currents will change as shown, while the efficiency will be improved by reducing the current.

In addition, it must be mentioned that often the so-called thermal torque or root-mean-squared torque is also computed and analyzed at the dimensioning stage. This torque is calculated by simply computing the root mean square of the torque during the elevator's operating cycles, also taking into account the time when the elevator is stopped and therefore there is no current through the motor. This computed torque provides a view of the average torque required for the motor, depending on the usage of a specific elevator.

With regards to the converter, the DC bus capacitance is dimensioned to 10mF, which gives a maximum DC bus voltage ripple of around 5V in the most unfavorable case, when the highest power is being transmitted. This fact is seen in the following subsection, in simulation-based validations. On the other hand, the crowbar resistance is set to 8Ω. This resistance must be appropriately designed, taking into account the DC bus capacitance employed at the converter and also the power and energy generated by the elevator at the regenerative operation.

Finally, is has to be mentioned that this machine and converter design has been specially oriented for this elevator. In reality, manufacturers tend to use the same converter and machine designs for elevators with similar characteristics, in order to avoiding having too many different components. This means they keep manufacturing costs down, even if it also means they install oversized electric drives in some elevators.

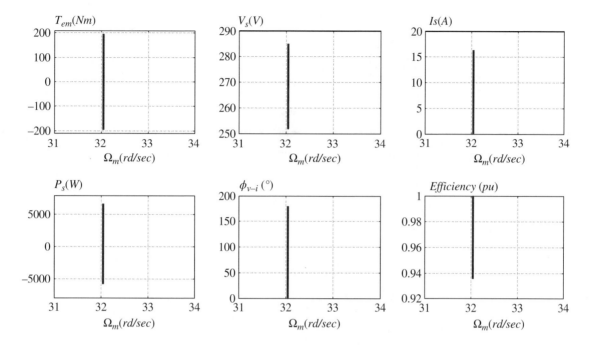

Fig. 8-50 *Real operation values of the 6.6kW PMSM when working at the elevator at constant speed with any of the possible loads (0–800kg)*

8.5 Computer-based simulation

This section illustrates how the model of the elevator seen in previous sections can be implemented in computer-based simulation software. The software used here is Matlab-Simulink. For simulation of the power electronic system (i.e. the electric drive), the library SimPower from Matlab is used. In this library, models of machines, converters, and so on are already implemented and well validated.

The simulation of the whole elevator system is useful and of interest for many purposes, for instance:

- validate developed control strategies;
- validate elevator system, machine and converter designs, in terms of appropriate dimensioning of rated load, speed, torques, voltages, switching frequencies, currents, established dynamics;
- validate fulfillment of safety and grid codes.

It must be mentioned that the developed model of the elevator system is just a first approach, which goes into a reasonable level of detail. For studies that require further detail, such as those that look at the thermal behavior of machines and converters, loss model, loss minimizations for increased efficiency, vibration problems, and so on, additional more sophisticated methods would be necessary, which is beyond the scope of this book.

Hence, by using the illustrative example of elevator dimensioned and described in the previous subsection, this section shows how it can be implemented in Matlab-Simulink, showing some validating representative results.

First of all, Fig. 8-51 shows the general Simulink simulation model of the gearless elevator with a 2:1 roping arrangement. All the model parts studied comprising the elevator are implemented: the electric drive,

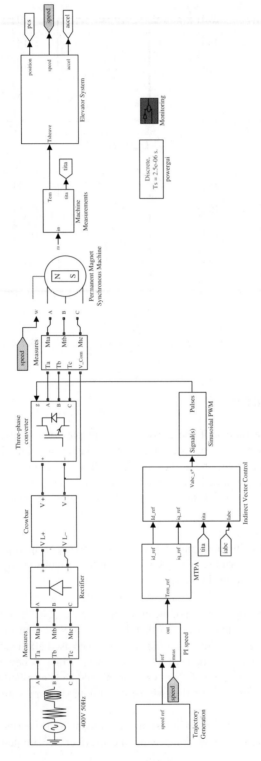

Fig. 8-51 General simulation model of the gearless elevator, with a 2:1 roping arrangement

composed by the multi-pole PMSM; the converter; the crowbar; the diode rectifier; and the grid model. On the other hand, it also implements the elevator system model and the control strategy of the elevator. The converter is switched to 2.5kHz switching frequency by a sinusoidal PWM with a third-harmonic injection. Note that often in reality a quite high switching frequency is employed in elevators in order to (among other objectives) reduce noise, as mentioned previously.

Most of the blocks corresponding to the drive are directly taken from models of the SimPower library already developed. The crowbar circuit implemented, which is controlled by a PI regulator switched at a constant switching frequency, is as described in Fig. 8-42(b).

The model of the elevator system is built by following the model equation (8.22) and (8.25), which is shown in Fig. 8-52. As it has been implemented, only rotatory movement magnitudes are provided. Note that the position, speed, and acceleration are provided by this elevator system model. The electromagnetic torque produced by the electric machine is directly transmitted to the drive sheave, controlling the movement of the elevator. Then the speed calculated for the elevator system model of Fig. 8-52 is transmitted as input to the electric machine model.

The control of the drive is carried out by an indirect vector control of PMSM, as studied in Chapter 3. Most of the blocks are described in that chapter. The speed reference trajectory is created simply using a look-up table where the jerk trajectory is generated, and then the acceleration and speed references are calculated by using integrators. The block diagram is shown in Fig. 8-53. Note that this trajectory generation can be implemented by different methods and this is only an illustrative example and not representative of something done in the real world.

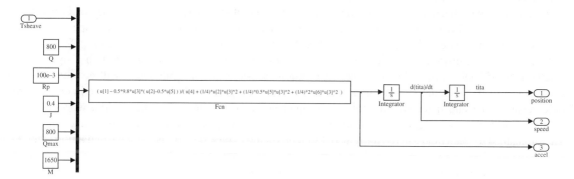

Fig. 8-52 *Elevator system model*

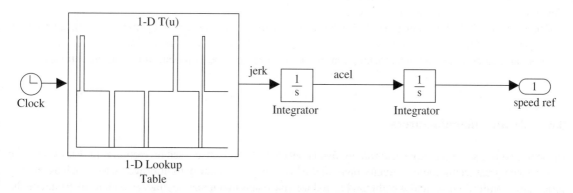

Fig. 8-53 *Trajectory generation for speed control of the elevator*

Thus, once the models are implemented in Simulink blocks and after tuning all the regulators, the elevator is driven in one floor elevation with a maximum load of 800kg. The most representative magnitudes comprising the elevator itself and the electric drive are shown in Fig. 8-54. The speed reference generation has been created according to the time, jerk, and accelerations specified in the previous subsection. The trajectory is composed of the commonly used acceleration, deceleration, and approximation. The zone where the elevator travels at constant speed has been shortened, for easier visualization of all the variables. It can be seen that the torque created by the drive and applied to the sheave is able to maintain the speed, acceleration, and jerk requirements imposed. For that purpose, the vector control dynamic capacity is very suitable. It will be seen that the torque remains positive all the time, which means that the drive is working as a motor also all the time. This fact is accordingly seen at the power P_m provided by the machine, which also remains positive throughout the simulation. As mentioned in the previous subsection, during the acceleration process (up to 2.32 seconds after the movement has been started), the torque and consequently, power, stator current, and stator required voltage exceed their rated values. For this purpose, the machine must be accordingly designed in order to not produce any malfunctioning of the drive. Stator currents i_{ds} and i_{qs} are equivalent to the behavior of the torque. Even at very low speeds, when the car approaches the floor, the current consumption is high due to the necessity of a high torque. The stator voltage reference created automatically by the vector control is of a different amplitude and frequency (the amplitude and frequency are dependent on what speed the elevator is traveling). The DC bus voltage behavior depends on the stator voltage and currents consumed every instant. Since in this experiment no generator operation is required, the crowbar is not activated at any time. Finally, the grid voltage and current behavior show that the higher the power demand is, the higher the amplitude of grid current required.

For an easier visualization of some of the variables, a zoom of this experiment is represented in Fig. 8-55. There, details of ripples at torques, powers, and voltages can be distinguished as well as the shape of the sinusoidal varying magnitudes, like voltages and currents, for a time range of between 2.15 and 2.25sec.

In a different simulation experiment, the same one-floor elevation is carried out but with a significantly smaller load of 200kg. Fig. 8-56 shows details of the most interesting magnitudes. It will be seen now that with a lower load the required torque to the drive is different compared with the previous experiment, in terms of lower amplitude and also with time intervals being negative, which means that the machine works as a generator regenerating power. This fact can also be seen at the power shape P_m, which is negative for a long period. Therefore, the DC bus voltage behavior is also different, owing to the effect of the crowbar activation when P_m is negative, which also provokes an absence of current consumption from the grid during regeneration, or crowbar activation, owing to the increase of the DC bus voltage.

In a similar experiment, Fig. 8-57 shows a descending floor trajectory with a load of 200kg in the car. It will be noted that this experiment is not symmetric in respect to the previous experiment, because in this case the third quadrant of motor operation is the dominant one. But for a certain time when the power is negative, the elevator is working in the fourth quadrant.

Finally, to conclude the simulations, Fig. 8-58 shows the most extreme case of operation in terms of regenerative operation, which is the elevator descending at a rated load of 800kg. Here the dominant quadrant is the fourth quadrant and, as seen, almost not consumption of current from the grid is produced during the entire journey.

8.6　Elevator manufacturers

In the last 10 years, the elevator sector has evolved considerably, as is discussed in Section 8.11. Elevators have experienced great technological development in the last few years, mainly to improve factors such as: energy efficiency, comfort, noise, and weight, and to reduce space and maintenance time, as well as to eliminate the machine room altogether, etc.

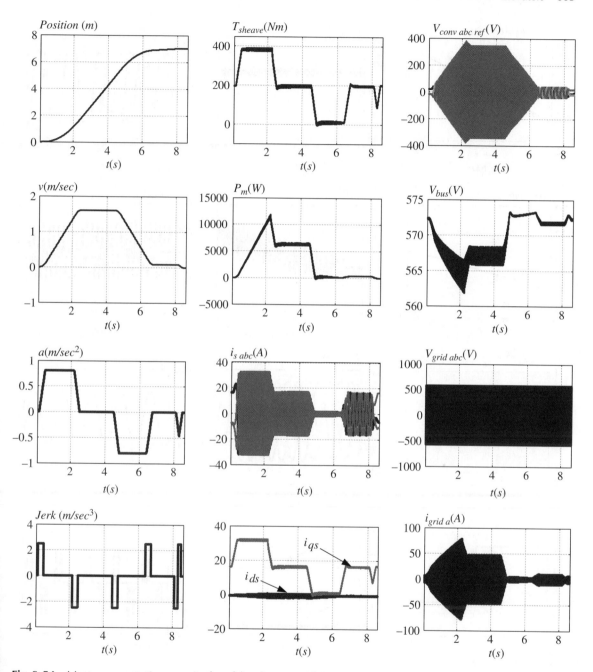

Fig. 8-54 *Most representative magnitudes of the elevator with a 2:1 roping arrangement, elevating one floor with a maximum load of 800kg*

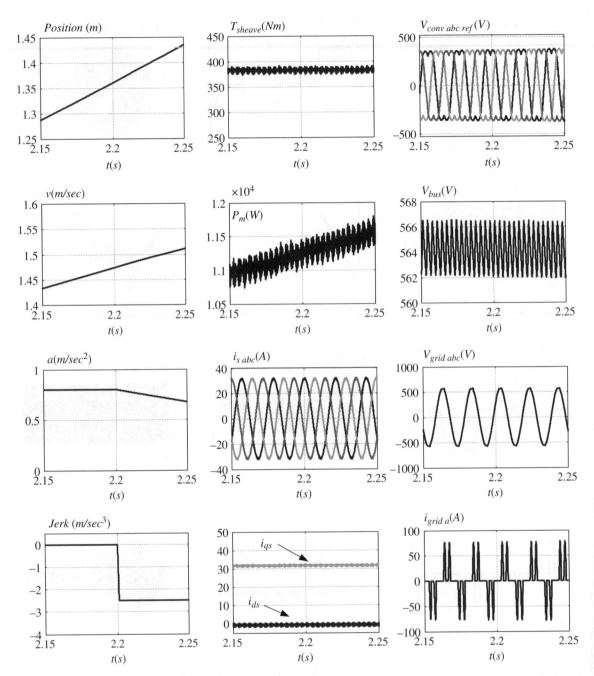

Fig. 8-55   *Zoom of the most representative magnitudes of the elevator with a 2:1 roping arrangement, elevating one floor with a maximum load of 800kg*

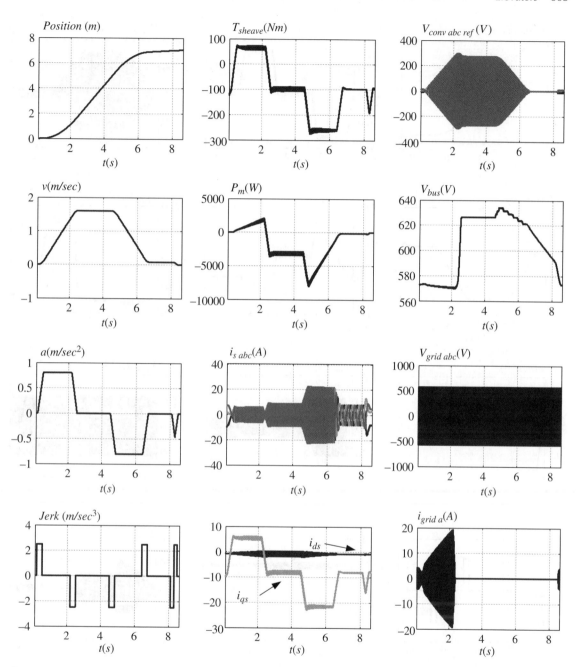

Fig. 8-56   *Most representative magnitudes of the elevator with a 2:1 roping arrangement, elevating one floor with a maximum load of 200kg*

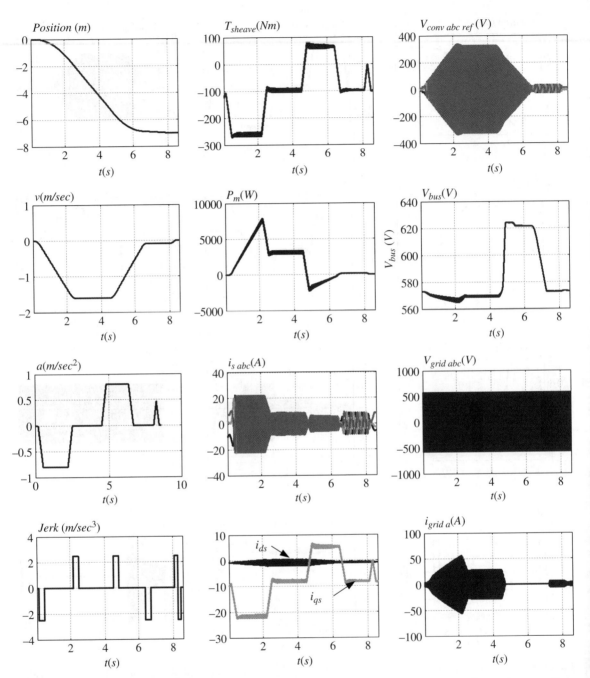

Fig. 8-57   *Most representative magnitudes of the elevator with a 2:1 roping arrangement, descending one floor with a maximum load of 200kg*

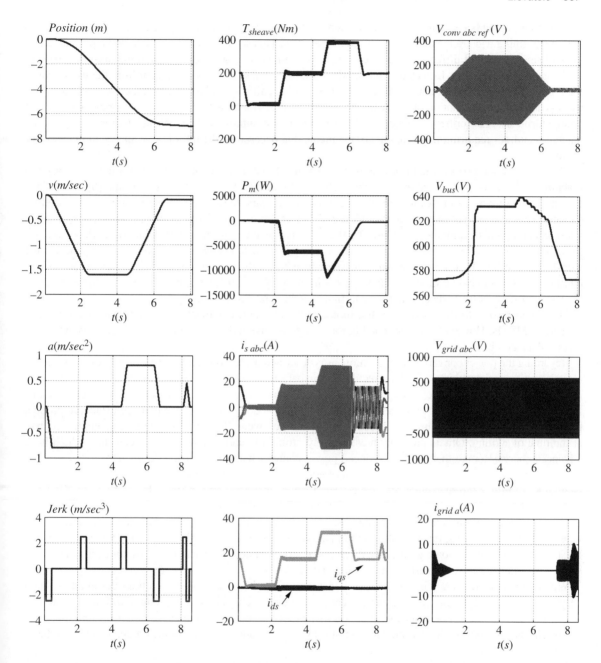

Fig. 8-58 *Most representative magnitudes of the elevator with a 2:1 roping arrangement, descending one floor and maximum load of 800kg*

Therefore, as can been seen in previous sections in this chapter, from the electric drive point of view, it is possible to find a wide range of different types of elevators. This section does not seek to make a detailed and meticulous overview of different types of elevators. Instead, it gives some examples of the most modern and well-established elevators dedicated to residential buildings. These elevators are taken from the usual manufacturers who operate in this market. Basically, these are machine-room-less residential elevators that do not use a gearbox. The elimination of the gearbox normally obliges manufacturers to use a 2:1 or even 4:1 roping arrangement, as is described in Section 8.2.5, mainly as a compromise between not requiring big torque machines and not requiring large roping arrangements that necessitate the use of large spaces in the elevator cavity.

Thus, this section gives some information about the most international and well-established elevator manufacturers. These chosen companies are: Otis, ORONA, Schindler, Kone, and ThyssenKrupp [6]–[10]. Since they have been involved in elevator development for many years, these companies represent many different elevator types and technologies, from the oldest to the most sophisticated ones. Although we will focus on the latest technological developments, obviously, they also produce and install older technologies depending on the market or country where the specific elevator is going to be installed. However, like most companies in the world, they continually research and innovate to produce better elevators. From the technology point of view, all these manufacturers use similar technologies, with no very real difference between them. However, there are some specific or basic details that we can find and the next lines focus on these.

Thus, with respect to the traction electric motor of the elevator, practically all of the manufacturers have gear-less PMSMs. However, as mentioned before, they also use induction motors with a gearbox when the market demands it.

Regarding the power converter configuration, as is described in Section 8.4.4.1, they all employ the standard electric drive configuration with diode rectifier, crowbar, and classic two-level converters, which all share relatively similar characteristics. However, when the client or the market demands something special and different, such as energy storage, they are able to adapt to that need. However, as the main and standard product they employ the classic configuration based on diodes and a crowbar. Perhaps it is necessary to emphasize that, at the time of writing, Otis is probably most actively promoting and installing newer elevators which have advanced energy storage that enable a reduced power contract with the electric grid operator.

Finally, in terms of the sheave and rope technology developed, it is possible to say that manufacturers have been able to reduce the diameters of sheaves. Here again, manufacturers need to be singled out: Otis and Schindler also have traction tapes which enable the use of smaller-diameter sheaves.

Hence, Otis' use of flat ropes gives the necessary adherence while reducing the diameter of the sheaves. As noted in previous sections, this fact allows the manufacturer to also reduce the torque that the electric motor must produce for a given load, giving as a result a very efficient and compact electric motor design and, consequently, requiring fewer materials such as magnets.

With regards to ORONA, it is possible to say that it provides advanced PMSM solutions designed to obtain very competitive and optimized efficiencies. They can adapt effectively and in an optimized manner to different and specific market demands, such as, for instance, in some residential environments where the space for the elevator is much reduced forcing the use of gear-less machines with 1:1 roping arrangements, and therefore high motor torques.

With regards to Schindler, its distinction arises from the use of induction motors as traction electric motors. The company actively employs gearless induction motors and drive sheaves of a small diameter, thanks to the employment of traction tapes. Thus, among other benefits, this means the manufacturer is in some ways not directly affected by the market particularities of magnets. However, with these types of induction motors, the efficient designs force the use of low pole numbers and probably results in less compact machines.

Otherwise, Kone is characterized by its use of axial flux PMSMs. In general, these types of motors present higher diameters than radial flux PMSMs, but with smaller longitudes. This configuration of motor enables probably the best reduction of space in the market, since it allows an ingenious disposition of the machine in the cavity of the elevator, which is patented. Moreover, this optimized space reduction also allows the use of high-ratio roping arrangements (higher than the rest of the competition's). This even avoids in some cases, the necessity to use counterweights. Obviously, these ingenious advances are accomplished thanks to complex and not easy to install suspension systems, but they are affordable for this manufacturer. In addition, the new UltraRope technology has recently been unveiled. It is an advanced form of rope oriented toward elevators in high buildings.

Finally, ThyssenKrupp is a well-known company mainly characterized for being a pioneer of high-speed elevators. Probably its most competitive solutions are offered to the premium market segment of long distances and high and special performances.

To conclude it must also be mentioned that there exist other, newer elevator manufacturers that currently enjoy strong market penetration, especially in emerging countries with particular demands for high performance, for example to travel long distances, at high speeds, and with heavy loads.

8.7   Summary of the most interesting standards and norms

The construction and installation of electric elevators in Europe are governed by UNE-EN81-1:2001 + A3:2010, a regulation entitled: *Safety Rules for the Construction and Installation of Elevators. Part 1: Electrical elevators.* In North America there is the American Society of Mechanical Engineers and Canadian Standards Association standard ASME A17.1-2013/CSA B44-13, *Safety Code for Elevators and Escalators*.

In Europe, vertical transport (elevators, freight elevator, car elevators, etc.) is regulated by different Directives dependent on whether the devices reach a top speed of 0.15m/sec, either electrical or hydraulic.

In Spain, for instance, regulations EN 81.2 (hydraulic system) and EN 81.1 (electrical system) apply. These regulations originate from Directive of Elevators: Directive 95/16/CE of the European Parliament and of the Board, of 29 June 1995: On the approximation of the legislations of the member states relative to the elevators.

A great interest exists throughout the industry worldwide for the new procedures EN 81-20 and EN 81-50 since they are intended to substitute the current ones (EN 81-1 and EN 81-2) on elevators. This is because these constitute the most important change to the regulation of elevators in the last 20 years and their introduction implies that all the current components of elevators will expire when the new procedures are in place.

It is not only a remodeling of the regulation of the electrical elevators part 1 of EN-81 or of the hydraulic ones of EN 81-2 but also the design and installation of elevators (EN 81-20) and the description of the inspections, tests, calculations and types of tests of components of elevators (EN-81.50).

With regards to the new regulations (EN 81-20 and EN 81-50), the most important aspects concern:

- pit and well;
- machine space;
- door floor and cabin access;
- elevator car;
- roping arrangement;
- acceleration rate, security elements, and non-desired movements;
- guide rails;
- shock-absorbers;
- electrical installation;
- controllers;
- mechanical and hydraulic traction units.

8.8   Door opening/closing mechanism

The door opening/closing mechanism is also a system that requires an electric drive, conceptually, of similar characteristics to the electric drive of the elevator itself. Fig. 8-59 illustrates a commercial example of this mechanism.

As noted in previous sections, there are a wide range of different types of doors that can be used for an elevator. The weight of the doors typically is in the range of 40 to 60kg, but can be up to 100kg in some models. In general, the so-called door operator is the device in charge of controlling the door's movement and is at the same time governed by the maneuver. Physically, the door operator is typically located in the roof of the car, or in the doors' supporting structure. From the motor generating the rotation, the movement is transmitted to the doors by a mechanical arrangement of toothed belts and sheaves, as can be appreciated in Fig. 8-59.

The electric drive which activates the door is depicted in Fig. 8-60. The electric motor in charge of moving the door is typically a PMSM, with similar technology to the elevator's vertical motor described in Section 8.4.3. The motor is specially designed to be compact, enabling the quick, silent, and accurate opening and closing of the doors, without the need for a gearbox. However, some manufacturers and models also employ induction motors for this mechanism.

The power consumption of the machine occurs when the elevator is stopped; therefore, it does not augment the power consumption of the elevator, because the car and the doors are not moved simultaneously.

The electric drive which moves the door mechanism is characterized by performing at 110V and the power is about 150W, having a rated speed of 600rpm. The motor used to traction the door is a synchronous one and has several numbers of poles (between 5 and 10 poles pairs). The control employed in the door's electric drives

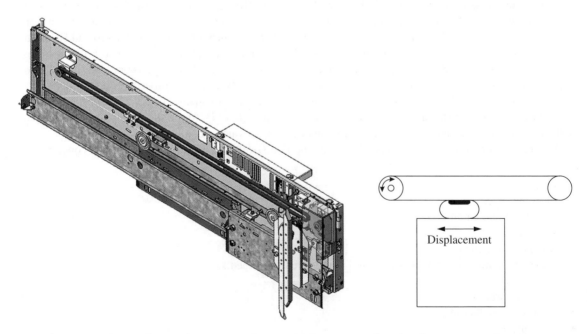

Fig. 8-59   *Close and open door system (Source: ORONA. Reproduced with permission of ORONA)*

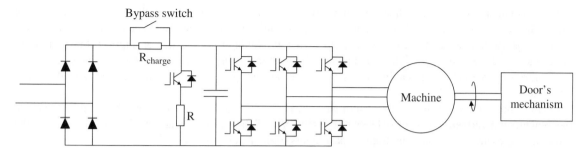

Fig. 8-60 *Electric drive for the door opening/closing mechanism*

could be a vector control or a V/F control. The former uses an encoder in order to control motor speed. The vector control strategy employed is equivalent to the vertical transport control strategy of the elevator described in Section 8.4.2. And here also a specific trajectory generation strategy is typically used to move the doors properly. Nowadays, modern electronics components allow technicians to develop a more complex strategy of control, which improves the drive performance and gives more information about the cycle life in order to receive better maintenance. This is possible because the consumption of the drive is known more precisely and reports any damage to the door and so the companies can act in advance.

The required energy is supplied from the single-phase AC grid. This voltage is then rectified by a diode-based rectifier. The electric drive incorporates a DC bus capacitor's pre-charge circuit, typically locating the resistance at the DC bus itself as depicted in Fig. 8-60. Also, a crowbar circuit is also added for deceleration and for stopping the door.

In general, the horizontal door opening/closing mechanism is not physically modeled as with the vertical movement in Section 8.3. It can be said that the physics describing the movements of the doors together with its mechanism is more complex than the vertical movement and often there are several types of doors that would also add complexity to the model. Consequently, the door mechanism and its electric drive are typically adjusted by the commissioning engineer, following simple intuition control basics.

The electric drive incorporates an encoder to control the door's movement. If the door finds an obstacle, this is detected by the control, and the door is stopped and returns to its original position. Then, it starts moving again, but upon reaching the detected obstacle it slows down to pass over it. Illustratively, we can say that the norms indicate that the force employed to stop the door must be lower than 150N. The kinetic energy of the total system is lower than 10J, and the force for unlocking the doors must be lower than 300N. Keeping the doors opened must be done with a force of around 80N.

There exist several operating modes of the doors, such as the "learning mode", which is employed for the first ensemble of the system. It allows the cycles of opening and closing the doors to be accurately defined. There are some other operating modes, such as the "rescue mode" and "forced closing".

8.9   Rescue system

The rescue operation takes place when it is necessary to evacuate persons who have remained enclosed in the car following a failure of the elevator. Before certain breakdowns, the elevator stops instantly, which means it can remain stationary between floors, and it is because of this that it is necessary to make a rescue that can be manual or automatic.

If it is a manual rescue, a specialized technician or the building's administrator will access the control box of the machinery and will press the button related to dealing with the issue having followed the instructions on the door of the cupboard. At this time, with help of a UPS (uninterruptible power supply), the brakes are deactivated and the windings of the machine are short-circuited. In this way, with the non-balanced forces experienced by the cabin, the cabin moves up to the nearest floor, the doors are opened, and the car is evacuated.

On the other hand, when the rescue is automatic, the elevator is provided with a more powerful UPS to open the brakes, and furthermore it also feeds the regulator and the maneuver automatically carrying out the sequence of rescue to evacuate the trapped passengers to the nearest floor.

8.10   Traffic

There are many traffic controlling algorithms for buildings with multiple elevators. These depend on the type of maneuver installed at the elevator, and on the number of elevators at the building. In this sense, systems of elevators could be duplex, triplex, or multiplex that incorporate coordinated algorithms of traffic.

All traffic technologies seek to optimize passenger waiting times. Depending on the segment to which the elevator is destined—residential, offices, hospitals, malls—it is necessary to adapt the performance of the elevator to the needs of its passengers.

Hereby, there are programmed algorithms that attend to the calls from floor or cabin according to a certain priority. In addition, they might attend the first call first or seek to optimize waiting times by stopping at the nearest call first.

8.11   Historical evolution

This section examines the technological evolution of elevators from their beginnings to the present day. First of all, a synthesized summary of the most relevant historic events related to the historical evolution of elevators is given. After that, the last 30 years of evolution are focused on, and more specifically various drive configurations, as well as details of the most interesting elevator installations.

Hence, Table 8-5 depicts the most relevant events related to elevators from our ancient history until today. It will be seen that the idea of the elevators predates Christ. However, it can be said that in 1853 Elisha Otis gained a significant improvement in elevators, inventing the concept of the safety elevator and preventing the fall of the car. It can be said that this invention facilitated the mass introduction of elevators into buildings, improving their performance and safety year after year.

Thus, since the invention of the safe elevator, the evolution has been characterized by improvements in comfort, capacity to reach greater heights and speeds, becoming increasingly efficient, and reducing the required space in the building.

Thus, now the most interesting developments related to the elevator have been presented, the next subsections provide more details of the electric drives configurations employed in the last 30 years.

It can be said that the elevators drive technology, from a historical perspective, of the past few years can be classified according to the schema shown in Fig. 8-61. Analyzing today's elevator manufacturers' products, it is possible to say that most of them still have four configurations of drives for their various elevators so they can adapt what they offer to the specific and different needs of the market.

Table 8-5 *The evolution of elevators*

Year	Development
3rd century BC	There is already evidence of primitive elevators operated by human, animal, or water wheel power
1000 AD	In 1000, the Book of Secrets by al-Muradi in Islamic Spain describes the use of an elevator-like lifting device
19th century	In 1846, Sir William Armstrong introduces the hydraulic crane and in the early 1870s hydraulic machines begin to replace the steam-powered elevator
1852	Elisha Otis introduces the safe elevator that includes a device to prevent the car falling if the ropes are broken. He demonstrates his invention in New York in 1854 in a death-defying presentation
1857	The first Otis passenger elevator is installed in New York City
1868	Waygood manufactures its first hydraulic elevator
1872	First geared electric elevator is installed in New York
1880	First electric elevator is built by inventor Werner von Siemens
1874	J. W. Meaker patents a method which permits elevator doors to open and close safely
1918	New York City adopts its first "elevator rules"
1924	In the US, automatic control of the elevator is introduced, which avoids the presence of a personal operator permanently remaining in the elevator
1980s	Microprocessors are introduced to elevators for the first time, substituting older maneuvers based on contacts
1985	Variable frequency control based on power electronic converter is introduced to the elevator, increasing comfort levels for passengers
1995	Machine-room-less elevators appear, being a good advance for the architects of buildings
1997	Gear-less elevator installation becomes generalized, increasing efficiency and comfort
2002	Self-propelled elevator is developed, with the drive being integrated in the car itself. Commercially it is not a successful step but marks a future tendency to be considered
2002	Non-conventional traction ropes are introduced, with the objective of reducing the drive sheaves and, therefore, the volume of motors
2005–2016	Efforts of manufacturers to: • reduce drive sheaves; • improve traction ropes; • be more efficient, have less volume, and optimize electric drive and elevation systems; • introduce energy storage systems and renewable energy generation at the elevator

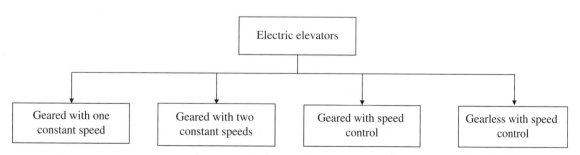

Fig. 8-61 *Electric elevator configurations used today (the trend in advanced countries is for gearless concepts)*

8.11.1   Geared with one constant speed

The most primitive and simple drive configuration was based on a classic, reliable, and simple squirrel cage asynchronous machine (induction machine), as illustrated in Fig. 8-62.

The most representative characteristics are summarized as:

- AC asynchronous machine for traction directly connected to the electric grid.
- The traction drive incorporates a gearbox, so the electric machine rotates at high speeds of around 1000rev/min, while the drive sheave, owing to the gearbox effect, rotates at lower speeds.
- The electric machine, as it is directly connected to the grid, is supplied at constant voltage amplitude and frequency, resulting in a constant speed of operation. It can work as a motor or as a generator (regenerating the energy to the grid).
- It incorporates a flywheel coupled to the axis, in order to mechanically control the acceleration and deceleration values of the elevator. This flywheel compensates the inertias of the mechanical elements (machine, gearbox, drive sheave, and equivalent car inertia itself) coupled to the axis and the transient provoked by the electric machine startups (note that this elevator concept does not incorporate an electric control of the speed). The flywheel is typically located in the low speed shaft, adjusted depending on the load of the elevator in order to obtain a defined comfort level by actuating basically on the acceleration and deceleration moments, where the torque produced by the machine is very high since there is no speed control.
- The elevating or descending mode is determined by the phase sequence abc or acb, which is changed by the contactors.
- Since there is no speed control, it can be said that it presents poor performances in: comfort, accuracy at the stop, and energy consumption. However, it is very cheap.
- It is possible to say that it is falling into disuse in advanced countries.

8.11.2   Geared with two constant speeds

The elevator configuration in this case is exactly the same as with one constant speed drive and illustrated in Fig. 8-62. The main characteristics are summarized as:

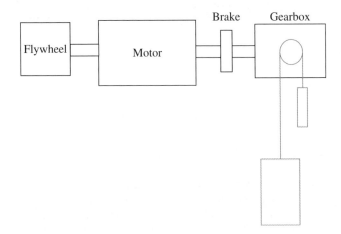

Fig. 8-62 *Elevator configuration based on a geared fixed-speed traction drive*

- The electric machine used for traction is also asynchronous but specially designed with two three-phase windings at the stator, with two different numbers of poles at each winding.
- It is also directly connected to the grid and therefore supplied at constant voltage amplitude, and frequency. Therefore, the achievable constant speeds are two, depending on the winding which is being supplied at each moment (no simultaneous supply is possible).
- Therefore, for instance in one typical machine design, the high-speed winding presents two pole pairs, which provide 1500rev/min with 50Hz of supplying voltage. On the other hand, the low-speed winding designed with eight pole pairs, provides 375rev/min rotational speed.
- Thanks to this two-speed operation, the elevator runs at a constant high speed for most of the journey, but when one sensor detects that it is approaching the floor the high-speed winding is deactivated and the low-speed winding is supplied, reaching the destination floor at low speed and so increasing comfort performance.
- The rest of the working principles, features, and characteristics of this elevator are very similar to the previous presented one, but it can be said that the main advantage compared to one speed of operation is the improved comfort characteristic, owing to the low-speed approach to the end of the journey.

8.11.3 Geared with speed control

The next step of the electric drive's evolution incorporates the speed control by means of power electronic conversion. By using the same machine concept of asynchronous machine with one three-phase winding, the incorporation of a power electronic converter allows (enabled, permitted) to incorporate (include, perform) speed control of the elevator. The elevator configuration is depicted in Fig. 8-63. The concept is very similar to the previous presented elevators, including the gearbox but eliminating the flywheel, thanks to the speed being controlled by the electric drive.

The main characteristics of this elevator concept are summarized as:

- The electric drive configuration is shown in Fig. 8-64. It is composed of the same converter, crowbar, and rectifier topology as the most modern elevator configurations studied in previous sections (same levels of power, voltage, and current). However, as mentioned before, it uses a simple and classic squirrel cage

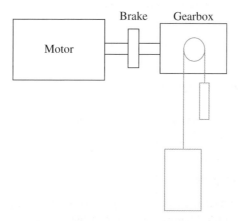

Fig. 8-63 *Elevator configuration based on a geared elevator with speed control*

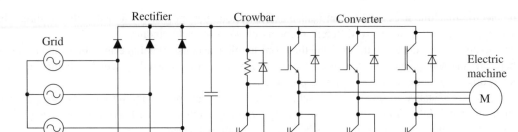

Fig. 8-64   *Electric drive of a modern geared elevator with speed control*

asynchronous machine, whose efficient design obliges one to use 1-, 2-, or 3-pole pairs (3000, 1500, or 1000rev/min at 50Hz), consequently obliging one to employ a gearbox.

- By the incorporation of the power electronic converter, it is possible to supply the machine at different voltage amplitude and frequency, allowing speed control to be performed.
- Compared to older models, improved performances are obtained—including better energy consumption, better comfort and accuracy, better reliability, and improved maintenance—thanks to the introduction of variable speed capability. However, the cost of the elevator itself is increased.
- It must be mentioned that the evolution of the electric drive technology has also affected the evolution of electric drives used in elevators, first, for instance using open loop or scalar controls, with converter topologies more simples and not using IGBTs as controlled semiconductors.

8.11.4   Gearless with speed control

Finally, this last generation of elevators is based on the inclusion of versatile high pole number PMSMs, eliminating the need for a gearbox (reducing the space required and moving toward the machine-room-less-concept) and increasing the energy efficiency of the elevator, owing to the utilization of, in principle, more-efficient electric machines. The electric drive configuration is equivalent to the two previously seen elevator families (Fig. 8-64) and the dimensioning and performance has been studied in previous sections of this chapter, so is therefore not repeated here.

8.12   New trends and future challenges

As with previous chapters, some new trends and future challenges are covered in this section. In the world of elevators, some challenges are presented.

8.12.1   Reducing the required space

With the increase of the cost of the square meter, the utilization of the space in the buildings turns into a challenge that the elevators designers must approach. This is because it is necessary to make elevators in which the ratio loads to transporting versus space occupied in the shaft it is the possible maximum. According to this premise, the elevators of the future will have to achieve that configurations in the shaft which occupy the least possible space. In addition, it will be necessary for an optimized distribution of the mechanical components; nearby guides of counterweight and cabin, systems of sheaves with limited dimensions and even the elimination of the counterweight.

8.12.2   Improving energy efficiency

In a future in which energy resources will be scarce, it will be necessary to try to use that energy in the most-efficient manner. In the case of the elevator, during the cycle of stopped or functioning in favor of load, it is possible to store or to recover the energy that has up until now been vanishing via heat in the resistances. This regenerated energy is that needed for moving the load. And it might be used in the electrical circuit of the rest of the building (lighting, heating, domestic appliances, computers, etc.) or stored in the system of the elevator to be able to be used during the cycle of traction and to reduce thereby the contracted power or the energy demanded for the network.

In addition, integration is well known in the field of designing buildings to use renewable resources. For example, solar panels or small wind generators included in a building's design might supply energy to the elevator.

8.12.2.1   *Returning the energy to the electric supply: Developing different possibilities*

Back-to-back converters consist of two converters joined by the DC bus. In this way, all the energy is transferred by the power electronics converters. Compact converter topologies allow smaller drives to be used because only the regenerative energy passes through the reversible part of the drive.

The best topology is the four-quadrant converter. This kind of converter allows the energy to flow through the electronic components not only in regenerative mode but also in motor traction.

8.12.2.2   *Energy storage*

Storage in batteries of different compositions, such as nickel metal hydride and lithium-ion, will become necessary. Ultracapacitors' principal disadvantage is the loss of long-term energy. This means that ultracapacitors are not the best choice of storage device to be used for energy systems based on renewable resources (such as solar and wind).

8.12.2.3   *Increasing reliability of components*

Designing components in order to decrease production and maintenance costs, and thereby increasing the feasibility and enlarging the cycle of life of the product and improving its efficiency are key aims.

The elevator is a way of transporting people that demands reliability, hardiness, and a lengthy cycle life. For this reason, it is necessary to design elevators with a high component reliability. This will allow manufacturers to develop products that are increasingly competitive in terms of cost and hardiness, thus guaranteeing the quality of the product from the design phase.

Maintenance will increasingly be based on monitoring and telematics. This will mean it becomes necessary to design applications that allow maintenance personnel to know at the push of a button what the maintenance history of the elevator has been, on top of what elements must be replaced at the time.

Finally, the mechanical aspects that try to meet the main exposed challenges are relative to the cable. They are necessary cables that allow elevators that cover major trips, cabins with more load and high speeds.

The current technology is limited to 200–300mm, although there are already manufacturers who set out to design technologies of carbon fiber to reach trips of up to 1000m.

In these heights of buildings, it becomes necessary to also solve the system of guided systems—and not only from the point of view of the design and application stages but also from the assembly stage. Owing to the lengths of trips, it is necessary to present solutions that allow the assembly of the guided system in a reliable way.

The latest issue in elevator design is multiple cages using the same shaft with the intention to build very tall buildings. There are many challenges ahead in making this work. Cabs have to be very light (think carbon fiber) because linear induction motors are expensive. Elevators will not limit the height of buildings and genuine vertical cities will grow ever higher.

References

[1]   Strakosch GR. *The Vertical Transportation Handbook*, 4th edition. Chichester: John Wiley & Sons, 2010.
[2]   Miravete A, Larrodé E. *Vertical Transport Book*. Barcelona, Spain: Editorial Reverté, 1995.
[3]   Sul K. *Control of Electric Machine Drive Systems*. New York: John Wiley & Sons, Inc., 2011.
[4]   www.mayr.com.
[5]   Almandoz G. *Advanced Design Methodology Oriented to the Application for Multipolar Permanent Magnet Machines*. PhD thesis, University of Mondragon, 2008.
[6]   www.otis.com.
[7]   www.ORONA-group.com.
[8]   www.schindler.com.
[9]   www.kone.com.
[10] www.thyssenkrupp-elevator.com.

Index

Note: page numbers in *italics* refer to figures; page numbers in **bold** refer to tables.

Power Electronics and Electric Drives for Traction Applications, First Edition. Edited by Gonzalo Abad.
© 2017 John Wiley & Sons, Ltd. Published 2017 by John Wiley & Sons, Ltd.